World Sustainability Series

Series Editor

Walter Leal Filho, European School of Sustainability Science and Research, Research and Transfer Centre "Sustainable Development and Climate Change Management", Hamburg University of Applied Sciences, Hamburg, Germany

Due to its scope and nature, sustainable development is a matter which is very interdisciplinary, and draws from knowledge and inputs from the social sciences and environmental sciences on the one hand, but also from physical sciences and arts on the other. As such, there is a perceived need to foster integrative approaches, whereby the combination of inputs from various fields may contribute to a better understanding of what sustainability is, and means to people. But despite the need for and the relevance of integrative approaches towards sustainable development, there is a paucity of literature which address matters related to sustainability in an integrated way.

Notes on the quality assurance and peer review of this publication

Prior to publication, the works published in this book are initially assessed and reviewed by an in-house editor. If suitable for publication, manuscripts are sent for further review, which includes a combined effort by the editorial board and appointed subject experts, who provide independent peer-review. The feedback obtained in this way was communicated to authors, and with manuscripts checked upon return before finally accepted. The peer-reviewed nature of the books in the "World Sustainability Series" means that contributions to them have, over many years, been officially accepted for tenure and promotion purposes.

Luciana Brandli · Francisco Dalla Rosa ·
Rudi Petrorius · Lucas Veiga Ávila ·
Walter Leal Filho
Editors

The Contribution of Life Cycle Analyses and Circular Economy to the Sustainable Development Goals

Springer

Editors
Luciana Brandli
Post Graduate Program in Civil and Environmental Engineering
Passo Fundo University
Passo Fundo, Rio Grande do Sul, Brazil

Francisco Dalla Rosa
Post Graduate Program in Civil and Environmental Engineering
University of Passo Fundo
Passo Fundo, Brazil

Rudi Petrorius
Department of Geography, Science Campus
University of South Africa
South Africa, South Africa

Lucas Veiga Ávila
Post Graduate Program in Production Engineering, Post Graduate Program in Administration and Accounting Sciences
University of Santa Maria
Santa Maria, Brazil

Walter Leal Filho
European School of Sustainability Science and Research
Hamburg University of Applied Sciences
Hamburg, Germany

ISSN 2199-7373 ISSN 2199-7381 (electronic)
World Sustainability Series
ISBN 978-3-031-85299-2 ISBN 978-3-031-85300-5 (eBook)
https://doi.org/10.1007/978-3-031-85300-5

This Springer imprint is published by the registered company Springer Nature Switzerland AG
The registered company address is: Gewerbestrasse 11, 6330 Cham, Switzerland

Preface

In an era characterised by unprecedented environmental challenges and a pressing need for sustainable solutions, the convergence of life cycle analyses (LCA) and circular economy principles emerges as a promising pathway to achieving the Sustainable Development Goals (SDGs). The intricate relationship between our economic activities and their environmental impacts has never been clearer, prompting a critical re-evaluation of how we design, produce, consume, and dispose of goods.

This book aims to explore and describe the integral role that LCA and circular economy frameworks play in advancing sustainability. Life cycle analysis provides a comprehensive tool for assessing the environmental impacts of products and services throughout their entire life span, from raw material extraction to end-of-life disposal. Meanwhile, the circular economy embodies a transformative approach to resource management, emphasising the importance of rethinking waste, extending product lifecycles, and fostering resource efficiency.

This book is structured into two parts:

Circular Economy Approaches
Life Cycle Approaches

Both parts present case studies, theoretical frameworks, and innovative practices that exemplify the synergistic potential of LCA and circular economy initiatives. Contributors from diverse fields share their insights and experiences, illustrating how these concepts can be operationalised to contribute meaningfully to the SDGs.

The collective journey towards sustainability is not only about mitigating environmental degradation but also about creating including compilation of knowledge and research, we aspire to inspire action, foster collaboration, and propose solutions that align with the vision of a more sustainable future for our planet and its inhabitants.

We thank all authors for their valuable contributions. We hope this book will provide an inspiration for further works in this field.

Passo Fundo, Brazil — Luciana Brandli
Passo Fundo, Brazil — Francisco Dalla Rosa
South Africa, South Africa — Rudi Petrorius
Santa Maria, Brazil — Lucas Veiga Ávila
Hamburg, Germany — Walter Leal Filho

Contents

Circular Economy Approaches

Circular Economy and Life Cycle Assessment in Extraction Process and Transformation of Essential Oils for Reaching Sustainable Development Goals

Saleth Causil and Aída Luz Villa

Abstract This chapter presents a bibliographic review of the environmental analysis reported for processes of extraction and transformation of essential oils. Results are contrasted with the principles of circular economy and sustainable development goals. Several LCA studies of extraction and/or catalytic transformation of essential oils, published between 2009 and 2023, are analyzed. The use of environmental tools as Life Cycle Analysis allow to identify preventive and corrective actions for mitigating the potential environmental impacts of the process in industries such as the extraction and transformation of essential oils. In this way, more sustainable processes that follow circular economy principles can be achieved, using forest biomass residues and waste from juice industries, for example. Those actions not only promote the reduction of negative environmental impacts and the overexploitation of land, but also mitigate poverty, hunger, and inequalities.

Keywords Circular economy · Biomass · Catalysis · Essential oils · LCA · Sustainable development

1 Introduction

Today, the societies face the tough challenge of guarantee future generations the environmental conditions necessary for the normal development of life. This has become a common purpose for humanity, manifested in important international agreements and materialized in the legislation of each country (Sonnemann et al. 2018). The work of the sciences has subscribed to this dynamic, which, based on the growing

S. Causil · A. L. Villa (✉)
Environmental Catalysis Research Group, Universidad de Antioquia, Medellin, Colombia
e-mail: aida.villa@udea.edu.co

S. Causil
e-mail: saleth.causil@udea.edu.co

L. Brandli et al. (eds.), *The Contribution of Life Cycle Analyses and Circular Economy to the Sustainable Development Goals*, World Sustainability Series,
https://doi.org/10.1007/978-3-031-85300-5_1

awareness of the importance of environmental protection and the possible environmental impacts associated with production processes that can be evaluated by life cycle assessment (LCA), has increased interest in the development of methods to better deal with these impacts (Kassim and Simoneit 2005).

The current industrial economic model "extract-manufacture-dispose", known as linear economics (ECV 2009) that is based on resource extraction, production of goods and services, and, then disposal once their use is completed, seems to be changing nowadays. The concepts of circular economy and life cycle encourage a more sustainable pace of production and consumption, and help to make more efficient use of limited financial and natural resources (Clark et al. 2016). By optimizing processes and production chains, it is possible to contribute to Sustainable Development Goals (SDGs) related to responsible consumption and production, no poverty, and climate action.

The use of essential oils as a source of more added products is according to the circular economy because materials than are obtained from agro-industrial wastes are transformed into sources of new opportunities and trends for economy and society. The inclusion of environmental impact assessment tools in the conceptualization and design stages of chemical processes contributes to the improvement and optimization of projects, in the technical and economic components, and identifying better options in terms of raw materials, catalysts, solvents, and synthesis routes. One of these tools is the life cycle assessment (LCA). In this chapter, several scientific articles published between 2009 and 2023 that are related to LCA analysis of essential oil extraction and catalytic transformation are compared. The comparison is respect to the boundaries, functional units, life cycle inventories, environmental impact assessments, and interpretations of LCA; the analysis is interpreted also in terms of the circular economy and SDGs.

2 Generalities of Essential Oils

As much as for human beings as for the biosphere in general, plants are beings of superior importance; from them we obtain products such as oxygen, food, wood, ornamental products, medicines, and essential oils, which are a mixture of volatile components that are products of the secondary metabolism of plants (Ríos 2015). These oils are mixtures that generally contain terpenic hydrocarbons (mono-, di- and sesquiterpenes); in small proportion, other common constituents are oxygenated derivatives of hydrocarbons, benzenoids, fatty acids, amino acids, nitrogenous and sulfur compounds. They are extracted by steam distillation or physicomechanical processes such as pressing or milling (Pino Alea 2015). Essential oils should not be confused with vegetable oils, which are composed of mixtures of lipids that are not necessarily volatile. Essential oils can be found in different parts of plants as roots, flowers, leaves, fruits, seeds, bark, and wood; so, they are the source of the smell of that part of the plant. Since ancient times they have been used in the cosmetic and food industry as flavorings and fragrances; currently the spectrum of applications

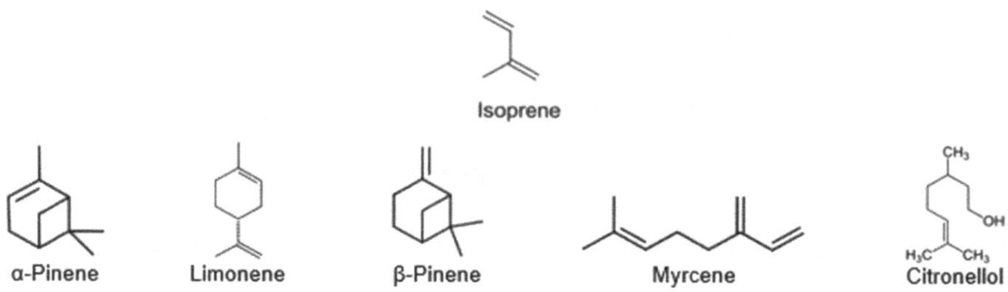

Fig. 1 Isoprene and examples of terpenes and their isoprene units

of essential oils, and their derivatives, has increased strongly, having uses in sectors such as chemical, pharmaceutical, agro-industrial, energy and fuels, materials (Can Baser and Buchbauer 2016).

The main components of essential oils contain several isoprene (2-methyl-1,4-butadiene) units connected in a head-to-tail way, Fig. 1. There are a wide variety of terpenes because the presence of stereogenic centers in the molecule, and the diverse molecules with several functional groups that can be obtained (Silvestre and Gandini 2008), via isomerization, hydroformylation, oxidation, dehydrogenation, polymerization, hydrogenation, hydration, condensation, hydroformylation, cyclization, oxidation, rearrangement, and ring contraction/enlargement (Rubulotta and Quadrelli 2019).

3 Life Cycle Assessment of Extraction and Transformation of Essential Oils

The environmental impact assessment must be evaluated in any human activity that requires the use of natural resources; it is necessary to identify, evaluate and describe the environmental impacts that these activities will produce in its environment, if it is carried out or modified (if it has already been implemented). The purpose is to prevent, mitigate, correct, and compensate the environmental impacts (Ott et al. 2012). There is a variety of methods used to assess impacts that are implemented depending on the goal or scope of analysis. In general, LCA is currently the most widely used method, due to its applicability to many processes, products, and services. Besides quantifying environmental impacts, the LCA method allows to track the environmental impacts and assess them from a systems perspective, identifying strategies for improvement without burden shifting (Hellweg and Cannals 2014). The LCA method has become a leading tool in the business world, to identify and understand environmental risks, and manage the opportunities for improvement associated with the studied systems.

The relevance of biomass utilization as a new source of hydrocarbon derivatives of interest in the food, cosmetic and pharmaceutical industries, justify quantify environmental impacts associated to the chemical production processes. One alternative green process is the catalytic transformations of essential oils (mainly the terpenic compounds they contain), because minimize waste treatment, decrease production costs, and separation and purification stages of the process (Kukhta et al. 2011). Therefore, this section presents an analysis of reports about life cycle assessment of processes that involve essential oils, its components and obtained products from their transformations; the relevance of this environmental analysis technique and the subsequent use in areas of circular economy and sustainable development are highlighted.

For selecting the papers for the review, a literature search was performed in the Google Scholar search engine and several data bases (ScienceDirect, Scopus, Springerlink, ACS Publications) limited to the English and Spanish languages. The selected keywords "essential oils", "LCA", "extraction", "catalysts", "circular economy", "life cycle assessment", "terpenes", "citrus", "orange wastes", "sustainability", "agriculture", "biomass", "sustainability goals", and "SDGs", were used; those key words were combined using the Boolean operator "AND". No specific period was selected in the research; however, selected reports were published after 2005, especially after 2014. The abstracts, introduction and conclusions sections were initially screened, and the rest of the paper was analyzed for selected articles. Some additional literature from the article's reference section was also analyzed.

Reports were the LCA methodology is applied for essential oils extraction or catalytic transformation, in general include the entire production process, from planting, going by the harvesting, extraction, processing and chemical transformation, to its market distribution. Table 1 presents the identified relevant 20 LCA studies of extraction and/or catalytic transformation of essential oils, published between 2009 and 2023, including information of data sources, methodology impact assessment, software, functional unit, system boundaries and scenarios.

From the analyzed papers, it is observed that the most of the LCA reports are about extraction of essential oil from citric fruits, perhaps because it can be extracted from peels that are wasted from juice industries and because limonene is its main component that has applications for the food, pharmaceutical, and cosmetic industries. Each of the analyzed papers starts with the general objective of evaluating the environmental impact of the process, some of the papers related to transformation of essential oils present the environmental evaluation together with kinetic, catalytic (Ravelli et al. 2011; Gallego et al. 2021), economic (Iñigo et al. 2023), or techno-economic studies (Moncada et al. 2014; Gwee et al. 2020).

Although there is a wide range of selected scopes, the most common is the "cradle to gate" that includes the stages of –plant cultivation (planting, growing, maintenance, harvesting)—transport—extraction—processing—transformation (catalyst synthesis, energy production, synthesis of the product of interest, purification of the product of interest, waste treatment)—packaging/distribution, because the complexity of analyzing the final use of the product (essential oils and derivatives have endless applications). A LCA that includes the agricultural steps is desirable for identifying the environmental impacts associated also to the plant growing and, in this way, to make the appropriate modifications to the process for reaching a more sustainable essential oil extraction. Furthermore, if the residues generated in the process can

Table 1 Main characteristics of the LCA reports of essential oil extraction or catalytic transformed

Source of the essential oil/catalytic transformation	Scope	Functional unit	Data source	Environmental impact categories (software)	References
Extraction of essential oils					
Italian citrus	Cradle to gate • Plant cultivation • Transport • Extraction and processing of products (pectin, juices, and oils) • Packaging • Distribution	1 kg of extracted essential oil	Literature On-site Measurements Inventory Databases (Ecoinvent and European Platform on LCA)	IPCC 2001 GWP100 CML 2 baseline 2000 (SimaPro®)	Beccali et al. (2009)
Citronella (*Cymbopogon winteriana*), Lemongrass (*Cymbopogon citrus*)	Gate to gate The extraction process of essential oils with different technologies • Extractive steam distillation • Solvent extraction • Supercritical fluid extraction	1 kg of extracted essential oil	Laboratory-scale data Literature Simulations (Aspen Plus®)	The Waste Reduction Algorithm (WAR) Inter Panel on Climate Change (IPCC)	Moncada et al. (2014)
Macroalga *Ochtodes secundiramea*	Cradle to gate • Raw material production • Transport of raw materials • Biotechnological processes • Extraction of oils	700 mg terpenes* (1 production batch) *Myrcene and derivatives	Literature and models On-site Measurements Inventory Databases (Ecoinvent®)	CML 2 baseline 2001 (SimaPro®)	Pérez-López et al. (2016)

(continued)

Table 1 (continued)

Source of the essential oil/catalytic transformation	Scope	Functional unit	Data source	Environmental impact categories (software)	References
Oregano (*Origanum vulgare*), Rosemary (*Rosmarinus officinalis*)	Gate to gate The extraction process of essential oils with different technologies • Extractive steam distillation • Solvent extraction • Supercritical fluid extraction	1 kg of extracted essential oil	Laboratory-scale data Literature Simulations (Aspen Plus®)	The Waste Reduction Algorithm (WAR) Inter Panel on Climate Change (IPCC)	Moncada et al. (2016)
Dragonhead (*Dracocephalum moldavica* L.)	Cradle to gate • Plant cultivation • Transport of raw materials • Essential oi extraction	1 kg of extracted essential oil	On-site Measurements Inventory Databases (Ecoinvent®)	IMPACT 2002+ (SimaPro®)	Maham et al. (2018)
Macroalgae	Cradle to gate • Plant cultivation • Transport of raw materials • Milling and hydrolysis • Fermentation • Essential oi extraction	1 t of refined single-cell oil	Laboratory-scale data Literature Simulations Inventory databases (Ecoinvent)	ReCiPe 2016 (Brightway LCA software)	Parsons et al. (2019)

(continued)

Table 1 (continued)

Source of the essential oil/catalytic transformation	Scope	Functional unit	Data source	Environmental impact categories (software)	References
Ylang-Ylang (*Cananga Odorata*)	Cradle to gate • Plant cultivation • Transport • Essential oil extraction	1 l of oil	Interviews On-site Measurements Literature Inventory databases (Ecoinvent®)	ReCiPe 2016 midpoint IMPACT 2002 + (SimaPro®)	Lambert (2019)
Aquilaria sinensis	Cradle to gate • Plant cultivation • Transport of raw materials • Milling • Essential oi extraction	1 kg of unrefined oil	Laboratory-scale data Literature Simulations Inventory databases (GaBi)	CML 2001 (GaBi)	Gwee et al. (2020)
Citrus waste	Cradle to gate • Raw materials obtention • Pre-treatment • Extraction • Purification • Anaerobic digestion	100 kg/h of citrus waste	Laboratory-scale data Simulations (Aspen Plus) Inventory databases (Ecoinvent®)	Cumulative Energy Demand (CED) CML 2001 (SimaPro®)	Santiago et al. (2020)
Calendula (*Calendula Officinalis*)	Cradle to gate • Calendula crop activities • Transport • Essential oil extraction	1 g of essential oil	Interviews On-site Measurements Literature Inventory databases (Ecoinvent®)	(SimaPro®)	González-Aguirre, Solarte-Toro and Cardona (2020)

(continued)

Table 1 (continued)

Source of the essential oil/catalytic transformation	Scope	Functional unit	Data source	Environmental impact categories (software)	References
Orange peel waste	Cradle to gate • Plant cultivation • Transport of raw materials • Orange manufacturing • Extraction and processing of products from orange peel (animal feed, essential oil, and citric acid)	1 t of orange peel waste	Literature Simulations (ProsimPlus®) Inventory databases (Ecoinvent®)	EF life cycle impact assessment method (SimaPro®)	Teigiserova et al. (2022)
Argentine lemon	Cradle to gate • Plant cultivation • Transport of raw materials • Lemon pretreatment • Extraction and processing of products (juices, essential oil, and biofuel/biogas)	1 t of each product	Interviews Technical Reports Literature Inventory databases (Ecoinvent®)	The International Reference Life Cycle Data System (ILCD) method (SimaPro®)	Machin et al. (2022)
Orange waste	Gate to gate • Orange pretreatment • Juice production • Essential oi extraction • Pectin production	1 kg of pectin	Literature Inventory databases (Ecoinvent®)	ReCiPe 2016 midpoint (SimaPro®)	da Costa et al. (2022)

(continued)

Table 1 (continued)

Source of the essential oil/catalytic transformation	Scope	Functional unit	Data source	Environmental impact categories (software)	References
Rosemary	Cradle to gate • Raw materials obtention • Transport • Extraction of oils. Three essential oil extraction processes, based on hydro-distillation (HD), HD followed by lyophilization (L) and supercritical fluid extraction (SFE)	1 g of aromatic extract and 1 g of α-pinene	Laboratory-scale data Literature Inventory databases (Ecoinvent®)	Cumulative Energy Demand (CED) Inter Panel on Climate Change (IPCC) (SimaPro®)	Moura et al. (2022)
Lemon and by-product production in Argentina	Cradle to gate • Plant cultivation • Transport of raw materials • Lemon pretreatment Extraction and processing of products (juices, essential oil, and others)	1 t of each final product	Interviews On-site Measurements Literature Inventory databases (Ecoinvent® and Agri-footprint)	ReCiPe 2016 midpoint (SimaPro®)	Iñigo et al. (2023)
Lippia origanoides	Cradle to gate • Raw materials obtention • Extraction step • Separation and purification of products	1 kg of extracted oleoresin 1 kg of isolated pinocembrin	Laboratory-scale data Inventory databases (Ecoinvent® and Agri-footprint)	ReCiPe 2016 midpoint (SimaPro®)	Arias et al. (2023)
Extraction—catalytic transformation of essential oils					

(continued)

Table 1 (continued)

Source of the essential oil/catalytic transformation	Scope	Functional unit	Data source	Environmental impact categories (software)	References
Rose oxide from β-citronellol	Cradle to gate • Production of raw materials • Transport • Synthesis • Separation and purification • Waste treatment	–	Literature Inventory databases (Ecoinvent®)	Eco-Indicator 99 (SimaPro®)	Ravelli et al. (2011)
Vanillin from depolymerization of lignin (by-product of paper production)	Cradle to gate • Raw materials obtention • Catalyst synthesis • Terpenes separation • Catalytic transformation of terpenes • Separation and purification of products	1 g of vanillin	Laboratory-scale data Literature Inventory databases (Ecoinvent®)	TRACI (SimaPro®)	Isola et al. (2018)

(continued)

Table 1 (continued)

Source of the essential oil/catalytic transformation	Scope	Functional unit	Data source	Environmental impact categories (software)	References
Nopol from β-pinene (turpentine oil component)	Cradle to gate • Raw materials obtention • Transport • Catalyst synthesis • Terpenes separation • Catalytic transformation of terpenes • Separation and purification of products	1 kg of purified nopol	Literature Simulations (Aspen Plus) Inventory databases (Ecoinvent®)	ReCiPe 2016 midpoint (Umberto LCA+)	Gallego et al. (2021)
Verbenone from α-pinene (turpentine oil component) Carvone from *d*-limonene (orange peel oil component)	Cradle to gate • Raw materials obtention • Transport • Catalyst synthesis • Terpenes separation • Catalytic transformation of terpenes • Separation and purification of products	1 kg of each product (carvone and verbenone)	Laboratory-scale data Literature Simulations (Aspen Plus) Inventory databases (Ecoinvent® and Agri-footprint)	ReCiPe 2016 endpoint (SimaPro®)	Becerra et al. (2022)

be used as source of added value compounds, the circular economy perspective is emphasized. Due to the objectives of each study and the stages involved, there is no consensus about the selected functional unit, which may vary at the author's discretion.

The environmental impacts of these processes are generally reflected in all categories of the LCA, highlighting the category of acidification, eutrophication, ecotoxicity, land use and biotic/abiotic resources depletion. Most of the work focuses on already established processes, so data collection (input–output/mass and energy balances) is done with in site measurements or through questionnaires sent to farmers, plant managers, and distributors; this information is mostly complemented with inventory databases (mainly Ecoinvent®).

The use of machinery in the agricultural stages (land preparation, sowing, maintenance and harvesting) is a challenge for the LCA implementation when they are included as entries in the inventory. In many cases it is necessary to take measurements *in situ* or with a real base of those equipments and this is a relevant point because this machinery uses fossil fuels with a high environmental impact load. In general, the papers converge in the agricultural stage (land preparation, planting, maintenance, and harvesting) as the phase with the greatest contribution to the environmental impacts evaluated by LCA. This is because the use of fertilizers, pesticide treatment and herbicides contribute with considerable quantities of sulfur, nitrogen, and phosphate compounds, that are directly responsible for the environmental impacts of acidification, eutrophication (Morelli et al. 2018), ozone formation. Furthermore, the irrigation systems involve the use of quantities of fresh water extracted from the surface and underground water bodies, which is reflected in the environmental impact of water consumption.

Another important hotspot is due to the use of solvents and reagents (mostly non-green) and energy consumption in processes that involve stages of catalytic transformations of the extracted essential oils (Ravelli et al. 2011), and which are converted into products with greater added value: ethyl acetate as a solvent in the process of obtain nopol from β-pinene (Gallego et al. 2021), acetone as a solvent and TBHP as a oxidant agent in the process of obtain verbenone from α-pinene and carvone from d-limonene (Becerra et al. 2022), CO_2 in extraction process of oregano (*Origanum vulgare*), rosemary (*Rosmarinus officinalis*) (Moncada et al. 2016).

The waste management and distribution stages represent a crucial issue to reduce environmental impacts (Bisinella et al. 2024), but since most essential oils have countless applications, it would be necessary to evaluate a considerable number of waste generation and disposal scenarios, which would lead to a complex analysis work; therefore, in all the analyzed papers a cradle to gate scope is presented, where the waste is presented only as "system outputs" with a generic final disposition established in the LCA analysis software. The results allow concluding in most of the cases, that the impacts associated with transportation are not very relevant compared to the other stages of the process as plant cultivation or catalytic transformation stages.

Regarding the papers that address the catalytic transformation stage of essential oils or their components (Ravelli et al. 2011; Isola et al. 2018; Gallego et al. 2021;

Becerra et al. 2022), it is evident how a synergy is generated between catalysis and environmental evaluation that allows the development or improvement of chemical processes. It is possible to make decisions or choices (type of catalyst, solvents, or reaction conditions) not only considering parameters such as yield, conversion, or purity, but also including the analysis of the associated environmental impacts. A notable aspect of some of the analyzed papers is that the environmental analysis is framed within feasibility studies (Gwee et al. 2020; Moncada et al. 2016), evaluating the process or product of interest technically, economically, and environmentally. In this way, the authors obtain scenarios that are technically possible, economically beneficial, and environmentally friendly.

The cases presented in Table 1 of extraction and transformation processes of terpenes, show how environmental analysis allows the development or adaptation of new technologies or the improvement of existing processes to produce of essential oils and their derivatives, considering current guidelines focused on green chemistry, and sustainability. The four principles of eco-friendly extraction include renewable plant materials, alternative and safe solvents, lower energy consumption, and bio- and agro-refining industries (Li et al. 2023). LCA tool generates environmental information of interest on raw materials, products, by-products, waste, stages and/or routes associated with the biorefineries where the biomass management processes are carried out (Clark and Deswarte 2015). Those results allow decision-making regarding improvements or changes, where the performance of the process is not affected, and in turn, the impact on the environment is reduced. Other aspects to be considered in the comparison of essential oil extraction processes are related to the energy requirements, time, equipment costs and extraction efficiency for evaluating profitability of the plant.

4 Towards a Circular Economy: LCA and Improvement of Essential Oil Production Processes

There is a current great interest in solving contemporary environmental problems, which are the result of the historical and permanent action of human beings on natural resources. A large extent, the current economic model is based on business activities inherited from the time of the Industrial Revolution (ECV 2009), under the assumption that there will be a constant and economically viable supply of natural resources, the economic development that has taken place since then has been accompanied by the thinking—now almost outdated—that consists of "extract—manufacture—dispose"; what is known as linear economy (Morseletto 2023). But, given that the resources of the planet are a limited asset, when humanity follows a linear economy model to satisfy its needs, the characteristics of the Earth are modified and generates several problems on a planetary scale, Fig. 2.

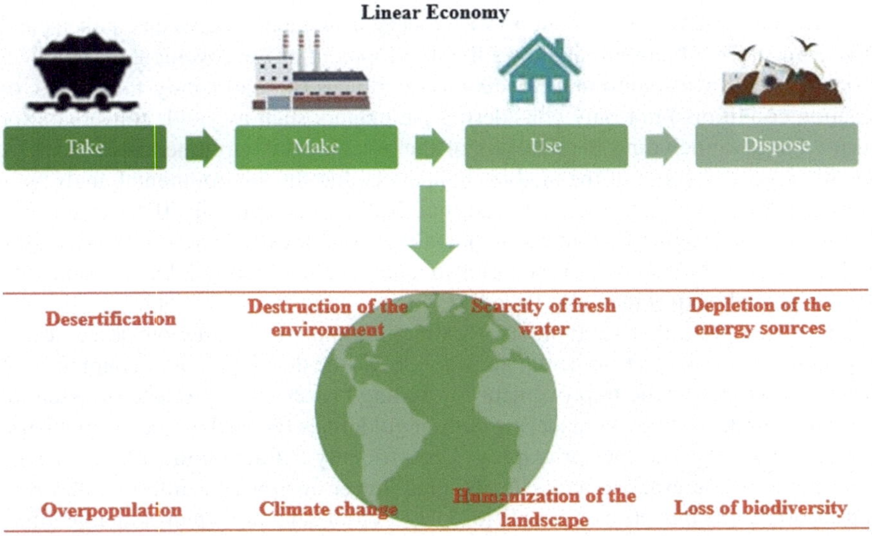

Fig. 2 Linear economy and its effects on the Earth

Consequently, the linear economy approach based on the extraction of resources, the production of goods and services, and, once their use is complete, the elimination of waste, is being displaced by the approach where the search for efficiency, use of resources and moderation are the guiding principles, the circular economy. In this "green" production/economy approach, the generation of value is prioritized, preventing it from leaking out of the production chain or system; it favors the recovery of materials and prevents unnecessary losses of resources from being perpetuated. It seeks to reduce waste, recover resources at the end of a product's useful life and channel them back into production, significantly reducing pressure on the environment (Liu and Kringos 2024). That is, the circular economy seeks to close the life cycle of products, promoting reuse, recycling, and regeneration of resources. This transition towards a more sustainable economy is essential to guarantee a better tomorrow for future generations.

The bioeconomy concept relates the circular economy with the processes associated to biomass; it is related to the consumption and production of goods and services derived from the sustainable transformation of biological resources and the biomass waste generated in the processes associated with their transformation, production and consumption, emulating biological principles (Martinez 2020).

Therefore, circular economy of processes associated with biomass, involve recycling and valorization of renewable agro-industrial waste into a mix of new products with added value, such as feed, food, chemicals, energy and fuels (Chaudhary and Singh 2024). Circular bioeconomy has as basic principles (Muscat et al. 2021): take care of agro-ecosystems, consuming only the required products avoiding their wasting, preferring biomass sources as raw materials, using and recycling secondary products from agro-ecosystems, using renewable energy and reducing energy consumption.

Considering the characteristics of the circular bioeconomy and that essential oils are a type of biomass generated from agricultural and/or forestry activity (Tsagaraki et al. 2017), it is adequate to consider the processes of obtaining and transforming these oils as bioeconomic processes that meet the condition of reuse and recycling of agro-industrial wastes. For example, turpentine oil is a byproduct generated in the paper industry, essential oils as citrus essential oils are byproducts of the food industry that generally result as secondary streams of other processes. The use in essential oils transformation of LCA analyses and their results, allow the optimization of these processes, through the identification of environmental impacts that permits the improvement of the process and the incorporation within the guidelines of the circular bioeconomy. From the analysis of the cases shown in Sect. 3 about environmental impacts evaluated by LCA of essential oil extraction and catalytic transformation processes, is possible to establish the following common points with the circular bioeconomy.

- Materials that are obtained from agro-industrial wastes are transformed into sources of new opportunities and trends for economy and society. A key challenge of the last decades is the development of efficient processes to convert biobased renewable feedstocks into fine chemicals, pharmaceuticals, food additives, bioenergy, biofuels, and polymers that are currently obtained from the petrochemical industry.
- The use of catalysts allows to produce goods and services while reducing material resource and energy use. It is known that the use of clean catalytic technologies, especially with heterogeneous catalysts is the key to rapid chemical transformation and can help to maximize the use of the available feedstocks. Therefore, the use of catalysts and optimal reaction conditions (pressure, temperature, pH, reaction time, among others) has become a crucial tool in biomass transformation processes in biorefineries.
- The inclusion of environmental impact assessment tools in the conceptualization and design stages of chemical transformations contributes to the improvement and technical and economic optimization of processes, because it allows to identify better options in terms of raw materials, catalysts, solvents, synthesis routes, and water/energy consumption.

Even though any plant where biomass processing is carried out is considered a biorefinery, these in themselves cannot be classified as examples of a circular bioeconomy. To achieve this, strategies must be developed or implemented that enable biomass to be transformed into energy, heat, chemicals, and other valuable materials, maximizing the value of biomass and minimizing waste (Jong and Jungmeie 2015), and therefore environmental impacts. In other words, and for the case addressed in this chapter, to bring the processes of extraction/transformation of essential oils to the level of circular bioeconomy, the environmental studies are vital, in addition to implementing the actions or improvements that arise on already established processes or biorefineries or performing technical–economic studies since the planning stage of the process.

5 Essential Oil Extraction and Transformation: Opportunities of a Sustainable Process

In recent decades, concern for the competitiveness of nations has significantly increased, especially in developing countries, where there is a growing awareness of the importance of this factor for economic growth and the improvement of the life quality of population (Boom et al. 2018). Companies and governments must inexorably improve productivity in the existing economic sectors in each territory by increasing product quality, adding desirable characteristics, improving production technology, and increasing production efficiency.

The United Nations has promoted strategies for global benefit with the formalization of the Sustainable Development approach (UN 2015) which corresponds to a guide about how we should live today if we want a better future, addressing present needs and assuring the opportunity of future generations to meet theirs. In 2015, all members of the United Nations adopted the 2030 Agenda for Sustainable Development. It includes a plan which will allow to get the peace and prosperity of everybody in the planet. At their core are the 17 SDGs (UN 2015), Fig. 3, which require the action of all countries in a cooperative way.

Previous sections showed that it is possible to establish the benefits and opportunities that are generated by linking the life cycle assessment to the study of the

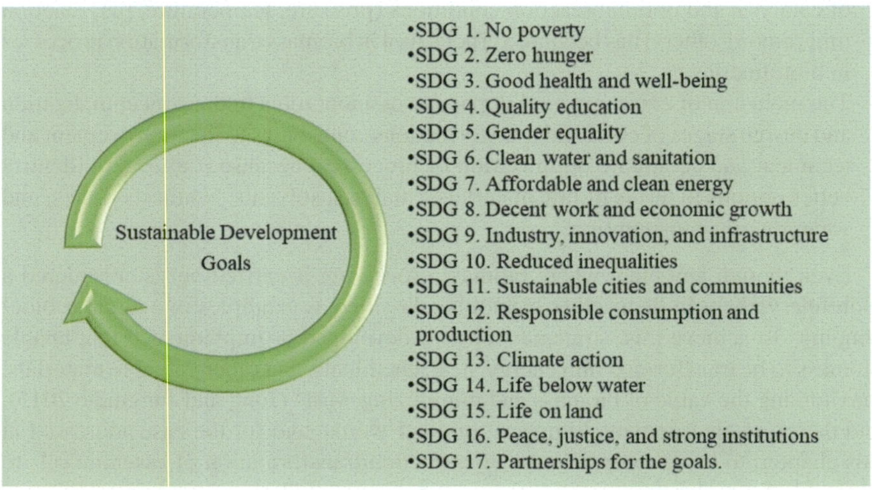

Fig. 3 Sustainable development goals

processes of extraction and transformation of essential oils. The transformation of wastes gives a great opportunity to get sustainable procedures that allow their conversion into high added value products (Chaudhary and Singh 2024). The main objective of these strategic routes (Processes—LCA—circular economy) is to be long-term sustainable policies that will give environmental and socioeconomic opportunities. To illustrate this, some of the common characteristics and methodological recommendations to be implemented derived from the LCAs, listed in Sect. 3 are shown, and their relationship with the SDGs is presented.

- One aspect that has become evident in many of the papers is the study of replacing traditional raw materials or routes in the catalytic processes by biodegradable raw materials to mitigate environmental impacts. In this sense, the replacement of raw materials with high environmental impact, such as petrochemicals, by biomass is a strategy that helps achieve the **SDGs 11–15**: implementation of sustainable processes and protection of biodiversity.
- Many of the articles integrate economic analyzes or suggest them as a complementary analysis. With this integration, companies seek to generate both economic and environmental benefits that enable them to grow financially and innovate in their processes, while maintaining environmental standards. That is, contributing to achieve the **SDGs 8, 9, 12,** and **13**.

Given the complexity of the study of catalytic processes, many authors incorporate into the methodology, together with LCA, the calculation of the so-called **green chemistry metrics**, metrics that measure aspects of a chemical process related to green chemistry principles (Sheldon 2018). These metrics facilitate the quantification of the environmental performance of chemical processes, and also allow evaluating the effect of changes or improvements made to chemical processes. Metrics as E factor, atom economy, mass intensity, solvent intensity, and carbon economy, contribute to quantify specific aspects of catalytic synthesis in terms of sustainability. Contributing to reducing pollution, that is, acting together with the **SDGs 6–7, 13–15**.

- The efficient use of energy is another aspect that is relevant and is present in most of the analyzed works. After the impact of greenhouse gas emissions, the effect of energy consumption is one of the most relevant. Its analysis is carried out simultaneously with the LCA, commonly using the Cumulative Energy Demand methodology. Strategies to minimize energy consumption and search for new energy sources are related to the **SDGs 7, 11,** and **13**.
- In general, the stage with the greatest contribution to environmental impacts is plant cultivation. In this sense, one of the aspects to improve is the reduction of the use of fertilizers and pesticides or using green products, to reduce damage to ecosystems, contributing to **SDGs 6, 7,** and **13–15**.
- The use of solvents, in the extraction/purification stage and/or in the catalytic transformation stage, constitutes another relevant aspect, and the transition towards the use of green solvents or the implementation of technologies that optimize the use of solvents are some of the most frequent aspects of these works. With this strategy,

a considerable reduction of environmental impacts that affect bodies of water and the environment is reached and contribute to reach **SDGs 6,** and **13–14**.

It was illustrated that following the recommendations obtained from LCA studies associated with the essential oil agroindustry, is possible to achieve the SDGs, for establishing a more economical and environmentally friendly process. Through innovation, knowledge transfer and research, it will be possible to get a mature stage of the essential oil agroindustry with a focus towards circular bioeconomy and sustainability.

6 Conclusions

This chapter shows the importance and how the use of environmental study tools can contribute to achieving the goals of Sustainable Development. This, using a study strategy: Process—LCA—Principles of Circular Economy, applied to the extraction and transformation processes of essential oils and their derivatives. The study of the production processes of essential oils, and their environmental evaluation through LCA allows us to show how the linear economy, approach that is based on the extraction of resources, the production of goods and services, and, their disposal once their use is complete, seems to be changing in recent years. Through the results of the LCA, regarding to propose preventive and corrective actions to mitigate the potential environmental impact, it is possible to get more out of the money invested (generate more wealth, expand access to wealth, health, and security, and reduce environmental impacts) by improving processes and production chains, obtaining positive results in terms of time, money, materials and impact on the environment. The life cycle assessment promotes a more sustainable pace of production and consumption and help to take advantage of limited financial and natural resources more efficiently.

In this review the environmental impacts were analyzed not only for the essential oil extraction, but also for some transformations of some essential oils for obtaining more added value compounds. The review of the selected papers for this chapter permit us to establish concrete examples of how the essential oil sector has been migrating from a linear economy to a circular economy; some of these examples correspond to the use of forest biomass residues from the paper production process to produce terpenes and derivatives, and the use of waste from the citrus juice industry to produce essential oils and pectin. In general, the studied reports are examples of the transition from a linear economy to a circular economy in production processes that involve the use of biomass, where the authors seek to reduce waste, recover resources at the end of a product's useful life and channel them back into production. This transition requires changes in the way we produce and consume biological resources, as well as in how we manage agro-industrial wastes. It is necessary to encourage reuse, recycling, and regeneration of resources, as well as promote the design of more durable and sustainable products. Furthermore, the collaboration of governments, companies and consumers is required to drive this transition towards a more circular economy. The responsibility falls on everyone, both individually and collectively, to adopt more sustainable practices and look for alternatives to the linear economy.

As biomass is being used as a source of hydrocarbons, it is a challenge to identify for each of the process steps, the sustainability of the process for concluding the pertinence of the process as an alternative to traditional ways of obtaining products from non-renewable sources. It is necessary to research about transformations of more components of essential oils by clean processes as heterogeneous catalytic transformations and to evaluate the environmental impacts of those processes. Furthermore, it is necessary a most integrated use of the agroindustrial wastes for extracting not only the essential oil, but also for obtaining products from other generated materials for working with a circular economy criterion. For example, extraction residues could be used to produce energy carriers, fertilizers, catalysts, and as source of bioactive compounds. On-site measurements will help to obtain a more realistic results of LCA and those data would contribute to increase the library of information about extraction of different essential oils, that grow in different parts of the World and that are obtained by different ways.

Finally, it is possible to establish how environmental analyzes of processes of obtaining and transforming essential oils generate information for taking actions that aim to contribute to the objectives of sustainable development. The improvements promote the process optimization and contribute to reducing the pollution of bodies of water and productive lands, enhancing air quality, and increasing availability of resources. Those actions lead to attacking poverty, hunger, and inequalities, while reducing climate change and the overexploitation of land, contributing to the improvement of life in cities and presenting innovations of rural agricultural processes. The results reached with essential oils can be extrapolated to sustainability studies of vegetable oils, and in general to processes that involve the use of biomass.

Acknowledgements The authors thank financial support from Universidad de Antioquia and the Ministry of Science, Technology and Innovation, the Ministry of Education, the Ministry of Industry, Commerce and Tourism, and ICETEX, Programme Ecosistema Científico-Colombia Científica, from the Francisco José de Caldas Fund, Grant RC-FP44842-212-2018.

References

Arias J, Muñoz F, Mejia J, Kumar A, Villa A, Martinez J, Stashenko E (2023) Advances in sample preparation simultaneous extraction with two phases (modified supercritical CO_2 and CO_2-expanded liquid-) to enhance sustainable extraction/isolation of pinocembrin from Lippia origanoides (Verbenaceae). Adv Sample Prep 6(100059). https://doi.org/10.1016/j.sampre.2023.100059

Beccali M, Cellura M, Ludicello M, Mistretta M (2009) Resource consumption and environmental impacts of the agrofood sector: Life cycle assessment of Italian citrus-based products. Environ Manag 43(4):707–724. https://doi.org/10.1007/s00267-008-9251-y

Becerra J, González J, Villa A (2022) Comparison by life-cycle assessment of alternative processes for carvone and verbenone production. Molecules 27(17):5479. https://doi.org/10.3390/molecules27175479

Bisinella V, Schmidt S, Varling A, Laner D, Christensen T (2024) Waste LCA and the future. Waste Biomass Valorization 174:53–75. https://doi.org/10.1016/j.wasman.2023.11.021

Boom E, Rodríguez E, Buelvas L (2018) Estrategias de desarrollo empresarial a partir de la extracción de aceites esenciales. Espacios 39(47):17. https://www.revistaespacios.com/a18v39n47/18394717.html

Can Baser K, Buchbauer G (2016) Handbook of essential oils, 2nd edn. CRC Press. https://doi.org/10.1201/b19393

Chaudhary S, Singh B (2024) Pineapple by-products utilization: progress towards the circular economy. Food Human 2:100243. https://doi.org/10.1016/j.foohum.2024.100243

Clark J, Deswarte, F (2015) The biorefinery concept: an integrated approach. In: Clark J, Deswarte F (eds) Introduction to chemicals from biomass, 2nd edn, pp 1–29. Elsevier. https://doi.org/10.1002/9780470697474.ch1

Clark J, Farmer T, Herrero-Davila M, Sherwood J (2016) Circular economy design considerations for research and process development in the chemical sciences. Green Chem 18:3914–3934. https://doi.org/10.1039/C6GC00501B

da Costa J, Maranduba H, de Sousa S, de Almeida J, Brito L (2022) Environmental performance of orange citrus waste as raw material for pectin and essential oil production. Food Bioprod Process 135:165–177. https://doi.org/10.1016/j.fbp.2022.07.008

ECV (2009) La economía lineal: Riesgos y consecuencias. https://economiacircularverde.com

Gallego L, Hasenstab C, Alarcon E, Villa A (2021) Identification by life cycle assessment of the critical stage in the catalytic synthesis of nopol using heterogeneous catalysis. Sustain Prod Consum 27:23–34. https://doi.org/10.1016/j.spc.2020.10.017

González-Aguirre J, Solarte-Toro J, Cardona C (2020) Supply chain and environmental assessment of the essential oil production using Calendula (*Calendula Officinalis*) as raw material es Gonz. Heliyon 6:e05606. https://doi.org/10.1016/j.heliyon.2020.e05606

Gwee Y, Yusup S, Tan R, Yiin C(2020) Techno-economic and life-cycle assessment of volatile oil extracted from *Aquilaria sinensis* using supercritical carbon dioxide. J CO_2 Util 38:158–167. https://doi.org/10.1016/j.jcou.2020.01.002

Hellweg S, Cannals L (2014) Emerging approaches, challenges, and opportunities in life cycle assessment. Science 344:1109–1113. https://doi.org/10.1126/science.1248361

Iñigo M, Díaz G, Machado D, de Nucci P, Paz D, Arena P (2023) Life cycle assessment of energy saving and circular economy strategies: contributions to the environmental footprint lemon and by-product production in Argentina. Clean Technol Environ Policy. https://doi.org/10.1007/s10098-023-02693-y

Isola C, Sieverding L, Numan A, Rajappagowda E, Boakye E, Raynie D, Smirnova A, Stone J (2018) Vanillin derived from lignin liquefaction: a sustainability evaluation. Int J Life Cycle Assess 23(9):1761–1772. https://doi.org/10.1007/s11367-017-1401-0

Jong E, Jungmeie G (2015) Biorefinery concepts in comparison to petrochemical refineries. In: Pandey A, Höfer R, Taherzadeh M, Nampoothiri K, Larroche C (eds) Industrial biorefineries and white biotechnology. Elsevier, pp 3–34. https://doi.org/10.1016/B978-0-444-63453-5.00001-X

Kassim T, Simoneit B (2005) Environmental impact assessment: principles, methodology and conceptual framework. In: Kassim T, Williamson K (eds) Environmental impact assessment of recycled wastes on surface and ground waters. Springer, pp 1–57. https://doi.org/10.1007/b98263

Kukhta N, Vasilenko I, Kostjuk S (2011) Room temperature cationic polymerization of β-pinene using modified $AlCL_3$ catalyst: toward sustainable plastics from renewable biomass resources. Green Chem 13(9):2362–2364. https://doi.org/10.1039/C1GC15593H

Lambert M (2019) Cradle-to-gate life cycle analysis (LCA) of Ylang-Ylang complete essential oil destined for cosmetic use from central region, Ghana Using the Software SimaPro. Université

Libre de Bruxelles Institut. https://mem-envi.ulb.be/Memoires_en_pdf/MFE_18_19/MFE_Lambert_18_19.pdf

Li Q, Putra NR, Rizkiyah DN, Abdul Aziz AH, Irianto I, Qomariyah L (2023) Orange pomace and peel extraction processes towards sustainable utilization: a short review. Molecules 28:3550. https://doi.org/10.3390/molecules28083550

Liu Z, Kringos N (2024) Transition from linear to circular economy in pavement engineering: a historical review. J Clean Prod 449:141809. https://doi.org/10.1016/j.jclepro.2024.141809

Machin L, Wheeler J, Mele F (2022) Life cycle assessment of the Argentine lemon and its derivatives in a circular economy context. Sustain Prod Consum 29:672–684. https://doi.org/10.1016/j.spc.2021.11.014

Maham S, Rahimi A, Smit D (2018) Environmental assessment of the essential oils produced from dragonhead (*Dracocephalum moldavica* L.) in conventional and organic farms with different irrigation rates. J Clean Prod 204:1070–1086. https://doi.org/10.1016/j.jclepro.2018.08.348

Martínez C (2020) Economía circular y bioeconomía, pilares del desarrollo sostenible. Revista Colombiana de Biotecnologia 22(2):3–5. https://doi.org/10.15446/rev.colomb.biote.v22n2.93351

Moncada J, Tamayo J, Cardona C (2016) Techno-economic and environmental assessment of essential oil extraction from Oregano (*Origanum vulgare*) and Rosemary (*Rosmarinus officinalis*) in Colombia. J Clean Prod 112:172–181. https://doi.org/10.1016/j.jclepro.2015.09.067

Moncada J, Tamayo J, Cardona C (2014) Techno-economic and environmental assessment of essential oil extraction from Citronella (*Cymbopogon winteriana*) and Lemongrass (*Cymbopogon citrus*): a Colombian case to evaluate different extraction technologies. Ind Crops Prod 54:175–184. https://doi.org/10.1016/j.indcrop.2014.01.035

Morelli B, Hawkins T, Niblick B, Henderson A, Golden H, Compton J, Cooter E, Bare J (2018) Critical review of eutrophication models for life cycle assessment. Environ Sci Technol. American Chemical Society 52(17):9562–9578. https://doi.org/10.1021/acs.est.8b00967

Morseletto P (2023) Sometimes linear, sometimes circular: states of the economy and transitions to the future. J Clean Prod 390:136138. https://doi.org/10.1016/j.jclepro.2023.136138

Moura B, Monteiro H, Mata T, Martins A (2022) Life cycle energy and carbon emissions of essential oil extraction from Rosemary. Energy Rep 8(3):291–297. https://doi.org/10.1016/j.egyr.2022.01.063

Muscat A, Olde E, Ripoll-Bosch R, Zanten H, Metze T, Terneer C, Ittersum M, Boer I (2021) Principles, drivers and opportunities of a circular bioeconomy. Nature Food 2:561–566. https://doi.org/10.1038/s43016-021-00340-7

Ott K, Mohaupt F, Ziegler R (2012) Environmental impact assessment. In: Chadwick R (ed) Encyclopedia of applied ethics, pp 114–123. https://doi.org/10.1016/B978-0-12-373932-2.00345-8

Parsons S, Allen M, Abeln F, McManus M, Chuck C (2019) Sustainability and life cycle assessment (LCA) of macroalgae-derived single cell oils. J Clean Prod 232:1272–1281. https://doi.org/10.1016/j.jclepro.2019.05.315

Pérez-López P, Jeffryes C, Agathos S, Feijoo G, Rorrer G, Moreira M (2016) Environmental life cycle optimization of essential terpene oils produced by the macroalga Ochtodes secundiramea. Sci Total Environ 542:292–305. https://doi.org/10.1016/j.scitotenv.2015.10.045

Pino Alea J (2015) Aceites Esenciales, Editorial Universitaria. ISBN 978-959-16-2556-4

Ravelli D, Protti S, Neri P, Fagnoni M, Albini A (2011) Photochemical technologies assessed: the case of rose oxide. Green Chem 13(7):1876–1884. https://doi.org/10.1039/C0GC00507J

Ríos J (2015) Essential oils: what they are and how the terms are used and defined. In: Preedy V (ed) Essential oils in food preservation, flavor and safety. Elsevier, pp 3–10. https://doi.org/10.1016/B978-0-12-416641-7.00001-8

Rubulotta G, Quadrelli E (2019) Terpenes: a valuable family of compounds for the production of fine chemicals. In: Albonetti S, Perathoner S, Quadrelli E (eds) Horizons in sustainable industrial chemistry and catalysis. Elsevier, pp 215–229. https://doi.org/10.1016/B978-0-444-64127-4.00011-2

Santiago B, Moreira M, Feijoo G, Gonzalez-García S (2020) Identification of environmental aspects of citrus waste valorization into D-limonene from a biorefinery approach. Biomass Bioenerg 143:105844. https://doi.org/10.1016/j.biombioe.2020.105844

Sheldon R (2018) Metrics of green chemistry and sustainability: past, present, and future. ACS Sustain Chem Eng 6:32–48. https://doi.org/10.1021/acssuschemeng.7b03505

Silvestre A, Gandini A (2008) Terpenes: major sources, properties and applications. In: Belgacem M, Gandini A (eds) Monomers, polymers and composites from renewable resources. Elsevier, pp 17–38. https://doi.org/10.1016/B978-0-08-045316-3.00002-8

Sonnemann G, Gemechu E, Sala S, Schau E, Allacker K, Pant R, Adibi N, Valdivia S (2018) Life cycle thinking and the use of LCA in policies around the world. In: Hauschild M, Olsen S, Rosenbaum R (eds) Life cycle assessment: theory and practice. Springer, pp 429–463. https://doi.org/10.1007/978-3-319-56475-3_18

Teigiserova D, Hamelin L, Tiruta-Barna L, Ahmadi A, Thomsen M (2022) Circular bioeconomy: life cycle assessment of scaled-up cascading production from orange peel waste under current and future electricity mixes. Sci Total Environ 812:152574. https://doi.org/10.1016/j.scitotenv.2021.152574

Tsagaraki E, Karachaliou E, Delioglanis I, Kouzi E (2017) D2.1 Bio-based products and applications potential. https://www.bioways.eu

UN (2015) Sustainable development. https://sdgs.un.org/

Circular Economy Framework as a Tool for the Construction of Prospective Scenarios for the Energy Management of a Higher Education Institution

N. G. Faitani, D. N. Lopez de Munain, and R. O. Bielsa

Abstract To improve their institutional commitment and internal campus operations, higher Education Institutions (HEIs) have developed new tools and implemented programmes to achieve the Sustainable Development Goals. In this sense, energy management became one of the focal points of institutional management to improve the environmental footprint and simultaneously contribute to the reduction of institutional expenditure. In this work, a diagnosis and next energy analysis of an administrative building of the Universidad Nacional de General Sarmiento (Buenos Aires, Argentina) is carried out to determine the management trend between 2015 and 2022. The indicators used for the analysis are those provided by the ReSOLVE and DISRUPT frameworks of the Circular Economy, adapted to an institutional context. The information for the energy diagnosis is based on a survey of the structures of the building taken as a case of analysis. Through the construction of the trend scenarios, the replacement of LED lights and the virtualisation of administrative activities improved the Institution's energy footprint. Results obtained after the construction of the future scenarios showed that the following actions are relevant for energy care and reduction programmes: application of automatic switch-on and switch-off devices; collaborative work; and the institutionalization of improvement and environmental management programmes.

Keywords Energy management · Circular economy · ReSOLVE · DISRUPT · Higher Education Institutions · Index

N. G. Faitani
Consejo Nacional de Investigaciones Científicas y Técnicas—CONICET, Buenos Aires, Argentina

N. G. Faitani · D. N. Lopez de Munain
Universidad Nacional de José Clemente Paz, José Clemente Paz, Buenos Aires, Argentina

N. G. Faitani · R. O. Bielsa (✉)
Universidad Nacional de General Sarmiento, Los Polvorines, Buenos Aires, Argentina
e-mail: rbielsa@campus.ungs.edu.ar

L. Brandli et al. (eds.), *The Contribution of Life Cycle Analyses and Circular Economy to the Sustainable Development Goals*, World Sustainability Series,
https://doi.org/10.1007/978-3-031-85300-5_2

1 Introduction

Higher Education Institutions (HEIs) are key agents for the socio-economic development and environmental care of its regions through knowledge creation (research), knowledge transfer (teaching) and internal and external community development (social action activities). They have a particular responsibility in training future professionals and putting knowledge and ideas into practice (Žalėnienė and Pereira 2021).

HEIs therefore play a key role in supporting sustainable development as they ease the processes through which sustainability innovations can flourish (Stephens et al. 2008). This can be achieved by reducing the environmental footprint of the institution, engaging strongly with communities, and strengthening the institutionalization of environmental actions (Faitani et al. 2021). Furthermore, when it comes to greening a university campus, it starts by applying strategies about waste management and energy management (Leal Filho et al. 2015), which is extremely important as a starting point for university institutions embarking on the path towards sustainability. Furthermore, an exploratory study conducted by Leal Filho et al. (2023) found that HEIs have a broad focus on sustainability research and investigate the greening of their campuses as the third most researched topic. Regarding energy management in university buildings, a study by Lange Salvia and Londero Brandli (2020) mentions that research is the main area that universities have been using to contribute to sustainable development in the energy context.

The Circular Economy (CE) is presented as a key strategy for the improvement of campus performance. HEIs are also considered strategic actors to support the transition to a more circular economy (Schröder et al. 2020). Furthermore, CE should be applied in all pillars of HEI functioning, i.e. education, research and operations (Lozano 2006). CE actions should be interlinked across all pillars of functioning, supported by the governance of the institution, as it influences and guides the decisions made in the areas of functioning of a HEI (Faitani 2023). CE helps to solve complex problems, as it studies structures from a systemic perspective, to understand their internal workings and find the most efficient paths or flows for the use of resources and waste and their reinsertion (CIEC 2019). Similarly, the research conducted in HEIs is a key point for renewing the proposal of CE actions or environmental issues in general. The strong presence of the CE research pillar is an advantage within HEIs, as it can be the link between them and the public or private governmental sectors (Faitani et al. 2022). Figure 1 shows the interrelationship of governance and the three pillars of HEI functioning.

The implementation of CE requires systemic and holistic innovation among stakeholders, including HEIs. HEIs propose improvements in production processes through the creation and implementation of policies to ensure that the solid waste generated on university campuses has a sustainable destination for next reuse (Faitani et al. 2021).

According to Nunes et al. 2018, CE can act in three ways within HEIs. The first aim is to build material flows where the introduction of CE would help those flows

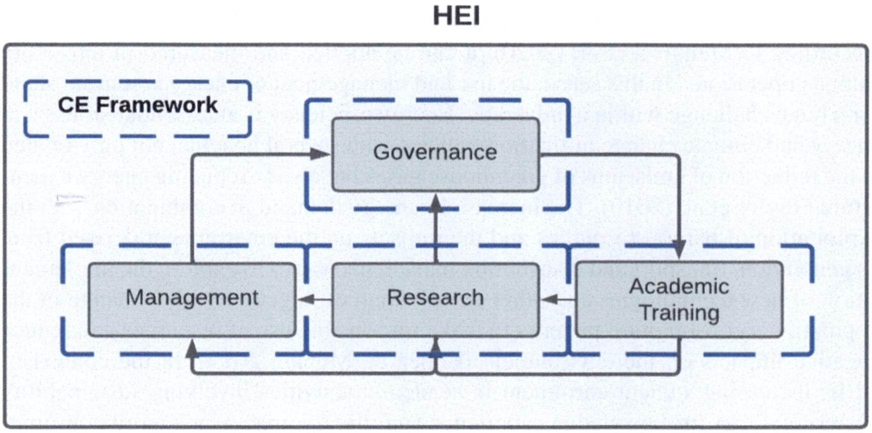

Fig. 1 Interrelation of CE decisions from the governance to the operational pillars of HEIs. *Source* Based on Faitani (2023)

by decreasing the input of new products and the negative outputs of the system. The second aim is to raise awareness in the whole university community. The third aim is to transform the HEI into a catalyst for the application of CE in different economic activities. However, universities need to advance their sustainability plans to meet all these aims, and fully contribute to a circular economy. Furthermore, they need to create the enabling policy framework for this.

Two theoretical frameworks of the Circular Economy were created to generate and implement strategies in business settings. The ReSOLVE framework was created by the Ellen McArthur Foundation (2019). It takes the basic principles of circularity and applies them to six actions: Regenerate, Share, Optimize, Circularise, Virtualise and Exchange. On the other hand, the DISRUPT theoretical framework, which was created by the Circle Economy group. This framework considers these basic principles: Design for the future, Incorporate digital technology, Sustain and preserve what is already there, Rethink business models, Use waste as resources, Prioritize regenerated resources, Team up to create joint value (Circle Economy 2020).

The principles in both theoretical frameworks are used as indicators for the application of CE in companies. Their application in university contexts is not commonly used, and only some examples of application have been presented by Mendoza et al. (2019, 2020). In both works, the usefulness of the tool to reach significant findings to evaluate and prioritise the implementation of CE strategies is highlighted. In the case of the DISRUPT framework, no work has been found that has been applied in HEI contexts. Moreover, the insights provided by CE are often substantially associated with recirculation in material terms, so its application, particularly with reference to the ReSOLVE and DISRUPT frameworks, to the specific study of energy management and more broadly to other pillars of HEI operations is a novel aspect.

The implementation of CE strategies in HEIs takes place in diverse ways. According to Mendoza et al. (2020), it can be applied and measured in university campus operations. In this sense, the use and management of energy resources stand for a huge challenge within a university. Energy efficiency is an essential strategy in energy and climate change mitigation policies with several benefits, not only related to the reduction of emissions of greenhouse gases but also to reducing energy expenditure (Taylor et al. (2010). The increase in energy demand in combination with the exploitation of natural resources and the impacts on the environment derived from its generation, transport and distribution make it necessary to explore the implementation of new technologies and other energy sources, together with a change in the population's consumption patterns to make responsible use of resources and reduce negative impacts on the environment (Lopez de Munain 2017). In the context of HEIs, increasing student enrolment in academic activities involving sustainability is a critical part for knowledge economies and direct impacts on energy consumption with consequences on environmental, ethical, and financial aspects (Mtutu and Thondhlana 2016). The study of the energy consumption patterns, and their characteristics should also considered factors such as the diverse types of use of equipment and spaces, the surface area of the HEI, seasonality, and the use of energy from renewable sources (Leal Filho et al. 2019a, b).

Frameworks used for the study of energy resource management include ISO 50001 standard which provides guidelines to improve energy use through the development of an energy management system (EnMS) (ISO n. d.). This framework is used for the study of energy use and consumption in works such as Diaz et al. (2023) applied to the Autonomous University of the West, Colombia; Ocampo Battle et al. (2019) on Federal University of Itajubá, Brazil; Pinzón Casallas et al. (2014) which, together with the implementation of energy indicators, applies this methodology to the Alejandro Suárez Copete Building of the Francisco José de Caldas District University. Against this background and considering the role of HEIs as an element in the training of professionals and their link with the community, the inclusion of strategies and actions related to energy management and its rational and efficient use in their facilities is of utmost importance (Lange Salvia and Londero Brandli 2020), as well as innovation in relation to the theoretical-methodological frameworks used for this purpose.

Once these measurements have been taken to identify, evaluate and prioritise the application of CE solutions in university contexts, it is important to assess the information to improve energy management in the institution's buildings. In this sense, being able to understand how the system will function in the future is of great strategic interest for the allocation of the university budget. Strategic foresight is considered a formalized approach that uses a combination of qualitative and quantitative tools to derive multiple scenarios of future alternatives, supported by a comprehensive analysis and incorporating possible actions and their consequences (Bradfield et al. 2005).

In Argentinean universities, the inclusion of sustainability and particularly the optimisation of energy consumption on university campuses, is contemplated through the resolutions of the National Interuniversity Council (Resolution No. 1195/2016), which proposes a Sustainable University Infrastructure Programme, which must also be implemented under the Strategic Plan for Environmental Management and Risk Prevention approved by the same council (Resolution No. 1054/2015). Both resolutions aim for HEIs to reduce the impacts generated by university campus activities. Furthermore, in 2023, a new resolution declares of national interest the National Sustainability Strategy for Argentine Universities (Resolution No 1766/2023). This strategy was proposed by a Network of Argentinean Universities for Environmental Management and Social Inclusion (UAGAIS, Spanish acronyms), which includes universities from all over the country, and particularly the HEI taken as a case study in this paper. According to Rokicki (2020), in terms of energy policy, the circular economy is closely related to increasing energy efficiency, developing renewable energy sources, and reducing greenhouse gas emissions. Argentinian HEIs have an enabling regulatory framework to implement activities aimed at minimising the use of energy resources and maximising the use of renewable energy sources and increasing energy efficiency. However, the limitations for the implementation of the strategies are related to the economic budget, institutional governance, and the lack of institutionalisation of the actions (Nunayon et al. 2021; Amaral et al. 2020).

In this work, a diagnosis and later energy analysis of an administrative building of the Universidad Nacional de General Sarmiento (Buenos Aires, Argentina) is conducted to determine the management trend between 2015 and 2022. The indicators used to conduct the analysis are those provided by the ReSOLVE and DISRUPT frameworks of the Circular Economy, adapted to an institutional context. The information for the energy diagnosis is based on a survey of the structures of the building taken as a case of analysis. The future scenario is built through the construction of the baseline scenario, where some aspects that are recognised in the baseline scenario as key points for the application of Circular Economy criteria in energy management are improved.

2 Method

This work takes as a case study the Universidad Nacional de General Sarmiento (UNGS) which is in the municipality of Malvinas Argentinas, located in the Metropolitan Region of Buenos Aires (RMBA), Argentina. Figure 2 shows the location of the HEI taken as a case study.

RMBA and UNGS were chosen as a case study institution and area because RMBA is the main urban conglomerate of Argentina. RMBA is a typical megacity of Latin America, with its centre in the Ciudad Autónoma de Buenos Aires (Buenos Aires City, CABA), and 40 surrounding municipalities of the Province of Buenos Aires (Fig. 2). Its geographic gross product is 40% of the national gross domestic product. It has 40% of the country's industries, with plants for the manufacture of cars, food, bovine leather, textiles and paper and cardboard as well as shopping malls and large supermarkets that are considered generators of commercial and industrial waste. About 23.5% of the country's total population lives in RMBA (INDEC 2023), where

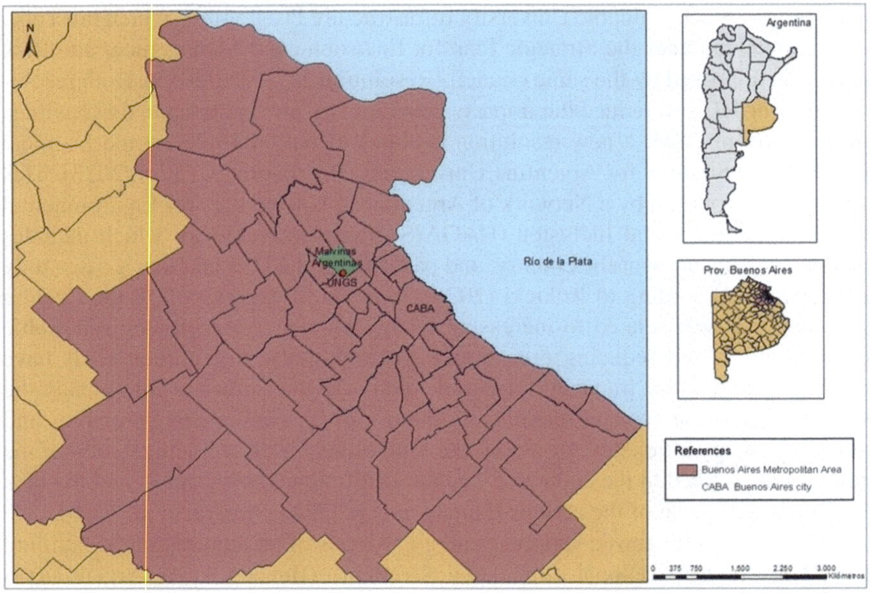

Fig. 2 Location of the National University of General Sarmiento and Buenos Aires Metropolitan Region. *Source* Geographic Information System Laboratory of the National University of General Sarmiento

the student population of national universities reaches 350 thousand students, which is 17% of the student population nationwide (SPU 2021). Most of these RMBA universities are "young" institutions because they are 30 years old on average or less. They were created in the decade of 1990 to offer high education opportunities to thousands of students that could not access the traditional University of Buenos Aires due to typical problems of megacities (long time transportation, crowded classes, etc.). The national universities in Argentina are not tariffed so many of these HEI supply a nearby and accessible alternative to university education. HEI's main goal is to become a space for research, teaching, and learning, providing opportunities for people in RMBA to improve their social and economic development.

The case study strategy is useful in exploratory investigation to seek evidence through data collection and compare the result to theory for the examining and understanding of contemporary events (Yin 2014).

Likewise, the building corresponding to Module 1 (M1) within the UNGS Campus was taken as a case study (Fig. 3). This building is where the administrative activities of the University are carried out, so it has the most stable population in terms of attendance and hours of use of the site; the work schedule is 8 h per day from Monday to Friday. In addition, emphasis was placed on the operations of M1, those operations that require the use of electrical power to understand its performance against this criterion.

The method is divided into four steps (Fig. 4):

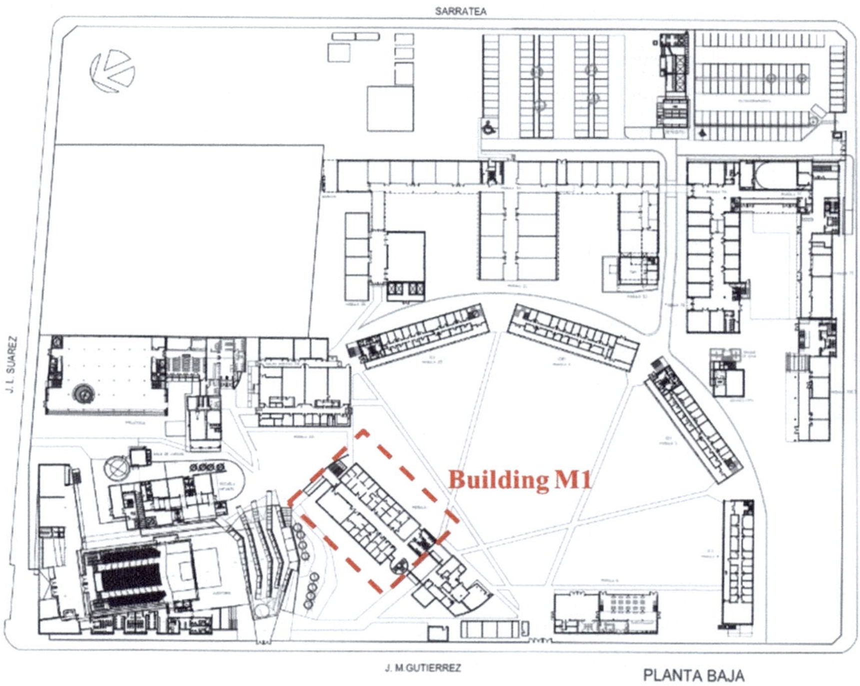

Fig. 3 Location of the Module 1 building in the UNGS Campus. *Source* Modified from plan provided by UNGS Administration Secretariat

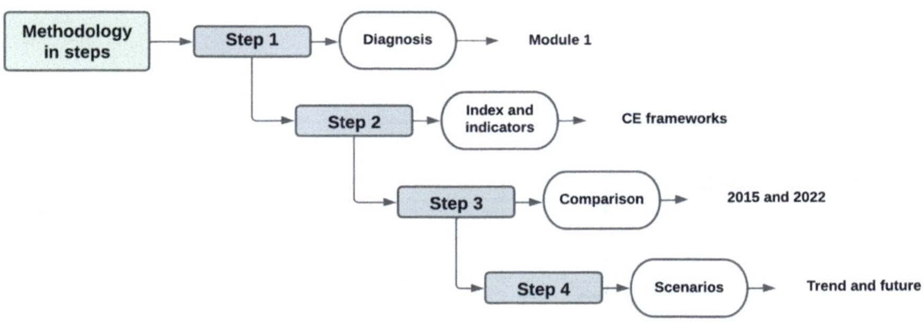

Fig. 4 Methodological scheme

Step 1—Diagnosis of the use of electrical equipment in the M1 building.
Step 2—Creation, adaptation, and application of CE indicators for the generation of an energy management index.
Step 3—Comparison of scenarios for the year 2015 and 2022.
Step 4—Generation of scenarios (Trend and Future).

Table 1 Types of equipment and examples of each of them

Types of electrical equipment	Examples of equipment
Heating, ventilating and air conditioning equipment	Air conditioners, fans, cookers, etc
Operational equipment	PCs, printers, routers, etc
Lighting equipment	Lamps, fluorescent tubes, LED lights, etc

2.1 Step 1—Diagnosis

First, a diagnosis of energy consumption was conducted, particularly the use of electrical energy, considering the type of equipment used and the installed power (in kW). Table 1 shows the types of equipment to be identified and examples of equipment.

2.2 Step 2—Index and Indicators

Secondly, although this paper uses different circular economy frameworks, only the management pillar, which stands for campus operations, will be emphasized to address the tangible activities related to energy use and management of the university campus. In addition, the activities that are implemented from the governance of the institution related to the CE will be considered. In the following Fig. 5, it is possible to see the edges that are specifically addressed in this work.

A proposal for indicators was generated based on the CE ReSOLVE and DISRUPT frameworks (Fig. 6). These indicators are presented in Table 2.

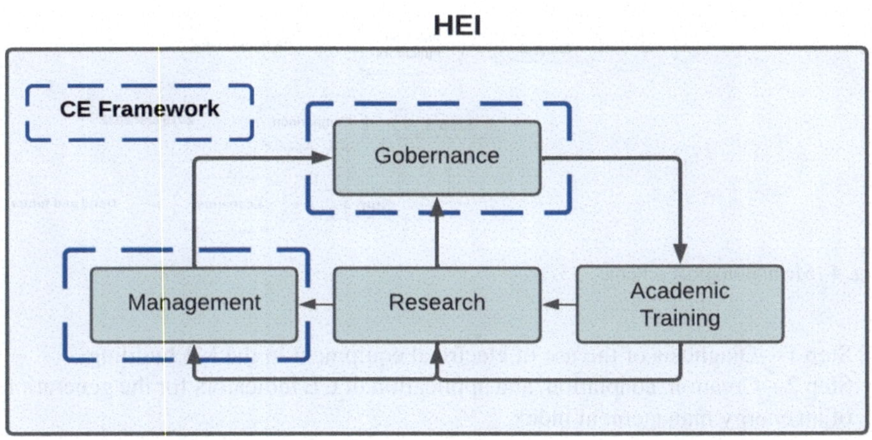

Fig. 5 Issues to be analysed, management and governance of HEIs

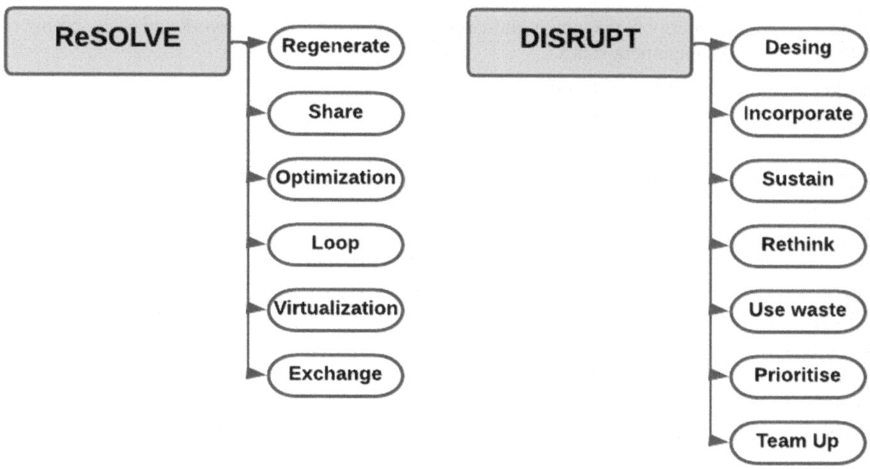

Fig. 6 ReSOLVE and DISRUPT circular economy frameworks

Once each indicator is scored, all assigned scores are added together. The following Eq. 1 stands for the Energy Management Index according to the CE (EMIce), composed of the six energy management indicators described in Table 2:

$$EMIce = EMI1 + EMI2 + EMI3 + EMI4 + EMI5 + EMI6 \tag{1}$$

Thus, the total score that EMIce can reach is a maximum of 18 and its minimum score is 0.

2.3 Step 3—Comparison

Thirdly, the performance of energy management in Building M1 in the years 2015 and 2022 was compared by applying the proposed indicators.

2.4 Step 4—Scenarios

Finally, the prospective scenario of a tendency character was elaborated based on the result of the comparison of the years 2015 and 2022. Then, the future scenario with the improvement in all aspects of energy management is considered in the proposed indicators (Fig. 7).

Table 2 Proposed energy management indicators based on the CE framework and their scoring to form the Energy Management Index

Energy management index—CEs (EMIce)			
Indicators	Description	Adequacy	Score
EMI 1 Use of renewable energies	Use of renewable energies to generate electricity	This electrical energy must be used for the operations of the M1 The collection of renewable energy can be in the M1 or another unit of the HEI	0 = Null 1 = Incipient 2 = Partial 3 = Total
EMI 2 Virtualization of administrative tasks	Use of virtualisation tools or software for administrative activities	Administrative activities that take place in M1. The tool can be used in the whole HEI campus or in any part of it	
EMI 3 Automating functions	Use of automation tools	Processes can be applied to switching on lights, shutting off water supply, automatic watering, etc	
EMI 4 Energy efficiency in processes	Processes requiring energy efficiency in their structural components	The processes can be applied to lighting, air conditioning or operating equipment	
EMI 5 Rethinking existing structures	Use of existing structures for the implementation of energy efficiency or energy use improvement strategies	The structures on which the M1 is based do not have any CE or energy-saving perspectives. In this sense, strategies are sought that can take advantage of existing structures. For example, green roofs, double glazing, orientation of desks, etc	
EMI 6 Teamwork and institutionalization of activities	CE activities with an emphasis on energy use in HEIs should be institutionalised or accompanied by institutional governance	The institution should institutionalise CE and energy management activities. The M1 community should be aware, engaged, and willing to report any suggestions or queries	

3 Results and Analysis

The M1 building has 2 floors and a total of 103 offices (see Fig. 8). As it is previously mentioned, it employs 184 employees who work eight hours a day from Monday to Friday. It has no environmental or energy management plans. In this regard, it only has the University Statute that incorporates environmental issues and sustainability (UNGS Statute 2017). Three main points could be found in this statute: establishment of a healthy environment, training of students with visions towards sustainability, and incorporation of environmental research as a priority. These points are related to the three pillars of an HEI: education, research, and management (Lozano

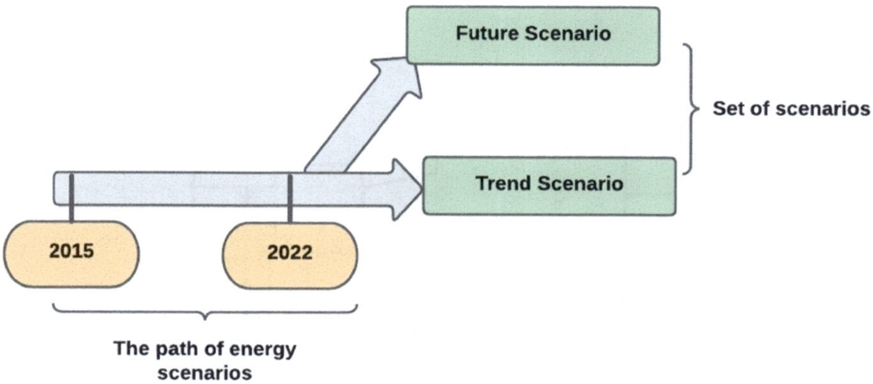

Fig. 7 Prospective energy management scenarios

2006). However, the statute does not provide for tangible strategies but only establishes fundamental rules for the organisation and functioning of an HEI. It is a basic document in which the necessary considerations for the development of university activities are written down and detailed in an orderly manner and respecting the rights and obligations of the community of the HEI. However, it does not show the environmental footprint the institution should adopt in the face of external needs.

The diagnostic results show that operational equipment stands for the largest amount of installed power (Fig. 9). This result is consistent with the type of activity conducted in Building M1. Among the equipment surveyed are PCs, printers, and others such as intercoms, coffee machines, refrigerators, and water dispensers. Concerning air conditioning equipment, split and low-silhouette air conditioners are installed, and to a lesser extent fans and cookers. Finally, regarding lighting equipment, there are mainly fluorescent tubes in the corridors and inside the offices, with an incipient replacement by LED luminaires.

Regarding the survey of activities to score the energy management indicators generated from the CE frameworks, the activities listed in Table 3 are shown in each year.

After surveying the activities that were conducted in 2015 and 2022, it is found that the UNGS tends to increase activities aimed at energy efficiency, which improves energy management in the institution and in the proposed Circular Economy indicators.

The survey indicates that neither the use of renewable energies nor strategies for automating the on/ off switching of lights are considered.

On the other hand, improvements in energy efficiency indicators have been verified. Although some offices are not extensively used by M1 administrative staff, they are occasionally used for meetings. The virtualisation strategies were intensified and reinforced after the COVID-19 pandemic and the environmental policy has been under development since 2020. Unfortunately, environmental policy and plan have not yet been finalised.

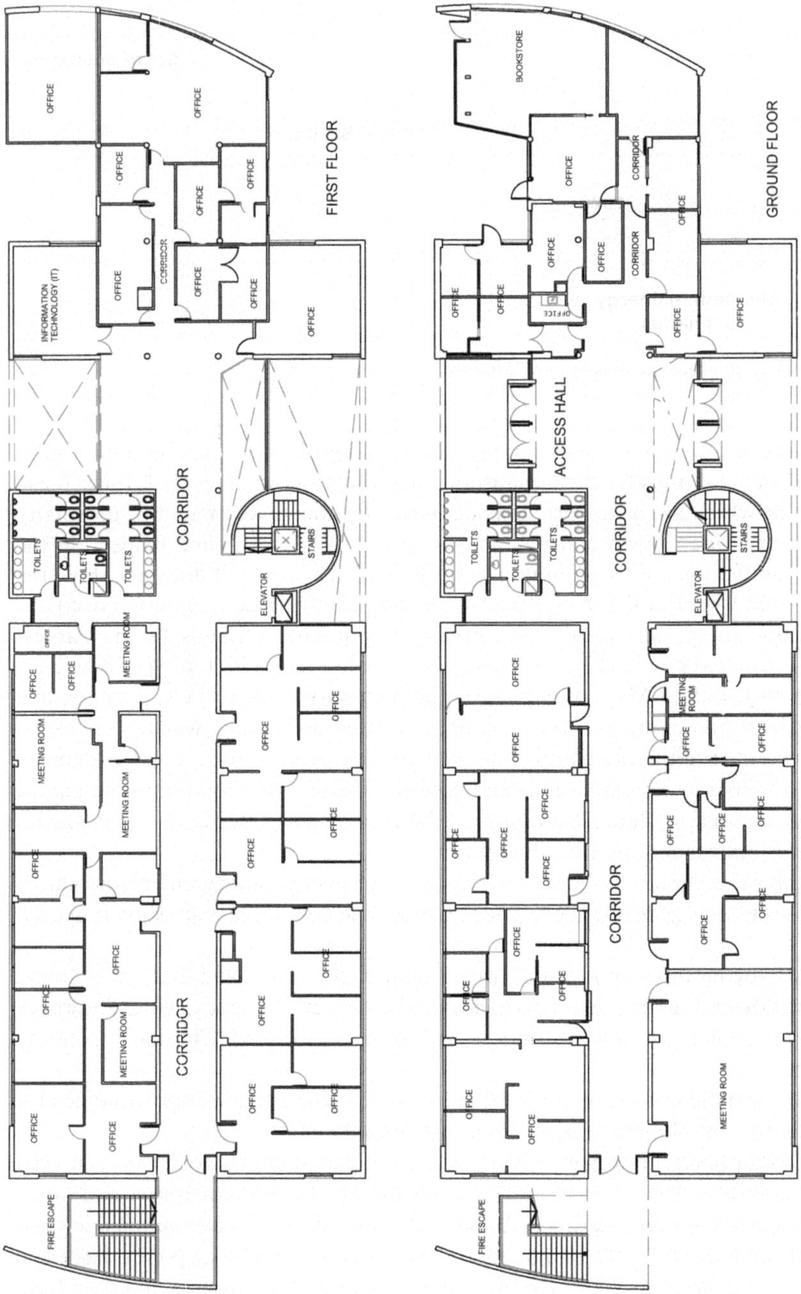

Fig. 8 Plan of the M1 building in UNGS Campus. *Source* UNGS Secretariat of Administration

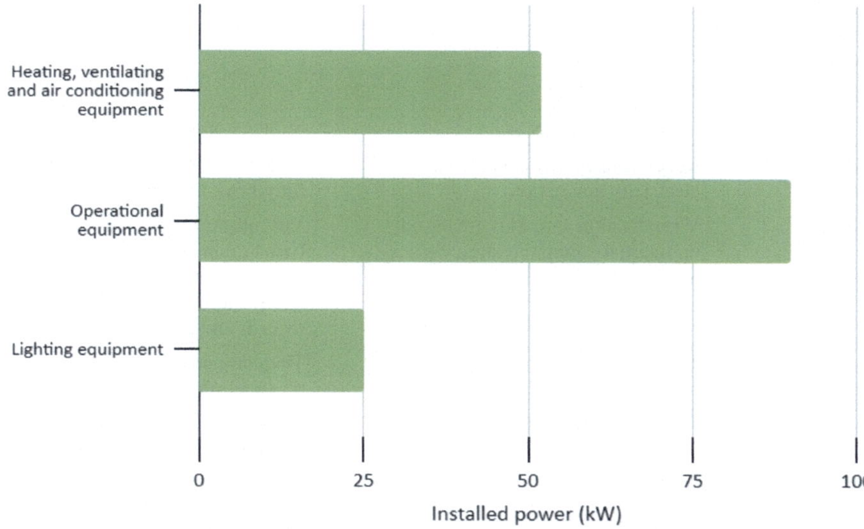

Fig. 9 Comparison of installed electrical power by type of equipment in M1. *Source* Based on Lopez de Munain (2017)

Following this analysis, the future scenario is created as the points for improvement and implementation of the EMIce strategies were identified. Table 4 summarises the activities to be implemented in a prospective future scenario.

Finally, the results obtained are summarised in Fig. 10, which shows the improvement in EMI 1–6 between the measured (2015 and 2022) and proposed (future) scenarios.

The EMIce in the year 2015 yields a total of 4, while in the year 2022 it yields a total of 9 points. In the latter year, the index reaches 50% of the maximum value. The improvement is largely due to the implementation of strategies driven by the institution's governance.

The EMIce for the future scenario reaches a total of 13 points, remaining below the achievable maximum. However, a greater institutional commitment is needed in this scenario, as the implementation of environmental policy requires the creation and designation of a specific budget and roles to apply strategies such as automatic functions.

The set-up of the future scenario aims to support the proportion of actions implemented between 2015 and 2022, i.e. those of the baseline scenario. Therefore, it is not intended to create a desirable scenario, where 100% of the score would be obtained, but rather a future scenario that is realistic to the behaviours of the HEI taken as a case study.

Table 3 Scoring assigned for each indicator of the trend scenario

Trend scenario				
Indicators	2015		2022	
	Score	Strategy or activity	Score	Strategy or activity
EMI 1	0	Renewable energy is not used for electricity generation	0	Renewable energy is not used for electricity generation
EMI 2	1	Virtualisation activities are based on internal communication via mail	3	Virtualisation activities are based on internal communication and administrative processes such as records and contracting
EMI 3	0	There are no automatic light on/ off functions	0	There are no automatic light on/ off functions
EMI 4	2	The replacement of conventional luminaires with LED lights is progressing	3	The replacement of LED lights throughout the M1 is completed and new, more energy efficient fixtures are bought
EMI 5	0	There is no evidence of the use of existing structures to reduce energy use	1	Meeting offices are created in places where one of the walls is a large window to take advantage of natural light. These are not offices that are commonly used for 8 h a day
EMI 6	1	The replacement of LED lights is carried out to save the monetary expenditure associated with energy consumption. The strategy was promoted by the university's administration secretariat	2	The replacement is completed and the guidelines for an environmental policy, which includes energy efficiency in the institution's buildings among its pillars, are being drawn up

Table 4 Scores assigned for each Future scenario indicator

Indicators	Future scenario	
	Score	Strategy or activity
EMI 1	1	Renewable energy is used for electricity generation in an incipient way, to power the luminaire in an M1 office
EMI 2	3	Virtualisation activities are based on internal communication and administrative processes such as records and contracting and become a catalyst for other buildings
EMI 3	1	The lights in the corridors of the M1 have motion sensors for automatic switching on and off
EMI 4	3	The replacement of LED lights throughout the M1 is completed and new, more energy efficient fixtures are purchased
EMI 5	2	Offices are created in places with large windows to take advantage of natural light
EMI 6	3	An environmental policy is created and approved, one of the pillars of which is energy efficiency in the institution's buildings

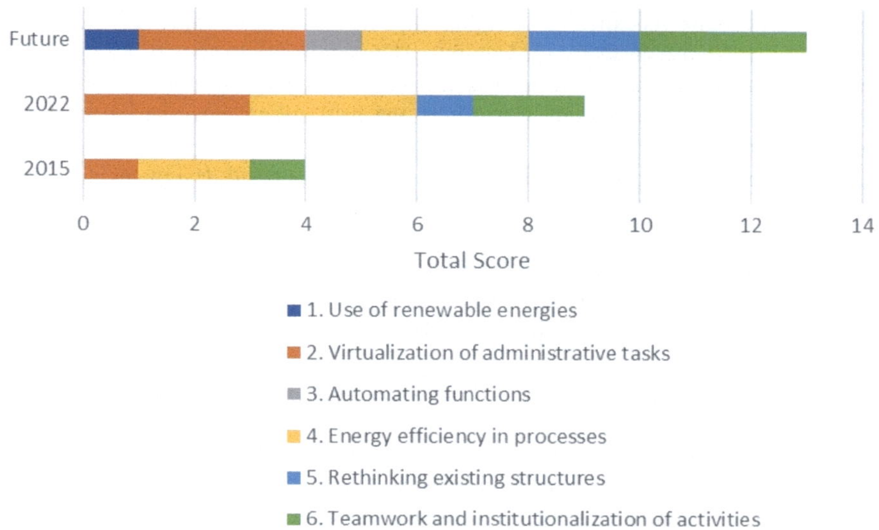

Fig. 10 Comparison of EMIce of measured and proposed scenarios for M1 by indicator

4 Conclusions

The proposal of indicators from the Circular Economy framework is useful for the elaboration of diagnostics and energy management scenarios in HEIs. The indicators have a dual purpose, which increases the amount of information provided. Similarly, the proposal of indicators broadens the view of CE, as in many areas it is only related to waste management and recovery and not to energy management. Likewise, several studies have detailed the study of ISO 50001 Standard for the study of energy management in institutions and organisations and not from the perspective of CE.

The results of the application of the proposed indicators make it possible to find the points to be strengthened for efficient energy use, such as the implementation of automation strategies and the creation of environmental policy. They also relate closely to the principles on which both CE frameworks were founded. ReSOLVE and DISRUPT aim for CE to be a driver for new building designs for the future, incorporating digital technology and for teams to keep working together for the environmental improvement of the organisation. This makes up several applicable strategies for prospective scenarios to improve the environmental performance of the university.

Environmental policy is closely related to HEI governance. The lack of institutionalisation of CE, energy management or any activity to enhance the environmental sustainability of the university campus is a constraint to the sustainability of the institution and generates a diffuse environmental image (Faitani 2023; Leal Filho et al. 2019a, b; López de Munain 2017). Specifically, there is a tendency to conduct actions associated with energy management in the case of UNGS. In the Future scenario, a better performance is expected including aspects that are still under development and other innovative aspects for the Campus. This requires the a specific budget allocation with the aim of the actions are sustained over time and monitored to recognise potential areas for improvement. Therefore, the institutional governance must act accordingly and propose an environmental policy that delimits clear goals, shows the environmental footprint that the institution will adopt and designates an office that brings together the activities for the sustainability of the campus with the specific monetary budget.

It is unknown whether the new environmental policy foreseen for UNGS will include among its priorities the activities or strategies proposed in the future scenario, as it has not been assembled or publicly consulted with the university community. Public consultation is also relevant as it is essential that the environmental policy covers all specific areas and includes the input of all specialists in the field and the campus stakeholders.

Among the limitations of this study, it can be mentioned that it was only carried out in a single module of the entire university campus. Another source of error to consider is that the information for the survey was collected in a fragmented way, as there is no office that brings together the environmental management or energy management of the institution, so the information could be lost and/or incomplete. On the other hand, the origin of the applied ReSOLVE and DISRUPT frameworks were created for corporations and not for Higher Education Institutions. Although the relevant adaptation was carried out, it could be possible to add other reference frameworks such as the aforementioned ISO 50001 standard, the Leed certification (Leed n.d.) or the recent ISO 59004 standard on Circular Economy. It is expected that the proposal of the Energy Management Index in the framework of the Circular Economy will be replicated throughout the UNGS Campus, to know the complete situation, as well as in other HEIs. It is also expected that this index will be adjusted to possible findings of other strategies aimed at the implementation of the CE and the improvement of energy management that may be found in other buildings on the university campus. It is also planned to adjust this index in view of possible findings of other strategies aimed at the application of CE and the improvement of energy management that may be found in other buildings on the university campus. Finally, it is intended to investigate in the future the presence of energy efficiency from the framework of the Circular Economy in the other pillars of operation of Higher Education Institutions not considered in this work, namely education and research.

5 Recommendations

We believe it is important to further explore the practical implications of the results for decision-making in HEIs, providing specific recommendations for managers and decision-makers in each area. The recommendations are intended for HEIs in contexts such as the university studied.

Primarily, institutional governance is essential to contribute and support the complete process of improving energy management to achieve the results or reach the future scenario in the short term and thus improve the environmental quality of the HEI and its surroundings. Institutional governance as the highest governing body of the HEI should create and distribute a budget for this line of management and for its implementation, maintenance, and continuous improvement. At the other hand, renewable energy generation demand the installation of energy harvesting devices such as solar panels or solar collectors. Although this requires a structural change in energy collection, pilot tests or small-scale installations can be test and adapted. Additionally, a similar scheme can be considered for the installation of automation tools. In these cases, the feasibility and cost mitigation study should be understood as part of the transformation process towards renewable energies. These feasibility studies should involve responsible for infrastructure, administration, and governance. HEI researchers can also provide comparative studies to determine the best applicable option.

Another relevant topic to increase energy efficiency is the procurement of equipment, that should be established by a specific institutional decision. In this case, HEIs have decision-making councils that should create basic guidelines and establish sustainable procurement criteria for the purchase of new, energy-efficient equipment by means of energy labelling. Similarly, the destination of discarded equipment should be known, so that it can be sent to sustainable destinations or used by cooperatives or associations that reuse some components to minimise waste generation.

For expanding the use of virtualisation tools for documents and administrative procedures, technological support for this purpose is essential. This requires secretaries and area coordinators to promote the practice among office staff and to increasingly reduce the use of paper.

Existing building structures have enormous potential for adaptation to nature-based strategies or solutions. Green roofs or gardens can be implemented to minimise the use of energy for air conditioning in buildings. In addition, it increases the biodiversity of the area and decreases the heat island effect in urban areas (Leveratto et al. 2023).

Finally, training for administrative and support staff, teachers, researchers, and students should be mandatory for the maintenance of good practices in energy efficiency and the good use of spaces by managing resources appropriately and responsibly. This establishes institutional recognition of the actions carried out and the proposed joint work of the entire university community to achieve a common goal. The actions mentioned above, such as the installation of solar panels, automation, green roofs, can be used as concrete examples of sustainable actions within university campuses and transferred to teaching. These actions also have repercussions outside the university institution, as they improve its environmental footprint and commitment and transmit good practices that can be replicated by other institutions or neighborhoods surrounding the HEI.

Acknowledgements Thanks are due to the Secretariat of Administration of the National University of General Sarmiento, which supplied the ground plan and provided information for the survey.

The authors also thank the National Agency for the Promotion of Research, Technological Development and Innovation of Argentina (ANPCyT), that financed the research project PICTO UNGS 2021-003 "Sustainability of university campuses in the Buenos Aires metropolitan area: current practices and management models" carried out at the National University of General Sarmiento, to which this paper belongs.

References

Amaral AR, Rodrigues E, Gaspar AR, Gomes Á (2020) A review of empirical data of sustainability initiatives in university campus operations. J Clean Prod 250:119558. https://doi.org/10.1016/J.JCLEPRO.2019.119558

Bradfield R, Wright G, Burt G, Cairns G, Van Der Heijden K (2005) The origins and evolution of scenario techniques in long range business planning. Futures 37(8):0–812. https://doi.org/10.1016/j.futures.2005.01.003

Centre for Innovation and Circular Economy (CIEC) (2019) Circular economy and public policies: State of the art and challenges for the construction of a policy framework for the promotion of circular economy in Latin America. Recovered from: https://bit.ly/3vUJy7s

Circle Economy (2020) Disrupt framework: 5fdccf51df0c5ef3ed6382ad_20201215—frameworks—disrupt—A4.pdf (website-files.com)

Diaz MC, Mira MV et al (2023) A methodology to analyze significant energy uses and energy consumption for improving energy performance in higher education buildings. Int J Energy Econ Policy 13(6):S. 636–649. https://www.econjournals.com/index.php/ijeep/article/download/14974/7606/35071. https://doi.org/10.32479/ijeep.14974

Ellen MacArthur Foundation (EMF) (2019) Circular economy in cities. Recovered from: https://bit.ly/3vX5D5h

Faitani NG (2023) Environmental sustainability of a Higher Education Institution in the Buenos Aires suburbs. Postgraduate thesis—Ph.D. in Science and Technology. National University of General Sarmiento. In Spanish

Faitani N, Galvan L, Bielsa R (2021) Key social actors in the management of university waste for the recovery of materials. Case study: National University of General Sarmiento. ISBN: 978-99905-809-9-0. http://bitly.ws/AzBS

Faitani N, Galván L, Bielsa R (2022) The contribution of the circular economy to the fulfilment of the 2030 Agenda in Higher Education Institutions. In: Handbook of sustainability science in the future. https://doi.org/10.1007/978-3-030-68074-9_124-1

INDEC-Instituto Nacional de Estadística y Censos (2023) National census of population, households and housing 2022: provisional results. In Spanish

ISO (n. d.) ISO 50001. Energy management. ISO—ISO 50001—Energy management

ISO (n.d.) ISO 59004. Circular economy. ISO 59004—Circular economy

Lange Salvia A, Londero Brandli L (2020) Energy sustainability at universities and its contribution to SDG 7: a systematic literature review. In: Leal Filho W et al (eds) Universities as living labs for sustainable development. World sustainability series. Springer, Cham, pp 29–45. https://doi.org/10.1007/978-3-030-15604-6_3

Leal Filho W, Frankenberger Silva F, Salvia A, Shiel C, Paço A, Price E, Brandli LL, Rampasso IS, Anholon R, Quelhas OLG, Pretorius RW (2023) An overview of research trends on sustainability in higher education—an exploratory study. Int J Sustain High Educ 24(5):1161–1175. https://doi.org/10.1108/IJSHE-08-2022-0252

Leal Filho W, Salvia AL, Paço, A, Anholon R, Gonçalves Quelhas OL, Rampasso IS, Brandli LL (2019a) A comparative study of approaches towards energy efficiency and renewable energy

use at Higher Education Institutions. J Clean Prod 237:117728. https://doi.org/10.1016/j.jclepro.2019.117728
Leal Filho W, Will M, Lange Salvia A, Adomßent M, Grahl A, Spira F (2019b) The role of green and sustainability offices in fostering sustainability efforts at Higher Education Institutions. J Clean Prod 232:1394–1401, ISSN 0959-6526. https://doi.org/10.1016/j.jclepro.2019.05.273
Leal Filho W, Shiel C, do Paço A, Brandli L (2015) Putting sustainable development in practice: campus greening as a tool for institutional sustainability efforts. Sustain High Educ 1–19. https://doi.org/10.1016/b978-0-08-100367-1.00001-9
LEED® (n.d.) https://new.usgbc.org/
Leveratto MJ, González Otharán F, Fernández Bados O, Propati M, Reobo M, Satelier J (2023) Green roofs as a tool for heat island mitigation in urban areas of the City of Buenos Aires. Available in: Cubiertas verdes isla de calor CABA.pdf (buenosaires.gob.ar)
Lopez de Munaín D (2017) Energy management in Higher Education Institutions. The case of the UNGS campus. Degree thesis. In Spanish
Lozano R (2006) Incorporation and institutionalization of SD into universities: breaking through barriers to change. J Clean Prod 14(9–11):787–796. https://doi.org/10.1016/j.jclepro.2005.12.010
Mendoza J, Gallego-Schmid A, Azapagic A (2019) A methodological framework for the implementation of circular economy thinking in Higher Education Institutions: towards sustainable campus management. J Clean Prod 226:831–844. https://doi.org/10.1016/j.jclepro.2019.04.060
Mendoza J, Gallego-Schmid A (2020) Implementation of a circular economy at universities. In: Handbook of the circular economy. Chap 33. Edward Elgar Publishing Ltd. https://doi.org/10.4337/9781788972727.00043
Mtutu P, Thondhlana G (2016) Encouraging pro-environmental behaviour: energy use and recycling at Rhodes University, South Africa. Habitat Int 53:142–150. https://doi.org/10.1016/j.habitatint.2015.11.031
National Interuniversity Council. Resolutions: CE No 1054/15; 1195/16; 1766/23. Available in: NormativalCIN
Nunes B, Pollard S, Burgess P, Ellis G, de los Rios I, Charnley F (2018) University contributions to the circular economy: professing the hidden curriculum. Sustainability 10(8):2719. https://doi.org/10.3390/su10082719
Nunayon SS, Olanipekun EA, Famakin IO (2021) Investigating barriers to efficient electricity management practices in government-owned tertiary institutions: a case of Obafemi Awolowo University, Ile-Ife. Energy Efficiency 14:69. https://doi.org/10.1007/s12053-021-09978-5
Ocampo Batlle EA, Escobar Palacio José Carlos, Silva Lora EE, Martinez Reyes AM, Melian M, Morejon MB (2019) A methodology to estimate baseline energy use and quantify savings in electrical energy consumption in Higher Education Institution buildings: case study, Federal University of Itajubá (UNIFEI). J Clean Prod 244:118551. https://doi.org/10.1016/j.jclepro.2019.118551
Pinzón Casallas JD, Corredor RA, Santamaría F, Hernández JA, Trujillo CL (2014) Implementación de indicadores energéticos en centros educativos. Caso de estudio: Edificio Alejandro Suárez Copete- Universidad Distrital Francisco José de Caldas. Revista Escuela de Administración de Negocios 77:186–200. (3) (PDF) Implementación de indicadores energéticos en centros educativos. Caso de estudio: Edificio Alejandro Suárez Copete-Universidad Distrital Francisco José de Caldas (researchgate.net)
Rokicki T, Perkowska A, Klepacki B, Szczepaniuk H, Szczepaniuk EK, Bereziński S, Ziółkowska P (2020) The importance of higher education in the EU countries in achieving the objectives of the circular economy in the energy sector. Energies 13(17). https://doi.org/10.3390/en13174407
Schröder P, Albaladejo M, Alonso Ribas P, MacEwen M, Tilkanen J (2020) The circular economy in Latin America and the Caribbean. Opportunities for building resilience. Chatham House Energy, Environment and Resources Programme Research Paper. https://doi.org/10.13140/RG.2.2.13883.13603

SPU-Secretaría de Políticas Universitarias (2021) Argentine University Statistics Yearbook. In Spanish. https://www.argentina.gob.ar/educacion/universidades/informacion/publicaciones/anuarios

Stephens J, Hernandez M, Román M, Graham A, Roland W. Scholz (2008) Higher education as a change agent for sustainability in different cultures and contexts. Int J Sustain High Educ 9(3):317–338. https://doi.org/10.1108/14676370810885916

Taylor P, D'Ortigue O, Francouer M, Trudeau N (2010) Final energy use in IEA countries: the role of energy efficiency. Energy Policy 38(11):6463–6474. https://doi.org/10.1016/j.enpol.2009.05.009

University Statute. UNGS (2017). https://www.ungs.edu.ar/wp-content/uploads/2019/11/Estatuto-2017.pdf

Yin RK (2014) Case study research. Design and methods, 5th edn. Sage. Can J Program Eval. https://doi.org/10.3138/cjpe.30.1.108

Žalėnienė I, Pereira P (2021) Higher education for sustainability: a global perspective. Geogr Sustain 2(2):99–106, ISSN 2666-6839. https://doi.org/10.1016/j.geosus.2021.05.001

Circular Economy in Brazil: The Nexus of Public Policy, Extended Governance, and Social Innovation

Teresa Villac, Amanda Xavier, Maria Carolina Santos, and Gabriel Grasso Ribeiro

Abstract The chapter presents the relationships between the circular economy, public policies, extended governance, and social innovation in Brazil and the potential contribution to implementing the Sustainable Development Goals in the country. It examines the national constitutional foundations for circular economy and the legal, institutional framework addressing the theme in scattered laws and norms rather than in a unified public policy. The need for articulation of public policies in the territory, with innovation and social participation in networks, is presented as the potential for implementing the 2030 Agenda in the country. In this sense, the territory emerges as a platform for articulation, mobilization, and political convention, where solutions go beyond the limits of public action through coordination between different territorial actors and scales of action, not limited to administrative jurisdictions. In this regard, the chapter presents the Sustainability Network ReciclaPorto Rio experience, a Technical Cooperation Agreement between public institutions located in Rio de Janeiro aimed at developing joint initiatives and projects with inter-institutional action. In searching for a joint solution for the appropriate management of recyclable waste, the

T. Villac (✉)
PhD in Environmental Science, University of São Paulo (USP) Philosopher, USP, Attorney for the Union in the Legal Advisory Office of the Ministry of the Environment and Climate Change (MMA), São Paulo, Brazil
e-mail: teresavillacp@gmail.com

A. Xavier
Post-Doctorate in Management, Innovation, and Sustainability—Production Engineering (COPPE), Professor of the Production Engineering Program, COPPE (Federal University of Rio de Janeiro—UFRJ), Rio de Janeiro, Brazil
e-mail: amandaxavier@coppe.ufrj.br

M. C. Santos
Doctoral Candidate in Management and Innovation in the Production Engineering Program (COPPE/UFRJ), Rio de Janeiro, Brazil
e-mail: carolina.santos6688@gmail.com

G. G. Ribeiro
Undergraduate Student in Environmental Sciences, Federal University of the State of Rio de Janeiro (UNIRIO), Rio de Janeiro, Brazil
e-mail: grasso.gabriel@outlook.com

L. Brandli et al. (eds.), *The Contribution of Life Cycle Analyses and Circular Economy to the Sustainable Development Goals*, World Sustainability Series,
https://doi.org/10.1007/978-3-031-85300-5_3

ReciclaPorto Network conducted a Shared Public Call, enabling collectors' cooperatives. Such an initiative outlines paths for circular trajectories and social innovations that emerge from extended governance, involving public actors, collectors' cooperatives, and territorial solutions for resource decoupling. The chapter concludes by recognizing the correlation between different actors and spheres of society, enriched with articulating knowledge for the breadth of social benefits regarding the Circular Economy and the 2030 Agenda.

Keywords Public policies · Circular economy · Extended governance · Social innovation · 2030 agenda

1 Introduction

This chapter critically examines the intricate interrelationship between the circular economy, public policies, extended governance, and social innovation in the Brazilian context, particularly within the framework of Agenda 2030. Anchored in the robust framework of the Federal Constitution, Brazilian public policies underscore the complex legal and programmatic foundations that shape the nation's approach to the circular economy.

Considering the absence of a comprehensive national policy for the circular economy, along with the constitutionally defined mandates and the distribution of powers within Brazil's federative system, this analysis offers a nuanced exploration of theoretical contributions. It underscores the significance of territory and culture as essential elements in the progression of the circular economy in Brazil. Given the nation's vast geographical expanse, rich biodiversity, and profound cultural diversity, these contributions emphasize the necessity of a participatory approach that aligns with and effectively advances the objectives of the 2030 Agenda.

Moreover, this chapter elucidates the ReciclaPorto Rio Sustainability Network initiatives—a consortium of public institutions in Rio de Janeiro engaged in collaborative projects that epitomize extended governance among public entities and cooperatives of recyclable material collectors.

2 Public Policies

The contextualization of the Circular Economy (CE) in Brazil and the associated implementation challenges demands a comprehensive examination through the lenses of public policies. The Legislative Branch is responsible for enacting laws that delineate and direct the policies required to safeguard the rights enshrined in the Federal Constitution. These statutes often need further elaboration through detailed regulations issued by Executive Decrees to ensure the effective implementation of public policies. Concurrently, the Judiciary upholds the principle of universal access to justice through a dual-level jurisdictional system, comprising initial adjudication by a single judge and appellate review by a collegiate court. Furthermore, it ensures access to two superior courts—the Superior Court of Justice and the Supreme Federal Court—in cases involving alleged violations of federal or constitutional matters.

2.1 Circular Economy: A Constitutional Analysis

In the context of public policies, a comprehensive examination of the domains pertinent to the circular economy in Brazil demands an in-depth analysis of the legislative and implementation competencies. These competencies are delineated and apportioned among the federative entities—namely, the Union, States, and Municipalities—under the Federal Constitution (CF) stipulations.

In the realm of implementation, the Union bears the primary responsibility for matters about the circular economy. This encompasses the development and execution of comprehensive national and regional plans aimed at territorial organization and socio-economic development. Additionally, the Union oversees the management of electricity services and facilities and coordinates with states to utilize hydroelectric potential. The Union's oversight extends to air navigation, aerospace, and airport infrastructure, alongside the administration of railway and waterway transportation services that link Brazilian ports to national borders or traverse state boundaries. Moreover, the Union is entrusted with managing interstate and international passenger road transport services, regulating maritime, riverine, and lacustrine ports, and establishing guidelines for urban development, including housing, basic sanitation, and urban transportation. It also formulates principles and procedures for the national transportation system and explores nuclear services and facilities. This encompasses the state monopoly over the research, mining, enrichment, reprocessing, industrialization, and commerce of atomic minerals and their derivatives (Article 21, FC).

Consequently, the Union is pivotal in formulating and implementing public policies fostering the circular economy, particularly emphasizing energy, logistics, and infrastructure.

At the municipal level, the focus is on local interests, which demand the adoption of measures that stimulate the circular economy within their jurisdictions. This involves the organization and provision of public services, either directly or through concessions (Article 30, Paragraph 1, FC).

States, in turn, are competent to implement public policies in areas not expressly reserved to the Union by the Federal Constitution. They can establish metropolitan regions, urban agglomerations, and micro-regions to integrate, organize, plan, and execute public functions of shared interest, creating a potential multiplier effect for actions aimed at the circular economy in multiple municipalities (Article 25, FC).

The Brazilian federal system includes an autonomous unit called the Federal District, where the federal government is seated. The district is located in Brasília and holds the constitutional competencies of both States and Municipalities (Article 32, Paragraph 1, FC).

In health, culture, education, science, technology, research, and innovation, the implementation of circular economy public policies falls under the shared competence of the Union, States, the Federal District, and Municipalities. The circular economy also encompasses environmental protection, pollution control, forest conservation, flora and fauna protection, agriculture and livestock, housing construction, and basic sanitation. The social dimension of sustainability is addressed within this framework through the shared competence to combat poverty and marginalization, thereby promoting the social integration of disadvantaged sectors (Article 23, FC).

Currently, Brazil lacks widely disseminated incentives for advancing the circular economy despite its extensive constitutional competencies, which empower the implementation of public policies across a diverse array of sectors and thematic areas.

Having delineated the constitutional competencies in the context of the circular economy, legislative competencies play a crucial role in its enhancement in Brazil, particularly concerning potential tax immunities and exemptions in strategic economic sectors under the concurrent legislative competencies over tax law shared among the Union, States, and the Federal District (Article 24, FC). Under this framework of concurrent competence, the Union establishes general norms, while the States and the Federal District exercise supplementary legislative authority.

Legislation on production and consumption, forestry, nature conservation, soil defense, natural resource protection, environmental protection, and pollution control falls within the concurrent jurisdiction of the Union, States, and the Federal District (Article 24, FC). This legislative framework facilitates the expansion of the circular economy through integrated public policies that adopt a systemic perspective, promoting sustainable development. Furthermore, it addresses the social dimension by advancing economic inclusion, reducing inequalities, ensuring social integration, and fostering a sense of social belonging. Studies by Santos (2014) underscores the feasibility of integrating circular economy initiatives related to recycling with the active participation of the community of waste pickers.

The circular economy encompasses a wide range of sectors, including civil, commercial, agriculture and livestock, maritime, aeronautical, space, water, energy, information technology, telecommunications, broadcasting, postal services, transportation, ports, inland navigation, maritime navigation, aerospace navigation, traffic, transportation, mining, natural resource management, metallurgy, and nuclear activities. Within these domains, the legislative competence is shared among the Union, States, and the Federal District (Article 24, FC), thereby offering substantial potential for the extensive dissemination of circular economy public policies through interfederative collaboration. Nonetheless, the formulation and implementation of unified policies present significant challenges due to Brazil's vast geographical expanse, the diverse economic sectors predominant in each State, and the unique characteristics stemming from the nation's mega biodiversity. Furthermore, transportation logistics pronounced social inequality, and disparities in access to fundamental human rights contribute to the complexity. These factors influence elected representatives' legislative priorities and decisions, who must navigate their respective territories' distinct environmental, social, and economic contexts.

Furthermore, given that the Union holds exclusive legislative competence over credit policy and both interstate and international commerce (Article 22, Federal Constitution), it is imperative to integrate and strengthen the circular economy framework. This will strategically position Brazil at the vanguard of global economic relations.

The most pertinent constitutional provisions for the circular economy are encapsulated in Article 225 of the Federal Constitution (CF), which states, "Everyone has the right to an ecologically balanced environment, an asset for the common use of the people and essential to a healthy quality of life, imposing on the Public Power and the community the duty to defend and preserve it for present and future generations." This is further reinforced by Article 170, CF, which asserts, "The economic order, founded on the valorization of human work and free enterprise, aims to ensure everyone a dignified existence, according to the dictates of social justice," adhering to the principles of social justice and environmental protection.

The Brazilian Federal Constitution embeds sustainable development within its environmental, social, economic, and cultural dimensions. Consequently, sustainability emerges as a guiding principle for both national public policies and private sector initiatives (Freitas 2014).

2.2 *Circular Economy and Brazilian Legislation*

In Brazil, the circular economy concept lacks a dedicated legislative framework. Therefore, it must be evaluated within the context of various existing statutes to discern pertinent provisions.

Brazilian legislation is exceptionally progressive in environmental conservation and associated policies. The National Environmental Policy (Law 6.938), enacted in 1981, has recognized the interdependence of environmental, social, and economic dimensions. This policy underscores the importance of governmental intervention, strategic planning, and environmental education. Concerning the circular economy (CE) and its nexus with innovation and social engagement, the policy mandates the diffusion of environmental management technologies, the dissemination of environmental data and information, and the fostering of public awareness regarding the imperative of sustaining environmental quality and ecological balance.

The National Environmental Policy requires the identification of priority zones for governmental intervention, establishing environmental quality criteria and standards, and regulating natural resource utilization and management. Both public and private entities are required to comply with these stipulations.

Moreover, the National Environmental System, established by this statute, serves as a framework for defining, coordinating, and promoting collaborative initiatives related to the circular economy across federal, state, and municipal governments, including the Federal District.

In 2006, Decree 5.940 mandated the segregation of recyclable waste by federal public administration bodies and its allocation to associations and cooperatives of recyclable material collectors. This decree imposed the responsibility on institutions to implement the Coleta Seletiva Solidária (Solidary Selective Collection) program and to conduct public solicitations to facilitate the participation of these cooperatives. These cooperatives are pivotal social actors in advancing the circular economy through managing and recycling waste materials.

In 2022, the Solidary Selective Collection program was superseded by the Coleta Seletiva Cidadã (Citizen Selective Collection) through Decree 10.936. While the mandate for federal public administration bodies to segregate reusable and recyclable waste and prioritize its allocation to associations and cooperatives of collectors persists, the policy's failure to remunerate these collectors constitutes a notable deficiency. Recognizing these collectors as environmental agents underscores the imperative for compensation to honor the dignity of their labor and their entitlement to financial remuneration and to acknowledge their contributions to environmental conservation.

Although not explicitly designated as such, the circular economy principles are inherently integrated within Brazil's National Solid Waste Policy (Law 12.305, 2010), the nation's seminal legislation addressing solid waste management. Before this comprehensive policy, waste disposal was regulated by a fragmented array of regulations specific to individual waste categories. The enactment of a national policy represented a significant legislative advancement.

The National Solid Waste Policy aligns closely with the principles of the circular economy, as the following analysis will elucidate and illustrate in Fig. 1.

Furthermore, a bill (PL 1874/2022) currently under consideration in the National Congress aims to establish the National Circular Economy Policy. Still under deliberation, this legislative initiative seeks to elevate the topic into the public discourse, fostering comprehensive dialogue and social participation in the public policy process (Secchi 2014). The bill represents a pivotal window of opportunity (Kingdon and Stano 1984) for advancing the circular economy, advocating for a systemic approach to circular economy as public policy, encompassing environmental benefits, social inclusivity, and mandatory obligations for the productive and business sectors.

The primary challenges confronting the circular economy in Brazil encompass the effective institutionalization of governance frameworks that fundamentally embrace sustainability (Cader and Villac 2022), coupled with a scarcity of socially participatory initiatives from their inception.

To address these challenges, two critical categories—territoriality and culture—are proposed as foundational components of socio-environmental public policies, as elucidated in preceding studies by Villac (2024). Given their intrinsic relevance to the circular economy, these categories can substantially enrich the discourse, extending an initially developed model for sustainable public procurement.

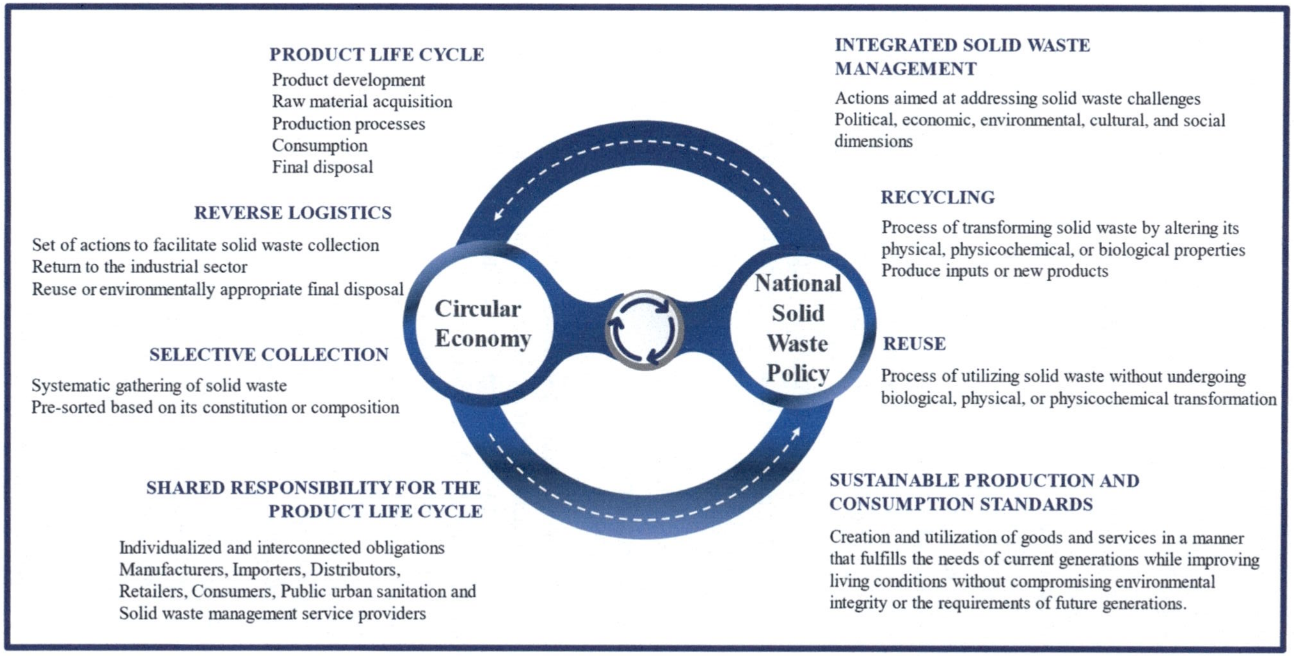

Fig. 1 Circular economy in the National Solid Waste Policy. *Source* Prepared by the authors

Drawing on the theoretical contributions of Santos (2009), which conceptualize space as an immaterial locus of action and reflection, public policies must integrate territoriality as an inseparable element within a democratic framework, considering geographical, human, socioeconomic, climatic, and environmental characteristics. Thus, "the field of human and political geography cannot be overlooked in implementing sustainable governance that is also inclusive and participatory" (Villac 2024).

In addition to territoriality, integrating culture into public policies presents a formidable challenge. A significant obstacle lies in the fact that "organizational culture, both public and private, as well as societal culture, does not prioritize sustainability as a core value. There remains a pressing need for educational initiatives for the workforce and citizens, advancing the principles of environmental education to be transversal, inclusive, and critical" (Villac 2024).

Notably, the absence of a national public policy on the circular economy does not preclude the development of related initiatives. Adopting collaboration acts as a propelling and multiplying force for such actions.

As proposed by Poteete et al. (2011), implementing collaborative actions in the stewardship of common-pool resources through collective governance constitutes a highly effective strategy for preserving shared resources, including environmental assets.

3 Extended Governance

The current disequilibrium in local ecosystems, driven by an economic paradigm centered on the mass production of material goods and a logic of financial performance, further exacerbated by escalating global interdependence, underscores the prevailing governance crisis and the disorder in territorial relations.

Territory constitutes the primary instance of governance, encompassing articulation, mobilization, and political convention, particularly concerning the concept of 'living in society.' Society is invariably situated within a territorial dimension, enriched by human action and sociocultural diversity (Santos and Silveira 2021). It is characterized by multiple cultures that enhance places, transforming them into spaces imbued with identity, relational dynamics, and historical significance (Augé 1992).

A systemic interrelationship among social, cultural, institutional, economic, and environmental domains is evident. However, as productivity drivers shift toward technological advancements and the economies of virtual platforms and scale, territory loses its role as a space for meeting and coexistence. Instead, territory is perceived as a constraint within the logistical frameworks of enterprises (Xavier et al. 2024).

This new context reveals disorder across all facets of social life, highlighting the inherent vulnerabilities in current organizational and governance models within these emerging economies, thereby contrasting territorial perspectives. Territory becomes confined within the boundaries of supply chains or sectoral silos, constructed on the foundations of labor division, financialization, and privileged information. In response, short supply chains and circular economic models emerge as transitional movements, seeking sustainable solutions for both the economy and territories.

While relevant, these circular initiatives face significant challenges in aligning decision-making with horizontal and transversal mechanisms for feedback and reflection. The lack of dedicated forums for discussing genuine work and consumption experiences within the territory weakens social ties, hindering the establishment of cooperation as a driving force of the economic model and obstructing the co-construction of more integrated solutions (Xavier et al. 2024; Costa and Xavier 2023).

The phenomenon of 'unnecessary circulations' (Santos and Silveira 2021), resulting from production activities geared toward export, engenders redundant exchanges involving heavy infrastructure and mobilizing vehicles and services to sustain such circulation, thereby imposing an undue burden on society. This scenario, arising from the challenge of decoupling economic growth from the consumption of finite resources and waste generation, reveals a movement predominantly driven by material flows within purely productive spatial circuits, detached from decentralized logics anchored in immaterial resources (e.g., intellectual capital, information, intangible heritage). The concentration of directive functions in specific locales underscores the spaces that exert control (Santos and Silveira 2021) as production coordinators, often overlapping with labor divisions typically governed by the pressures of international relations. This territorial fragmentation reflects a centralization of regulatory power, where commanding forces are exercised by private enterprises and public authorities, confined to unilateral benefits directed toward financialization and external relations rather than to the territory itself.

Within this context, novel challenges emerge regarding integrating the Sustainable Development Goals (SDGs) into regulatory frameworks. This demands carefully considering the roles of formal institutions and informal coordination among local actors in regulating territories and fostering local development processes. The governance paradigm emphasizes the necessity of both horizontal and vertical coordination of public action, encompassing independent authorities, territorial collectives, and civil society components (Fuini and Pires 2015).

Discourses on SDG monitoring highlight the critical importance of governance and the alignment of commitments to these goals with public policies across short-, medium–, and long-term horizons. However, the 2030 Agenda lacks explicit guidance on the interactions among the SDGs and methodologies for their evaluation. This gap demands new research on political innovation, monitoring, and systemic approaches that consider territorial constraints (Rocha et al. 2023).

In this transition, regulatory bodies will increasingly mandate integrating ESG (environmental, social, and governance) practices within productive systems, significantly shaping consumer perceptions. Consequently, organizations must prioritize stakeholder management concerning ESG practices, as these profoundly impact the communities in which they operate.

Creating more effective interfaces with societal innovations requires fundamentally rethinking public policy design. For this new governance paradigm, it is imperative (Mazzucato 2022): to overcome the silo mentality among various public institutions; centralize public procurement budgets to enhance bargaining power with suppliers; cultivate leadership capable of integrating competing groups and communicating purpose to society; achieve organizational and operational effectiveness through decentralization and delegation of authority; contextualize the development and financing of technology within social and political frameworks; and ensure the systematic and organized involvement of civil society organizations from the project evaluation stages onward.

Extended governance hinges on constructing collective solutions through shared management, engaging key stakeholders within the territory, and delineating their strategic decisions, responsibilities, and the boundaries of effects governed by corporate practices (Costa and Xavier 2023).

It is imperative to emphasize that the development of territorial solutions is significantly enhanced when governance prioritizes work activities. Work, as a transformative activity that provides goods, services, and integrated solutions, constitutes a critical source of value creation. The territory is where work materializes, and its recognition and development are contingent upon governance relationships (Xavier et al. 2024).

As a regulatory mode, governance establishes foundational social relationships and mechanisms for value accumulation. In this context, it is essential to incorporate elements of territorial governance (Dallabrida 2006) and inter-organizational networks (Albers 2005). Critical characteristics of territorial governance include acknowledging the territory's historical dynamics; implementing decentralized management, conciliation, and mediation processes practiced by pertinent local leaders and representatives; and fostering socio-territorial projects resulting from agreements and conventions among territorially organized societal segments as networks of power. Network models can take various forms: shared governance without a formal and exclusive administrative structure, governance led by a primary organization, governance through an administrative network organization, and hybrid models (Provan and Kenis 2008).

These diverse approaches reveal several limitations, such as low trust levels among actors (participants and managers); limited plurality and heterogeneity, often confined to private and governmental agents; a lack of integration among businesses and supply chains, resulting in competitive or misaligned solutions; metrics based on immediate and intentional effects that can be confusing; the absence of a political vision of the territory, treating it as an exogenous agent; and unilateral benefits that risk hegemony in external (and international) relations without equitable distribution of risks and rewards within and for the territory.

To meet the SDGs and the imperatives of circularity, governance must be transformative, altering power dynamics to emancipate stakeholders who embody transformative sustainability values. Governance becomes transformative if the approaches are integrative, inclusive, adaptive, and pluralistic. Additionally, these approaches must be implemented collectively, operationalized in context-specific ways, and

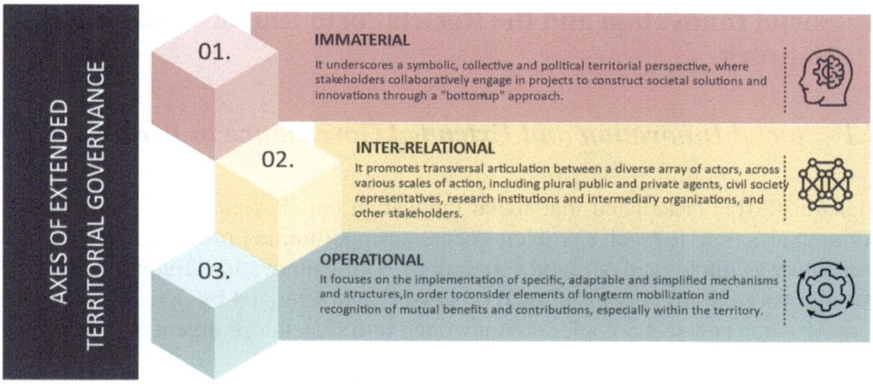

Fig. 2 Axes for extended territorial governance. *Source* Prepared by the authors

focused on addressing the indirect factors underlying sustainability issues, such as social change (Visseren-Hamakers et al. 2021).

To establish extended governance that effectively addresses contemporary challenges and fulfills the Sustainable Development Goals (SDGs), it is imperative to transcend traditional boundaries of public action. This demands the integration of three fundamental axes—immaterial, inter-relational, and operational—as illustrated in Fig. 2.

In this context, the critical role of public agencies as mediators between the State and society in advancing the general well-being of the population becomes evident. The formulation of sustainable public policies through extended governance frameworks serves as a strategic asset for governmental actions, anchored in the principle of legality, thereby ensuring universal applicability and mandatory enforcement. Moreover, it is crucial to underscore that public authority is guided by the principles of equity, justice, impartiality, and equal rights, reinforcing its intersection with society and its territorial connections. Beyond this political and territorial perspective, it is imperative to design streamlined and context-specific mechanisms and structures capable of engaging multiple stakeholders in long-term participation, recognizing mutual benefits—particularly for the territory—and the collaborative efforts of all involved.

4 Social Innovation and the ReciclaPorto Rio Sustainability Network

4.1 Social Innovation and Extended Governance in Brazil

Manzini (2017) articulated that social innovation entails changes in individual or communal actions to resolve problems or create opportunities, prioritizing behavioral shifts over technological or market alterations. These innovations frequently emerge through "bottom-up" organizational processes, characterized by periods of intense activity when new technologies are integrated into societies or urgent or widespread issues demand solutions.

Numerous scholars have extensively examined the significance of social innovation over recent decades (Rocha et al. 2023; Manzini 2017; Bignetti 2011; Baumgarten 2007, 2008). It is crucial for advancing sustainability and inclusive practices, particularly in vulnerable areas like Brazil, where it addresses complex social problems by disrupting entrenched power structures.

Brazil's public administrative model, traditionally characterized by structural rigidity, often results in stagnation in actions and decisions. Political appointments to high administrative positions, frequently prioritizing political loyalty over technical competence and rational decision-making, have been criticized by Abrucio (1977) for their tendency towards depoliticization. Franzese (2021) highlights these challenges, advocating for less hierarchical management approaches emphasizing partnerships, shared responsibilities, and fostering public governance to strengthen interinstitutional and sectoral relationships. He posits that network management is particularly effective in overcoming such barriers, offering a more flexible and adaptive model.

In the realm of organizational networks, Provan and Kenis (2008) assert that delineating shared governance is imperative for engaging participants in mutual support and collective actions, including conflict resolution. Governance underscores the involvement of institutions, their collaborative structures, and the authority to disseminate information and implement joint programs.

Bessa and Doetzer (2020) further accentuate the positive impact of shared governance, highlighting the enhancement of collective work through dialogue and opportunities for innovation and experiential sharing. This approach bolsters participants' self-esteem and fosters environments conducive to joint action, dialogue promotion, inter-agency coordination, and the appreciation of complementary roles among involved actors.

The ReciclaPorto Rio Sustainability Network epitomizes social innovation through shared governance among governmental entities. Initiated in 2017 and formalized in 2019 via a Technical Cooperation Agreement, its thematic areas of action are guided by the A3P—Public Administration Environmental Agenda, as depicted in Fig. 3.

The management style represents a significant advancement in Brazilian public administration by integrating interinstitutional and multidisciplinary practices, encompassing waste management and social inclusion. This model epitomizes

Fig. 3 ReciclaPorto Rio sustainability network. *Source* Prepared by the authors

extended governance, fostering robust collaboration between public entities and civil society organizations. Initially comprising six federal entities, the network expanded to encompass 14 federal and state institutions by 2023. This growth underscores the appeal and efficacy of a less hierarchical, more participatory administration, as articulated by Franzese (2021). This flexible and horizontal governance model facilitates awareness campaigns and training initiatives, promoting sustainable practices and empowering vulnerable communities.

Following its formalization, the Network initiated the first Shared Public Call and joint selective collection in 2019, engaging waste pickers' cooperatives and enhancing the institutional management of recyclable waste, as delineated in Decree 5.940/2006. Despite the revocation of the mandate to allocate recyclable waste from public institutions to waste pickers' cooperatives (Decree 10,936/2021), ReciclaPorto maintained exclusive allocations to these cooperatives in 2023 through the implementation of the 2nd Shared Public Call, benefiting nine qualified cooperatives.

The initiatives undertaken by the ReciclaPorto Rio Network exemplify extended governance, linking public actors with recyclable material waste pickers' cooperatives across Brazil. This approach strengthens collaborative networks between institutional representatives and local associations, ensuring efficient and sustainable operations. The continuous interaction among these groups facilitates effective partnerships in the public sphere, promotes the transfer of knowledge and resources, and fosters the development of more inclusive and effective public policies. This governance model is characterized by its inclusive and participatory nature, ensuring broad and impactful societal engagement.

4.2 Interconnection with the Circular Economy and the SDGs

Santos et al. (2023) assert that the model established by the ReciclaPorto Network through Shared Public Calls not only adheres to current legislation but also promotes environmental sustainability, bolsters the local economy, enhances the quality of life for workers, and advances social justice. This model aligns with various targets of the Sustainable Development Goals (SDGs) outlined in the UN's 2030 Agenda and stages of the circular economy cycle, as illustrated in Fig. 4.

The shared management of recyclable waste ensures proper disposal with traceability, lending credibility to its final destination (SDG 12, directly impacting Target 12.5 and tangentially related to Target 12.6). It strengthens the recycling chain in Rio de Janeiro, facilitating the reintegration of waste into the production cycle (SDG 8, directly impacting Target 8.4 and tangentially related to Targets 8.3 and 8.2) and mitigates risks of soil and water contamination. The shared initiative enhances the attractiveness of Waste Pickers' Cooperatives by aggregating segregated waste from multiple institutions along the same collection route. This strategy reduces transportation costs, maximizes the efficiency of kilometers traveled to collect larger quantities

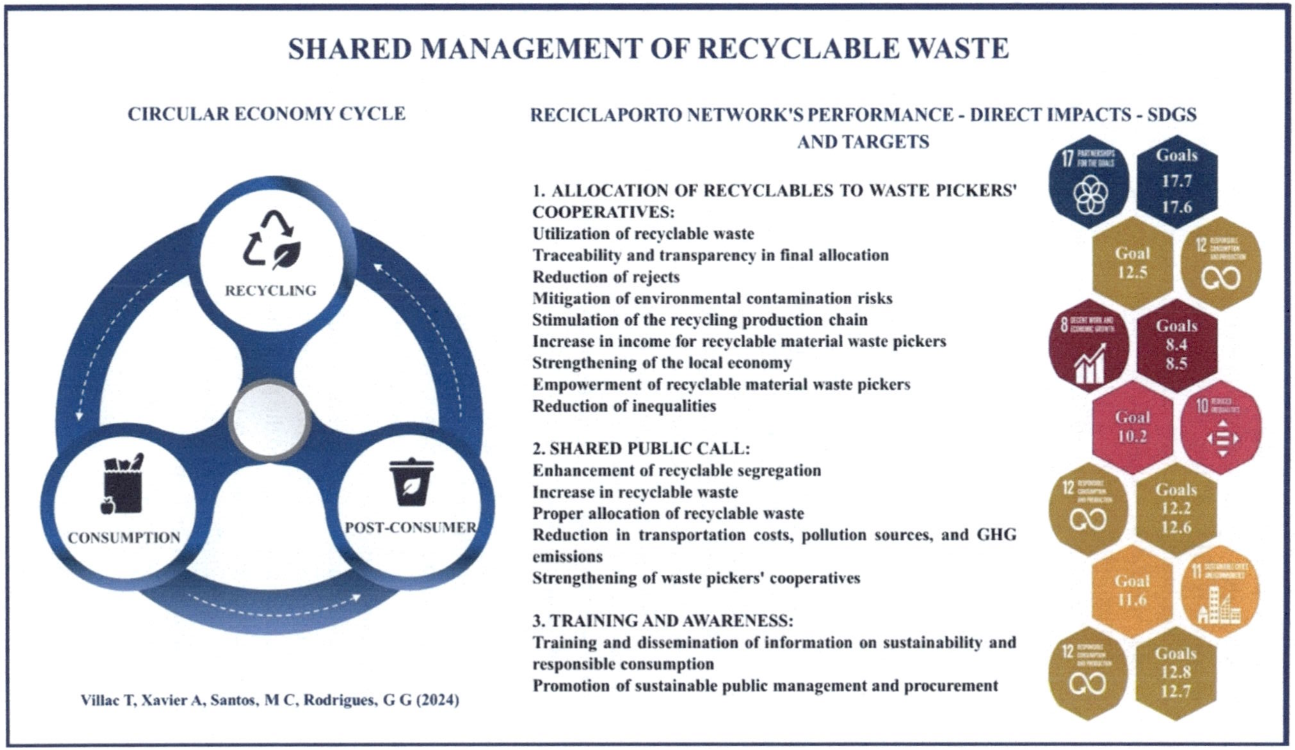

Fig. 4 Shared management of solid waste with SDGs and circular economy. *Source* Prepared by the authors

of waste, mitigates pollution sources and reduces greenhouse gas emissions (SDG 9, Target 9.5 and SDG 13, Target 13.3).

Promoting and supporting waste pickers' cooperatives in Brazil not only elevates their income levels but also significantly contributes to dignified employment and fortifies the local economy (SDG 8, Target 8.4). Consequently, this indirectly aids in attaining SDG 1 by contributing to poverty eradication. Strengthening these cooperatives is instrumental in mitigating social inequalities by fostering the social inclusion of waste pickers, a demographic predominantly composed of women from vulnerable populations (SDG 5, Target 5.5; SDG 10, Target 10.2). The strategic alignment of ReciclaPorto with multiple Sustainable Development Goals, notably SDG 17 (Target 17.17) on promoting effective partnerships and SDG 11 (Target 11.6) on empowering vulnerable groups and communities, underscores its comprehensive contribution to the sustainability and resilience of local territories.

From a circular economy perspective, the efficient management of recyclable waste during post-consumption and recycling stages enables the transformation of discarded materials into new resources, thereby minimizing reliance on virgin raw materials in production processes. The initiatives undertaken by the ReciclaPorto Network facilitate this transformation by ensuring the collection, segregation, and proper disposal of recyclable waste generated by participating institutions, which are then channeled to qualified waste pickers' cooperatives through collaborative public calls. This methodology mitigates the environmental impact associated with the extraction of new resources. It enhances waste valorization, strengthening the role of cooperatives within the recycling production chain and promoting the efficient reintegration of materials into the production of new goods.

In the broader context of the Circular Economy, the actions of ReciclaPorto generate substantial indirect impacts across various stages, including raw material extraction, design, manufacturing, and distribution. In the raw materials phase, emphasis is placed on the judicious use of natural resources. The network's initiatives on conscious consumption and appropriate waste disposal practices contribute to reducing the use of virgin raw materials. Circular economy design principles aim to create products based on sustainable foundations. ReciclaPorto actively disseminates knowledge demonstrating the efficacy of recycling and material reuse practices in alignment with these principles. Regarding manufacturing processes, the network focuses on minimizing waste and maximizing production efficiency, ensuring the reintegration of recyclable materials into the production cycle. In the realms of distribution and transportation, the emphasis is on energy efficiency and reducing greenhouse gas emissions. ReciclaPorto aims to reduce the carbon footprint associated with sustainable procurement and optimize transportation routes utilized by waste pickers' cooperatives by advocating for conscious consumption and effective waste management.

4.3 Future Perspectives of the ReciclaPorto Network

The collaborative implementation process has significantly bolstered interinstitutional cooperation, strengthening ReciclaPorto. As Rindfleisch (2000) elucidates, interinstitutional cooperation materializes through collective actions to achieve desirable outcomes. Joint training sessions have been instituted to refine each agency's Citizen Selective Collection procedures and augment the quantity and quality of recyclable waste. These sessions, conducted by representatives of accredited waste pickers' cooperatives and participating institutions, engage members of the Selective Collection Committee, cleaning teams, contract inspectors, and personnel responsible for waste management. Owing to their positive outcomes, these initiatives have been integrated into the strategic work plan for periodic execution over the next five years.

In response to institutional demands and operational exigencies within the territory, ReciclaPorto is poised to extend its network into the interior regions of the state of Rio de Janeiro. This expansion aims to optimize recyclable waste management through joint allocation efforts and reinforcing connections among local actors. Preliminary studies reveal that certain regions are serviced by only a single registered waste pickers' cooperative. Consequently, ReciclaPorto intends to initiate programs to support the establishment of new cooperatives, thereby contributing to the formalization and training of waste pickers. Additionally, it seeks to structure a Regional Shared Public Call and forge academic partnerships focused on initiatives that bolster waste pickers' cooperatives, the recycling production chain, and the local economy, particularly in areas with higher vulnerability indices.

Simultaneously, efforts are being undertaken to overcome administrative hurdles to facilitate the remuneration of waste pickers' cooperatives for their services in alignment with the network's objectives and Decree 11.414, 2023, which addresses the integration and coordination of public administration actions, projects, and programs aimed at promoting and safeguarding the human rights of recyclable material waste pickers. Within this framework, a proposal for contracting a waste pickers' cooperative is considered by a network entity. This proposal envisions the development of consultancy services aimed at enhancing selective collection procedures, performing gravimetric analyses of recyclable waste, evaluating the content of awareness campaigns, and conducting training and mentorship activities for the workforce.

Integrating shared governance and management within a territorialized network substantially enhances public administration by fostering the convergence of actions and the optimal deployment of resources and instruments across diverse entities. This approach respects and harmonizes the unique cultures of all stakeholders, thereby amplifying the impact of implemented initiatives. Facilitating coordinated actions among institutions mitigates the risks of inefficiency and ineffectiveness that often plague isolated efforts. Moreover, this model epitomizes how extended governance can advance more sustainable and efficacious public administration practices, serving as an exemplary and replicable paradigm.

5 Conclusions

Despite constitutional competencies to implement public policies across various domains, Brazil lacks comprehensive incentives for advancing a circular economy. A primary challenge lies in institutionalizing participatory governance frameworks that embed sustainability. A territorial perspective is essential, integrating cultural, political, and situational dimensions into the foundations of socio-environmental public policies, which are intrinsically linked to promoting and disseminating the principles of a circular economy.

The ReciclaPorto Rio Sustainability Network exemplifies the potential of social innovation through shared governance among governmental and territorial actors. This initiative represents a significant advancement in Brazilian public administration, aligning with SDG 17 by fostering interinstitutional and multidisciplinary partnerships. Dialogical practices and collective creation spaces enhance actor complementarity, involving civil society and emphasizing the territory as a sphere of action. This approach supports environmentally responsible, economically viable, and social work in alignment with the 2030 Agenda and the SDGs.

Encouraging and promoting effective partnerships across public, public–private, private, and civil society sectors hinges on resource mobilization strategies derived from these collaborations.

Extended and transformative governance requires symbolic axes within the territory to engage actors over the long term, imbuing the territorial project with substantive meaning. This inter-relational approach connects various action scales (civil society, public and private enterprises, collectives, and associations) to foster social innovations. Operational elements mobilize mechanisms of conduct, convention, articulation, and evaluation grounded in experiential feedback and the recognition of practical work. Thus, the territory evolves into a locus and an endogenous agent in creating, producing, and capturing value and wealth, mobilizing meanings, situated innovations, and intersections with local needs and expectations.

Recognizing the role of the state and the interplay among diverse societal actors and action spheres is crucial for governance models capable of integrating the SDGs and the circular economy. This paradigm positions the territory at the center, not merely as a recipient of externalities from local and global economic activities, but as a foundational pillar for value creation.

References

Abrucio FL (1977) The impact of the managerial model on public administration: a brief study on recent international experience. Cadernos ENAP, vol 10, p 52. ISSN: 0104-7078

Albers S (2005) The design of alliance governance systems. Springer Fachmedien Wiesbaden, Wiesbaden

Augé M (1992) Non-places—introduction to an anthropology of supermodernity. Éditions du Seuil, Paris

Baumgarten M (2007) Science & technology in the semi-periphery and social innovation: inequalities, excellence, and competitiveness. In: Maciel ML, Albagli S (eds) Information and development: knowledge, innovation, and social appropriation. UNESCO/IBICT, Brasília, pp 271–295

Baumgarten M (2008) Science, technology, and development—networks and social innovation. Parcerias Estratég 26:101–123

Bessa FLBN, Doetzer GD (2020) Sustainable development goals 16 and 17—the creation of the Sustenta Paraná network and the challenges of networked work. In: Villac T, Bessa FLBN, Doetzer GD (eds) Brazilian public management: sustainable networked innovation. Editora Fórum, Belo Horizonte

Bignetti LP (2011) Social innovations: an incursion through ideas, trends, and research focuses. Ciên Soc Unisinos 47(1):3–14

Cader R, Villac T (2022) Governance and sustainability: an essential link in Brazil. 1st reprint. Editora Fórum, Belo Horizonte

Caron A (2007) Social innovation and the role of industry. In: Farfus D, Rocha MCS (eds) Social innovations. FIEP—Federation of Industries of the State of Paraná, Curitiba, pp 85–116

Costa C, Xavier A (2023) Creativity, value co-creation, and reflexivity: how the functional and cooperative economy's approach enhances service design performance. In: Service design and innovation conference, pp 1077–1091

Dallabrida VR (2006) Territorial governance: institutional density and social capital in the territorial development management process. Santa Cruz Sul Edunisc 1:1–19

Franzese C (2021) The potential of network management in public administration. GV Executivo 20(2):49. Available at: https://periodicos.fgv.br/gvexecutivo/article/view/83960/79537. Accessed on 17 Sept 2023

Freitas J (2014) Sustainability—the right to the future. Editora Fórum, Belo Horizonte

Fuini LL, Pires ELS (2015) The dimensions of territorial governance: preliminary delineation of theoretical and morphological aspects. In: Saquet MA, Sposito ES (eds) Territories and territorialities: theories, processes, and conflicts. 1st edn. Expressão Popular, São Paulo

Kingdon JW, Stano E (1984) Agendas, alternatives, and public policies. Little, Brown, Boston

Manzini E (2017) Design: when everybody designs—an introduction to design for social innovation. Unisinos, São Leopoldo

Mazzucato M (2022) Mission economy: a moonshot guide to changing capitalism. Portfolio-Penguin

Poteete AR, Ostrom E, Janssen MA (2011) Working together: collective action, common goods, and multiple methods. Senac, São Paulo

Provan K, Kenis P (2008) Modes of network governance: structure, management, and effectiveness. J Public Administration Research and Theory, Oxford, UK 18(2):229–252

Rindfleisch A (2000) Organizational trust and interfirm cooperation: an examination of horizontal versus vertical alliances. Mark Lett 11(1):81–95

Rocha K, Xavier A, Santos MC, Rodrigues-Ferreira A (2023) Analysis of the integration and contribution of the SDGs to equitable global results in the light of service design: a scoping review. In: Service design and innovation conference, pp 1353–1372

Santos M (2009) The nature of space: technique and time. reason and emotion. Edusp, São Paulo

Santos MCL (ed) (2014) Design, waste & dignity. Olhares; CNPq, São Paulo

Santos M, Silveira ML (2001, 2021). Brazil: territory and society at the beginning of the 21st century. Record, 22nd edn. São Paulo

Santos MC, Ribeiro GGL, Bertolossi C, Xavier A (2023) Shared public call—a network of public institutions co-creating solutions for recyclable waste management. In: Hagge A, Asti A, Ramos AP et al (eds) Anais do CIRS: 1st international congress on solid waste in Búzios. Atena, Ponta Grossa, PR, pp 55–66

Secchi L (2014) Public policies: concepts, analytical frameworks, and practical cases. Cengage Learning

Villac T (2024) Governance, territoriality, and culture: assumptions for the advancement of sustainable public procurement based on Law No. 14,133/21. Available at: https://www.novaleilicitacao.com.br/2024/11/08/governanca-territorialidade-e-cultura-pressupostos-para-o-avanco-das-contratacoes-publicas-sustentaveis-a-partir-da-lei-n-o-14-133-21/. Accessed on 10 Nov 2024

Visseren-Hamakers IJ, Razzaque J, McElwee P, Turnhout E, Kelemen E, Rusch GM, Zaleski D (2021) Transformative governance of biodiversity: insights for sustainable development. Curr Opin Environ Sustain 53:20–28

Xavier AF, Valle WAD, Souza MAD, Duarte FJDCM, Lima FDPA (2024) Searching for a sustainable economy: work, cooperation, and territorial solutions. Work, p 77

Teresa Villac Ph.D. in Environmental Science, University of São Paulo (USP) Philosopher, USP. Attorney for the Union in the Legal Advisory Office of the Ministry of the Environment and Climate Change (MMA). Coordinator of the National Chamber of Sustainability of the General Counsel of the Union (Advocacy-General of the Union).

Amanda Xavier Post-Doctorate in Management, Innovation, and Sustainability—Production Engineering (COPPE). Professor of the Production Engineering Program at COPPE (Federal University of Rio de Janeiro—UFRJ). Founder and coordinator of the Advanced Center for Sustainability, Local Ecosystems, and Governance (Casulo—COPPE). Director of Planning and Institutional Development at COPPE (Federal University of Rio de Janeiro—UFRJ).

Maria Carolina Santos Doctoral Candidate in Management and Innovation in the Production Engineering Program (COPPE/UFRJ). Master of Production Engineering (COPPE/UFRJ). Senior Science and Technology Analyst at the Ministry of Science, Technology, and Innovation (MCTI). President of the Sustainability Commission at the National Institute of Technology (INT—MCTI). Co-founder and Coordinator of the Sustainability Network ReciclaPorto Rio.

Gabriel Grasso Ribeiro Undergraduate Student in Environmental Sciences at the Federal University of the State of Rio de Janeiro (UNIRIO). Intern at the Sustainability Commission of the National Institute of Technology (INT—MCTI). Active participant in projects: Movimento é Vida, Agroecology for All, and the Economic and Social Development Panel of the Itaocara Region (Movement is Life, Agroecology for All, and the Economic and Social Development Panel of the Itaocara Region).

Circular Economy in the Textile Sector: Driving Sustainable Development Goals

Elda Rodrigues Steinhorst Kraetzig, Lucas Veiga Ávila, and Marceli Adriane Schvartz

Abstract This chapter investigates the integration of the Circular Economy (CE) with the 17 Sustainable Development Goals (SDGs) in the textile sector, through a Systematic Literature Review (SLR) that analyzed 188 relevant articles. The research highlights how the adoption of circular practices can positively impact all the SDGs, from reducing waste to promoting social and economic justice. The bibliographic analysis identified predominant keywords such as "sustainable consumption", "circular fashion" and "circular economy", and highlighted influential articles such as Towards sustainability? Forest-based circular bioeconomy business models in Finnish SMEs and Trends in the fashion industry: the perception of sustainability and circular economy. It also mapped key academic sources, such as Sustainability and the Journal of Cleaner Production, and highlighted key organizations and authors. The chapter not only contributes to academic advancement, but also offers practical insights for the textile industry and policymakers, suggesting a more integrated and interdisciplinary approach to promoting sustainable development and tackling environmental and social challenges. The research highlights gaps in the literature, pointing to new research opportunities and reinforcing the importance of CE as a strategic tool to transform the textile sector.

Keywords Circular economy · Sustainable development goals · Textile industry · Sustainability · Sustainable business models

E. R. S. Kraetzig (✉) · L. V. Ávila · M. A. Schvartz
Federal University of Santa Maria, Santa Maria, Brazil
e-mail: Kraetzig.elda@gmail.com

L. V. Ávila
e-mail: lucas.avila@ufsm.br

M. A. Schvartz
e-mail: schvartz.marceli@gmail.com

L. Brandli et al. (eds.), *The Contribution of Life Cycle Analyses and Circular Economy to the Sustainable Development Goals*, World Sustainability Series,
https://doi.org/10.1007/978-3-031-85300-5_4

1 Introduction

The textile industry, although a fundamental part of the global economy, faces strong criticism due to its significant environmental impacts. The intensive use of natural resources and the disproportionate generation of waste characterize an unsustainable production model that demands urgent reformation (Niinimaki et al. 2020; Albuquerque et al. 2021; Brydges 2021).

In response to these pressing challenges, the circular economy emerges as a promising alternative, proposing a paradigm in which waste is minimized, resources are optimized, and development is truly sustainable (EMF 2019; Gallego-Schmid et al. 2020). According to Sharma et al. (2021), waste management and the circular economy play a crucial role in various Sustainable Development Goals (SDGs).

The Sustainable Development Goals (SDGs) are a collection of 17 global goals established by the United Nations General Assembly in 2015 as part of the 2030 Agenda for Sustainable Development. They were adopted by all UN Member States at a historic summit held in September 2015 in New York, with the aim of promoting peace, prosperity, and sustainability for all people and the planet (United Nations 2015).

The SDGs offer a novel approach to global governance through goal setting, but their success depends on institutional factors and research communities' involvement (Biermann et al. 2017). Integrating Sustainable Development Goals requires greater attention to interlinkages across sectors, societal actors, and countries, resulting in better outcomes and synergies (Stafford-Smith et al. 2016).

As highlighted by the Ellen MacArthur Foundation (2019) and the United Nations Environment Programme (2018), the circular economy, if adopted on a large scale, has the potential to play a significant role in achieving the SDGs. This approach can address urgent contemporary challenges such as biodiversity loss, resource scarcity, pollution, and waste, thus contributing to the progress of the SDGs.

The 17 SDGs address a broad range of sustainable development issues, including the eradication of poverty, environmental protection, promotion of health and well-being, quality education, gender equality, clean water and sanitation, affordable and clean energy, decent work and economic growth, reduced inequalities, sustainable cities and communities, responsible consumption and production, climate action, life below water, life on land, peace, justice, and strong institutions, and partnerships for the goals (United Nations 2015).

The circular economy plays a prominent role in promoting global sustainability by encouraging the reuse, recycling, and regeneration of materials. Important studies (Belmonte-Ureña et al. 2021; Dantas et al. 2021; Khalifa et al. 2022; Zibunas et al. 2022; Thakker and Bakshi 2023) have begun to explore the relationship between the circular economy (CE) and the SDGs. Several studies (Chen et al. 2020; Fatimah et al. 2020; Cecchin et al. 2020; Kayikci et al. 2022; Khalifa et al. 2022) that analyzed the CE in conjunction with the SDGs observed a positive correlation between these concepts.

Although the topic has gained popularity, the joint approach to the circular economy (CE) and the SDGs is still limited. Various studies mention the SDGs in their analyses of the CE, but few directly explore the interconnection between the CE and the SDGs (Govindan 2023). The textile and fashion industry impacts the environment and human health and is associated with all 17 SDGs, despite rarely being linked to the entire SDG framework (Thakker and Sun 2023).

Considering the concerning impacts of the textile sector, the SDGs demonstrate even greater relevance for this sector (Cobbing et al. 2023). According to Aramendia-Muneta et al. (2022), circular fashion stands out as one of the sectors where the commitment to the SDGs is most necessary, given its global nature and significant growth in terms of consumption. In this context, this chapter aims to examine the potential of the circular economy in the textile sector and its potential in achieving the Sustainable Development Goals (SDGs) through a systematic literature review.

2 Circular Economy in the Textile Sector and Contributions to the Sustainable Development Goals (SDGs)

The circular economy has emerged as a prominent strategy to address global environmental and economic challenges, aiming for sustainability along the entire value chain (Schroeder et al. 2019). In the textile sector, known for its substantial environmental impact due to the intensive use of natural resources and significant waste production, the circular economy offers innovative approaches to mitigate waste, optimize resource management and promote sustainable development (Peters et al. 2019; Niinimaki et al. 2020). Given this, the circular economy in the textile sector has the potential to contribute to each of the SDGs.

2.1 Contributions to SDG 1 (Poverty Eradication)

According to Cai and Choi (2020), SDG 1 (Poverty Eradication) receives less attention in research related to the circular economy and the textile sector. However, the circular economy in this sector contributes significantly to the creation of new jobs and economic opportunities, including "green jobs" in areas such as sustainable design, recycling and waste management (Cuc et al. 2015; Cappellieri et al. 2019).

Jia et al. (2020) also reaffirm the importance of circular economies in the textile and clothing industry, stating that they create a viable relationship between ecosystems and economic growth. Therefore, if it generates economic growth, it contributes to SDG 1. Chaudhary and Shamsi (2020) state that adopting a circular economy in textiles can reduce post-consumer textile waste, generate employment and create a sustainable environment.

According to Saha et al. (2021), the circular economy in the textile and clothing industry correlates with sustainability performance (economic, environmental and social). In other words, economic performance is integral. According to Filho et al. (2019) textile recycling offers socio-economic benefits, including reduced waste, improved environmental impact and increased employment in the clothing sector. Recycling and waste management in the textile sector offer a number of applications, such as the use of natural fibers and composite materials, which have proven valuable in the textile industry.

2.2 Contributions to SDG 2 (Zero Hunger and Sustainable Agriculture)

The circular economy in the textile sector plays an important role in achieving SDG 2 (Zero Hunger and Sustainable Agriculture) by positively impacting both food security and sustainable agriculture. The implementation of circular practices, such as waste reduction and material recycling, contributes to reducing the adverse environmental impacts associated with textile production, which in turn can mitigate the effects of climate change (Leal Filho et al. 2022). Mitigating climate impacts helps promote the stability of agricultural production, increasing food security by reducing the vulnerability of crops to extreme weather changes (United Nations 2021; Li et al. 2023).

In addition, the circular economy in the textile sector generates employment opportunities and fosters economic growth, which can have positive effects on reducing hunger. Creating new jobs in areas such as sustainable design, recycling and waste management (Chaudhary and Shamsi 2020) not only contributes to the eradication of poverty, but also improves the purchasing power of poor populations, helping to ensure access to food (Jia et al. 2020).

By implementing circular economy techniques, the textile sector not only reduces the negative environmental impact of its operations, but also contributes to the creation of a more inclusive and sustainable economic cycle, benefiting both food security and sustainable agriculture.

2.3 Contributions to SDG 3 (Good Health and Well-Being)

The circular economy in the textile sector can significantly contribute to SDG 3 (Good Health and Well-being) through various strategies that promote public health and safety in the workplace (Agrawal et al. (2021). As stated by Furferi et al. (2022), the adoption of circular economy policies can lead to the reduction of waste production and the regenerative use of materials, such as wool, in addition to improving the management of the end-of-life of products. These practices not only minimize the environmental impact, but also reduce the risks associated with the accumulation of waste and pollutants, promoting a healthier environment for the communities surrounding the factories.

In addition, the circular economy addresses the reduction of the use of toxic chemicals in textile production. Koszewska (2018) highlights that the transition to circular models implies significant changes in production and consumption processes, resulting in a decrease in the use of harmful substances and an improvement in working conditions. By adopting safer and less polluting production methods, the textile sector not only protects the health of workers but also contributes to reducing health problems related to exposure to toxic chemicals, in line with SDG 3.

Implementing circular economy practices, such as recycling materials and improving working conditions in garment factories, is intrinsically linked to promoting a safer and healthier work environment. These practices not only ensure a better quality of life for workers, but also have a positive impact on general public health by reducing the risks associated with pollution and the use of harmful chemicals, strengthening the connection with SDG 3 and also with SDG 8, on Decent Work and Economic Growth (Ellen MacArthur Foundation and Circular Fiber Initiatives 2017).

2.4 Contributions to SDG 4 (Quality Education)

The integration of the circular economy in the textile and apparel sector can significantly contribute to SDG 4, which aims to ensure inclusive, equitable and quality education and promote lifelong learning opportunities for all. The LAB University of Applied Sciences has been a pioneer in creating teaching methods and learning environments aimed at promoting circular economy expertise in the textile sector, through several national and international projects (Heikinheimo et al. 2022).

A notable example is the Design4Circle project, which developed an e-learning platform to train designers in eco-innovation, empowering them to adapt to a circular textile and fashion industry (Avadanei et al. 2021). Furthermore, research by Mansour et al. (2020) demonstrates the potential of using natural waste as sources for recycling eco-friendly textile dyes, promoting the circular economy and sustainability in arts education.

Collaborative learning projects, as highlighted by Whitehill et al. (2022), can generate value for all stakeholders and contribute to the circular economy by encouraging innovation and sustainability in manufacturing. These initiatives exemplify how education can be a catalyst for the transition to a circular economy by equipping future professionals with the skills and knowledge needed to promote sustainable practices.

2.5 Contributions to SDG 5 (Gender Equality)

The circular economy in the textile sector plays a crucial role in promoting gender equality, in line with SDG 5. Historically, the inclusion of women as wage workers in Eritrean textile factories between the 1950s and 1970s not only transformed the factory environment into a more gender-responsive space, but also expanded the role of women beyond formal work (Fusari 2023). This example illustrates how female participation in the textile industry can bring about significant changes in social and economic dynamics.

In many societies, textile production directly influences social status, with men often assuming the role of primary producers in non-state economies (Jolie 2014). In the Indian context, women account for almost half of the workforce in the textile sector, with high rates of female ownership and employment in some states (Swami and Padma 2015).

Implementing circular economy practices in the textile sector can help address inequalities by fostering a more inclusive work environment. The use of recycled materials and the adoption of sustainable production processes can open up new employment and entrepreneurship opportunities for women, helping to reduce the wage gap and improve working conditions.

In addition, training and skills development programmes focused on sustainable practices have the potential to empower women to take on leadership roles in the industry, strengthening their economic and social role. Thus, the circular economy in the textile sector not only advances environmental sustainability but also plays an important role in promoting gender equality, a key aspect of sustainable development.

2.6 Contributions to SDG 6 (Clean Water and Sanitation)

Circular economy practices can also contribute to water conservation (SDG 6), and the percentage of recycled water is a significant indicator in measuring CE performance (Jia et al. 2020). There is evidence in the case study by Neto et al. (2021) in a textile industry that the implementation of Cleaner Production practices, such as water reuse and reduction of toxic effluents, led to a significant reduction in environmental impacts and a return on investment in 29 months.

Projects such as the AquaFit4Use project can contribute to SDG 6. The AquaFit4Use project successfully separated wastewater based on pollution levels, enabling a reduction of more than 40% in freshwater consumption in the European textile finishing industry. During the textile manufacturing process, there is water waste, intensive energy use, application of toxic chemicals, emission of solid waste and water pollution (Shirvanimoghaddam et al. 2020).

Therefore, there is a need to contribute to SDG 6 (Clean Water and Sanitation). To this end, the textile sector has implemented several solutions, such as atmospheric plasma technology developed by SENAI CETIQT. This technology applies electrical discharges to gases at room temperature to treat material surfaces, minimizing the use of chemical reagents and eliminating the production of effluents. This promotes water savings and reuse, reducing both consumption and water pollution (Senai 2024).

2.7 Contributions to SDG 7 (Affordable and Clean Energy)

The circular economy (CE) in the textile sector can significantly contribute to Sustainable Development Goal (SDG) 7, which aims at affordable and clean energy. CE, as described by Geissdoerfer et al. (2017), enables resource input, waste, emissions, and energy leakage to be minimized by slowing, closing, and tightening material and energy loops. CE and sustainability-driven innovation can help companies transition to a cleaner energy future by promoting customer value and considering environmental and social aspects (Koval et al. 2022). It is essential that the sector invests in renewable energy sources and energy efficiency to decarbonize the energy matrix, reduce costs (Quantis, 2018; Radulescu et al. 2018), and achieve SDG 7 (Affordable and Clean Energy).

The Ellen MacArthur Foundation (2013) highlights that, in addition to focusing on resources, CE encourages the production and expansion of the use of renewable energy and the end of the use of toxic chemicals. The toxicity of natural resources prevents them from being safely returned to the environment, which is mitigated by the EC (European Environment Agency 2016).

The stages of the product life cycle with the greatest environmental impacts in the textile industry include the production of synthetic fibers, finishing processes, yarn manufacturing, and the washing and drying of clothes in the use phase (Koszewska 2018). Therefore, the adoption of circular economy practices can improve energy efficiency and reduce the environmental impact of these processes, contributing to SDG 7 by promoting sustainable energy use and reducing emissions.

2.8 Contributions to SDG 8 (Decent Work and Economic Growth)

The circular economy in the textile sector plays a significant role in advancing SDG 8, which aims to promote decent work and economic growth. This approach can transform the textile industry by aligning business practices with the principles of social and economic sustainability.

Entrepreneurial leadership is essential to achieving sustainable development goals, including Goal 8. In the context of the Pakistani textile industry, such leadership can foster fairer work environments and promote economic growth, which are central elements of SDG 8 (Shehri et al. 2022). In this sense, fostering entrepreneurial leadership in the textile circular economy can be a lever for SDG 8.

In addition, the quality of working life, a vital aspect of decent work, is influenced by factors such as fair pay, healthy working conditions, and work-life balance. These conditions, observed in the textile sector of Coimbatore, are essential to ensure that workers have a work environment that respects their dignity and promotes their development (Valarmathi and Bhalakarishnan 2013). Improving these conditions is essential to balance the work and personal lives of workers, as evidenced in the Indian textile industry (Gangwani et al. 2020).

Managing quality of work is also crucial for long-term stability in the textile industry. By addressing basic requirements such as stewardship, quality of work and team harmony, the industry can achieve better economic and social outcomes, contributing to Goal 8 (Purushothama 2013).

The circular economy is one of the measures proposed by Thakker and Sun (2023) as a solution to promote decent work and economic growth (SDG 8). As mentioned as a solution for SDG 1, the circular economy generates employment opportunities, which can boost economic growth (Cuc et al. 2015; Cappellieri et al. 2019).

2.9 Contributions to SDG 9 (Industry, Innovation and Infrastructure)

Channeling innovations in textiles and fashion to be sustainable from start to finish is essential to advance the Sustainable Development Goals (SDGs), especially SDG 9 (Industry, Innovation and Infrastructure). These innovations not only support sustainable industrialization, but also adapt environmentally friendly methods and methodologies to reduce environmental impact. Digital innovation models, such as on-demand production involving the manufacturer and retailer to create personalized products, as well as virtual 3D customization of cutting, printing and assembly of parts, significantly boost SDGs 9 and 12 by promoting resilient infrastructure and sustainable consumption and production practices (Larsson 2018).

The combination of the Circular Economy and Industry 4.0 directly benefits the achievement of Sustainable Development Goals, particularly SDG 9 and SDG 12. The circular economy, promoting sustainable development and resource management, can revolutionize the textile industry and contribute to achieving SDG 12 (Gabriel and Luque 2019).

According to Govindan (2023), several studies, such as those carried out by Menton et al. (2020), Pizzi et al. (2020) and van Zanten and van Tulder (2020), highlight that all Sustainable Development Goals (SDGs) are interconnected, influencing each other both directly and indirectly. This is true across several SDGs, and one example of clear interconnection is SDGs 1, 8 and 10.

This interconnection between SDG 1 (No Poverty), SDG 8 (Decent Work and Economic Growth) and SDG 10 (Reduced Inequalities) is particularly evident. For example, promoting inclusive and sustainable economic growth (SDG 8) can significantly contribute to reducing poverty (SDG 1) by generating quality jobs and ensuring equal opportunities for all. At the same time, policies that encourage social and economic inclusion can help reduce inequalities (SDG 10), creating a virtuous cycle where progress on one SDG drives progress on the others.

2.10 Contributions to SDG 10 (Reduced Inequalities)

The circular economy in the textile sector can contribute to SDG 10 by promoting economic and social equity in several ways. By creating inclusive jobs (Cuc et al. 2015; Cappellieri et al. 2019) and empowering vulnerable communities with circular economy skills, the textile sector helps to reduce inequalities by providing equitable opportunities for economic and social development. These jobs can be created in disadvantaged communities, providing opportunities for people who might otherwise be excluded from the traditional labour market.

Equipping vulnerable communities with circular economy-specific skills not only increases their chances of employment, but also strengthens local economic development and social justice, also encompassing SDG 16 (Peace, Justice and Strong Institutions). The circular economy can facilitate business models such as product rental and exchange, which can reduce the initial cost for low-income consumers. These models allow people to access sustainable products without having to purchase expensive items.

2.11 Contributions to SDG 11 (Sustainable Cities and Communities)

The circular economy in the textile sector can play a key role in promoting more sustainable cities and communities, as per SDG 11. Responsible management of textile waste contributes to reducing the volume of urban waste, promoting a cleaner environment and reducing pressure on cities' waste management systems (Campello 2020). By adopting circular production practices, such as reusing and recycling materials, it would be possible to reduce the amount of waste sent to landfills, minimizing negative environmental impacts and improving the quality of life in urban areas.

In addition, implementing textile waste collection and sorting systems in communities facilitates efficient waste management, promoting environmental awareness and education among citizens. These practices encourage active community participation in waste management and promoting a more sustainable life cycle for textile products. For example, recycling programs and drop-off points for used or end-of-life clothing encourage consumers to participate in the circular economy, helping to reduce the sector's environmental impact.

The circular economy could also promote the development of new sustainable infrastructure, such as recycling facilities and upcycling centers, which could create jobs and economic opportunities in urban areas. These centers would not only contribute to the local economy but would also help to reduce pressure on natural resources and reduce the environmental impacts associated with conventional textile production. Therefore, by integrating circular economy practices, the textile sector can help build more resilient and sustainable cities, aligning with the goal of promoting cleaner and more environmentally responsible communities.

2.12 Contributions to SDG 12 (Responsible Consumption and Production)

There are many possibilities for responsible textile production that meet SDG 12 (Araújo and Vieira 2017; Schroeder et al. 2019), such as the use of waste generated by agricultural and food production that can effectively generate eco-friendly fibers and fabrics. For example, fabrics made from orange and citrus peels through innovative and patented methods, such as blending orange fibers and lyocell, illustrate the transformation of waste into valuable resources (Orange Fiber Srl 2022).

Similarly, vegetable leathers made from cereals, pineapple, cactus, corn, apple peels and flowers through carbon–neutral processes offer sustainable alternatives to animal leather, contributing to the protection of biodiversity (Hirsh 2020). Furthermore, the use of algae to make foam represents a nature-based, eco-friendly solution for textiles and fashion (Veerah 2022). Bacterial cellulose, found in Kombucha, has the potential to replace fabrics in the textile industry, contributing to environmental and social sustainability and aligning with the Sustainable Development Goals, more specifically, SDG 12 (Provin et al. 2021).

2.13 Contributions to SDG 13 (Climate Action)

The authors Leal Filho et al. (2022) highlight mitigation practices needed in the textile sector, some of which encompass the scope of other SDGs that ultimately contribute to SDG 13. For example, the implementation of regenerative models for agriculture, which are beneficial for climate action, is also connected to SDG 15, which deals with life on land. These models promote soil recovery and biodiversity, reducing the need for textile products based on non-renewable fibers and, consequently, reducing environmental impact.

In addition, the circular economy encourages innovation in recycling technologies and the development of biomaterials, which are essential to reducing dependence on virgin materials and reducing greenhouse gas emissions (Choudhary et al. 2015). These advances are aligned with SDG 9, which emphasizes industry, innovation, and infrastructure. Phasing out the use of coal and transitioning to renewable energy sources are practices highlighted in the context of the circular economy and are directly linked to SDG 7, which promotes affordable and clean energy.

These strategies not only reduce the carbon footprint of the textile sector, but also contribute to the climate resilience of communities by reducing the environmental burden and promoting a more sustainable production and consumption system. In short, the circular economy not only helps reduce greenhouse gas emissions and resource consumption, but also drives innovations that are key to addressing today's climate challenges.

2.14 Contributions to SDG 14 (Life Below Water)

The textile sector presents environmental challenges that directly impact SDG 14, which seeks to conserve and sustainably use the oceans and marine resources. Fabrics composed of polylactic acid and cellulose/plastic blends, for example, do not biodegrade rapidly in marine waters, which suggests that, although widely promoted as sustainable alternatives, they are inadequate for mitigating plastic pollution in the oceans (Royer et al., 2023). Furthermore, effluents from the textile industry, when not specifically treated, have a negative impact on the quality of water and aquatic ecosystems. Studies show that these effluents can alter benthic communities, increasing density but decreasing species diversity, compromising the ecological integrity of aquatic habitats (Gómez et al., 2008). Likewise, inadequate treatment of textile wastewater can deteriorate river water quality and affect biological communities, especially fish, although environmental conditions recover some distance from treatment plants (Colin et al., 2016). The circular economy, unlike the linear model of production and consumption, proposes a more sustainable approach, minimizing the environmental impacts of the textile sector. The reuse of materials, design for recycling, and waste reduction directly reduce pollution, especially with regard to the use of fabrics and the disposal of textile waste. In addition, the circular economy offers solutions to the problem of synthetic microfibers, promoting the reengineering of fabrics and the improvement of filtration systems in washing machines, as well as advanced wastewater treatment techniques (Weis and De Falco, 2022). Thus, the adoption of circular practices in the textile sector can not only reduce plastic pollution and improve water quality, but also contribute to the conservation of marine ecosystems, aligning with the objectives of SDG 14 and promoting the sustainability of oceans and water resources (Araújo and Vieira, 2017; Schroeder et al. 2019).

2.15 Contributions to SDG 15 (Life on Land)

The circular economy in the textile sector contributes to SDG 15 (Araújo and Vieira 2017; Schroeder et al. 2019) by promoting practices that minimize the exploitation of natural resources and combat deforestation associated with the production of textile fibers. By reusing and recycling materials, the industry reduces pressure on natural habitats and biodiversity, reducing the need for new raw material extraction and the degradation of terrestrial ecosystems. Reusing fibers and reducing waste not only helps conserve natural resources, but also promotes practices that prevent the expansion of agricultural areas for fiber production, which in turn combats deforestation and soil degradation.

In addition, producing fabrics from agricultural waste, such as fruit peels and other by-products, not only reduces the environmental impact of textile production, but also promotes the sustainable management of agricultural resources (Orange Fiber Srl 2022). This practice helps preserve soil quality and prevent desertification, contributing to the health and sustainability of terrestrial ecosystems. The circular economy, by encouraging the efficient use of resources and the reduction of waste, also supports reforestation and restoration initiatives for degraded areas, aligning with the SDG 15 targets for the conservation and sustainable management of natural resources and the protection of life on land.

2.16 Contributions to SDG 16 (Peace, Justice and Strong Institutions)

The circular economy in the textile sector can play a significant role in promoting SDG 16) by encouraging more responsible and transparent business practices. Adopting circular models requires a deep commitment to social and environmental responsibility, which can lead to more robust and ethical governance within the industry. By integrating circular economy principles, companies become more aware of the social and environmental impacts of their operations, thus promoting the protection of human rights and better working conditions.

In addition, the circular economy contributes to the creation of more transparent and traceable supply chains, allowing greater visibility into production practices and the origin of materials. This increased transparency helps to combat exploitation and injustice throughout the value chain, ensuring that business practices are carried out in a fair and ethical manner.

The traceability promoted by the circular economy also facilitates the identification and mitigation of risks associated with the violation of labor and environmental rights, contributing to a fairer and more peaceful business environment. By strengthening governance practices and promoting greater accountability in business operations, the circular economy aligns with SDG 16 by contributing to more effective and fair institutions, promoting a more ethical and responsible textile sector.

2.17 Contributions to SDG 17 (Partnerships for the Goals)

The circular economy in the textile sector depends on collaboration between governments, companies, NGOs and other stakeholders to be effective. Partnerships between different sectors and countries are essential to share knowledge, resources and technologies that facilitate the transition to a more circular and sustainable economy.

These collaborations strengthen the capacity to implement sustainable practices globally, meeting the targets of SDG 17. In addition, the development of public policies that encourage the circular economy and the creation of international cooperation networks can accelerate the adoption of sustainable practices in the textile sector, promoting more equitable and sustainable development.

According to Koszewska (2019), the transition to a circular economy presents significant challenges for the textile and clothing industries, requiring significant changes in production and consumption models. In this sense, Table 1 lists the 17 Sustainable Development Goals (SDGs) and addresses how the textile sector faces challenges in each of them. Additionally, we highlight the contributions of the circular economy in the textile sector to the achievement of each SDG.

Table 1 clearly illustrates how the textile sector faces multiple challenges in relation to the SDGs and how the circular economy can offer relevant solutions to these problems. The information presented highlights the importance of adopting sustainable and innovative practices to advance towards global sustainable development goals. Each contribution identified demonstrates the ability of the textile sector to not only mitigate negative impacts but also to drive positive change. The Methodology section will then address how the Systematic Literature Review was structured.

Method

The search was carried out in the Web of Science Core Collection database on May 9, 2024, using the keywords ((((TS = (SDG)) OR TS = (SUSTAINABLE DEVELOPMENT GOAL)) AND TS = (CIRCULAR ECONOMY)) AND TS = (TEXTILE*)) OR TS = (FASHION). Initially, 205,962 results were obtained without any filter applied. A preliminary analysis was carried out to identify the number of publications per year since 1946, and it was observed that the largest number of publications is concentrated in the last five years. Based on this and with the aim of carrying out a systematic review of the current literature, we decided to limit our analysis to publications from the last five years. After applying the filter to include only publications from the last five years, this number was reduced to 38,584. Then, a document type filter was applied, where only articles and reviews were selected, resulting in 33,026 publications.

Furthermore, a filter was applied to include only open access publications, totaling 18,721 publications. Category and research area filters were also applied, resulting in 4,204 publications. Subsequently, the titles and abstracts of the selected publications were reviewed to determine their relevance to the scope of the research. After this review, 188 articles that were directly aligned with the theme of the study were selected. These articles were considered for detailed analysis and inclusion in the

Table 1 Challenges facing the textile sector and its contribution to the sustainable development goals (SDGs) through the circular economy

Challenges of the textile sector	Sustainable development goals (SDGS)	Contributions of the textile sector
– Low wages for sector employees (Franco, 2017)	1—Eradication of poverty	– Creation of new "green jobs" in areas such as sustainable design, recycling, and waste management (Cuc et al. 2015; Cappellieri et al. 2019)
– Resource wastage (EMF 2019), unsustainable agricultural practices, and the influence of climate change on food security (Leal Filho et al. 2022)	2—Zero hunger	– Reuse, recycling, and waste management reduce climate impacts that lead to food insecurity. Therefore, they are a measure to achieve SDG 2 (United Nations 2021; Leal Filho et al. 2022; Li et al. 2023)
– Use of toxic chemicals and poor working conditions (Ellen MacArthur Foundation 2017) – High-risk health and safety conditions and prolonged working hours (Almanza and Corona 2020)	3—Good health and well-being	– Reducing the use of toxic chemicals and improving working conditions in garment factories can contribute to the health and well-being of workers (MacArthur Foundation and circular fibers initiatives 2017)
– Lack of training and education programs in sustainable practices in the sector	4—Quality education	– Implementation of training programs in circular economy practices. This may include courses on design for circularity and recycling methods for the staff, as well as extension courses for the community on clothing customization, proper textile disposal, more sustainable fabrics, and the environmental impact of the textile sector
– Gender inequality in the distribution of leadership opportunities and training in the textile sector	5—Gender equality	– Promoting training and capacity-building in new skills related to circular economy and providing leadership opportunities in the textile industry can promote gender equality
– High water consumption and water pollution	6—Clean water and sanitation	– Atmospheric plasma technology to promote water savings and reuse, reducing both consumption and water pollution (Senai 2024)
– Dependency on non-renewable energy sources in textile production	7—Affordable and clean energy	– Investment in renewable energy sources and energy efficiency (Quantis 2018; Radulescu et al. 2018)

(continued)

Table 1 (continued)

Challenges of the textile sector	Sustainable development goals (SDGS)	Contributions of the textile sector
– Poor working conditions and high informality in the business (Amaral et al. 2019) – Low wages, child labor, slavery, long working hours (Franco 2017)	8—Decent work and economic growth	– Creation of new jobs and economic opportunities, including "green jobs" in areas such as sustainable design, recycling, and waste management (Cuc et al. 2015; Cappellieri et al. 2019)
Low level of technological innovation and outdated infrastructure in textile production (Araujo Galvão et al. 2018; Kirchherr et al. 2018)	9—Industry, innovation, and infrastructure	– Business models aligned with SDGs and digital transformation (Leal Filho et al. 2023) – Innovations in textiles, such as the use of agricultural waste to produce eco-friendly fibers and fabrics, exemplified by the case of Orange Fiber Srl. which patented methods to create fabrics from orange peels and citrus fruits combined with lyocell, demonstrate the transformation of waste into valuable resources (Orange Fiber Srl. 2022)
Economic and social inequalities in access to jobs in the textile industry	10—Reduced inequalities	– Generating inclusive jobs (Cuc et al. 2015; Cappellieri et al. 2019) and empowering vulnerable communities with skills for the circular economy, the textile sector helps reduce inequalities, providing equitable opportunities for economic and social development
Lack of knowledge about proper disposal of textile waste, lack of textile waste collection points, ineffective management of urban and industrial waste	11—Sustainable cities and communities	– Sustainable management of urban waste
– Fast fashion business model (Amaral et al. 2019) – Consumer awareness (Araujo Galvão et al. 2018)	12—Responsible consumption and production	– Promotion of workshops on conscious consumption, customization, upcycling, incentive through discounts for receiving used clothes, etc. – Producing more with less, using renewable energy sources, circular water solutions, natural dyes, recycled yarns in production (Amaral et al. 2019)

(continued)

Table 1 (continued)

Challenges of the textile sector	Sustainable development goals (SDGS)	Contributions of the textile sector
– High greenhouse gas emissions and industrial waste (Leal Filho et al. 2022)	13—Climate action	– Design for circularity, waste elimination, and substitution of materials with renewable, low-carbon substances, or secondary materials, as alternative inputs for new production (EMF 2019, Amaral et al. 2019)
– Ocean pollution by microplastics and chemicals	14—Life below water	– Reducing pollution from plastics and chemicals (Amaral et al. 2019)
– Deforestation and degradation of terrestrial ecosystems for fiber production	15—Life on land	– Decreasing the use of natural resources and reducing deforestation for fiber production preserve terrestrial ecosystems
– Lack of transparency and corporate responsibility (Thakker and Sun 2023) – Regulatory policies (Araujo Galvão et al. 2018)	16—Peace, justice, and strong institutions	– Responsible and transparent business practices (Thakker and Sun 2023)
– Low awareness and collaboration among stakeholders (Amaral et al. 2019)	17—Partnerships for the goals	– Collaboration among governments, businesses, and NGOs strengthens the global partnerships necessary to implement circular economy practices in the textile industry

systematic review of the literature on the circular economy in the textile sector and its contributions to the Sustainable Development Goals (SDGs). The results of the sample of 188 articles analyzed can be seen in the next results and discussion section.

Results and Discussions

In this section, the sample of articles selected for review will be presented and discussed in terms of the evolution of the topic and bibliometric data provided with the aid of VOSviewer software (Van Eck and Waltman 2021). First, the analysis of the evolution of the number of studies on the topic over the years is presented in Fig. 1.

Figure 1 reveals a significant increase in the number of articles from 13 to 41 from 2020 to 2021, representing a growth of about 215%, followed by a continuous increase to 61 in 2022, indicating persistent interest and research on the topic. However, between 2022 and 2023, there was a slight decrease to 59 articles, suggesting a possible stabilization in academic production, despite the still high interest. The sharp drop to 14 articles in 2024 should be interpreted considering that the year has not yet ended and has not gone through a full semester.

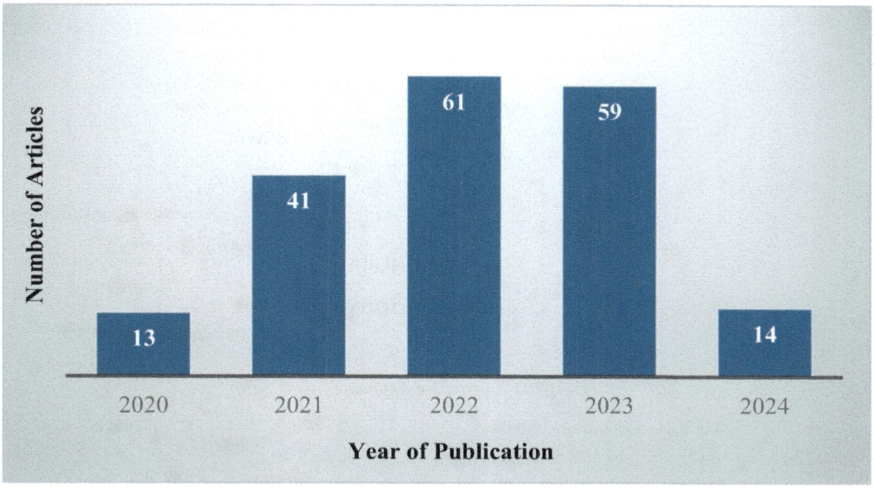

Fig. 1 Evolution of the topic over the past five years

Next, bibliometric analyses were conducted using VOSviewer software, a widely used text mining tool in scientific mapping, especially for visualizing large networks in bibliometric analyses (Van Eck and Waltman 2010, 2020). The co-occurrence analysis of keywords was performed, considering a minimum of 5 occurrences per keyword, totaling 62 keywords to be analyzed. This analysis explores the proximity of terms in a text, allowing the identification of the main research themes (Arita 2017). The generated terms are presented in Fig. 2.

The co-occurrence analysis of the keywords resulted in three distinct clusters. Cluster 1 (Red) emphasizes consumer behavior and the collaborative economy. Among the keywords in this cluster, "consumption" and "sustainable fashion" stand out both in occurrences and link strength, reflecting the central importance of these concepts in research.

Cluster 2 (Green) focuses on business models and implementation challenges. The keywords "Circular economy" and "sustainability" have the highest occurrence and link strength in this cluster. Cluster 3 (Blue) focuses on circular fashion practices and environmental impacts. "Circular fashion" and "textile" are the keywords that stand out the most in terms of occurrence and link strength within this cluster, indicating the centrality of these issues in the analyzed research.

Subsequently, the analysis of bibliographic coupling and citation of documents was carried out, presented in Fig. 3. For better understanding, a bibliographic coupling link is a link between two items that cite the same document (van Eck and Waltman 2020). In the process of parameterizing the analysis, no minimum number of citations was included, aiming to encompass all documents and not exclude recently published documents that may not have had time to be cited. Thus, it was parameterized to include all 188 documents in the map. However, of the 188, only 186 were connected and only the 186 connected documents were analyzed.

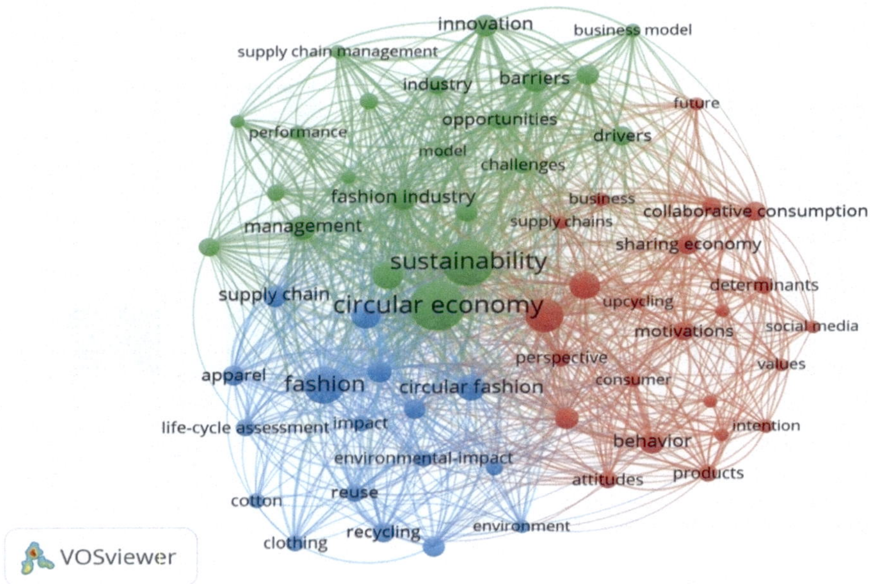

Fig. 2 Co-occurrence of keywords

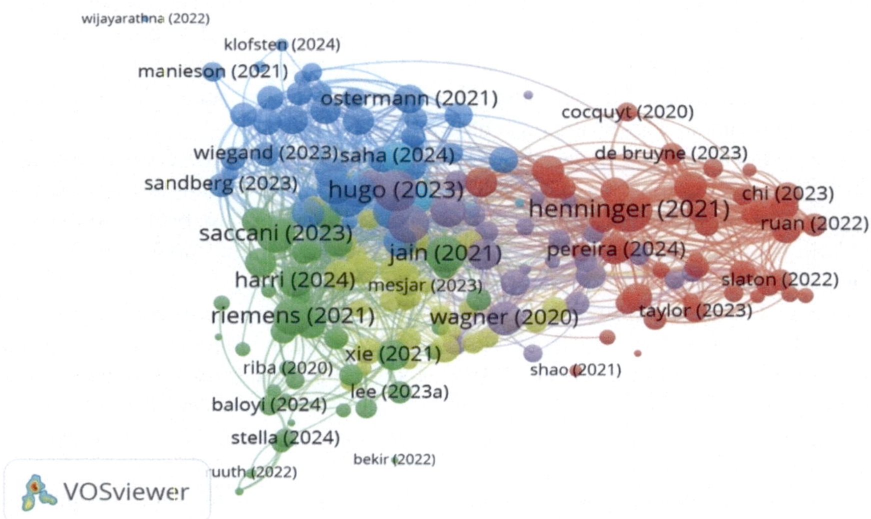

Fig. 3 Bibliographic coupling of documents

Figure 3 illustrates the total link strength, which refers to the influence of each theme/author/country in the research. The greater the total link strength, the greater the impact of the topic in the respective field (Leal Filho et al. 2024). The five most relevant documents are: Henninger et al. (2021), de Aguiar Hugo et al. (2021, 2023), Jain et al. (2021), and Riemens et al. (2021). In Cluster 1 (Red), themes such as circular and sustainable fashion, clothing reuse, and digital platforms are addressed.

Clothing reuse through the purchase of second-hand clothes is part of a circular economy, as opposed to new clothing items (Silva et al. 2021; Kasavan et al. 2021). Additionally, according to Corboş et al. (2023), recent studies show an increasing need to understand consumer motivations and behavior regarding second-hand clothing purchases, with a focus on sustainability and the circular economy. In this context, a significant number of studies address the consumption of second-hand clothing within the circular economy. Notable works include Henninger et al. (2021), Poldner et al. (2022), Arrigo (2022), Johnson and Plepys (2021), and Mora-Sojo (2023).

Cluster 2 (Green) focuses on textile waste recycling and digital transformation. This cluster includes one of the most relevant articles for this review, titled "Sustainable Development Goals for Textiles and Fashion" by authors Thakker and Sun (2023), which addresses the textile sector's contribution to all 17 Sustainable Development Goals, but unlike the present study, does not directly focus on the circular economy. Additionally, another important article in this cluster is titled "How digitalization transforms the traditional circular economy to a smart circular economy for achieving SDGs and net zero" by author Govindan (2023), which covers the circular economy, textile sector, and SDGs that is the focus of this study but differs by specifically focusing on digitalization.

In Cluster 3 (Blue), sustainable fashion practices are explored. Key documents include Teixeira et al. (2023), Colucci and Vecchi (2024), Hellström and Olsson (2024), Jimenez-Fernandez et al. (2023), and Ostermann et al. (2021). In this cluster, one of the most relevant articles for this review is titled "Circular Fashion: Cluster Analysis to Define Advertising Strategies" by Aramendia-Muneta et al. (2022), which connects circular fashion with the SDGs, but the focus is on advertising strategies for circular fashion.

Cluster 4 (Yellow) discusses sustainable fashion, environmental impacts, and design strategies. The documents with the highest total link strength are by Degenstein et al. (2023), Piller (2023), Bailey et al. (2022), Eppinger (2022), and Benkirane et al. (2022). One of the most relevant articles for this review in this cluster is titled "Textile-apparel manufacturing and material waste management in the circular economy: A conceptual model to achieve sustainable development goal (SDG) 12 for Bangladesh" by Akter (2022), which presents a conceptual model led by the circular economy to achieve SDG 12 in Bangladesh. Unlike this review, the focus is regional and not global, and the SDGs and the circular economy are not addressed generally but are focused only on SDG 12 and material waste management.

Cluster 5 (Purple) addresses sustainable fashion practices, including upcycling and consumer participation. Influential documents include de Aguiar Hugo et al. (2021, 2023), Wagner and Heinzel (2020), Rotimi et al. (2021), and De Wagenaar et al. (2022). Cluster 6 (Light Blue) deals with sustainable business strategies in the fashion industry, including reviews on the circular economy. Relevant documents are Saha et al. (2024), Barletta et al. (2024), D'Adamo and Colasante (2022), and Clark (2022).

This analysis reveals a variety of approaches within the circular economy in the textile industry, highlighting the importance of innovation and awareness in promoting sustainability. However, it is noted that few studies directly address the relationship between the circular economy, the textile sector, and the 17 SDGs, confirming Govindan's (2023) argument that few studies address this relationship. The five most influential documents are presented in Table 2.

The first article, "Towards sustainability? Forest-based circular bioeconomy business models in Finnish SMEs" by D'Amato et al. (2020), investigates how Finnish SMEs in the forestry sector implement circular bioeconomy business models. The study identifies challenges and opportunities, contributing to the theory and practice of sustainable circular bioeconomy.

The second article, "Trends in the Fashion Industry. The Perception of Sustainability and Circular Economy: A Gender/Generation Quantitative Approach" by Gazzola et al. (2020), analyzes how sustainability and circular economy influence the perception of fashion among young consumers. The research reveals attitudes across different generations, offering insights for brands wishing to engage this audience.

Table 2 Top five most cited documents

Document title	Author	Year	Source	Number of citations on May 9, 2024
Towards sustainability? Forest-based circular bioeconomy business models in Finnish SMEs	D'amato, Dalia; Veijonaho, Simo; Toppinen, Anne	2020	Forest policy and economics	155
Trends in the fashion industry. the perception of sustainability and circular economy: a gender/ generation quantitative approach	Gazzola, Patrizia; Pavione, Enrica; Pezzetti, Roberta; Grechi, Daniele	2020	Sustainability	144
Closing the loop on take, make, waste: investigating circular economy practices in the swedish fashion industry	Brydges, Taylor	2021	Journal of cleaner production	83
Sustainability through online renting clothing: circular fashion fueled by instagram micro-celebrities	Shrivastava, Archana; Jain, Geetika; Kamble, Sachin S.; Belhadi, Amine	2021	Journal of cleaner production	71
Consumers' value and risk perceptions of circular fashion: comparison between secondhand, upcycled, and recycled clothing	kim, inhwa; jung, hye jung; lee, yuri	2021	Sustainability	61

The third article, "Closing the loop on take, make, waste: Investigating circular economy practices in the Swedish fashion industry" by Brydges (2021), studies the implementation of circular economy practices in the Swedish fashion industry. The analysis identifies gaps and challenges in adopting circularity strategies throughout the supply chain. The fourth article, "Sustainability through online renting clothing: Circular fashion fueled by Instagram micro-celebrities" by Shrivastava et al. (2021), investigates the use of online second-hand clothing rental platforms. The study reveals that the influence of Instagram micro-celebrities promotes the adoption of these platforms, driving circular fashion.

The fifth article, "Consumers' Value and Risk Perceptions of Circular Fashion: Comparison between Secondhand, Upcycled, and Recycled Clothing" by Kim et al. (2021), examines consumers' perceptions of value and risk regarding secondhand, upcycled, and recycled clothing. The results highlight the importance of emotional value and the role of individuality in consumer attitudes. These studies demonstrate the growing relevance of the circular economy in the textile sector, offering varied perspectives on promoting sustainable practices and reducing environmental impact. Next, the bibliographic coupling analysis of sources is presented in Fig. 4.

In the process of parameterizing the bibliographic coupling analysis of sources, no minimum number of citations was included, aiming to encompass all documents and not exclude recently published documents that may not have had sufficient time to be cited. The bibliographic coupling analysis conducted in VOSviewer resulted in the identification of six main clusters of sources (Fig. 4), each representing different thematic focuses and approaches in academic literature. The top five sources with the highest total link strength are: Sustainability, Journal of Cleaner Production, Sustainable Production and Consumption, Business Strategy and the Environment,

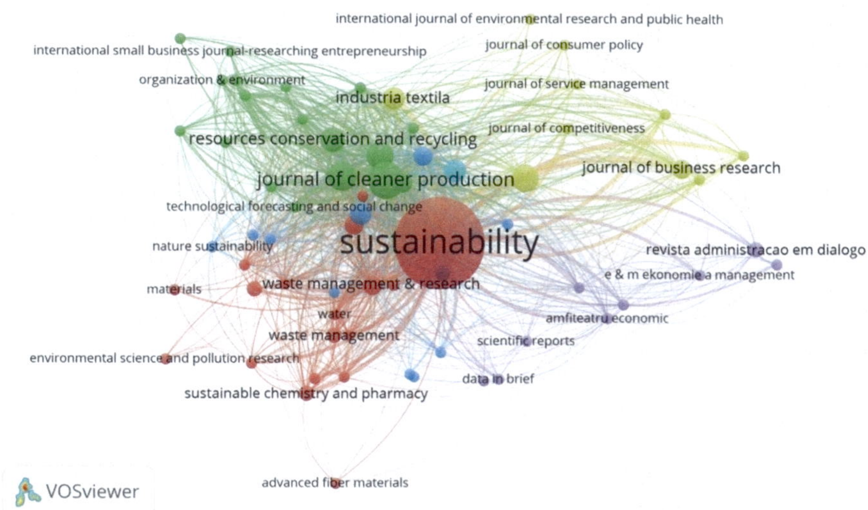

Fig. 4 Bibliographic coupling of sources

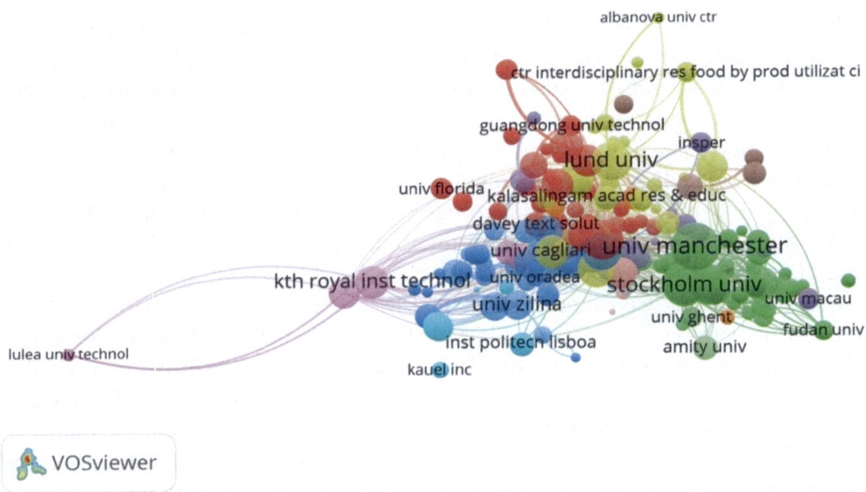

Fig. 5 Bibliographic coupling of organizations

and Journal of Fashion Marketing and Management. The top five most cited sources are: Sustainability, Journal of Cleaner Production, Forest Policy and Economics, Journal of Business Research, and Business Strategy and the Environment. Cluster 1 (Red) is dominated by sources focused on sustainability and waste management, while Cluster 2 (Green) explores business strategies and sustainable management. In Cluster 3 (Blue), the focus is on sustainable development and environmental responsibility. Cluster 4 (Yellow) encompasses publications related to trade, marketing, and management in the fashion industry, while Cluster 5 (Purple) covers economics, administration, and business innovation. Finally, Cluster 6 (Light Blue) focuses on social economy issues and cooperation, as well as possible topics related to economics and sustainability. Next, the bibliographic coupling (Fig. 5) analysis of organizations is presented.

The analysis was based on 310 interconnected items, where the top five organizations with the highest total link strength are: Manchester, Stockholm, Technology Sydney, Otago, and Lund. The top five most cited organizations are: Helsinki, Insubria, Manchester, Stockholm, and Technology Sydney. Organizations with the highest total link strength, such as Manchester, Stockholm, Technology Sydney, Otago, and Lund, not only have a robust network of connections but also play a key role in facilitating collaboration and knowledge exchange among different institutions and researchers.

Their prominent presence in the network suggests that they are centers of excellence in research on circular economy in the textile industry, possibly leading innovative projects, publishing high-quality works, and promoting strategic partnerships. On the other hand, the most cited organizations, such as Helsinki, Insubria, Manchester, Stockholm, and Technology Sydney, indicate that their contributions and research are widely recognized and referenced by the academic community. This suggests that these institutions have produced outstanding works that influence and shape discussions and theoretical development in the field of circular economy in the textile industry. Next, the bibliographic coupling analysis of countries is presented.

The analysis of the data reveals an interesting overview of the distribution of research related to circular economy in the textile sector among different countries. The total strength of the link and the connection between the countries indicate the intensity of the relationships among different research centers. Countries with a high total strength of link, such as the United States, China, Germany, and Italy, demonstrate a robust network of collaboration and knowledge exchange in research on circular economy in the textile industry.

This strong connection between countries suggests that research in this area is often conducted in partnership with institutions and researchers from different parts of the world. This may result in a broader exchange of ideas, methodologies, and best practices, leading to faster and more comprehensive advancements in the field of circular economy in the textile industry.

Part of the prominence of the United Kingdom in the results may have occurred because it has a Circular Economy Action Plan (CEAP), demonstrating a commitment to the theme (Sharpe et al. 2022). Additionally, the presence of emerging countries such as Brazil, India, and South Africa in the research network indicates a growing participation and contribution of these countries to the development of knowledge in this field. This suggests significant potential for international collaboration and resource sharing to further drive research on circular economy in the textile industry on a global scale. Next, in Fig. 6, the bibliographic coupling of authors with a minimum of 5 documents is presented.

In the analysis of Fig. 7 of bibliographic coupling of authors, "McNeill, Lisa S." stands out with a total link strength of 607, suggesting strong integration and influence in the research network. "McQueen, Rachel H." and others like "Bressanelli, Gianmarco", "Saccani, Nicola", and "Visintin, Filippo" also show high connectivity, reflecting their significant positions in the network. "Brydges, Taylor" stands out for the highest number of citations, indicating their significant influence and innovative contribution to the field.

Other authors, such as "Cai, Liling" and "Ji, Xiaofen" with a total link strength of 323, demonstrate strong collaboration presence, although "Cai, Liling" has no registered citations yet. "De Nadae, Jeniffer" and "Colucci, Mariachiara" also stand out for their significant collaborations. Authors like "Potdar, Balkrushna" and "D'Itria, Erminia" exhibit strong connection and influence in their co-authorship networks, facilitating knowledge exchange and innovation.

3 Conclusions

This chapter explored the connections between the circular economy (CE) and the 17 Sustainable Development Goals (SDGs) through a systematic literature review (SLR) of 188 articles. The findings demonstrate that circular practices in the textile

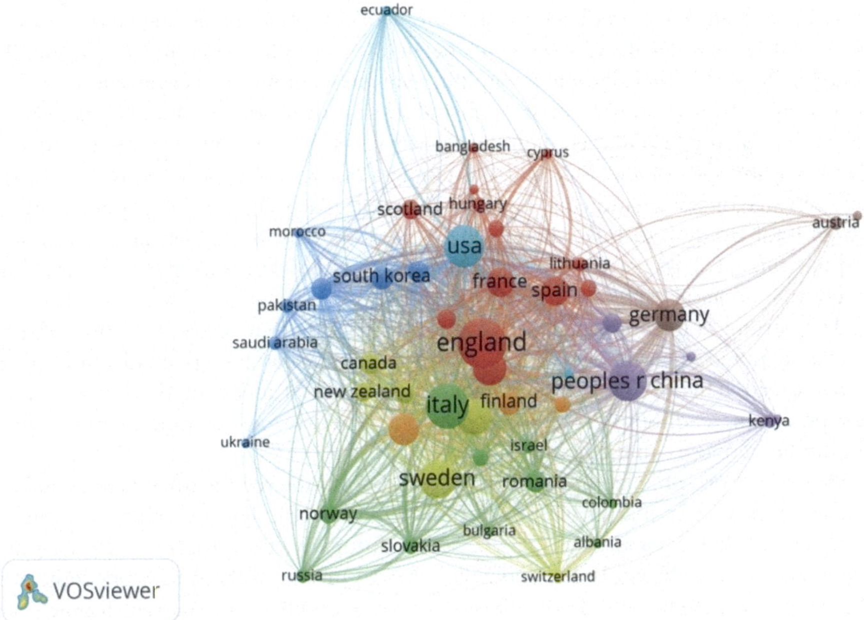

Fig. 6 Bibliographic coupling of countries

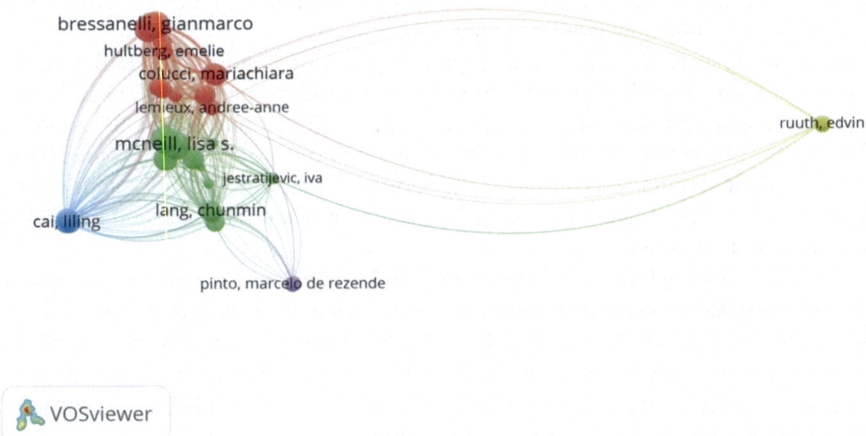

Fig. 7 Bibliographic coupling of authors

industry significantly contribute to the advancement of the SDGs, addressing both environmental and social challenges. Additionally, bibliographic coupling analysis provided valuable insights, identifying key themes, influential articles, primary sources, contributing organizations, prominent authors, and leading countries in this research area.

Despite these advancements, there remains a notable gap: the lack of comprehensive studies addressing the contribution of CE to all 17 SDGs simultaneously. This chapter attempts to fill part of this gap by proposing a holistic view, but further research is needed to deepen this integration across various sectors and economic contexts.

The implications of this research are significant for academia, industry, and society. For scholars, it underscores the necessity of an interdisciplinary approach to applying CE for broader sustainable development. For the textile industry, it highlights the opportunity to lead the transition to more sustainable practices, meeting consumer and regulatory expectations. For society, this study illustrates that adopting CE practices is essential for achieving the SDGs, improving quality of life, reducing inequality, and preserving natural resources for future generations.

However, several limitations must be acknowledged. The bibliometric analysis is based on a sample of articles that may not encompass all relevant research due to access restrictions and keyword selection. Additionally, the keyword co-occurrence and bibliographic coupling analyses may not fully capture emerging themes' nuances due to their reliance on textual data and quantitative metrics. The preliminary decrease in the number of articles in 2024 also warrants caution in interpreting recent trends.

Future studies should adopt a more integrated approach to explore the relationship between the circular economy and all 17 SDGs. Interdisciplinary research connecting CE with climate change, social justice, and technological innovation is crucial. Examining the implementation of CE in various regional and sectoral contexts can offer specific insights into adapting these practices for maximum socioeconomic and environmental benefits.

Longitudinal impact analysis is another promising area, evaluating the evolution of CE practices over time and their long-term sustainability consequences. Mixed-method studies combining qualitative and quantitative analysis can deepen the understanding of factors facilitating or hindering the transition to a circular economy. Furthermore, exploring the role of public policies and regulations in promoting CE, and how governments can collaborate with industry and academia, is essential for fostering an environment conducive to sustainable innovations.

References

Agrawal R, Wankhede VA, Kumar A, Luthra S (2021) Analisando os obstáculos à adoção da economia circular no setor automobilístico: redução de resíduos e perspectivas ambientais. Estratégia Empresarial e Meio Ambiente 30(2):1051–1066

Akter MMK, Haq UN, Islam MM, Uddin MA (2022) Textile-apparel manufacturing and material waste management in the circular economy: a conceptual model to achieve sustainable development goal (SDG) 12 for Bangladesh. Clean Environ Syst 4:100070. https://doi.org/10.1016/j.cesys.2022.100070

Albuquerque SF, Santos MSF, Neto JMM (2021) Pre-consumption textile waste management in the clothing industry in Teresina/PI. Sustain Debate 12(3):39–50. https://doi.org/10.18472/SustDeb.v12n3.2021.40474

Almanza AM, Corona B (2020) Using social life cycle assessment to analyze the contribution of products to the sustainable development goals: a case study in the textile sector. Int J Life Cycle Assess 25:1833–1845. https://doi-org.ez47.periodicos.capes.gov.br/10.1007/s11367-020-01789-7

Amaral WAN et al (2019) Moda circular no Brasil. Laudes Foundation. https://www.laudesfoundation.org/pt/results/publicacoes-pdf/moda-circular-no-brasil.pdf

Aramendia-Muneta ME, Ollo-López A, Simón-Elorz K (2022) Circular fashion: cluster analysis to define advertising strategies. Sustainability 14(20):13365. https://doi.org/10.3390/su142013365

Araújo MG, Vieira AO (2017) A economia circular pode ser solidária. Política Nacional de Resíduos Sólidos: Implementação e 54

Arita HT (2017) Multisite and multispecies measures of co-occurrence and co-diversity overlap. Ecography 40:709–718. https://doi.org/10.1111/ecog.01942

Arrigo E (2022) Digital platforms in fashion rental: a business model analysis. J Fash Mark Manag 26(1):1–20. https://doi.org/10.1108/JFMM-03-2020-0044

Avadanei M, Beļakova D, Martinez R, Souto R, Sivevska N, Mouazan E (2021) e-learning platform for eco-design in textile and fashion sectors in the direction to a circular textile. e-Learn Softw Educ. https://doi.org/10.12753/2066-026x-21-150

Bailey K, Basu A, Sharma S (2022) The environmental impacts of fast fashion on water quality: a systematic review. Water 14(7):1073. https://doi.org/10.3390/w14071073

Barletta M, D'Adamo I, Garza-Reyes JA, Gastaldi M (2024) Business strategy and innovative models in the fashion industry: clothing leasing as a driver of sustainability. Bus Strat Environ 33(5):4730–4743

Belmonte-Ureña LJ, Plaza-Úbeda JA, Vazquez-Brust D, Yakovleva N (2021) Circular economy, degrowth and green growth as pathways for research onsustainable development goals: a global analysis and future agenda. Ecol Econ 185:107050

Benkirane R, Thomassey S, Koehl L, Perwuelz A (2022) A new longevity design methodology based on consumer-oriented quality for fashion products. Sustainability 14(13):7696. https://doi.org/10.3390/su14137696

Biermann F, Kanie N, Kim RE (2017) Global governance by goal-setting: the novel approach of the UN Sustainable Development Goals. Curr Opin Environ Sustain 26–27:26–31. https://doi.org/10.1016/j.cosust.2017.01.010

Brydges T (2021) Closing the loop on take, make, waste: investigating circular economy practices in the Swedish fashion industry. J Clean Prod 293:126245. https://doi.org/10.1016/j.jclepro.2021.126245

Cai Y-J, Choi T-M (2020) A United Nations sustainable development goals perspective for sustainable textile and apparel supply chain management. Transp Res Part E: Logistics Transp Rev 141:102010. https://doi.org/10.1016/j.tre.2020.102010

Campello M (2020) Economia circular, sustentabilidade e logística: uma combinação para melhorar o planeta, as pessoas e os negócios. CONVIBRA.ORG. https://convibra.org/congresso/res/uploads/pdf/artigo21794_20201318.pdf

Cappellieri A, Borboni E, Tenuta L, Testa S (2019) Economia circular no mundo da moda. In: Tejerina B, Miranda de Almeida C, Perugorría I (eds) Anais da Sociedade Internacional de Compartilhamento de Conferências. Bilbao, Espanha, pp 112–125

Cecchin A, Salomone R, Deutz P, Raggi A, Cutaia L (2020) Relating industrial symbiosis and circular economy to the sustainable development debate. Industrial symorosts for the circular economy: operatonal experiences, best pracuces and obstacles to a collaborative business approach

Chaudhary A, Shamsi MA (2020) Need for post-consumer textile waste management through the circular economy. Int J Modern Trends Sci Technol

Chen Z, Ming X, Zhou T, Chang Y, Sun Z (2020) A hybrid framework integrating rough fuzzy best-worst method to identify and evaluate user activity-oriented service requirement for smart product service system. J Clean Prod 253

Choudhary A, Sarkar S, Settur S, Tiwari MK (2015) Um modelo de otimização sensível ao mercado de carbono para logística reversa integrada. Int J Prod Econ 164(6):433–444

Clark JH (2022) Using green chemistry to progress a circular fashion industry. Curr Opin Green Sustain Chem 38:100685. https://doi.org/10.1016/j.cogsc.2022.100685

Cobbing M, Wohlgemuth V, Vicaire Y (2023) Zona de perigo Greenwash. Greenpeace

Colin N, Maceda-Veiga A, Flor-Arnau N, Mora J, Fortuño P, Vieira C, Prat N, Cambra J, De Sostoa A (2016) Ecological impact and recovery of a Mediterranean river after receiving the effluent from a textile dyeing industry. Ecotoxicol Environ Saf 132:295–303. https://doi.org/10.1016/j.ecoenv.2016.06.017

Colucci M, Vecchi A (2024) Tackling climate change with end-of-life circular fashion practices—remade in Italy with Amore. Br J Manag 35:1157–1179. https://doi.org/10.1111/1467-8551.12806

Corboş R, Bunea O, Triculescu M (2023) Towards sustainable consumption: consumer behavior and market segmentation in the second-hand clothing industry. Amfiteatru Economic, Suppl Special Issue 25:1064–1080. https://doi-org.ez47.periodicos.capes.gov.br/10.24818/EA/2023/S17/1064

Cuc S, Gîrneatə A, Iordənescu M, Irinel M (2015) Sustentabilidade ambiental e socioeconômica através da reciclagem têxtil. Industria Textila 66:156–163

D'Adamo I, Colasante A (2022) Survey data to assess consumers' attitudes towards circular economy and bioeconomy practices: a focus on the fashion industry. Data Brief 43:108385. https://doi.org/10.1016/j.dib.2022.108385

Dantas TET, de-Souza ED, Destro IR, Hammes G, Rodriguez CMT, Soares SR (2021) How the combination of circular economy and industry 4.0 can contribute towards achieving the sustainable development goals. Sustain Prod Consumption 26:213–227

de Aguiar Hugo A, de Nadae J, da Silva Lima R (2021) Can fashion be circular? A literature review on circular economy barriers, drivers, and practices in the fashion industry's productive chain. Sustainability 13(21):12246. https://doi.org/10.3390/su132112246

de Aguiar Hugo A, de Nadae J, da Silva Lima R (2023) Consumer perceptions and actions related to circular fashion items: perspectives of young Brazilians on circular economy. Waste Manag Res 41(2):350–367. https://doi.org/10.1177/0734242X221122571

De Araujo Galvão GD, Nadae J, Clemente DH, Chinen G, De Carvalho MM (2018) Circular economy: overview of barriers. Procedia CIRP 73:79–85. https://doi.org/10.1016/j.procir.2018.04.011

Degenstein LM, McQueen RH, Krogman NT, McNeill LS (2023) Integrating product stewardship into the clothing and textile industry: perspectives of New Zealand stakeholders. Sustainability 15(5):4250. https://doi.org/10.3390/su15054250

de Wagenaar D, Galama J, Sijtsema SJ (2022) Exploring worldwide wardrobes to support reuse in consumers' clothing systems. Sustainability 14(1):487. https://doi.org/10.3390/su14010487

Ellen MacArthur Foundation (2013) Towards the circular economy Vol. 1: economic and business rationale for an accelerated transition. https://ellenmacarthurfoundation.org/towards-the-circular-economy-vol-1-an-economic-and-business-rationale-for-an

Ellen MacArthur Foundation (2017) A new textiles economy: redesigning fashion's future. Ellen MacArthur Foundation. https://www.ellenmacarthurfoundation.org/a-new-textiles-economy

Ellen MacArthur Foundation (EMF) (2019) Completando a figura: como a economia circular ajuda a enfrentar as mudanças climáticas* (v. 3). Disponível em https://emf.thirdlight.com/file/24/34jSrJc344.wq9z34q-R3P-DB5B/Completando-a-figura%20Como-a-economia-circular-ajuda-a-enfrentar-as-mudanc%CC%A7as-%20clima%CC%81ticas.pdf. Acesso em 20 mai. 2024

Eppinger E (2022) Recycling technologies for enabling sustainability transitions of the fashion industry: status quo and avenues for increasing post-consumer waste recycling. Sustain: Sci Pract Policy 18(1):114–128. https://doi.org/10.1080/15487733.2022.2027122

European Environment Agency (2016) Circular economy in Europe—developing the knowledge base (European environment agency report No 2/2016). In Publication Office of the Euopean Union (Issue 2)

Fatimah YA, Govindan K, Murniningsih R, Setiawan A (2020) Industry 4.0 based sustainable circular economy approach for smart waste management system to achieve sustainable development goals: a case study of Indonesia. J Clean Prod 269:122263. https://doi.org/10.1016/j.jclepro.2020.122263

Filho W, Ellams D, Han S, Tyler D, Boiten V, Paço A, Moora H, Balogun A (2019) A review of the socio-economic advantages of textile recycling. J Cleaner Prod. https://doi.org/10.1016/J.JCLEPRO.2019.01.210

Franco MA (2017) Circular economy at the micro level: a dynamic view of incumbents struggles and challenges in the textile industry. J Clean Prod 168:833–845. https://doi.org/10.1016/j.jclepro.2017.09.056

Furferi R, Volpe Y, Mantellassi F (2022) Circular economy guidelines for the textile industry. Sustainability. https://doi.org/10.3390/su141711111

Fusari V (2023) Accounting for textile industry and labor dynamics in Eritrea (1956–1975): adding gender to the equation. Northeast Afr Stud 22:130–191

Gabriel M, Luque M (2019) Sustainable development goal 12 and its relation to the textile industry. The UN sustainable development goals for the textile and fashion industry. https://doi.org/10.1007/978-981-13-8787-6_2

Gallego-Schmid A et al (2020) Links between circular economy and climate change mitigation in the built environment. J Clean Prod 260:1–14. https://doi.org/10.1016/j.jclepro.2020.121115

Gangwani S, Sharma S, Sheikh S (2020) Effect of working conditions on work life balance of workers in textile industry. Int J Innovative Technol Exploring Eng. https://doi.org/10.35940/ijitee.e2259.049620

Gazzola P, Pavione E, Pezzetti R, Grechi D (2020) Trends in the fashion industry. The perception of sustainability and circular economy: a gender/generation quantitative approach. Sustainability 12(7):2809. https://doi.org/10.3390/su12072809

Geissdoerfer M, Savaget P, Bocken NMP, Hultink EJ (2017) A economia circular—um novo paradigma de sustentabilidade. J Clean Prod 143:757–768

Gómez N, Sierra M, Cortelezzi A, Capítulo A (2008) Effects of discharges from the textile industry on the biotic integrity of benthic assemblages. Ecotoxicol Environ Saf 69(3):472–479. https://doi.org/10.1016/j.ecoenv.2007.03.007

Govindan K (2023) How digitalization transforms the traditional circular economy to a smart circular economy for achieving SDGs and net zero. Transp Res Part E: Logistics Transp Rev 177:103147. https://doi.org/10.1016/j.tre.2023.103147

Heikinheimo L, Karlström L, Cura K (2022) Development of textile circular economy education in various networks in Finland. Adv Sci Technol 113:163–168. https://doi.org/10.4028/p-d5e8p5

Hellström D, Olsson J (2024) Let's go thrift shopping: Exploring circular business model innovation in fashion retail. Technol Forecast Soc Change 198:123000. https://doi.org/10.1016/j.techfore.2023.123000

Henninger CE, Brydges T, Iran S, Vladimirova K (2021) Collaborative fashion consumption—a synthesis and future research agenda. J Clean Prod 319:128648. https://doi.org/10.1016/j.jclepro.2021.128648

Hirsh S (2020) These companies are making vegan leather out of plants instead of plastic. Retrieved from green matters: https://www.greenmatters.com/p/vegan-leather-made-from-plants

Jain V, O'Brien W, Gloria TP (2021) Improved solutions for shared value creation and maximization from used clothes: streamlined structure of clothing consumption system and a framework of closed loop hybrid business model. Clean Responsible Consum 3:100039. https://doi.org/10.1016/j.clrc.2021.100039

Jia F, Yin S, Chen L, Chen X (2020) The circular economy in the textile and apparel industry: A systematic literature review. J Cleaner Prod. https://doi.org/10.1016/j.jclepro.2020.120728

Jimenez-Fernandez A, Aramendia-Muneta ME, Alzate M (2023) Consumers' awareness and attitudes in circular fashion. Clean Respons Consum 11:100144. https://doi.org/10.1016/j.clrc.2023.100144

Johnson E, Plepys A (2021) Product-service systems and sustainability: analysing the environmental impacts of rental clothing. Sustainability 13(4):2118. https://doi.org/10.3390/su13042118

Jolie R (2014) Exploring textile traditions, gender shifts, and social capital in the American southwest. North Am Archaeologist 35:375–403. https://doi.org/10.2190/NA.35.4.f

Kasavan S, Yusoff S, Guan NC, Zaman NSK, Fakri MFR (2021) Global research trends on textile waste from 2005 to 2020 using bibliometric analysis. Environ Sci Pollut Res 28:44780–44794

Kayikci Y, Kazancoglu Y, Gozacan-Chase N, Lafci C, Batista L (2022) Assessing smart circular supply chain readiness and maturity level of small and medium-sized enterprises. J Bus Res 149:375–392

Khalifa AA, Ibrahim AJ, Amhamed AI, El-Naas MH (2022) Accelerating the transition to a circular economy for net-zero emissions by 2050: a systematic review. Sustainability 14(18):11656

Kim I, Jung HJ, Lee Y (2021) Consumers' value and risk perceptions of circular fashion: comparison between secondhand, upcycled, and recycled clothing. Sustainability 13(3):1208. https://doi.org/10.3390/su13031208

Kirchherr J, Piscicelli L, Bour R, Kostense-Smit E, Muller J, Truijens A, Hekkert M (2018) Barriers to the circular economy: evidence from the European Union (EU). Ecol Econ 150(April):264–272. https://doi.org/10.1016/j.ecolecon.2018.04.028

Koszewska M (2018) Circular economy—challenges for the textile and clothing industry. Autex Res J 18:337–347. https://doi.org/10.1515/aut-2018-0023

Koszewska M (2019) Circular economy in textiles and fashion—the role of a consumer. In: Circular economy in textiles and apparel. Elsevier, pp 183–206. https://doi.org/10.1016/B978-0-08-102630-4.00009-4

Koval V, Arsawan I, Suryantini N, Kovbasenko S, Fisunenko N, Aloshyna T (2022) Circular economy and sustainability-oriented innovation: conceptual framework and energy future avenue. Energies. https://doi.org/10.3390/en16010243

Larsson JK (2018) Digital innovation for sustainability—experiences based on projects in textile value. Res J Textile Apparel 22(4):370–389. https://doi-org.ez47.periodicos.capes.gov.br/10.1108/RJTA-02-2018-0016

Leal Filho W, Trevisan LV, Rampasso IS, Anholon R, Dinis MAP, Brandli LL, Sierra J, Salvia AL, Pretorius R, Nicolau M, Eustachio JHPP, Mazutti J (2023) When the alarm bells ring: why the UN sustainable development goals may not be achieved by 2030. J Clean Prod 407:137108. https://doi.org/10.1016/j.jclepro.2023.137108

Leal Filho W, Schmidberger I, Sharifi AR, Vargas V, Rampasso IS, Dibbern T, Liakh O, Aina YA, Trevisan LV, Mbah MF, Anholon R, Kozlova V (2024) Design thinking for sustainable development: a bibliometric analysis and case study research. J Clean Prod 455:142285. https://doi.org/10.1016/j.jclepro.2024.142285

Leal Filho W et al (2022) An overview of the contribution of the textiles sector to climate change. Front Environ Sci 10

Li C et al (2023) What influences the climate entrepreneurship? Chinese-based evidence. Front Environ Sci 10

Mansour H, Hilal N, Alhajri S, Al-Yahyai F, Al-Amri M (2020) The education of artistic culture in the Sultanate of Oman through the multidisciplinary integration of graphic design and ecological textile printing. Part 1: standardization of extraction and dyeing with natural waste products. Energy Rep 6:933–939. https://doi.org/10.1016/j.egyr.2019.12.020

Menton M, Larrea C, Latorre S, Martinez-Alier J, Peck M, Temper L, Walter M (2020) Environmental justice and the SDGs: from synergies to gaps and contradictions. Sustain Sci 15:1621–1636. https://doi.org/10.1007/s11625-020-00789-8
Mora-Sojo MC, Krych K, Pettersen JB (2023) Evaluating the current Norwegian clothing system and a circular alternative. Resour Conserv Recycl 197:107109. https://doi.org/10.1016/j.resconrec.2023.107109
Neto G, Silva P, Tucci H, Amorim M (2021) Reuse of water and materials as a cleaner production practice in the textile industry, contributing to the blue economy. J Clean Prod. https://doi.org/10.1016/J.JCLEPRO.2021.127075
Niinimäki K, Peters G, Dahlbo H, Perry P, Rissanen T, Gwilt A (2020) Author's correction: the environmental price of fast fashion. Nat Rev Earth Environ 1:278
Orange Fiber Srl (2022) The first fabric from oranges. Retrieved from Orange Fiber: https://orangefiber.it/tencel-limited-edition-initiative/
Ostermann CM, Nascimento LdS, Steinbruch FK, Callegaro-de-Menezes D (2021) Drivers to implement the circular economy in born-sustainable business models: a case study in the fashion industry. Revista de Gestão 28(3), 223–240. https://doi.org/10.1108/REGE-03-2020-0017
Ovam. (n.d.) Seven messages about the circular economy and climate change. [S.I.]: OVAM
Peters GM, Sandin G, Spak B (2019) Perspectivas ambientais para a reciclagem de têxteis mistos na Suécia. ACS Sustain Chem Eng 7:11682–11690
Piller LW (2023) Designing for circularity: sustainable pathways for Australian fashion small to medium enterprises. J Fash Mark Manag: Int J 27(2):287–310. https://doi.org/10.1108/JFMM-09-2021-0220
Pizzi S, Caputo A, Corvino, A, Venturelli A (2020) Management research and the UN sustainable development goals (SDGs): a bibliometric investigation and systematic review. J Clean Prod 276:124033. https://doi.org/10.1016/j.jclepro.2020.124033
Poldner K, Overdiek A, Evangelista A (2022) Fashion-as-a-service: circular business model innovation in retail. Sustainability 14(20):1–18. https://doi.org/10.3390/su1420xxxx
Provin A, Cubas A, Dutra A, Schulte N (2021) Indústria têxtil e meio ambiente: o uso de celulose bacteriana na fabricação de biotêxteis pode contribuir para o setor? Tecnologias Limpas e Política Ambiental. https://doi.org/10.1007/s10098-021-02191-z
Purushothama B (2013) Work quality management in the textile industry. Woodhead Publishing India Pvt Limited. ISBN 978-1-78242-267-9, 200 p
Quantis (2018) Measuring fashion: environmental impact of global apparel and footwear study. [S.l.]
Radulescu CV, Popescu L, Badea CG (2018) Integrating sustainability into business: an essential need in the global context. Eur J Sustain Dev 7(4):223–235
Riemens J, Lemieux A-A, Lamouri S, Garnier L (2021) A Delphi-Régnier study addressing the challenges of textile recycling in Europe for the fashion and apparel industry. Sustainability 13(21):11700. https://doi.org/10.3390/su132111700
Rotimi EOO, Topple C, Hopkins J (2021) Towards a conceptual framework of sustainable practices of post-consumer textile waste at garment end of lifecycle: a systematic literature review approach. Sustainability 13(5):2965. https://doi.org/10.3390/su13052965
Royer S, Greco F, Kogler M, Deheyn D (2023) Not so biodegradable: polylactic acid and cellulose/plastic blend fabrics do not exhibit rapid degradation in marine waters. PLOS ONE 18. https://doi.org/10.1371/journal.pone.0284681
Saha K, Dey P, Papagiannaki E (2021) Implementing the circular economy in the textile and clothing industry. Bus Environ Strategy. https://doi.org/10.1002/BSE.2670
Saha K, Dey PK, Kumar V (2024) A comprehensive review of circular economy research in the textile and clothing industry. J Clean Prod 444:141252. https://doi.org/10.1016/j.jclepro.2024.141252
Schroeder P, Anggraeni K, Weber U (2019) The relevance of circular economy practices to the sustainable development goals. J Ind Ecol 23(1):77–95. https://doi.org/10.1111/jiec.12732

Senai C (2024) Solução inovadora para economizar água na indústria têxtil: conheça a tecnologia do plasma atmosférico. Por Jaqueline Cruz. Disponível em. https://senaicetiqt.com/solucao-inovadora-para-economizar-agua-na-industria-textil-conheca-a-tecnologia-de-plasma-atmosferico/

Sharma A, Singh RK, Kumar S, Bhattacharyya JK (2021) Urban solid waste management in the developing world with emphasis on India: challenges and opportunities. Rev Environ Sci Bio/technol 20(1):1–58

Sharpe S, Brydges T, Retamal M, Pugh R, Lavanga M (2022) Wellbeing Wardrobe. Disponível em. https://www.uts.edu.au/sites/default/files/2022-04/Wellbeing_Wardrobe_Report_long%20doc_FINAL%20final_21%20Mar%20v2.pdf

Shehri F, Rehman T, Mansoor M, Saif O (2022) Exploring the role of entrepreneurial leadership in decent work and economic growth: a sustainable development of textile industry of Pakistan. J Soc Sci Rev. https://doi.org/10.54183/jssr.v2i4.82

Shirvanimoghaddam K, Motamed B, Ramakrishna S, Naebe M (2020) Death by waste: fashion and textile circular economy case. Sci Total Environ 718:137317

Shrivastava A, Jain G, Kamble SS, Belhadi A (2021) Sustainability through online renting clothing: circular fashion fueled by Instagram micro-celebrities. J Clean Prod 278:123772. https://doi.org/10.1016/j.jclepro.2020.123772

Silva SC, Santos A, Duarte P, Vlačić B (2021) O papel do constrangimento social, sustentabilidade, familiaridade e percepção de higiene na experiência de compra de roupas de segunda mão. Int J Retail Distrib Manag 49:717–734

Stafford-Smith M, Griggs D, Gaffney O et al (2017) Integration: the key to implementing the sustainable development goals. Sustain Sci 12:911–919. https://doi.org/10.1007/s11625-016-0383-3

Swami M, Padma S (2015) Women in the informal/unincorporated textile sector of India: gendered relationship between proprietorship, employment and types of enterprises across states. Dev Econ: Microeconomic Issues Developing Econ eJ. https://doi.org/10.2139/ssrn.2813632

Teixeira TGB, Medeiros JFde, Kolling C, Ribeiro JLD, Morea D (2023) Redesign in the textile industry: proposal of a methodology for the insertion of circular thinking in product development processes. J Cleaner Product 397:136588. https://doi.org/10.1016/j.jclepro.2023.136588

Thakker V, Bakshi BR (2023) Ranking eco-innovations to enable a sustainable circular economy with net-zero emissions. ACS Sustain Chem Eng 11(4)

Thakker AM, Sun D (2023) Sustainable development goals for textiles and fashion. Environ Sci Pollut Res Int 30(46):101989–102009. https://doi.org/10.1007/s11356-023-29453-1. Epub 2023 Sep 4. PMID: 37667117; PMCID: PMC10567856

United Nations Environment Programme (2018) Building circularity into our economies through sustainable procurement. In: United Nations environment programme

United Nations (2015) Sustainable development goals. Retrieved from https://sdgs.un.org/goals

United Nations(2021) Climate change biggest threat modern humans have ever faced. World-renowned naturalist tells security council, calls for greater global cooperation

Valarmathi A, Bhalakarishnan H (2013) A study on quality of work life in textile sector in and around Coimbatore district. J Bus Manag Soc Sci Res 2:42–46

Van Eck NJ, Waltman L (2010) Software survey: VOS viewer, a computer program for bibliometric mapping. Scientometrics 84(2):523–538. https://doi.org/10.1007/s11192-009-0146-3

Van Eck NJ, Waltmann l (2020) Manual for vosviewer version 1.6.14

Van Eck NJ, Waltman L (2021) VOSviewer manual. Universiteit Leiden, Leiden

Van Zanten JA, Van Tulder R (2020) Beyond COVID-19: applying "SDGLogics" for resilient transformations. J Int Bus Policy. https://doi.org/10.1057/s42214-020-00076-4

Veerah (2022) Apple leather. Retrieved from Veerah: https://www.veerah.com/pages/organic-apple-leather

Wagner MM, Heinzel T (2020) Human perceptions of recycled textiles and circular fashion: a systematic literature review. Sustainability 12(24):10599. https://doi.org/10.3390/su122410599

Weis J, De Falco F (2022) Microfibers: environmental problems and textile solutions. Microplastics 1(4):43–58. https://doi.org/10.3390/microplastics1040043

Whitehill S, Hayles C, Jenkins S, Taylour J (2022) Engagement with higher education surface pattern design students as a catalyst for circular economy action. Sustainability. https://doi.org/10.3390/su14031146

Zibunas C, Meys R, Kätelhön A, Bardow A (2022) Cost-optimal pathways towards net-zero chemicals and plastics based on a circular carbon economy. Comput Chem Eng 162

EDGE Certification of Public Schools in Brazil: A Case Study

Camila Gregório Atem, Daniela Fonseca Mendes, Felipe Masanobu Hachiya, and Vanessa Costa Santos

Abstract The environmental impacts resulting from building construction persist throughout the project's lifecycle. One strategy for promoting sustainability in construction is adhering to green building certification programs, such as Leadership in Energy and Environmental Design (LEED) and Excellence in Design for Greater Efficiencies (EDGE). These certifications recognize buildings exhibiting high environmental performance, low consumption of natural and energy resources, and reduced greenhouse gas emissions. However, the certification process often involves complex assessment methods and requires compliance with criteria that may incur prohibitive costs, particularly for public buildings in Brazil. This paper aimed to apply and evaluate the EDGE certification process for a model public school developed by the Brazilian federal government. The applicability and usability of the EDGE application, which is available free of charge, were examined for three Brazilian municipalities with different climates. The findings suggested that the tool aids in designing solutions, particularly in addressing energy-related concerns.

Keywords EDGE certification · Green building · Public schools

C. G. Atem (✉) · D. F. Mendes · F. M. Hachiya · V. C. Santos
Center for Technology and Urbanism, State University of Londrina, Londrina, PR, Brazil
e-mail: camila.atem@uel.br

D. F. Mendes
e-mail: daniella.fonseca@uel.br

F. M. Hachiya
e-mail: felipe.hachiya@uel.br

V. C. Santos
e-mail: vanessacosta@uel.br

L. Brandli et al. (eds.), *The Contribution of Life Cycle Analyses and Circular Economy to the Sustainable Development Goals*, World Sustainability Series,
https://doi.org/10.1007/978-3-031-85300-5_5

1 Introduction

The environmental impacts resulting from building construction persist throughout the project's entire life cycle. Buildings were responsible for 32 Gt of CO_2 emissions in 2021, which represents about 30% of total emissions stemming from end-use sectors, including indirect emissions associated with energy usage and heating. Furthermore, buildings accounted for 30% of the global energy consumption (IEA 2022). With the growing impacts of climate change, the diminishing availability of natural resources, and the worsening of the energy crisis across all sectors of society, it becomes imperative to align building design with sustainable principles for construction, use, and maintenance.

Certification is a valuable strategy to verify that a product meets standard criteria (Chaterjee et al. 2022). For instance, green building certifications ensure that buildings are designed based on a life cycle design approach and satisfy quality criteria aimed at minimizing environmental impacts, thereby guiding the construction industry toward sustainable development goals (Klein et al. 2018). The most renowned green building certifications in the world are LEED (United States), BREEAM (United Kingdom), HQE (France), Green Star (Australia), CASBEE (Japan).

The certification process requires the application of complex assessment methods and demands compliance with criteria that often demand the use of computational tools for evaluation of building behavior (Bueno and Rossignolo 2013; Isimbi and Park 2022). Software such as EnergyPlus®, TRNSYS®, and Simapro® enable a reliable analysis of certification requirements. However, applying these computational tools to analyze design process or existing buildings is not an easy task. They require time and expertise, posing a barrier to incorporation into the workflow of designers, mainly in the early design phases (Chaterjee et al. 2022).

For professionals seeking to perform an in-depth assessment of the environmental impacts of buildings, there is a more accessible and simplified alternative: the Excellence in Design for Greater Efficiencies (EDGE) certification. The EDGE certification system was developed by the International Finance Corporation (IFC), a member of the World Bank Group. This global online certification platform allows designers to verify the efficiency in the use of energy and water, and the embodied energy of building materials, supporting the identification of opportunities for reducing resource consumption (EDGE 2022). Despite being a global certification, EDGE is adapted to local environmental and energy efficiency regulations. As a result, it has been successfully implemented in more than 150 countries (EDGE 2022).

The EDGE certification system encompasses different levels of sustainability. Level 1 certification requires a reduction of at least 20% in resource consumption in relation to the base case. Level 2 (advanced certification) can be attained by adopting energy efficiency strategies that result in at least 40% energy savings. All certification levels are evaluated by accredited certifiers listed on the platform.

Several case studies were conducted to illustrate the effectiveness of this tool in quantifying reductions in water, energy, and embodied energy and greenhouse gas emissions in buildings and assisting in the selection of design solutions (Kartikasari et al. 2018; Tarigan 2018; Robles et al. 2022; Isimbi and Park 2022). Setyowati et al. (2020) analyzed the potential of water saving strategies for a university campus in Indonesia. According to estimates from the EDGE app, the best efficiency scenario would reduce water usage by 37.86%, equivalent to an annual savings of 19,095.02 m^3 of water.

Other studies focused on discussing the advantages and disadvantages of the EDGE application and performed comparative analyses with other certification systems and simulation programs. Caldas (2020), for example, applied the EDGE method to a sustainable retrofit project of a Brazilian office building to assess whether the tool can quickly and effectively analyze improvement strategies and accurately determine embodied costs and consumption. The researcher concluded that the platform contributed in a practical way to decision-making on building strategies in the initial design phases. The EDGE app proved to be sensitive to the different climatic conditions of each region of the country. Nevertheless, the author pointed out limitations regarding the accuracy of cost estimates and the availability of construction strategies. The solutions provided by the app were deemed generic, overlooking envelope characteristics that are relevant to the Brazilian context.

Although the EDGE app provides data to guide designers in decision-making, it is important that these professionals understand the relevance of certification tools for project quality. Dlamini and Yessoufou (2022) developed a descriptive statistical study to assess the perceptions of professionals and residents about an EDGE-certified sustainable housing project developed in South Africa. The results showed that, although 95% of the professionals understood the concept of sustainable buildings, the majority had limited knowledge of the benefits that sustainable strategies can provide to civil construction. Furthermore, about 40% of the users were not aware of the energy efficiency measures implemented in the enterprise. This finding can be problematic, as numerous studies underscored the importance of conscious user behavior in achieving energy savings (Motherway et al. 2022; Corbos et al. 2023).

As a promoter of policies aimed at improving the quality of the school environment, the Ministry of Education of the Brazilian federal government published the Basic Infrastructure Parameters for Early Childhood Education Institutions (Brasil 2006). The document offers recommendations for the design, construction, and renovation of early childhood education units based on a replicable school unit model designed to meet high educational demands. The goals include resistance to vandalism, low cost, easy maintenance, harmonious relationship with the surroundings, thermal comfort, and sanitary quality (FDE 2006; FNDE 2005, 2006). This study investigated the applicability of the EDGE certification in a preschool building model. A case study was conducted in three different climates in Brazil to understand how the model can be adapted to enhance sustainability in these climatic contexts.

2 Methods

This case study evaluated a preschool project proposed by the National Program for Restructuring and Equipment Acquisition for the Public School Network of Early Childhood Education (Proinfância), supported by the National Education Development Fund (FNDE 2006). The chosen model was Type 2, which has a built-up area of 780.96 m^2 and can serve up to 188 children if operated in two shifts or 94 children if operated in a single full-time shift (Figs. 1 and 2).

Brazil is a country with continental dimensions. The analysis was performed for three municipalities across the country, each representing a different climate zone. The chosen municipalities were Curitiba, São Paulo, and Belém. Curitiba is the coldest capital of the country. São Paulo has mild winters and hot summers. Belém exhibits a typical hot-humid climate (Fig. 3).

A base project was developed for subsequent comparison with improved cases. The EDGE app supports the following building types: residential, hotel, office, hospital, retail, and education buildings. Education building assumptions are based on the type of educational facility (e.g., pre-school, university, or sports facility), as well as occupancy density and use hours. Because it is uncommon for the complete set of building parameters to be available at the design stage, EDGE provides default data to initiate the generation of a base case for each building. A unique base case is developed for each project using empirical data from actual buildings and current practices around the world.

Using the descriptive memorial and project plans, it was possible to fill in the required design information on the EDGE platform, such as type of construction, location, areas, systems, and materials (Tables 1, 2, 3, 4 and 5).

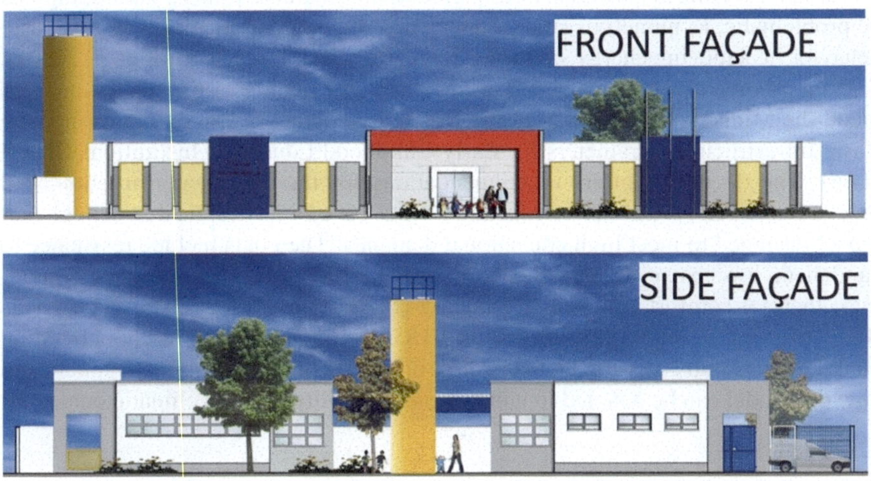

Fig. 1 Front and side façades of the model preschool (FDNE 2006)

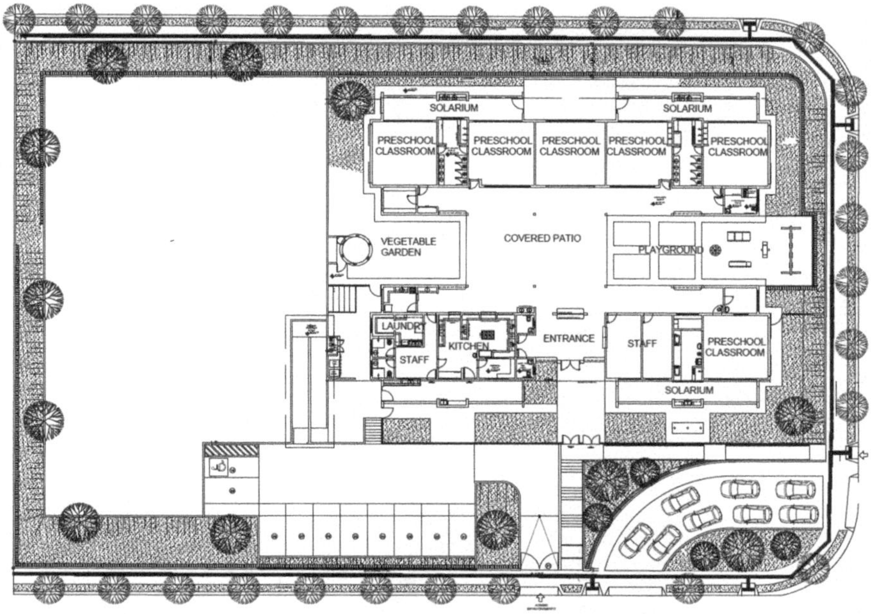

Fig. 2 Building plan of the model preschool (FDNE 2006)

3 Results and Discussion

Following the insertion of the physical and occupational characteristics of the building, the savings strategies for energy, water, and embodied energy of materials (improved case) were selected in the app. The base case was provided by the software, and the proposed strategies were compared to it. The estimated savings of the improved case are shown when the user selects specific technical measures. The difference in energy, water and materials consumption between the base case and the improved case defines whether the building meets EDGE certification criteria. In addition these consumption savings, EDGE also reports the estimated reductions in greenhouse gas emissions and operational costs (costs were not evaluated here).

The chosen measures for the improved case were based on the reality of public buildings in Brazil. The building was designed taking into consideration the country's budget. The energy efficiency measures are shown in Table 6, water efficiency measures in Table 7, and materials efficiency measures in Table 8.

The proposed strategies sought to reduce energy consumption via a reduction in glass area, reduction in solar radiation absorption via the roof and walls, use of shade devices, thermal insulation of the roof, use of natural ventilation throughout the building, smart energy metering, and generation of 50% renewable energy.

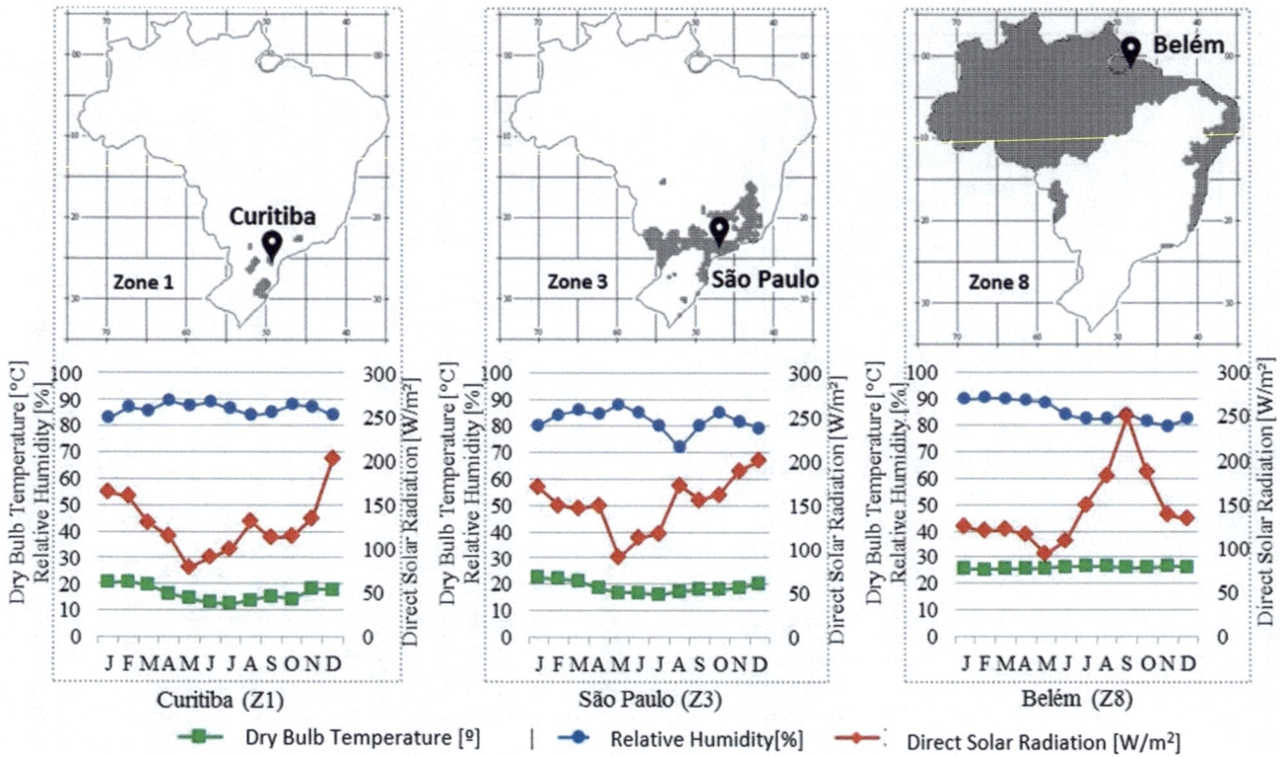

Fig. 3 Location and climate characteristics of the three municipalities included in the analyses (Nunes et al. 2020)

Table 1 General building parameters

Parameter	Input data
Gross internal area (m^2)	780.96
Working days	5
No. of holidays (days/year)	6
Hours of operation (h/day)	10
No. of floors above grade	1
Floor-to-floor height (m)	3
Occupancy density (m^2/person)	7.49

Table 2 Building areas

Room type	Input data (m^2)
Classrooms	178.12
Meeting rooms	16.20
Playrooms	38.4
Office/administration rooms	19.2
Restrooms	64.22
Cafeteria	121.05
Corridors	175.46
Staff rooms	11.15
Other space types	157.16

Table 3 Building dimensions

Orientation	Input data (m)
Northeast	40.8
Southeast	29.18
Southwest	36.9
Northwest	29.18

Table 4 HVAC parameters

Parameter	Input data
Building HVAC system	No
Ventilation	Natural ventilation

Table 5 Climate zones of the three evaluated municipalities

Municipality	ASHRAE climate zone
Curitiba	3A
São Paulo	2A
Belém	0A

Table 6 Energy efficiency measures

Measure	Item	Input data for base case	Input data for improved case
EEM01	Window-to-wall ratio	22%	20.6%
EEM02	Reflective roof: solar reflectance	45%	90%
EEM03	Reflective exterior walls: solar reflectance	45%	80%
EEM04	External shading/annual average shading factor	No shade	0.01
EEM05	Roof insulation	$U = 2.33$ W/m^2·K	$U = 0.62$ W/m^2·K
EEM06	Insulation of ground/ raised floor slab	$U = 0.49$ W/m^2·K	$U = 0.18$ W/m^2·K
EEM08	Insulation of exterior walls	$U = 1.86$ W/m^2·K	$U = 1.2$ W/m^2·K
EEM09	Window efficiency	$U = 5.75$ W/m^2·K, SHGC = 0.49, and VT = 0.7	$U = 5.62$ W/m^2·K, SHGC = 0.3, and VT = 0.43
EEM11	Natural ventilation	No	Yes/cross 30% opening
EEM18	Domestic hot water system	50% solar, 50% heat pump	100% solar
EEM22	Efficient lighting for internal areas	65 L/W	120 L/W
EEM31	Smart meters for energy	No	Yes
EEM33	Onsite renewable energy	No onsite renewable energy	50% of annual energy use

SHGC, solar heat gain coefficient; VT, visual transmittance

Table 7 Water efficiency measures

Measures	Item	Input data for base case	Input data for improved case
WEM01	Water-efficient showerheads	8 L/min	6 L/min
WEM02	Water-efficient faucets for all bathrooms	6 L/min, no aerators	6 L/min, with aerators
WEM04	Water-efficient closets for all bathrooms	6 L/flush (high volume flush)	Dual flush: 6 and 3 L/ flush
WEM06	Water-efficient bidet	6 L/min	5 L/min
WEM07	Water-efficient urinals	4 L/flush	2 L/flush
WEM08	Water-efficient faucets for kitchen sinks	8 L/min	8 L/min
WEM14	Rainwater harvesting system	No	200 m^2 catchment area
WEM17	Smart meters for water	No	Yes

Table 8 Materials efficiency measures

Measure	Input data for base case	Input data for improved case	Input data (%)
MEM01 Bottom floor construction	Concrete slab: in-situ reinforced conventional slab	Reuse of existing floor slabs	100
MEM03 Floor finish	Ceramic tiles	Linoleum sheet	100
MEM04 Roof construction	Concrete slab: in-situ reinforced conventional slab	Clay tiles of timber rafters	100
MEM05 Exterior walls	Concrete blocks: solid blocks of dense concrete	Compressed earth blocks	100
MEM06 Interior walls	Brick wall, solid brick (0–25% voids) with external and internal plaster	Cored bricks (25%–40% voids) with external and internal plaster	100
MEM07 Window frames	Aluminum	Aluminum	100
MEM08 Window glazing	8 mm single glazing	6 mm single glazing	100
MEM09 Roof insulation	No insulation	Air gap > 100 mm	100

The water savings strategies aimed to increase the efficiency of showers, faucets, water closets, and urinals. Furthermore, rainwater harvesting, and smart metering were proposed.

To reduce the embodied energy of materials, it was decided to reuse floor materials, apply clay tiles on the roof, and use compressed earth blocks on exterior walls.

General analysis of the improved case revealed that the building had different energy saving results according to the location. Nevertheless, location did not influence water efficiency or the embodied energy of building materials. The EDGE app considers climate differences and calculates their impact on energy consumption only. It, however, does not consider the regional contexts of building materials. There are large differences in procurement and distances traveled by raw materials across municipalities. Therefore, the embodied energy of materials should differ according to location in Brazil. The free Brazilian database, SIDAC, could be used in EDGE, allowing for a better approximation to the country's reality.

Figure 4 shows the savings in energy, water, and embodied energy of materials in the three evaluated municipalities. Belém, the hottest municipality, had the largest energy savings.

To obtain level 1 certification, buildings must achieve a minimum of 20% savings. Specifically for energy, this requirement would only be met with 100% use of renewable energy, in this case, solar energy (Fig. 5).

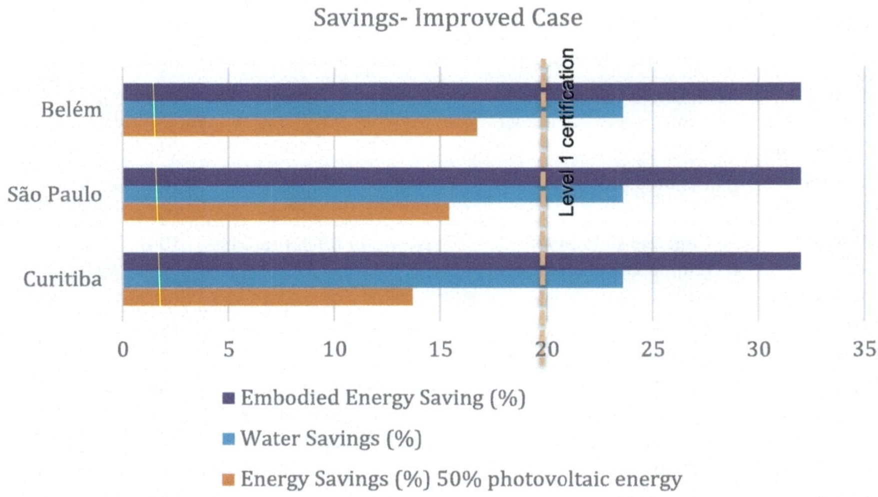

Fig. 4 Savings in energy, water, and embodied energy of materials in Belém, São Paulo, and Curitiba, Brazil (improved cases)

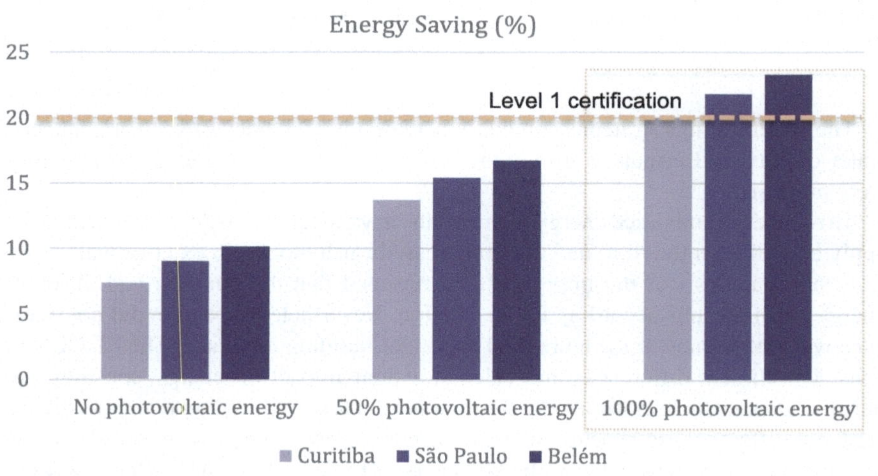

Fig. 5 Energy savings according to the relative use of renewable energy

According to Muta (2018), EDGE currently uses specific requirements from existing energy performance certifications in China (China Green Evaluation Label) and South Africa (SANS Building Regulations). The author suggested that energy savings seal criteria of the Brazilian National Program for Energy Conservation (ENCE, PROCEL) should be incorporated into the app, providing a better representation of the Brazilian reality, and making it easier to obtain the certification. The results would be better aligned with the country's energy efficiency policies.

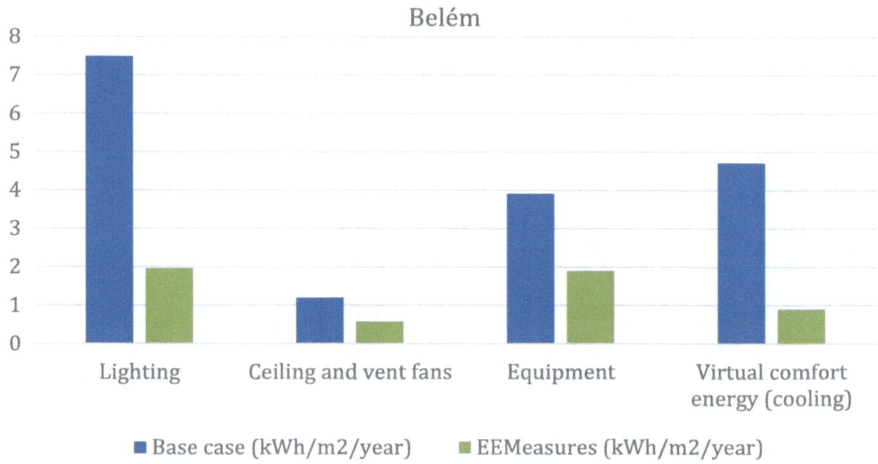

Fig. 6 Energy consumption of a model public preschool building in Belém, Brazil

Measures such as sunshade, natural ventilation, and reflective roof had little relevance in the results. Efficiency of the lighting system and renewable energy had the most relevant roles in energy savings. Energy consumption was found to stem mainly from cooking, and this variable could not be controlled by designers. Therefore, the analysis focused on measures related to the envelope, lighting, and equipment.

In Belém, artificial lighting had the largest impact on energy consumption (Fig. 6): the use of lamps with twofold efficiency in the improved case led to marked energy savings compared with the base case. Another important factor was virtual comfort energy, which represents the amount of energy required for future air cooling/heating systems. The improved case had a large decrease in this item, as only natural ventilation would be used.

In São Paulo (Fig. 7), the same behavior was observed for artificial lighting. However, the decrease in energy consumption associated with virtual comfort was lower than that observed in Belém, explained by the fact that São Paulo has a milder climate.

Curitiba (Fig. 8), which has the coldest climate among the three municipalities, had the lowest virtual comfort energy value.

3.1 *Evaluation of Passive Design Strategies*

Passive design is defined as "the design that maintains a comfortable temperature within the building using the climate and natural elements to get the optimum benefit and to reduce or eliminate the independence on mechanical systems for heating, cooling and lighting" (Altan et al. 2016). Some physical characteristics, such as thermal insulation and reflective properties of the walls and roof, were altered.

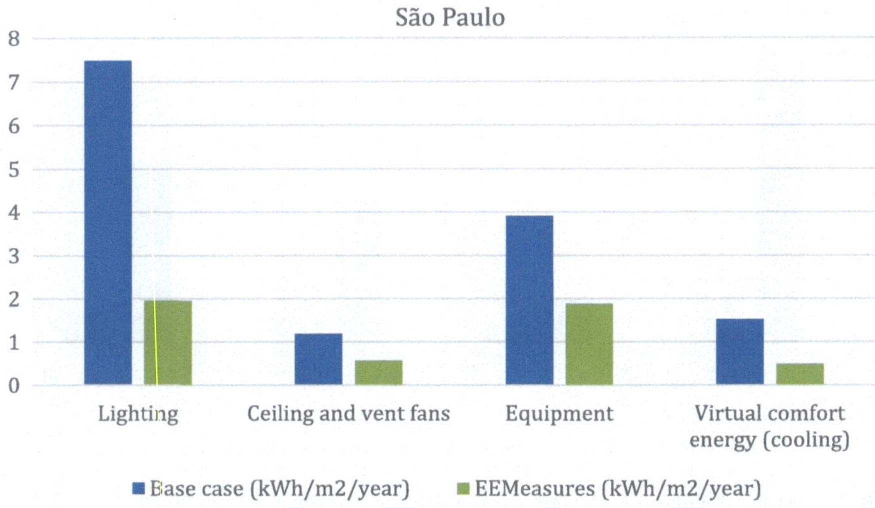

Fig. 7 Energy consumption of a model public preschool building in São Paulo, Brazil

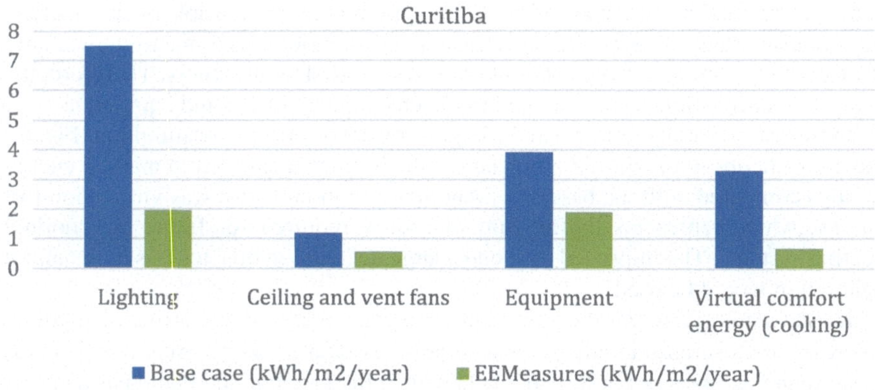

Fig. 8 Energy consumption of a model public preschool building in Curitiba, Brazil

Nevertheless, none had a significant change in energy savings, according to EDGE. According to the app, "EDGE was designed as a comparative software and not a design tool. Therefore, predicted results for energy, water and materials may vary from actual results" (EDGE 2022). With the addition of smart energy meters, the building had a savings of 3%. Shade devices and window-to-wall ratio had no impact on energy savings in any of the municipalities. This result does not seem to reflect the climate reality of São Paulo and Belém, which have hot summers.

The use of natural ventilation increased energy savings by 2%. However, there was no difference between cross-ventilation and single-sided ventilation. The embodied energy of the proposed building stems mainly from the structure, more specifically,

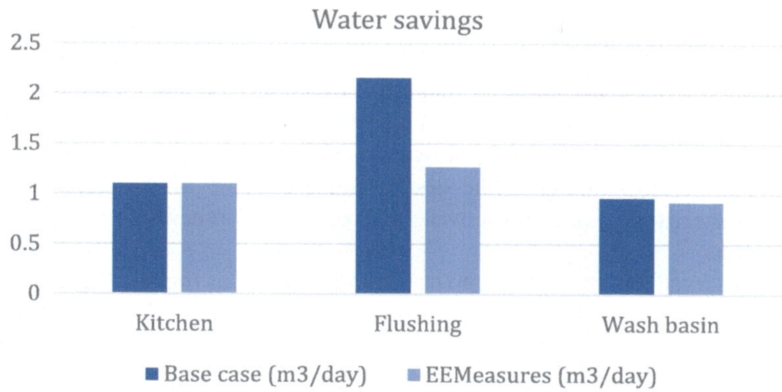

Fig. 9 Water savings in a model public preschool building

the foundation. Other materials, such as windows, roofing, and thermal insulation, do not have much impact. Compressed earth blocks were suggested, instead of concrete blocks, to decrease the environmental impact of building materials. This solution increased material savings by 8%.

The biggest impacts on water savings were associated with flushing and wash basin efficiency (Fig. 9). By controlling water flows, it was possible to achieve 20% water savings. Rainwater harvesting in a 200 m^2 area and with a tank capacity of 20 m^3 led to a water savings of 23.6%.

4 Conclusions

The objective of this study was to apply and evaluate the use of the EDGE app in a model public preschool developed by the federal government, replicated in three Brazilian municipalities with different climates.

After inputting the initial data (base case) and implementing the improvement strategies (improved case) for energy, water, and embodied energy savings, it was possible to identify the strengths and weaknesses of the application.

- The proposed measures only produced changes in energy savings, even though three different municipalities were tested.
- Generic strategies are available, such as using thermal insulation on the roof and façade. However, it was not possible to create new design strategies or detail specific construction systems that are typically used in Brazil.
- The software was useful for comparing solutions; however, adaptations are needed to better simulate the reality of the country. Energy savings seal criteria of the Brazilian National Program for Energy Conservation (ENCE, PROCEL) should be incorporated into the app. The public construction materials database SIDAC could also be used as a basis for the calculations.

- Passive design strategies have a minor impact on the energy savings results for the three cities. The significance of these solutions should be reconsidered, as in a country where most buildings are naturally conditioned, strategies such as natural ventilation and solar shading have greater importance and impact.
- In agreement with the literature, it was observed that complex measures, which require in-depth planning in the design and execution phases, generate more satisfactory results, such as rainwater harvesting and renewable energy (solar) generation.
- It was observed that the EDGE app has certain limitations regarding the base case simulation, as it does not allow selecting building materials. This option is only available for the improved case.
- It is a free software, with a friendly interface, available in several languages (English, Spanish, Portuguese, French, and Mandarin, among others). Thus, architects, engineers, and students can use the software to gain a better understanding of the impacts of their design decisions.
- The EDGE application necessitates minimal initial information at the outset of the simulation, thereby enabling its utilization in the early stages of architectural design. Although the results may not appear entirely accurate, it allows for the visualization of strategies and their impacts. The tool shows great potential for teaching purposes. This approach proves useful, especially considering that numerous energy efficiency evaluation tools are complex and require detailed information that is only available at the end of the design process.

References

Altan H, Hajibandeh M, Anissa K, Aouland KAT, Deep A (2016) Passive design. Chapter 8. In: Noguchi M (ed) ZEMCH: Toward the delivery of zero energy mass custom homes. Springer Tracts in Civil Engineering. https://doi.org/10.1007/978-3-319-31967-4_8

Brasil—Ministério da Educação (2006) Secretaria de Educação Básica. Parâmetros básicos de infraestrutura para instituições de educação infantil (2006). MEC, SEB, Brasília, p 45. https://sinapse.gife.org.br/download/parametros-basaicos-de-infra-estrutura-para-instituicoes-de-educacao-infantil-encarte-1

Bueno C, Rossignolo JA (2013) Análise dos sistemas de certificação ambiental de edifícios residenciais no contexto brasileiro. Risco Revista Pesquisa Arquitetura Urbanismo (Online) 17:6–22. https://doi.org/10.11606/issn.1984-4506.v0i17p6-22

Caldas LR (2020) Uso do excellence in design for greater efficiencies (EDGE) como Ferramenta de Apoio para Projetos de Retrofit Sustentável. Revista Gestão Gerenciamento 13. https://nppg.org.br/revistas/gestaoegerenciamento

Chatterjee D, Ray S, Kumar R (2022) Certification of green buildings. Wood Good 3(3). https://iwst.icfre.gov.in/Magazine/2022/Wood%20is%20Good%20-%20Vol.III,%20Issue%20III%20October-December%202022.pdf

Corbos A, Bunea O-I, Jiroveanu D-C (2023) The effects of the energy crisis on the energy-saving behavior of young people Razvan. Energy Strat Rev 49:101184. https://doi.org/10.1016/j.esr.2023.101184

Dlamini LN, Yessoufou K (2022) Residents and professionals' perspectives on energy and water consumption while transiting from conventional to sustainable housings in South Africa. Sustainability 14:4498. https://doi.org/10.3390/su14084498

EDGE—Excellence in Design For Greater Efficiencies (2022). https://www.edgebuildings.com

FDE—FUNDAÇÃO PARA O DESENVOLVIMENTO DA EDUCAÇÃO (2006) Arquitetura escolar paulista - estruturas pré-fabricadas. Diretoria de obras e serviços, São Paulo. https://www.fde.sp.gov.br/pagepublic/Interna.aspx?codigoMenu=255&AspxAutoDetectCookieSupport=1

FNDE – FUNDO NACIONAL DE DESENVOLVIMENTO DA EDUCAÇÃO (2005) Manual para Adequação de Prédios Escolares. 5ª Ed./Elaboração Carlos, Alberto Araújo Guimarães, Cláudia Maria Videres Trajano, Erinaldo Vitório, Rodolfo Oliveira Costa, Willamy Mamede da Silva Dias. Brasília: Fundescola/DIPRO/FNDE/MEC. https://www.gov.br/fnde/pt-br/acesso-a-informacao/acoes-e-programas/programas/par/manuais-par/VolumeIIIProjetosEd.EscolaresEnsinoFundamental.pdf

FNDE—FUNDO NACIONAL DE DESENVOLVIMENTO DA EDUCAÇÃO (2006) Orientação para elaboração de projetos de construção de centros de educação infantil. Cartilha Proinfancia. Brasília, pp 52–55. https://www.mprj.mp.br/documents/20184/1232335/volume_2elaboracao_de_projetos_ed._escolares_ed._infantil.pdf

International Energy Agency (IEA) (2022) World energy outlook 2022. https://iea.blob.core.windows.net/assets/830fe099-5530-48f2-a7c1-11f35d510983/WorldEnergyOutlook2022.pdf

Isimbi D, Park J (2022) The analysis of the EDGE certification system on residential complexes to improve sustainability and affordability. Buildings 12(10):1729. https://doi.org/10.3390/buildings12101729

Kartikasari FD, Tarigan E, Fransiscus Y, Lidyawati T (2018) Energy saving measures and potential of energy efficiency at the university of Surabaya, based on EDGE simulation. In: 5th International conference on information technology, computer, and electrical engineering (ICITACEE), pp 89–92. https://doi.org/10.1109/ICITACEE.2018.8576925

Klein M, Osterhage T, Müller D, Kosanović S, Hildebrand L (2018) Building certification systems and processes. In: Sustainable and resilient building design: approaches, methods and tools. TU Delft Open. https://doi.org/10.18154/RWTH-2018-230294

Motherway B, Klimovich K, Mooney E, Gelis C (2022) Empowering people to act: how awareness and behaviour campaigns can enable citizens to save energy during and beyond today's energy crisis. In: 7th Annual global conference on energy efficiency of IEA. https://www.iea.org/commentaries/empowering-people-to-act-how-awareness-and-behaviour-campaigns-can-enable-citizens-to-save-energy-during-and-beyond-today-s-energy-crisis

Muta LF (2018) Aplicação do software edge para identificar o nível de eficiência energética de uma edificação residencial unifamiliar. Revista Brasileira Engenharia Sustentabilidade 5(2):14–20. https://revistas.ufpel.edu.br/index.php/rbes/article/view/283

Nunes G, Moura JDM, Güths S, Atem C, Giglio T (2020) Thermo-energetic performance of wooden dwellings: benefits of cross-laminated timber in Brazilian climates. J Build Eng 32:101468. https://doi.org/10.1016/j.jobe.2020.101468

Robles VJF, Picó EC, Hosseini SMA (2022) Environmental performance assessment: a comparison and improvement of three existing social housing projects. Clean Environ Syst 5:2022. https://doi.org/10.1016/j.cesys.2022.100077

Setyowati DL et al (2020) Simulating water efficiency management at UNNES campus, Semarang, Indonesia using EDGE application. IOP Conf Ser Earth Environ Sci 485:012038. https://doi.org/10.1088/1755-1315/485/1/012038

Tarigan E (2018) Energy saving measures and simulation in the library building of University of Surabaya. J Teknol Rekayasa 3(1):63–70. https://doi.org/10.31544/jtera.v3.i1.2018.63-70

End-Of-Life Approach of the PEF Method Applied to Wooden Windows: An Italian Case Study

Elisabetta Palumbo, Sara Cortesi, and Valentina Castellani

Abstract The newly revised Energy Performance of Buildings Directive extends the evaluation of the energy performance of buildings to the entire life cycle. This means evaluating the embodied carbon of buildings, in addition to just the operational carbon covered by it to date. Windows typically account for 10–25% of the exposed area of buildings, resulting in more than 60% of its total energy loss. Due to the key role of windows in the performance of buildings over the whole life cycle, the evaluation of their environmental impacts through recognized and robust environmental methods should become part of building design practice. However, there are limited studies examining the embodied impacts of different types of windows with a life cycle approach. Furthermore, the majority of them focus on the upfront carbon, oversimplifying the end-of-life possibilities, often assuming incineration as the only available scenario. To support studies with a broader scope and a more accurate approach to end-of-life, the Product Environmental Footprint (PEF) method, developed by the European Commission, considers the whole life cycle of products and introduces the Circular Footprint Formula (CFF), a dedicated approach to model the end-of-life of products as well as their recycled content. The use of the CFF can support the assessment and comparison of circular economy actions, because it defines rules to model the use of secondary raw materials and of different end-of-life scenarios, including recycling, reuse, incineration (with electricity and heat recovery), and disposal, for both open- and closed-loop systems, in a consistent and reproducible way. Furthermore, LCA underpins PEF and is a pivotal method to support policy on transition to a circular economy design and the realization of respective SDGs (7, 9, 12 and 13). In June 2018, the Italian Ministry of Environment and Energy Security launched

E. Palumbo (✉) · S. Cortesi
University of Bergamo, Bergamo, Italy
e-mail: elisabetta.palumbo@unibg.it

V. Castellani
LCA Consultant, Milano, Italy

S. Cortesi
ENEA - Italian National Agency for New Technologies, Energy and Sustainable Economic Development, Rome, Italy

L. Brandli et al. (eds.), *The Contribution of Life Cycle Analyses and Circular Economy to the Sustainable Development Goals*, World Sustainability Series,
https://doi.org/10.1007/978-3-031-85300-5_6

the Made Green in Italy, a voluntary scheme that adopts the PEF method to evaluate and communicate the environmental impact of Made in Italy products. Between 2019 and 2023, the LIFE MAGIS project, co-funded by the LIFE Programme of the European Union, tested PEF and Made Green in Italy on a number of Italian product categories. This study presents the experience gained during the project working on the category of wooden windows, in particular highlighting the questions to ask and the choices to make when applying the CFF. Despite the limitations posed by the lack of an official PEFCR, also reflected in the lack of some application-specific parameters in Annex C, the experience proved that the application of the CFF to wooden windows can be a good method to assess the environmental implication of different EoL scenarios, including the implementation of CE actions. For example, for the wooden window and door sector, the growing recycling of materials such as wood and glass can be captured by the formula with higher values to be assigned to R1 while the higher recycled content, for example for glass, will be reflected by higher R2 values. This can help drive the sector towards more circular options, by recognizing and valorizing the efforts to reduce the use of virgin resources and to find alternative end-of-life options to landfill and incineration.

Keywords Product environmental footprint (PEF) · Life cycle assessment (LCA) · Wooden window · Circular footprint formula (CFF) · End-of-life (EoL) · Circular economy (CE)

1 Introduction

The implications of the scarcity of primary raw materials, the increasing global consumption of non-renewable resources, and the decreasing availability of landfilling areas are of increasing concern. In March 2020, the European Union (EU) published the new Circular Economy Action Plan (CEAP) (COM(2020)98 final) to promote circular economy processes, encourage sustainable consumption, and prevent waste, which are prerequisites to achieving the EU's 2050 climate neutrality target.

To transition to a CE, the building sector, which is one of the priority sectors in the European CEAP, should examine how CE principles can be incorporated into building projects. After focusing on energy efficiency for many years, the Architecture, Engineering, and Construction (AEC) sector has made great efforts to include the circularity concept in the evaluation of the environmental performance of building products. Even at the legislative level, the newly recast Energy Performance of Buildings Directive (EPBD) published on 24th April 2024 (Directive 2024/1275) has extended the evaluation of the energy performance of buildings to their entire life cycle. This means that in addition to the current assessment of operational carbon, evaluation of embodied carbon in buildings will also be performed, considering also the characteristics of materials used in the buildings and their production processes as well as end-of-life (EoL) scenarios.

More than 60% of a building's total energy loss is generated by windows, which typically represent 10–25% of its exposed area. Given the key role of windows in the performance of buildings over their whole life cycle, the evaluation of their environmental impacts through recognised and robust environmental methods such as the life cycle assessment (LCA) should be part of building design practices. LCA is a quantitative method used to support the implementation of CE principles in the AEC sector. By analysing the impacts on all life cycle phases, from the construction phase, which involves procuring materials, producing, and transporting, to the EoL phase, which involves dismantling, demolishing, and recycling, LCA evaluates the environmental impacts of the whole life cycle of buildings, considering both operational and embedded impacts.

Different papers have emphasised the importance of including EoL considerations in LCA to effectively implement CE strategies (Boje and Mack 2024; Cabeza et al. 2014; Dixit et al. 2012; Blengini and Di Carlo 2010).

However most studies have focused on upfront carbon, oversimplifying the EoL possibilities and often assuming incineration as the only available scenario. Despite the recent focus on LCA research on building reuse, academics and practitioners in the field have consistently focused on evaluating the environmental impacts of recycling (De Wolf et al. 2020).

Another recent trend connected with CE principles is highlighted by a recent paper that provides an overview of methodologies to evaluating the environmental impacts of reusable/recyclable building components/materials (Yang et al. 2024). This paper focused on the assessment of the environmental credits and loads associated with reusing modular components over multiple use cycles, highlighting that the determination of impact allocation methods is critical to avoid different assessment results.

By analyzing different methods of end-of-life allocation, the study asserts that the Circular Footprint Formula (CFF) was a strong incentive to prioritize designing for disassembly in the initial phase, according to Eberhardt et al. (2020).

According to Allacker et al. (2017), it is necessary for this method to be physically realistic and ensure a fair allocation of burdens and benefits between cycles, avoiding double counting and inconsistencies.

Although CE principles contribute to the reduction of environmental impact, their application in construction faces several challenges, including the lack of a standardised approach to guiding assessments in different EoL scenarios. In particular, there is a lack of clear, standardised guidance on how to model the EoL of building products with LCA, which leads to difficulties in properly comparing studies that include this phase.

The main reference document for the LCA of construction products in the context of Environmental Product Declarations schemes (EN15804:2019+A2, developed by CEN (2019)) defines requirements for modelling the burdens and credits associated with recycling of materials at the EoL by including them in a separate module (module D) that does not directly contribute to the calculation of the environmental impact associated with the product analysed.

To support studies with a broader scope and more diversified approach to EoL, the Product Environmental Footprint (PEF) method, an LCA-based methodology developed by the European Commission, analyses the whole life cycle of products and introduces the Circular Footprint Formula, a dedicated approach to consider burdens and credits associated with recycling (both as inputs of recycled materials and as the recycling of materials at EoL) and other EoL treatments. The aim of the CFF is to enable the assessments of possible EoL scenarios in a consistent and reproducible way, including recycling, reuse, incineration with electricity and heat recovery, and disposal, for both open- and closed-loop systems. Obrecht et al. (2021) examined five allocation methods and determined that the PEF and cut-off methods with module D were more favourable for accounting the effects of CE practices than the other methods because they consider both the benefits and burdens of recycling and reuse, thereby promoting the circular use of materials.

Although PEF is not widely spread in the construction sector, some studies have examined its use in this field and in three PEFCR developed at the EU level (decorative paints, hot- and cold-water supply pipes, and thermal insulation). Durão et al. (2020) recognised that it may be useful for achieving a harmonised approach to the building scale and for providing global assessment from construction to EoL. Passer et al. (2020) demonstrated the usefulness of the PEF method in assessing environmental performance in different climate and regulatory contexts by applying it to two office buildings. They provided recommendations for the advancement of benchmarking approaches and performance classes for various types of buildings. This latest work will support the definition of benchmarks and target values in light of the whole life cycle objectives proposed in the EPBD revision.

In June 2018, the Italian Ministry of Environment and Energy Security launched Made Green in Italy, a voluntary scheme that adopts the PEF method to evaluate and publish the environmental impacts of products made in Italy. Between 2019 and 2023, the LIFE MAGIS project, co-funded by the LIFE Programme of the EU, tested PEF and Made Green in Italy to study environmental impacts and support the definition of improvement options for several Italian product categories, including wooden windows. In the Italian window market, wooden windows account for approximately 25% of the windows and doors produced in Italy out of the three main materials (aluminium, wood, and PVC) (UNICMI 2023).

A PEF Category Rule (PEFCR), the document drafted, managed, and updated by a technical secretariat at the European level that guides the application of the PEF method to a specific type of product, was not available for wooden windows. During the LIFE MAGIS project, a PEFCR proposal for this product category, designed particularly for the Italian context, was developed by Inab–Institute of Sustainability in Civil Engineering of the RWTH Aachen University and Consorzio LegnoLegno, two of the project partners. This draft PEFCR was tested to study six products of three Italian companies. The test allowed to understand how well the document was suited for the specific context and offered suggestions for improvements. On the basis of the lessons learned from the LIFE MAGIS experience, the application of the CFF is presented below.

Therefore, starting with the explanation of the CFF (Sect. 2), the next section will demonstrate its use in a specific case, a wooden window (Sect. 3), and then discuss potential integrations and future development scenarios (Sect. 4).

2 Circular Footprint Formula

The CFF is part of the environmental footprint methods (PEF and Organisational Environmental Footprint [OEF]) recommended by the European Commission (EC 2020). The formula was developed to provide a comprehensive way to consider all the burdens and credits associated with different EoL scenarios that may happen within a product life cycle, taking into consideration the contribution of recycling to more than one life cycle (i.e. beyond the life cycle under analysis in the PEF study). The formula is composed of three parts: material, energy, and disposal (Fig. 1).

The first part of the CFF, 'material', refers to the input of recycled materials and the recycling of materials at EoL. The parameters considered for the material part are R1, Ev, *A*, Erecycled, Qsin/Qp, R2, ErecyclingEoL, E*v, and Qsout/Qp. R1 represents the share of recycled material in the product. For example, the product studied is made of aluminium, which is produced using 70% virgin material and 30% scraps. In this case, the R1 value for the input material aluminium is 0.3. The quality of the recycled material used in the analysed product in relation to the quality of the virgin material is accounted for (Qsin/Qsp).

In the same way, the quality of the recycled material produced after the EoL of the product by treating its material in relation to the quality of the virgin material is taken into account (Qsout/Qsp). For instance, the quality of a recycled metal is usually the same as that of the virgin metal (Qsin/QSout = 1), so 1 kg of recycled metal can substitute for 1 kg of virgin metal in a production process. In case of downcycling of the material, the qualities of the recycled and virgin materials are not the same (as it

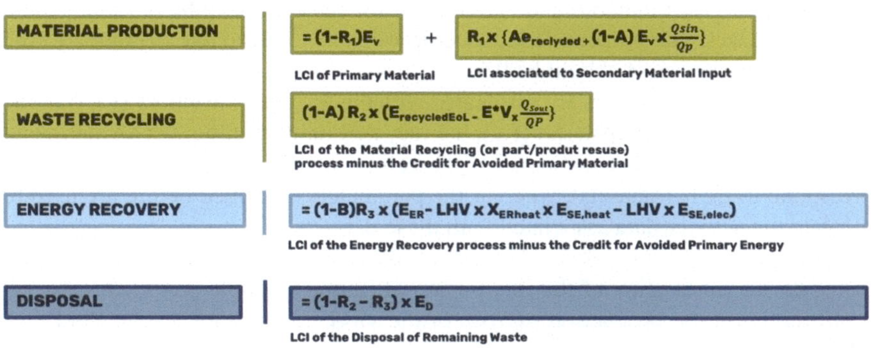

Fig. 1 The three parts of the circular footprint formula: material, energy, and disposal. Credit: E. Palumbo modified from Zampori and Pant (2019)

may be for paper or some types of plastics), so the resulting ratio will be lower than 1, which means that to substitute 1 kg of virgin material, more than 1 kg of recycled material is needed.

Parameter R2 represents the share of the material in the product that will be recycled (or reused) in a subsequent system. It must be measured from the output of the recycling plant (i.e. it should take into account the inefficiency of the collection and recycling systems and should be different from the share of the material collected separately from other waste streams). Company-specific R2 values shall be used whenever available (and justified in the PEF report). Default application-specific R2 values are available in Annex C of the PEF method.

The parameter *A* is the allocation factor of burdens (impact of the recycling process) and credits (avoided impact of the production of the substituted virgin material) of recycling between the two life cycles of the recycled material (i.e. the recycling of the material and the use of the recycled material as input). According to the PEF method, the *A* value can be between 0.2 and 0.8. This means that the 0–100 approach (i.e. no burdens and credits for the company that sends the material to recycling at EoL and 100% burdens and credits to the company that uses the recycled material) and the 100–0 approach (i.e. 100% burdens and credits to the company that generates the waste and send it to recycling and no impacts for the one that chooses to use a recycled material, in accordance with 'the polluter pays' principle) are not allowed in the PEF method, and that part of the burdens and credits shall always be allocated to the two cycles. The value of the *A* factor should reflect the market situation: 0.2 in the case of a low offer of recyclable materials and high demand (the recycling at EoL is incentivised through lower impact allocation); 0.8 in the case of a high offer of recyclable materials and low demand (the use of recycled materials is incentivised); and 0.5 in the case of equilibrium between offer and demand.

In the choice of the *A* value, the market situation should be considered for specific applications, not only for the material in general. The default application- and material-specific values for the parameter *A* are provided in Annex C of the PEF method. If the user of the PEF method cannot find an application-specific *A* value in Annex C for a given material and application, the material-specific *A* value should be used. In cases where the material-specific value is not available in Annex C, the value of the parameter *A* should be 0.5.

The described parameters are associated with the life cycle inventory (LCI) of the production of the virgin material (Ev), the recycled input material (Erecycled), and the recycling process (ErecyclingEoL). Therefore, the 'E' components of the CFF correspond to LCIs (datasets), whereas the parameters described before are used to calculate the activity data associated with these datasets.

More specifically, '(1 − R1)Ev' represents the impact associated with the production of the virgin fraction of the input material (in the example provided before, it would represent the production of 70% of the virgin aluminium input). Similarly, 'R1 × (AErecycled + (1 − A)Ev × Qsin/Qp)' represents the impact of the recycling process allocated to the current life cycle (AErecycled), plus a share of the impact generated during the original production of the material ('first life'), taking into consideration the quality ratio ((1 − A)Ev × Qsin/Qp)). In the example provided before, it would be associated with 30% of the recycled aluminium input.

Finally, the last part of the formula for materials, (1-A)R2 × (ErecyclingEoL − Ev × Qsout/Qp), represents the modelling of burdens and credits at the EoL. In fact, (ErecyclingEoL − Ev × Qsout/Qp) is the net impact calculated by subtracting the credits of avoided virgin material production (proportionally to the quality ratio) from the impact of the recycling process. The resulting net impact (or credit) is then multiplied by R2 to consider only the fraction of material recycled (compared with 100% of the material content in the product studied) and by (1 − *A*), which allocates only a share of the net impact to the life cycle under study (coherently with what has been done in the first part on recycled input).

The second part of the CFF, the 'energy' part, allows for calculating the net impact of the energy recovery processes (usually incineration with recovery of heat and/or electricity). The parameters used in this part of the formula are R3, lower heating value (LHV), B, XER.heat, and XER.elec.

Parameter R3 represents the share of the material used for energy recovery at EoL. The default R3 values for all the European countries are available in Annex C, based on Eurostat data. The EC recommendation specifies that these data are subjected to periodical update, but at the moment, the more recent version of Annex C dates back to 2020, so more recent data could be available on the Eurostat website. As in other contexts, LVH stands for 'lower heating value', which has to be considered in the calculation of the amount of heat and electricity produced by the conversion of the material.

Similar to the parameter *A*, the parameter *B* is an allocation factor of burdens and credits of the energy recovery process between the two systems (i.e. the one in which the energy recovery process occurs and the one in which the recovered heat or electricity is used as input). According to the PEF method, the parameter *B* should be 0. This means that 100% of the burdens and credits of energy recovery are attributed to the system studied (i.e. the system that recovers the energy). Therefore, this approach may incentivise the recovery of energy over the recycling of materials independently from their environmental impact.

Finally, XER.heat and XER.elec represent the efficiency of the heat and electricity recovery systems. When multiplied by the LVH of the material used to recover energy, they allow for calculating the amount of heat and electricity recovered.

As for the material part of the CFF, the described parameters are associated with the LCI of the energy recovery process (EER) and the avoided processes for generating heat (ESE.heat) and electricity (ESE.elec). As before, the 'E' components of the CFF correspond to LCIs (datasets), whereas the parameters described before are used to calculate the activity data associated with these datasets.

The structure of the energy part of the CFF is similar to that of the material part, where (EER − LVH × XER.heat × ESE.heat − LVH × XER.elec × ESE.elec) represents the net impact generated by the recovery of energy (burdens of the energy recovery process minus credits from the avoided generation of heat and electricity), which are then multiplied by (1 − B), which equals 1, and by the share of material in the product that is used for energy recovery (R3).

Finally, the disposal part of the CFF represents the impact generated by the disposal of the remaining fraction of the material, which is not recycled nor used for energy recovery. Its structure is simple: ED represents the LCI of the disposal process (e.g. landfilling), which is multiplied by (1 – R2 – R3), the remaining share of the material.

3 Application of the CFF to a Wooden Window

The case analysed was that of a wooden window produced in Italy. The window has been assessed taking into consideration the reference product described in Table 1.

The definition of the functional unit was a crucial aspect of the work, considering the wide range of application parameters for window fixtures, such as size, shape, and characteristics. According to the PEF guidelines, the functional unit is defined by answering the following four questions: what, how much, how well, and how long (Zampori and Pant 2019). Particular attention was given to the definition of the product performance and quality, which correspond to the two questions 'how well' and 'how long'. These questions are relevant to the comparability of products. In fact, comparing products necessitates including performance and quality in the functional unit (Pedersen and Remmen 2022).

Therefore, the functional unit used in this study was 1 m^2 of wooden window without its packaging that provided a thermal transmittance value (value U_w) of 1.3 W/m^2K and met the general requirements of the *Conformité Européenne* (EC conformity) marking declaration of conformity over a useful life span of 42 years.

As part of the PEF studies, the CFF was applied to evaluate the EoL phase of wooden windows, considering the four materials included in the product: pine wood, glass, aluminium, and EPDM rubber (Fig. 2).

Data about the EoL of windows are not collected systematically (e.g. as a product group in Eurostat records), so it is not easy to define a robust EoL scenario for the product analysed and for its constituent materials. In particular, the potential recyclability of materials per se is known (e.g. glass used as packaging is usually recycled quite easily), but the recyclability potential is related also to the possibility to easily separate materials at the EoL of the product and to the specific features of the material within the product (e.g. the type of preservation treatment applied to the wood components or the composition of the glass used to ensure specific thermal performance). Moreover, there can be design solutions that promote circular economy

Table 1 Characteristics of the wooden window type studied

Dimensions	1.23 m × 1.48 m
Frame thickness	70 mm
Double glazing	3 + 3 – 16 – 3 + 3
Wood species	Pine wood (*Pinus sylvestris*)
Thermal transmittance (U_w)	1.3 W/m^2K

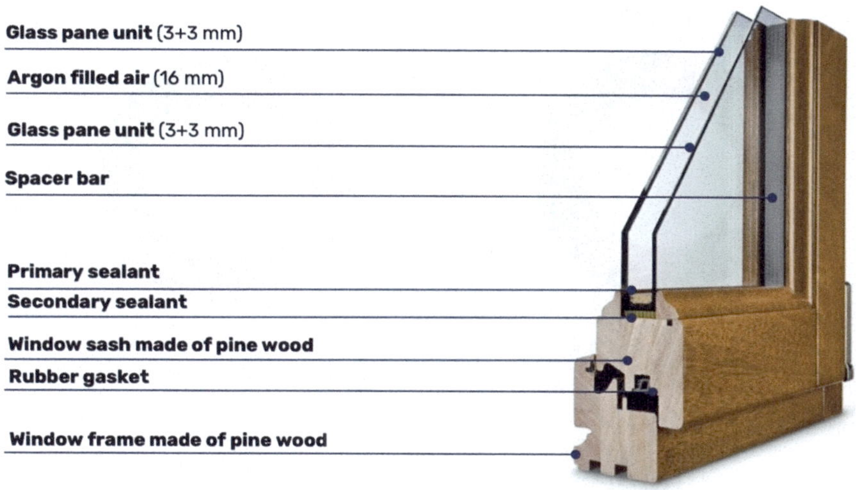

Fig. 2 Wooden window composition. Credit: E. Palumbo, modified by PM Serramenti

practices (including reuse, recycling or upcycling) at the EoL, but usually they are still a niche in the market, and are not captured by general statistics. Therefore, the EoL scenarios adopted for each component of window are shown in Fig. 3.

The application of the CFF and lessons learned are presented below.

The values of *A*, *B*, R1, R2, R3, Qsin, and Qsout, which were chosen by following the general indications of the PEF method and the specific indications, where available, and contained in Annex C are shown in Table 2.

The main limitation to the application of the CFF to wooden windows was the absence of an official PEFCR that is reflected in the absence of application-specific parameters in the Annex C, which was developed before the drafting of the PEFCR for windows in the LIFE MAGIS project (and not updated later). Therefore it was necessary to make some elaborations and assumptions based on the available parameters. The reasoning behind the choice of the parameters presented in Table 2 is illustrated below.

For wood, the Annex does not contain an application-specific value for the parameter *A*, given that it does not refer to the wood used in windows or construction but only to that used to make pallets, or a value for the generic material. Therefore, according to the indications of the PEF method, the intermediate value of 0.5 was assigned to the parameter.

For glass, *A* was assigned the value of 0.2, which is reported in the Annex for the generic material. For the same parameter, 0.2 was also used for aluminium, indicated both for the generic material and for its application in construction, for example, for doors and windows.

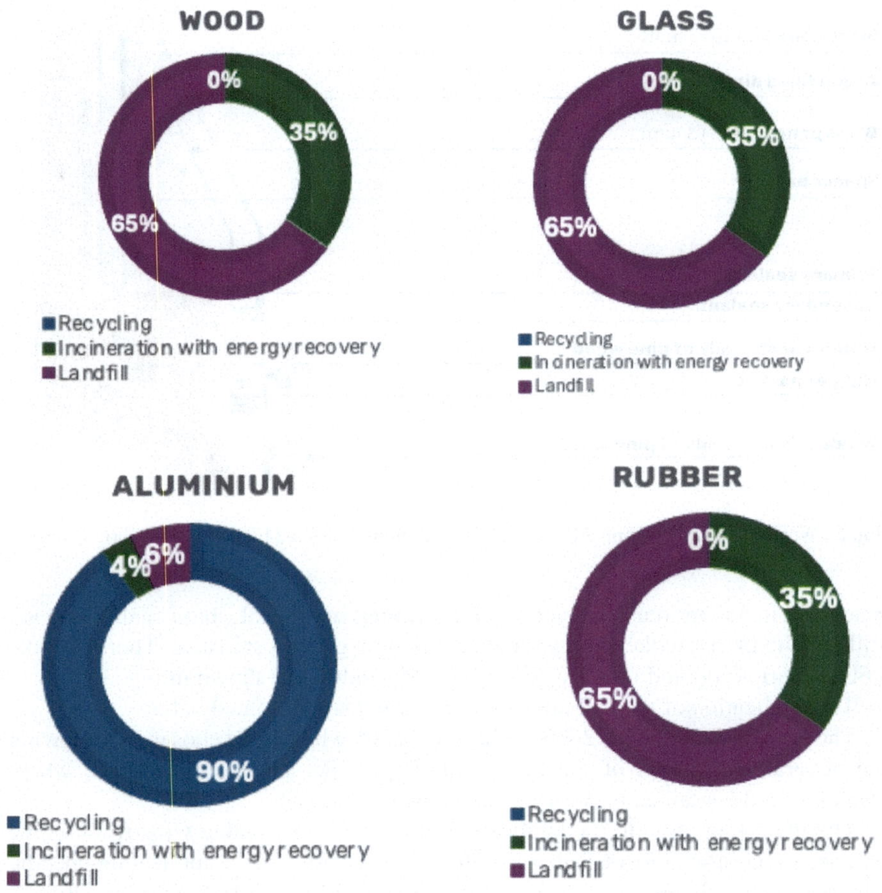

Fig. 3 Percentage contribution of the EoL scenarios adopted for each window component. Credit: E.Palumbo and S. Cortesi

Table 2 Values assigned to the CFF parameters for the materials contained in the studied windows

	A	*B*	R1	R2	R3	Qsin	Qsout
Pine wood	0.5	0	0	0	0.35	NA	NA
Glass	0.2	0	0	0	0.35	NA	NA
Aluminium	0.2	0	0	0.9	0.035	NA	1
EPDM rubber	0.5	0	0	0	0.35	NA	NA

NA: Not applicable

In the case of rubber, the Annex indicates an application-specific value for *A* for construction, but it is specific for water supply pipes. In this case, therefore, the intermediate value of 0.5 was chosen.

As indicated by the PEF method, the parameter B takes a default value of 0.

No specific information was available about the recycled content of glass and aluminium used in the windows, so the value of R1 provided in the Annex for the generic use of these materials was chosen. The value is 0 in both cases, which means that no recycled material content is considered. For wood and rubber, for which no suitable indications are available for the case studied, it was also considered that no recycled material was present, on the basis of the information collected from the producers during the project. As no quantity of recycled content was considered for any material, it was not necessary to define the Qsin parameter, that is, to define the quality of the incoming recycled material to be compared with the quality of the replaced virgin material (Qp parameter).

The value indicated for R2 by the Annex is equal to 0.9 for aluminium (generic use) and 0 for glass (generic use); that is, it is considered that 90% of the aluminium present in the window is recycled at the end of its life and that glass is not recycled.

These assumptions, particularly related to the glass component, represent the current scenario adopted by different manufacturers of glazing in their EPDs. However, they are not in line with the default values specified for wooden windows in EN 17213. In this standard, which is the standard that defines the product category rules (PCR) for Type III environmental declarations for windows and pedestrian doorsets, the EoL recycling scenarios for metals are 100% and 30% for glass.

For the remaining materials, wood and rubber, no company-specific values were available. As material- or application-specific guidance was not provided in Annex C for these materials, the parameter was set to 0.

The ratio between the parameters Qsout and Qsp for aluminium, the only material for which recycling was assumed, was set at 1 (based on Annex C); that is, the quality of the recycled aluminium was considered equal to that of virgin aluminium, which it is going to replace.

The percentage of material to be incinerated at EoL and thus the value of the R3 parameter were taken from Annex C, considering that its EoL takes place in Italy. Thus, R3 was calculated as 35% of what is not recycled, that is, 0.035 (35% of 10%) for aluminium and 0.35 (35% of 100%) for wood, glass, and rubber. The portion of material destined for landfill is that remaining after subtracting the recycled part and that sent for incineration (i.e. 9.75% for aluminium and 65% for wood, glass, and rubber).

A PEFCR drafted, managed, and updated at the European level for wooden windows would indicate the EF datasets for Ev, ErecyclingEoL, E*v, EER, ESE, heat, ESE, elec, and ED, to be used for all the products in the product category. As that was not available during the LIFE MAGIS project, it was decided that datasets from the Ecoinvent database would be used, despite recognising that this is a limitation for the application of the CFF, which can be implemented only in the foreground system (as described before) but not in the background datasets. The PEF method requires to indicate in the PEFCR the EF datasets to be used to model the EoL with the aim of ensuring consistencies in the modelling of products from different producers, i.e. to improve comparability of results. The authors of the present study made an analysis of the available ecoinvent datasets for the studied materials and selected those considered more representative for the product analysed, as shown in Table 2.

The documentation of the datasets used for EER provided the information necessary to define the value to assign to XER,heat and XER,elec in relation to the lower heating value (LHV), also indicated in the metadata of the datasets used.

Even if the ecoinvent 3.10 database was not developed to be used for modelling EoL according to the CFF, most of the datasets used in the model (and shown in Table 3) have a good correspondence to the process that we wanted to include in the model, with few exceptions. For instance, the dataset used to represent the recycling process for aluminium ("Aluminium scrap, post-consumer, prepared for melting {RER}| treatment of aluminium scrap, post-consumer, by collecting, sorting, cleaning, pressing") is based on recycling data of a mix of scraps coming from very diverse sources (e.g. motor block engines, used beverage cans, facade panels) and not only from the construction sector. The documentation of the dataset states "The actual activities associated with scrap preparation will depend on the scap nature [..] Some scap requires no preparation whatsoever." Therefore, the technological representativeness of the dataset in relation to the current study may be limited.

Another limitation is the non-availability of a specific dataset for the landfilling of rubber, which has been modelled using the dataset for the generic plastic mixture as a proxy. More in general, rubber may be a challenge for studies that include recycling of this material (which is not the case for the present study) because the ecoinvent database does not include a dataset for the recycling of rubber (component "ErecyclingEoL" in the CFF) and because it would require a careful analysis of the available material datasets to select the one which corresponds to the correct point of substitution. Rubber is presented here as an example, but it would be interesting to check potential limitations also for a wider set of materials, to assess if and how these limitations may influence the possibility to correctly model the environmental effect of circular economy solutions.

4 Conclusions and Future Trends

In order to transition to a CE, the building sector, one of the five priority sectors in the European CE package, should integrate CE principles into projects and evaluate their effectiveness in decreasing the overall environmental impact of buildings. The literature indicates that the EoL phase is crucial for CE, as it can have a significant impact on the overall environmental performance of a building. Several academic articles and case studies have emphasised the necessity of incorporating the EoL phase in LCA to achieve a more accurate and holistic evaluation of the environmental impact of a building throughout its entire life cycle.

Table 3 Ecoinvent datasets used for the CFF

Material	CFF parameter	Ecoinvent dataset
Wood	Ev	Sawlog and veneer log, softwood, measured as solid wood under bark {SE}| softwood forestry, pine, sustainable forest management | cut-off, U
	EER	Waste wood, untreated {RoW}| treatment of waste wood, untreated, municipal incineration | cut-off, U
	ESE,heat	Heat, central or small-scale, natural gas {Europe without Switzerland}| market for heat, central or small-scale, natural gas | cut-off, U
	ESE,elec	Electricity, medium voltage {IT}| market for electricity, medium voltage | cut-off, *U*
	ED	Waste wood, untreated {RoW}| treatment of waste wood, untreated, sanitary landfill | cut-off, *U*
Glass	Ev	Glazing, double, U < 1.1 W/m^2K, laminated safety glass {RER}| production | cut-off, U
	EER	Waste glass {CH}| treatment of waste glass, municipal incineration with fly ash extraction | cut-off, *U*
	ESE,heat	Heat, central or small-scale, natural gas {Europe without Switzerland}| market for heat, central or small-scale, natural gas | cut-off, *U*
	ESE,elec	Electricity, medium voltage {IT}| market for electricity, medium voltage | cut-off, *U*
	ED	Waste glass {CH}| treatment of waste glass, inert material landfill | cut-off, *U*
Aluminium	Ev	Alluminium alloy, AlMg3 {RER}| production | cut-off, *U*
	ErecyclingEoL	Aluminium scrap, post-consumer, prepared for melting {RER}| treatment of aluminium scrap, post-consumer, by collecting, sorting, cleaning, and pressing | cut-off, *U*
	E*v	Aluminium, primary, ingot {IAI Area, EU27 & EFTA}| aluminium production, primary, ingot | cut-off, *U*
	EER	Scrap aluminium {Europe without Switzerland}| treatment of scrap aluminium, municipal incineration | cut-off, *U*
	ESE,heat	Heat, central or small-scale, natural gas {Europe without Switzerland}| market for heat, central or small-scale, natural gas | cut-off, *U*
	ESE,elec	Electricity, medium voltage {IT}| market for electricity, medium voltage | cut-off, *U*
	ED	Waste aluminium {RoW}| treatment of waste aluminium, sanitary landfill | cut-off, *U*
Rubber	Ev	Synthetic rubber {RER}| production | cut-off, U
	EER	Waste rubber, unspecified {Europe without Switzerland}| treatment of waste rubber, unspecified, municipal incineration | cut-off, *U*

(continued)

Table 3 (continued)

Material	CFF parameter	Ecoinvent dataset
	ESE,heat	Heat, central or small-scale, natural gas {Europe without Switzerland}| market for heat, central or small-scale, natural gas | cut-off, *U*
	ESE,elec	Electricity, medium voltage {IT}| market for electricity, medium voltage | cut-off, *U*
	ED	Waste plastic, mixture {RoW}| treatment of waste plastic, mixture, sanitary landfill | cut-off, *U*

However, lack of a unique and comprehensive method for modelling the EoL of building products using LCA makes it difficult to properly compare studies that include this phase.

With the aim of providing harmonized rules for measuring the environmental footprint of products and organisations based on a life cycle approach (i.e. LCA) and allowing for more accurate and comparable results across different products and services and during the whole life cycle, the EC developed the PEF method. This method comprehensively assesses the environmental impact of a product throughout its entire life cycle, including all stages from raw material extraction, production, distribution, and use until the End of Life. In addition, it introduces the CFF, a formula for analysing the EoL scenario of a product as well as its recycled content in a standardised way.

Although CFF is not widely used in the construction sector, some studies have examined its use in this field. Eberhardt et al. (2020) applied the CFF to assess the environmental impacts of reusing building components over multiple life cycles.

Their findings suggested that the CFF prompted a strong incentive to prioritize disassembly design in the first cycle, as there was less impact assigned to that cycle.

Three LCA allocation rules, including Module D, Product Environmental Footprint (PEF), and Circular Footprint Formula, were used in a recent study by Yang et al. (Yang et al., 2024) to assess the environmental credits and burdens of reusing modular components across multiple use cycles. The authors assert that the CFF provides insights into the demand and supply dynamics of recycled materials by taking into account the market situation.

The case study described in this chapter discusses the application of the CFF formula to a specific family of construction products, also based on the experience gained during the LIFE MAGIS project, presenting the lesson learned.

The limitations posed by the lack of an official PEFCR, also reflected in the lack of some application-specific parameters in Annex C, imply flexibility in the choice of parameters and datasets by the study developers, thus reducing the comparability of results of studies carried out by different developers. Moreover, the use of ecoinvent datasets instead of EF-compliant ones may be critical for some materials especially in terms of technological representativeness and of the definition of the point of substitution. However, the case study on wooden windows demonstrated that the implementation of the CFF with ecoinvent datasets is anyway feasible and can ensure a sufficient level of detail and accuracy in the modelling. Despite the limitations, the CFF has proven to be a valid tool for valorising the efforts made by producers and designers to increasingly reduce the quantity of virgin resources used in construction products and organise EoL scenarios that minimise the use of landfill and incineration, favouring reuse and recycling, in line with European objectives. As is the intention of the European Commission, it is necessary that the CFF and the guidelines on its use evolve with the context they have to analyse, from a legislative, technical and market point of view.

For example, for the wooden window and door sector, the growing recycling of materials such as wood and glass can be captured by the formula with higher values to be assigned to R1 while the higher recycled content, for example for glass, will be reflected by higher R2 values.

In conclusion, the CFF can support the harmonisation of the analysis of EoL scenarios within environmental impact assessments, in combination with the LCA-based assessment systems already existing and widespread on the European market.

Acknowledgements The authors would also like to thank Stefano Mora (Legno Legno) and Patrizio Pesenti (PM Serramenti) for providing technical feedback on improving the work.

All the Authors contributed synergistically to the drafting of the article; however, the coordination for its drafting, the abstract, introductory paragraph are to be attributed to E. Palumbo; the paragraph 2 'Circular Footprint Formula' is to be attributed to V. Castellani; the paragraph 3 'Application of the CFF to a wooden window' to S. Cortesi; the paragraph 'Conclusion and future trends' to E. Palumbo and S. Cortesi.

Funding The study presented in this chapter was part of the activities of the LIFE MAGIS project (LIFE18 GIE/IT/000735). The LIFE MAGIS project has received funding from the LIFE Programme of the European Union.

The LCA survey and publication were financed by the Department of Department of Engineering and Applied Sciences of the University of Bergamo, through the project entitled 'Multicriteria Evaluation of Building Techniques and Systems for the Definition of Environmental Sustainability Strategies of Buildings'—Scientific Responsible Dr. Arch. E. Palumbo.

References

Allacker K, Mathieux F, Pennington D, Pant R (2017) The search for an appropriate end-of-life formula for the purpose of the European Commission environmental footprint initiative. Int J Life Cycle Assess 22(9):1441–1458. https://doi.org/10.1007/s11367-016-1244-0

Blengini GA, Di Carlo T (2010) The changing role of life cycle phases, subsystems and materials in the LCA of low energy buildings. Energy Build 42(6):869–880. https://doi.org/10.1016/j.enbuild.2009.12.009

Boje C, Mack N (2024) Combining building information model and life cycle assessment for defining circular economy strategies. Sustainability 16(11):4561. https://doi.org/10.3390/su16114561

Cabeza LF, Rincón L, Vilariño V, Pérez G, Castell A (2014) Life cycle assessment (LCA) and life cycle energy analysis (LCEA) of buildings and the building sector: a review. Renew Sustain Energy Rev 29:394–416. https://doi.org/10.1016/j.rser.2012.03.021

De Wolf C, Hoxha E, Fivet C (2020) Comparison of environmental assessment methods when reusing building components: a case study. Sustain Cities Soc 61:102322. https://doi.org/10.1016/j.scs.2020.102322

Directive 2024/1275/EC. Energy performance of buildings directive (recast)

Dixit MK, Fernández-Solís JL, Lavy S, Culp CH (2012) Need for an embodied energy measurement protocol for buildings: a review paper. Renew Sustain Energy Rev 16(6):3730–3743. https://doi.org/10.1016/j.enbuild.2009.12.009

Durão V, Silvestre JD, Mateus R, de Brito J (2020) Assessment and communication of the environmental performance of construction products in Europe: comparison between PEF and EN 15804 compliant EPD schemes. J Clean Prod 265:121580. https://doi.org/10.1016/j.jclepro.2020.121580

Eberhardt LCM, van Stijn A, Nygaard Rasmussen F, Birkved M, Birgisdottir H (2020) Development of a life cycle assessment allocation approach for circular economy in the built environment. Sustainability 12(22):9579. https://doi.org/10.3390/su12229579

European Committee for Standardization (CEN) (2019) EN 15804:2012+A2:2019. Sustainability of construction works—environmental product declarations—core rules for the product category of construction products. CEN, Brussels

European Commission (2020) A new circular economy action plan for a cleaner and more competitive Europe. COM(2020) 98 final. Publications Office of the European Union. https://doi.org/10.2779/085517. Available at: https://op.europa.eu/en/publication-detail/-/publication/9dc6aa01-39d2-11eb-b27b-01aa75ed71a1

Obrecht W, Röck M, Passer A (2021) End-of-life modelling of buildings to support more informed decisions towards achieving circular economy targets. Int J Life Cycle Assess 26:1830–1843. https://doi.org/10.1007/s11367-021-01906-4

Passer A, Lasvaux S, Allacker K, De Lathauwer D, Spirinckx C, Wittstock B, Kellenberger D, Gschösser F, Wall J, Wallbaum H (2020) Environmental product declarations entering the building sector: critical reflections based on 5–10 years experience in different European countries. Build Environ 182:107079. https://doi.org/10.1016/j.buildenv.2020.107079

Pedersen HB, Remmen A (2022) Sustainability assessment in product comparisons: incorporating performance and quality in the functional unit. J Clean Prod 340:130507. https://doi.org/10.1016/j.jclepro.2022.130507

UNICMI (2023) Serramenti e facciate continue: Linee guida per la progettazione e la certificazione. Unione Nazionale delle Imprese di Calcestruzzo Preconfezionato e dei Prodotti in Cemento

Yang Y, Zheng B, Luk C, Yuen K, Chan A (2024) Towards a sustainable circular economy: Understanding the environmental credits and loads of reusing modular building components from a multi-use cycle perspective. Sustain Prod Consumption 46:543–558. https://doi.org/10.1016/j.spc.2024.02.027

Zampori A, Pant R (2019) Product environmental footprint (PEF) guide for the construction sector. European Commission, Joint Research Centre (JRC)

Energy Optimization in Supply Chains: Paving the Way for a Circular Economy

Aliaksandr Novikau and Sami R. M. Musallam

Abstract Energy is an indispensable component of supply chains underpinning production, use, and disposal of goods and services. However, current energy consumption patterns are often unsustainable, leading to negative externalities such as pollution and greenhouse gases emissions. The circular economy paradigm has emerged as a promising solution to address the unsustainable use of resources, including energy. A cornerstone of the circular economy is minimizing waste and maximizing productivity, including energy efficiency. This study delves into the role of energy within the circular economy framework. The paper contends that transitioning to a circular economy necessitates a fundamental shift in our perception of energy use, its significance within the supply chain, and the adoption of innovative technologies and practices. Optimizing energy consumption within the supply chain can effectively reduce fossil fuel utilization and greenhouse gas emissions, lower operational costs, and enhance the efficiency of production processes. Embracing renewable energy sources can contribute to the decarbonization of energy supply and decrease the reliance on fossil fuels. Furthermore, integrating digital technologies, advanced analytics, and energy audits can play a crucial role in optimizing energy use and identifying energy-saving opportunities.

Keywords Energy · Energy efficiency · Supply chain · Energy management · Circular economy · Waste-to-energy

A. Novikau (✉)
Department of Political Science and International Relations, International University of Sarajevo, Ilidža, Bosnia and Herzegovina
e-mail: novikau@gmail.com

S. R. M. Musallam
Department of Economics and Management, International University of Sarajevo, Ilidža, Bosnia and Herzegovina

L. Brandli et al. (eds.), *The Contribution of Life Cycle Analyses and Circular Economy to the Sustainable Development Goals*, World Sustainability Series,
https://doi.org/10.1007/978-3-031-85300-5_7

1 Introduction

Material goods and services form the bedrock of human progress, and energy, the fundamental element required for their production, is critically essential for human development. Every stage of a supply chain—manufacturing, utilization, and final disposal—requires energy, often in significant quantities. The remarkable advances observed in manufacturing, transportation, thermal comfort, lighting, communication, and entertainment have been mainly driven by the exponential growth in energy consumption (Novikau 2022). Progress in energy use has happened in two major ways.

First, energy consumption has exhibited exponential growth. This phenomenon aligns with the broader trajectory of human civilization, which has been characterized by continuous development.

Second, energy use has also expanded in the number of energy sources used. The main surge in energy use that happened during the last two centuries has been caused by the transition from supply chains that relied on energy from firewood with occasional use of wind and hydropower to fossil fuels such as coal, oil, and, more recently, natural gas. New energy sources such as nuclear power were introduced in the last century. Renewable sources, such as wind, hydro, solar, and biofuel power, even known for centuries, have gained increased attention and widespread adoption in recent decades. Despite significant progress in renewables, the share of fossil fuels such as coal, oil, and natural gas in the global energy mix has remained around 80% for decades (IEA 2022).

Expansion in energy consumption has profoundly enhanced people's quality of life, providing access to a range of previously unavailable goods and services. However, this advancement has also brought to the forefront two of the most pressing challenges confronting the globe today—pollution and greenhouse gas emissions. Moreover, the excessive extraction of nonrenewable energy resources fueled by extensive energy consumption has become a pressing concern worldwide. Fossil fuel energy resources are finite, meaning that once they are depleted, they cannot be replenished. The current overdependence on nonrenewable energy sources presents a significant challenge to sustainability. Continued consumption at present rates raises concerns regarding resource depletion and potential scarcity for future generations. Unlike finite fossil fuels, renewable energy sources offer a significant advantage for future generations—accessibility. Wind, sun, geothermal heat, and tides are constantly replenished, ensuring a more sustainable and reliable energy source for the long term. However, harnessing this potential requires overcoming the hurdle of upfront costs. While renewable technologies are becoming more affordable, the initial installation of solar panels, wind turbines, or small hydropower systems can still be a barrier for low-income communities. To bridge this gap, innovative financing models, government subsidies, and community-ownership projects are crucial. By making renewable energy more accessible and affordable for all, we can ensure a future where clean energy benefits not only the environment but also uplifts communities by fostering local energy production, job creation, and energy independence.

Another solution to the issue of unsustainable resource utilization is the circular economy. This approach promotes the reuse and recycling of materials, which can help to conserve resources and minimize waste and pollution. The assumptions of the circular economy can be applied to various sectors and components of supply chains, including energy, which is the central focus of this paper. Acknowledging the critical role of energy in the supply chain and its substantial impact on human health and natural systems, this paper discusses how the circular economy principles can be implemented to improve energy management.

The paper is structured as follows. First, it offers a brief overview of the importance of energy in the supply chain and the negative externalities that energy use produces. After that, it briefly discusses how the principles of circular economy can be applied to energy management. Then, it examines the primary business and public policy strategies for energy management. The final section concludes with perspectives applying principles of circular economy in energy management.

2 Energy in the Supply Chain

Energy is a critical input for any production system, significantly impacting its productivity and competitiveness. Consequently, energy-related issues have become a central concern within business operations. Escalating energy costs and growing environmental issues surrounding energy use have garnered significant attention from businesses, governments, and academia.

Historically, energy use within supply chains has been framed predominantly through a logistics lens. This approach assumed that energy inputs would always be available and affordable. However, the numerous energy crises have demonstrated that supply chains are highly vulnerable to energy supply disruption and price volatility. Price volatility is a significant concern for energy system security as changes in energy prices generate serious business risks. Supply chain complexity significantly exacerbates risks caused by disruptions in energy supply or price surges (Mulhall and Bryson 2014).

The supply chain's vulnerability to energy shocks is particularly evident in industries that heavily rely on energy utilization. Within energy-intensive sectors, the expenses associated with energy can represent a substantial proportion, reaching up to approximately 15% of the added value and exceeding 20% for the sectors characterized by the highest energy intensity levels. Consequently, from a business standpoint, mitigating energy consumption within the supply chain has the potential to yield augmented profitability. Yet, notwithstanding the potential economic advantages, energy management often fails to receive sufficient consideration from corporations, even within energy-intensive sectors (Thollander and Ottosson 2010; Rudberg et al. 2013).

3 Externalities of Energy Use in Supply Chains

Although the transition to fossil fuels has allowed energy use more widely and intensively, it has also resulted in numerous negative externalities. Pollution is the most evident byproduct of energy use because fossil fuel combustion unleashes the chemicals stored in the fuel into the environment. As a result, air pollution, one of the most critical factors affecting human health, is mainly energy-related (Novikau 2022).

Furthermore, the combustion of fossil fuels leads to carbon dioxide and other greenhouse gases into the atmosphere, precipitating anthropogenic global climate change—a formidable peril to the environment and, ultimately, to human well-being and even existence. In 2021, there was a substantial increase in energy-related CO_2 emissions, resulting in a record-breaking annual rise of 36.6 Gt (IEA 2022).

The problems associated with fossil fuels and partially nuclear power and renewables are mainly related to the linear economic approach to their use. In the linear economy, resources are extracted from nature, turned into products, used for some time, and then disposed of as waste back to the environment. This approach is often characterized as a "take-make-dispose" model that leads to the depletion of natural resources, pollution, and waste accumulation.

Renewable energy sources are typically considered to have a closed energy cycle because they rely on energy from natural resources that are replenished naturally and do not need to be mined or extracted from underground geological formations like fossil fuels. For example, solar panels generate solar energy by converting sunlight into electricity. The solar energy is continuously replenished, making it renewable. Similarly, wind energy is generated by harnessing the power of the wind or water, which is also constantly filled through natural weather patterns.

In nuclear energy, there are two different ways to handle the nuclear waste generated by nuclear reactors. In an open fuel cycle, after the fission process, the byproduct is considered waste and usually deposited in geological long-term storage. The disposal and storage of nuclear fuel produced in an open fuel cycle pose significant challenges due to potential health and environmental risks for thousands of years (Macfarlane and Ewing 2006). In a closed nuclear fuel cycle, spent fuel is reprocessed using advanced technologies to recover valuable materials that could be further used for energy production. A closed fuel cycle can potentially reduce the amount of nuclear waste produced, but it needs to be implemented everywhere due to security and cost concerns (Graaf and Sovacool 2020).

Although alternative energy sources can replace fossil fuels, their current share in the energy mix is still relatively modest because the existing energy systems are designed for the use of fossil fuels and the political power of the fossil fuel industry.

4 Energy and Circular Economy

The circular economy emerges as a compelling alternative to the prevailing linear economic model. It presents itself as a potential solution to the challenges associated with intensive energy consumption within supply chains and the negative externalities arising from resource utilization (Kirchherr et al. 2017). The ideas of circular economy can be traced back to the 1960s, and in its essence, it pertains to the restoration of goods and materials using numerous strategies. These "R" strategies can include (van Buren et al. 2016; Fonseca et al. 2018):

(1) Refuse (avoidance of unnecessary use of materials or products by reducing their production and consumption),
(2) Reduce (minimization of resource use),
(3) Reuse (finding new uses for products or materials that would otherwise be discarded),
(4) Repair (fixing broken or damaged products instead of throwing them away),
(5) Refurbish (restoring products to their original condition, making them functional again),
(6) Remanufacture (creating new products from old products or their parts),
(7) Repurpose (finding new uses for materials that would otherwise be discarded),
(8) Recycle (breaking down materials and turning them into new products),
(9) Recover (extracting energy from waste that cannot be recycled or reused).

As such, a fundamental precept of the circular economy involves optimizing resource efficiency. By decreasing the consumption of primary materials, the circular economy simultaneously alleviates resource depletion and the accompanying environmental degradation risks while reducing energy consumption and pollution throughout the supply chain. Consequently, implementing circular economy principles and solutions holds the potential to attenuate energy consumption and pollution, all while sustaining economic growth. Not surprisingly, more and more countries have taken action to implement the ideas of the circular economy into practice (George et al. 2015). For instance, in 2015, the European Union launched an ambitious "Closing the Loop – An EU Action Plan for the Circular Economy" for the efficiency and environmental performance of energy-related products, bioenergy, and energy recovery from waste (European Commission 2015). In March 2020, the European Commission introduced the Circular Economy Action Plan as a cornerstone of the European Green Deal, an initiative to foster sustainable growth and development within the European Union. Embracing a circular economy framework is paramount to alleviating pressure on natural resources, facilitating the creation of sustainable jobs and economic growth, and achieving the European Union's 2050 climate neutrality goals. The action plan outlines a comprehensive set of initiatives encompassing all product life cycle phases. It strongly emphasizes product design, promotes circular economy processes, advocates for sustainable consumption, and endeavors to prevent waste and preserve resources within the EU economy for an extended period (European Commission 2020).

The idea of a circular economy can be seamlessly integrated into managing all resources employed within the supply chain, including energy. By effectively closing the loop between production and the end of a product's life cycle, valuable resources are continuously circulated, ensuring that useful energy remains within the economy for an extended period. Despite its immense potential to contribute significantly to achieving climate goals, the circular economy's role within the energy transition framework has garnered relatively limited attention among policymakers and business leaders (Fragkos 2022).

5 Business Strategies

The principles of the circular economy find application across three levels of energy management (Batista et al. 2018).

First, at the product level, the principle of circular economy is used in designing and manufacturing energy-efficient and environmentally sustainable products. It includes features that enhance energy efficiency, extend product lifespans, and make repair easier. Products can be created with modular, repairable, upgradable, and recyclable attributes, among others, to reduce energy consumption and promote sustainable energy management.

At the organizational level, circular economy principles advocate for restorative processes that minimize energy consumption and maximize resource utilization. These processes include energy-efficient methodologies, such as reducing energy use through process optimization, reusing energy, and implementing practices like repair, reconditioning, refurbishing, remanufacturing, and recycling. Such measures can yield substantial energy savings, reduced waste generation, and improved energy governance effectiveness.

Moreover, at the industrial level, circular economy principles can be operationalized through the sequential utilization of used materials and renewable resources among enterprises. It involves active participation in waste and byproduct synergy systems, resource and infrastructure sharing, as well as engagement in industrial symbiosis processes spanning diverse entities. By employing these methods, companies can reduce their environmental impact, enhance energy efficiency, and achieve greater energy sustainability.

Enhancing energy use within supply chains is possible through various means, yet firms often fail to implement such strategies fully. An initial step for companies is to adopt a comprehensive perspective toward their energy system. It includes centralizing energy planning to minimize the risk of sub-optimal energy system management and mitigate the likelihood of counterproductive initiatives being implemented across various locations. Applying measures to increase energy utilization at the corporate level is likely to yield more significant dividends than those arising from local initiatives (Rudberg et al. 2013).

There are two main strategies to improve energy use within supply chains. The first one focuses on supply-side management, such as investment in new energy generation facilities. The second one focuses on demand, specifically energy management (Thollander and Ottosson 2010). Demand-side management techniques such as energy-efficient products and processes, responsive energy consumption, and alternative energy technologies have received particular attention. These techniques can mitigate risks to supply chain stability and even generate additional revenues (Mulhall and Bryson 2014).

These practices may include an initial energy audit, support from senior management, continuous energy use monitoring, development of an energy policy, implementation of programs for energy-saving projects, and staff motivation and training (Thollander and Ottosson 2010).

5.1 Energy Efficiency

Energy efficiency is using energy resources to minimize energy waste and reduce energy consumption without compromising product or service quality. Through energy-efficient strategies, firms can balance energy conservation and output maximization, ensuring continuous production without compromising product or service quality.

In addition, energy efficiency measures have the potential to offer a range of non-energy benefits for a firm. These benefits can include increased other materials efficiency, an increase in the value of assets, better working conditions, improved product quality and increased productivity, improved organizational performance, better working conditions, reduced cost of regulatory compliance, contribution to greenhouse emissions reductions, and better public image (Kluczek and Olszewski 2017; Schubert et al. 2021).

Several barriers impede the implementation of energy efficiency measures. Access to capital is a prominent hindrance, as investments are frequently necessary for the dissemination of technological advancements in energy efficiency. Energy efficiency initiatives often fall short regarding investment and technology prioritization, exacerbating the challenge of securing the requisite funds for these projects. Bounded rationality, or the limitations on our capacity for entirely rational decision-making due to time constraints and a lack of readily accessible information, constitutes another impediment to the diffusion of energy-efficient technology. Investment decisions often hinge on inaccurate or incomplete information, leading to a reliance on traditional estimates and heuristics rather than rigorous data (Backlund et al. 2012).

However, a robust contention asserts that the concentration on energy efficiency is inherently flawed, as it does not culminate in an absolute diminution of total energy consumption at the national and global scales. Stated differently, optimizing energy efficiency at the micro level does not invariably translate into energy conservation at the macro level. While alterations in consumer conduct and lifestyle represent direct remedies for energy utilization predicaments, societal inclination typically leans towards technical or economic resolutions, some of which have been delineated previously. Consequently, the exclusive pursuit of energy efficiency proves insufficient in effectively tackling the difficulty of diminishing CO_2 emissions unless complemented by measures restricting consumption through regulatory or fiscal means. Thus, the paramount focus should be transitioning to non-fossil fuels, a more efficacious technical remedy than energy efficiency. The accentuation ought to center on curbing carbon dioxide emissions rather than curtailing energy usage, with the overarching objective of disentangling the trajectory of energy growth from CO_2 emissions. Adopting carbon–neutral renewables facilitates sustained energy expansion while concurrently curbing carbon emissions. Nevertheless, governments should advocate for energy efficiency grounded in economic and social welfare considerations. Eco-taxation mechanisms, such as a carbon tax, could be judiciously deployed to incentivize the integration of non-fossil fuels (Herring 2006).

5.2 *Extended Lifespan*

This approach endeavors to prolong the duration of a commodity and avert its disposal. It is predominantly associated with stages of product design, utilization, and upkeep. Various measures can be implemented, including the preservation, refurbishment, and reorganization of equipment to counteract the harmful effects of degradation and accommodate alterations in procedural parameters. It involves retrofitting or replacing outdated machinery with contemporary technologies, mitigating heat loss and energy wastage through adept heat management or insulation, harnessing spent heat, and refining process control to enhance both energy and material efficiency, as well as process productivity.

It is noteworthy that all circular economy strategies inherently require energy consumption. Remanufacturing, refurbishing, and repairing various products exemplify the implementation of circular economy principles with respect to energy use. These practices aim to conserve embodied energy by restoring old, worn-out, or malfunctioning products to a near-new condition or at least a functional state, thereby diverting them from disposal and the associated loss of the energy invested in their initial production. Remanufacturing, refurbishment, repair, and recycling require energy as any other industrial process. However, the total energy needed for these processes is usually a fraction of the energy required to manufacture new items from raw materials. For instance, recycling one ton of polyester garments requires 1.8% of the energy needed to manufacture an equivalent quantity of such items from primary materials. In contrast, recycling one ton of cotton clothing necessitates only 2.6% of the energy required to produce new cotton garments (Woolridge et al. 2006).

Concerns have arisen regarding the benefits of repairing or refurbishing old energy-inefficient devices, as these practices could extend their lifespan and delay their replacement with modern energy-efficient technologies. That, in turn, could lead to increased energy consumption over the long term. Several studies have demonstrated that remanufacturing typically saves energy over the life cycle of products that do not consume significant energy during use. However, for products with substantial energy requirements during use, such as washing machines, refrigerators, or air conditioners, especially those with significant generational improvements in energy efficiency, remanufacturing usually fails to achieve energy conservation (Kim et al. 2006; Gutowski et al. 2011; De Kleine et al. 2011). Hence, there is a dilemma regarding the optimal course of action, whether to persist with a given product or transition towards modern technologies possessing enhanced energy efficiency (van Loon et al. 2021).

Modular infrastructure is also a helpful strategy as it enables the circular economy by allowing the reuse and repurposing of functional modules. The strategy can help to address challenges related to energy transition and adaptation to future changes in energy demand. By enabling the reconfiguration and adaptation of energy systems, modular infrastructure can facilitate the integration of new technologies and energy sources and improve the resilience of energy systems. Furthermore, this approach can provide opportunities for innovative business models and new forms of collaboration between stakeholders (Mignacca et al. 2020).

5.3 *Digitalization*

The potential for digital technologies and advanced analytics to optimize energy use and identify energy-saving opportunities in the supply chain. Compelling empirical evidence suggests that digitalization can significantly enhance the efficiency of energy systems while simultaneously promoting energy conservation by end-users. Information technologies facilitate faster and more accurate analysis of power flows, enabling energy operators to balance loads more effectively. Moreover, digitalization permits the remote monitoring of distributed energy generation while facilitating the prediction of energy demand through an analysis of historical energy consumption. Furthermore, digitalization can enhance the energy efficiency of end users by providing real-time feedback on their energy consumption via smart meters. Dynamic tariffs, also known as real-time pricing, may also aid in shaping consumption patterns during peak hours, although some users may find them challenging to comprehend.

Additionally, information technologies enable users to compare their energy consumption with that of other households, potentially motivating them to conserve energy more effectively through social norms. Integrating digital thermostats, smart start features, and other digital functions in household appliances can further promote energy conservation. It has been estimated that this could yield energy savings of between 5 and 15% (Pearson 2011; Goldbach et al. 2018). At the same time, the digitalization of energy presents challenges in enhancing the accessibility of energy services. Underdeveloped countries and regions lacking advanced infrastructure and connectivity can face disadvantages in this regard.

5.4 Energy Recovery

The energy recovery strategic approach endeavors to forestall the disposal of resources throughout the product life cycle, with particular attention to increasing its lifespan. At the end of the product life cycle, the overarching objective is to systematically substitute disposal and landfilling practices with a systematic implementation of reuse and recycling. At this juncture, the cultivation of synergies and complementarities among companies and sectors assumes paramount significance. Applying industrial strategies can effectively address the complementary requirements of companies, aligning and harmonizing their production processes. Consequently, byproducts, waste materials, and emissions from one production process are repurposed as inputs for another.

Recovering energy from waste is often viewed as a most straightforward application of the ideas of circular economy to energy management (Fernando et al. 2022). For instance, the European Commission identifies three possible directions for energy-effective waste management (European Commission 2011).

- Reduce energy intensity of waste treatment,
- Increase the use of biodegradable waste for bioenergy and bioproducts,
- Ensure energy recovery of nonrecyclable waste.

Although ideal waste management mainly focuses on waste prevention, recycling, and material recovery, part of the waste for material recovery cannot be effectively applied. In this case, energy recovery can be viewed as a better solution than just the simple disposal of waste. A typical way for energy recovery of nonrecyclable waste is a thermal treatment that includes combustion, gasification, and pyrolysis of waste (Pan et al. 2015). The positive externalities of thermal waste treatment usually reduce the volume of nonrecyclable waste and decay prevention (Lombardi et al. 2015). The negative externalities of energy recovery from waste are similar to energy generation from fossil fuels, including pollution and greenhouse gas emissions.

The strategy for the utilization of energy derived from agricultural waste includes the transformation of organic residues from farms, encompassing crop residue, animal byproducts, and food waste, into energy through diverse processes such as anaerobic digestion, thermal conversion, and biorefining. Consequently, agricultural biomass waste holds the potential for conversion into clean energy, fostering the sustenance of business operations in developing nations and concurrently yielding supplementary income.

There are several energy challenges associated with the strategy. First, some conversion technologies may be expensive or not yet fully developed, making it challenging to scale up the process for widespread adoption. Second, collecting and transporting waste to processing facilities can be logistically challenging and may require significant investment in infrastructure. Third, depending on the region, there can be competition for agricultural waste as a feedstock for energy production and other uses, such as livestock feed or compost. Fourth, the economic viability of the energy from the agricultural waste strategy may depend on the availability of incentives, subsidies, and favorable policy environments to support the development of the industry. In addition, environmental challenges are associated with the energy from agricultural waste. Waste-to-energy was initially discussed mainly regarding negative externalities created by waste treatment facilities and how the risks were unjustly distributed among the population because such facilities were often located in poor communities (Malinauskaite and Jouhara 2019).

While converting agricultural waste into energy can help reduce greenhouse gas emissions, some conversion processes may still produce greenhouse gas emissions, primarily if fossil fuels are used. For instance, conversion technologies used to process agricultural waste, such as thermal conversion, can produce air pollutants and thus negatively affect human and planetary health. Disposing of waste byproducts from conversion technologies can also pose a risk to the environment if not appropriately managed (Fernando et al. 2022).

5.5 *Energy and Additional Revenue*

Applying circular economy principles to energy management can yield positive externalities within a company's supply chain. In other words, energy can become a byproduct of the primary business process. It can reduce dependence on external energy supply and potential revenue generation from selling energy surplus to other firms and communities. However, in practice, opportunities for additional revenue generation for companies not focused on energy generation are often limited. This limitation is usually due to the location of energy generation facilities, particularly their proximity to energy consumers. Nonetheless, in several cases, redesigning the supply chain to sell surplus energy on the market may be profitable, mainly if government incentives exist. However, this strategy involves risks related to high investment costs, fluctuating energy prices, and potential competition from energy consumers who may also adopt circular economy principles in their supply chains (Rudberg et al. 2013).

5.6 *Energy Audits*

The absence of comprehensive information regarding the costs and energy-saving advantages associated with industrial processes is identified as an impediment to implementing energy-efficiency measures within industrial settings, especially in small- and medium-sized companies where energy consumption has traditionally not been perceived as a substantial cost factor within the supply chain (Alhourani and Saxena 2009). Consequently, energy audits have been conducted proactively to surmount this efficiency gap (Kluczek and Olszewski 2017).

Numerous enterprises have embraced the practice of conducting energy audits to gain valuable insights into the energy consumption patterns across their entire supply chain, encompassing individual production processes and specific pieces of equipment. They were used in many production processes such as glass manufacturing (Li et al. 2010), food processing (Olayinka and Oladele 2013), plastic molding (Kluczek and Olszewski 2017), pharmaceutical industry (Bruni et al. 2023), and compressed air systems (Kluczek and Olszewski 2017; Uyan 2023).

The outcomes of these audits are contingent upon the precise definition of system boundaries, revealing many opportunities for enhancing industrial energy efficiency within the context of a circular economy. Energy audits effectively overcome obstacles such as time constraints, informational gaps, and uncertainties associated with the costs or risks to energy processes.

However, the heterogeneous nature of energy processes and the ensuing array of auditing methodologies pose challenges to the standardization of energy audits. Consequently, the efficacy of energy audits is contingent upon the proficiency of the auditor, rendering it a notably subjective undertaking. A compelling demand exists for all-encompassing and adaptable energy audit tools capable of accommodating divergent sectors, scales, and organizational structures. Pragmatic energy audits necessitate meticulous energy consulting services that articulate compelling cases for energy-saving investments. These audits can assist business decision-makers in assimilating energy-related advantages into their strategic deliberations. Moreover, energy audits facilitate the implementation of energy efficiency measures in diverse settings, refining energy management protocols and assessment methodologies. The reports derived from energy audits extend tangible benefits to enterprises with limited acumen in energy management, particularly during the planning or inaugural phases of implementation (Thollander et al. 2005; Kluczek and Olszewski 2017; Schubert et al. 2021).

5.7 Energy Modeling

Large-scale energy-economy models play a critical role in assessing the repercussions of climate policies on both energy and the economy. These models are frequently cited as essential tools as they provide quantitative analyses for consequential impact assessments. This is exemplified by benchmark studies conducted by the Intergovernmental Panel on Climate Change, which rely heavily on these models. Additionally, these models find application at the national level, where they are instrumental in evaluating the ramifications of climate policies. Energy system models assess the technical feasibility and associated costs of integrating renewable energy sources into the supply chain (Fragkos and Kouvaritakis 2018). Energy modeling has given comparatively less attention to the industry sector than other sectors such as oil and gas extraction, power generation, transport, or building. Hence, it is essential to comprehend and forecast alterations in industrial energy consumption with reasonable accuracy, which is a crucial task (Kesicki and Yanagisawa 2015).

6 Conclusion

The circular economy represents a promising approach to address the pressing issues of sustainable development, such as the depletion of natural resources and the environmental impact caused by energy use. Implementing principles of the circular economy into energy systems requires changes in the entire energy supply chain—from the extraction of energy resources to energy generation and consumption. It can change both business and consumer behavior with the active involvement of all stakeholders. A multitude of technological, financial, institutional, and regulatory barriers hinder the implementation of circular economy principles in energy management. Yet, in practice, the boundaries between them are blurred. Governments can play a crucial role in such a transformation by introducing public policies to change business and consumer behavior with a focus on circular economy practices. Governmental policies, either command and control or economic measures, can play a crucial role in alleviating these barriers and facilitating the transition to a circular economy.

Energy-efficient technologies and renewable energy sources are the essence of the circular economy principles for energy systems. On the one hand, they allow for a reduction in energy use, decreasing energy consumption and pollution. On the other hand, renewables, by definition, comply with the principles of the circular economy. Therefore, further investment in research and development of energy-efficient and renewable energy technologies is crucial for bringing the circular economy into energy systems.

At the same time, although the principles of the circular economy are promising, they should not be viewed as a universal solution for all the problems caused by the current linear "take-make-dispose" economic model. Circular economy solutions, such as energy generation from waste, can cause air pollution that potentially nullifies the advantages of such an approach. Therefore, the principles of the circular economy should not be blindly adopted and implemented into energy systems but rather scrutinized. Thus, it should be recognized that energy systems built on the circular economy approaches may not always be the optimal choice. Different sectors of the economy and different geographical regions may face unique energy needs and challenges that can affect how circular economy strategies are applied to energy systems.

References

Alhourani F, Saxena U (2009) Factors affecting the implementation rates of energy and productivity recommendations in small and medium sized companies. J Manuf Syst 28:41–45. https://doi.org/10.1016/j.jmsy.2009.04.001

Backlund S, Thollander P, Palm J, Ottosson M (2012) Extending the energy efficiency gap. Energy Policy 51:392–396. https://doi.org/10.1016/j.enpol.2012.08.042

Batista L, Bourlakis M, Smart P, Maull R (2018) In search of a circular supply chain archetype—a content-analysis-based literature review. Prod Plan Control 29:438–451. https://doi.org/10.1080/09537287.2017.1343502

Bruni G, Martini C, Martini F, Salvio M (2023) On the energy performance and energy saving potential of the pharmaceutical industry: a study based on the Italian energy audits. Processes 11:1114. https://doi.org/10.3390/pr11041114

De Kleine RD, Keoleian GA, Kelly JC (2011) Optimal replacement of residential air conditioning equipment to minimize energy, greenhouse gas emissions, and consumer cost in the US. Energy Policy 39:3144–3153. https://doi.org/10.1016/j.enpol.2011.02.065

European Commission (2011) Roadmap to a resource efficient Europe

European Commission (2015) Closing the loop—an EU action plan for the circular economy

European Commission (2020) A new circular economy action plan: For a cleaner and more competitive Europe

Fernando Y, Tseng M-L, Aziz N et al (2022) Waste-to-energy supply chain management on circular economy capability: an empirical study. Sustain Prod Consum 31:26–38. https://doi.org/10.1016/j.spc.2022.01.032

Fonseca L, Domingues J, Pereira M et al (2018) Assessment of circular economy within Portuguese organizations. Sustainability 10:2521. https://doi.org/10.3390/su10072521

Fragkos P (2022) Analysing the systemic implications of energy efficiency and circular economy strategies in the decarbonisation context. AIMS Energy 10:191–218. https://doi.org/10.3934/energy.2022011

Fragkos P, Kouvaritakis N (2018) Model-based analysis of intended nationally determined contributions and 2 °C pathways for major economies. Energy 160:965–978. https://doi.org/10.1016/j.energy.2018.07.030

George DAR, Lin BC, Chen Y (2015) A circular economy model of economic growth. Environ Model Softw 73:60–63. https://doi.org/10.1016/j.envsoft.2015.06.014

Goldbach K, Rotaru AM, Reichert S et al (2018) Which digital energy services improve energy efficiency? A multi-criteria investigation with European experts. Energy Policy 115:239–248. https://doi.org/10.1016/j.enpol.2017.12.036

Gutowski TG, Sahni S, Boustani A, Graves SC (2011) Remanufacturing and energy savings. Environ Sci Technol 45:4540–4547. https://doi.org/10.1021/es102598b

Herring H (2006) Energy efficiency—a critical view. Energy 31:10–20. https://doi.org/10.1016/j.energy.2004.04.055

IEA (International Energy Agency) (2022) World Energy Outlook 2022. IEA Publications

Kesicki F, Yanagisawa A (2015) Modelling the potential for industrial energy efficiency in IEA's world energy outlook. Energy Effic 8:155–169. https://doi.org/10.1007/s12053-014-9273-7

Kim HC, Keoleian GA, Horie YA (2006) Optimal household refrigerator replacement policy for life cycle energy, greenhouse gas emissions, and cost. Energy Policy 34:2310–2323. https://doi.org/10.1016/j.enpol.2005.04.004

Kirchherr J, Reike D, Hekkert M (2017) Conceptualizing the circular economy: an analysis of 114 definitions. Resour Conserv Recycl 127:221–232. https://doi.org/10.1016/j.resconrec.2017.09.005

Kluczek A, Olszewski P (2017) Energy audits in industrial processes. J Clean Prod 142:3437–3453. https://doi.org/10.1016/j.jclepro.2016.10.123

Li Y, Li J, Qiu Q, Xu Y (2010) Energy auditing and energy conservation potential for glass works. Appl Energy 87:2438–2446. https://doi.org/10.1016/j.apenergy.2010.02.029

Lombardi L, Carnevale E, Corti A (2015) A review of technologies and performances of thermal treatment systems for energy recovery from waste. Waste Manag 37:26–44. https://doi.org/10.1016/j.wasman.2014.11.010

Macfarlane A, Ewing RC (eds) (2006) Uncertainty underground: Yucca Mountain and the nation's high-level nuclear waste. MIT Press, Cambridge, Mass

Malinauskaite J, Jouhara H (2019) The trilemma of waste-to-energy: a multi-purpose solution. Energy Policy 129:636–645. https://doi.org/10.1016/j.enpol.2019.02.029

Mignacca B, Locatelli G, Velenturf A (2020) Modularisation as enabler of circular economy in energy infrastructure. Energy Policy 139:111371. https://doi.org/10.1016/j.enpol.2020.111371

Mulhall RA, Bryson JR (2014) Energy price risk and the sustainability of demand side supply chains. Appl Energy 123:327–334. https://doi.org/10.1016/j.apenergy.2014.01.018

Novikau A (2022) Energy, and human and planetary health: is the mutual link inevitable? In: Leal Filho W (ed) Handbook of human and planetary health. Springer International Publishing, Cham, pp 101–112

Olayinka OS, Oladele AT (2013) Energy audit of manufacturing and processingindustries in Nigeria: A case study of food processing industry and distillation & bottling company. Am J Energy Res 1:36–44. https://doi.org/10.12691/ajer-1-3-1

Pan S-Y, Du MA, Huang I-T et al (2015) Strategies on implementation of waste-to-energy (WTE) supply chain for circular economy system: a review. J Clean Prod 108:409–421. https://doi.org/10.1016/j.jclepro.2015.06.124

Pearson ILG (2011) Smart grid cyber security for Europe. Energy Policy 39:5211–5218. https://doi.org/10.1016/j.enpol.2011.05.043

Rudberg M, Waldemarsson M, Lidestam H (2013) Strategic perspectives on energy management: a case study in the process industry. Appl Energy 104:487–496. https://doi.org/10.1016/j.apenergy.2012.11.027

Schubert T, Breitschopf B, Plötz P (2021) Energy efficiency and the direct and indirect effects of energy audits and implementation support programmes in Germany. Energy Policy 157:112486. https://doi.org/10.1016/j.enpol.2021.112486

Thollander P, Ottosson M (2010) Energy management practices in Swedish energy-intensive industries. J Clean Prod 18:1125–1133. https://doi.org/10.1016/j.jclepro.2010.04.011

Thollander P, Karlsson M, Söderström M, Creutz D (2005) Reducing industrial energy costs through energy-efficiency measures in a liberalized European electricity market: case study of a Swedish iron foundry. Appl Energy 81:115–126. https://doi.org/10.1016/j.apenergy.2004.07.006

Uyan E (2023) Energy saving potential analysis of a short cycling industrial air compressor in a marine equipment manufacturing plant in Türkiye. Clean Responsible Consum 11:100147. https://doi.org/10.1016/j.clrc.2023.100147

van Buren N, Demmers M, van der Heijden R, Witlox F (2016) Towards a circular economy: the role of Dutch logistics industries and governments. Sustainability 8:647. https://doi.org/10.3390/su8070647

van Loon P, Diener D, Harris S (2021) Circular products and business models and environmental impact reductions: current knowledge and knowledge gaps. J Clean Prod 288:125627. https://doi.org/10.1016/j.jclepro.2020.125627

van de Graaf T, Sovacool BK (2020) Global energy politics. Polity, Cambridge, UK; Medford, MA

Woolridge AC, Ward GD, Phillips PS et al (2006) Life cycle assessment for reuse/recycling of donated waste textiles compared to use of virgin material: an UK energy saving perspective. Resour Conserv Recycl 46:94–103. https://doi.org/10.1016/j.resconrec.2005.06.006

Exploring the Circular Economy Route: Creating Value for End-of-Life Photovoltaic Equipment Through a Reverse Stream

Júlia Possebon Spellmeier, Carmen Brum Rosa, Paula Donaduzzi Rigo, Mario Fernando De Mello, and Marcelo Hoss

Abstract The market for solar energy harnessed through photovoltaic (PV) technology has experienced unprecedented growth worldwide in recent years. Consequently, the production and installation of photovoltaic equipment have increased at a comparable rate. Over time, such equipment reaches its end of life (EoL). The sustainability of this renewable energy source is questioned when the issue of EoL PV is considered, given that the current linear production model results in waste from equipment rich in materials due to the absence of reverse logistics and the implementation of CE principles in the sector. In Brazil, PV equipment does not yet have specific regulations for proper disposal, and it is classified as "general waste." Substances like lead and cadmium, present in PV modules, pose significant risks to both the environment and human health when improperly disposed of at the end of their use. Therefore, discussing studies on this subject is crucial for developing societal knowledge and awareness regarding this problem. This topic necessitates further research and dialogue and is recognized as a segment with substantial potential for progress. Thus, the present study aims to structure a CE pathway by mapping the relationship between EoL PV and the main stakeholders involved in its value chain, proposing a solution for the disposal of end-of-life photovoltaic equipment.

J. P. Spellmeier (✉) · P. D. Rigo
Production Engineering, Innovation and Competitiveness Center (NIC), Federal University of Santa Maria, Santa Maria, Brazil
e-mail: julia.possebon@acad.ufsm.br

C. B. Rosa · M. F. De Mello
Production and Systems Engineering Department, Innovation and Competitiveness Center (NIC), Federal University of Santa Maria, Santa Maria, Brazil
e-mail: carmen.b.rosa@ufsm.br

M. F. De Mello
e-mail: mario.mello@ufsm.br

M. Hoss
Production and Systems Engineering Department, Federal University of Santa Maria, Santa Maria, Brazil
e-mail: marcelohoss@ufsm.br

L. Brandli et al. (eds.), *The Contribution of Life Cycle Analyses and Circular Economy to the Sustainable Development Goals*, World Sustainability Series,
https://doi.org/10.1007/978-3-031-85300-5_8

Through this mapping, a reverse logistic will be established to promote the CE in the PV sector, involving all stakeholders identified in the mapped value chain, in order to close the existing gap in this market and suggest a second life for PV equipment. Consequently, this renewable energy source will potentially achieve greater sustainability throughout its life cycle.

Keywords Solar energy · End of life · Value chain · Circular economy

1 Introduction

Energy is an indispensable commodity for human existence, spanning its myriad sources. Among these, electrical energy emerges as one of the most favorable forms of energy supply for society and has established itself as a strategic resource for the development of numerous countries and regions (Maka 2022). Economic growth is increasingly leading to the depletion of natural resources. Intrinsically linked to this growth are energy consumption and the imperative to seek sustainable solutions capable of mitigating the detrimental effects of this depletion (Krishnan et al. 2023). The use of renewable sources for electricity generation is necessitated to support future demand (Simpson and Clifton 2017).

Solar energy, captured through photovoltaic (PV) technology, stands as one of the most promising and viable renewable energy sources to supplant fossil fuels. Its unlimited nature and potential to reduce the use of thermoelectric plants, which emit greenhouse gases, contribute significantly to environmental preservation (Bao et al. 2020). Beyond environmental benefits, solar energy, when harnessed through PV technology, offers considerable returns on investment, on par with those of fossil fuels, and has the capacity to meet global electrical energy consumption demands. Due to its expansive land area and tropical location, Brazil is identified as one of the countries with the most significant potential for implementing PV systems (Rigo et al. 2022). The levels of global solar irradiation incident on its territory (1500–2500 kWh/m^2)—with the highest irradiation observed in the northern region of Bahia State (6.5 kWh/m^2)—underscore the substantial potential for electricity generation from solar resources (INPE 2017). In this context, Brazil stands out due to its exponential growth in PV technology, which, according to the National Electric Energy Agency, is utilized across various classes (residential, commercial, rural, industrial) and generation modalities (distributed and centralized) (ANEEL 2022a).

However, paralleling the ascent of PV technology in the nation is the increase in the production and installation of PV equipment, growing at a comparable rate (Mahmoudi et al. 2019). Consequently, the sustainability of this energy source is questioned, especially when the end of life (EoL) of PV systems is emphasized. In the prevailing linear production model, these are generally deemed general waste (Klepa et al. 2020). It is important to note that the second life of PV equipment presents unique opportunities for new economic avenues (Sica et al. 2018). Nevertheless, in

Brazil, these opportunities are still in the rudimentary stages of exploration, attributed to the linear production chain in which this equipment is integrated.

These preliminary discussions culminate in the research problem explored in this study: To ensure sustainability within the PV value chain, how can we facilitate the transition to a circular economy (CE) in the sector? Consequently, this study aims to structure the CE pathway by mapping the interactions between EoL PV and the principal stakeholders involved in its value chain. The justification for this study is twofold: first, to solidify the CE within the PV segment so that this energy source inches closer to achieving environmental equilibrium throughout its lifecycle by transitioning its production chain to a circular production model. Second, to address the environmental and economic challenges posed by increased energy consumption through the paradigm shift from "photovoltaic waste" to a resource with a second life.

2 Circular Economy Principles and the End of Life of Photovoltaic Equipment

Exploiting finite natural resources through an unsustainable consumption and production pattern contributes to the depletion of Earth's resources (Martins and Castro 2020). Climate change, soil degradation, and the significant loss of biodiversity are currently observable effects of human activity, with the potential to destabilize the ecosystems that support its development (Hazen et al. 2020).

The economic model favored over the last two centuries, known for its depletion and degradation, is referred to as "linear" (Urbinati et al. 2017). It is based on the exploitation of virgin materials provided by nature to manufacture products that are ultimately discarded after use (take, manufacture, waste) (Salvi et al. 2023). In response to the negative externalities caused by the linear economy, the concept of the CE was developed, aiming to replace the concept of waste with a restorative approach (Di et al. 2023). The CE concept represents one of the most promising approaches for organizing sustainable economic activity in the future (Hazen et al. 2020).

The End-of-Life Management report from the International Renewable Energy Agency highlighted that, by the end of 2016, there was an expectation of 43.5–250 tons of waste from PV modules, with projections estimating accumulations of up to 8 million tons by 2030 (IRENA 2016). These figures point to the significant volume of PV waste a country needs to manage. At their EoL, PV modules are classified by the European Community Directive as waste electrical and electronic equipment (Ndzibah et al. 2022). In Brazil, electronic waste, such as the PV module, lacks specific regulations, being considered as "general waste" (Cucchiella et al. 2015), leading to considerable ambiguity and a lack of specific regulations PV for accelerating the collection, recovery, and recycling of solar PV waste (Sharma et al.

2021). As a result, much of the solar waste is treated as general electronic waste and inappropriately disposed of in landfills, as observed in Brazil (Jain et al. 2022).

Two waste segments are generally related to EoL PV: primary and secondary. The primary waste consists of PV modules, whereas the secondary flow includes system components such as inverters, cables, aluminum structures, and sometimes batteries (Okoroigwe et al. 2020). Despite the expected lifespan of 20–30 years for PV modules, substantial amounts of premature waste are also generated in this sector due to equipment damage during transportation, installation, and operation (Majewski et al. 2021). The International Renewable Energy Agency estimated that up to 80% of the PV waste stream may consist of products that were defective or failed during production, transport, or in the initial years of operation (IRENA 2016). Therefore, to minimize this high waste flow generated by the current linear production process, the CE offers a viable alternative to redesign the production process to avoid such waste.

This waste stream presents environmental and human health concerns due to hazardous elements in a PV system, such as lead and cadmium (Chowdhury et al. 2020). Besides environmental challenges, shortages in the supply of critical metals such as silicon, germanium, and lithium, used in solar panel manufacturing, also pose a significant resource challenge for this sector.

The solar PV waste management cycle currently follows an open-loop path (Sharma et al. 2021). However, this sector offers an opportunity to explore the '3Rs' concept of sustainable waste management, namely Reduce, Reuse, and Recycle, to close the open loop and introduce circularity to the life cycle of solar PV equipment (Farrell et al. 2020). The high recyclability potential (60–90%) of solar PV materials underscores the possibilities for recycling PV waste, thereby closing the material loop at its EoL (Santos and Alonso-García 2018; Gautam et al. 2021).

The CE prioritizes maintaining products at their original value for as long as possible over recycling (Tsanakas et al. 2020). Nevertheless, repair and reuse are valid and even preferred options for transitioning to a CE for PV components that have not reached their technical end of use, thus fostering their second life (Opstal and Smeets 2023). The second life of PV equipment can provide a sustainable solution to the current waste stream generated by the linear production model (Sica et al. 2018). Therefore, developing a circular production model is crucial to support future demand for PV systems (Cunha 2023). According to Farrell et al. (2020), adopting CE principles can help mitigate environmental factors such as industrial emissions and extend the useful life of PV equipment, thus reducing its waste stream.

Circular economy solutions encompass rethinking product/service design to enable, for example, efficiency gains, reduction of materials and energy required for production, extension of service life, long-term maintenance and repair, and initiatives for reuse, refurbishment, and remanufacturing (Amato and Korhonen 2021). Developing waste management that facilitates new business models, design thinking, and a more productive approach to consumption and production, all embody concepts of the CE (Di et al. 2023). Through this lens, the idea of a CE advances as a means to mitigate the impacts of unidirectional practices, seeking a restorative economic system that is less harmful to the environment (Tognato et al. 2023). Thus, implementing CE principles is increasingly recommended to achieve the Sustainable Development Goals (Saidani et al. 2019). Structuring the CE in the PV segment, with consideration for the three pillars of sustainability—social, environmental, and economic—is essential to ensure desired outcomes regarding sustainability and environmental balance in the sector.

3 Methodology

The methodological procedure was conducted in stages to understand the main aspects covered and achieve the proposed objective, as shown in Fig. 1.

To complete the initial stage, which involved constructing the theoretical framework of this research and identifying the problem of the study, we utilized the CAPES Journal Portal. This portal facilitated access to articles and Technical Notes related to PV solar energy. Our goal was to establish a theoretical base to understand the context of solar systems installed in Brazil and comprehend operations of the EoL phase within the country. In the second stage, we conducted a systematic literature review focusing on mapping the relationship between the EoL phase of PV systems and its primary stakeholders in the value chain. The systematic review protocol entailed exploring scientific studies on EoL practices worldwide. The development of the research question marked the starting point of this review: What are the primary stakeholders involved in the EoL PV value chain? To address this question, we selected two databases: SCOPUS and Web of Science. Preliminary readings facilitated the selection of keywords, resulting in the search phrase: TITLE-ABS-KEY ((“Solar PV” OR “Photovoltaic”) AND (“End of Life” OR “End of Use”)). Our search strategy involved the use of metadata fields—“Title,” “Abstract” and “Keywords”—with a document-type restriction to articles only, coupled with a time filter limited to the past five years and publications in English. The search in both databases yielded

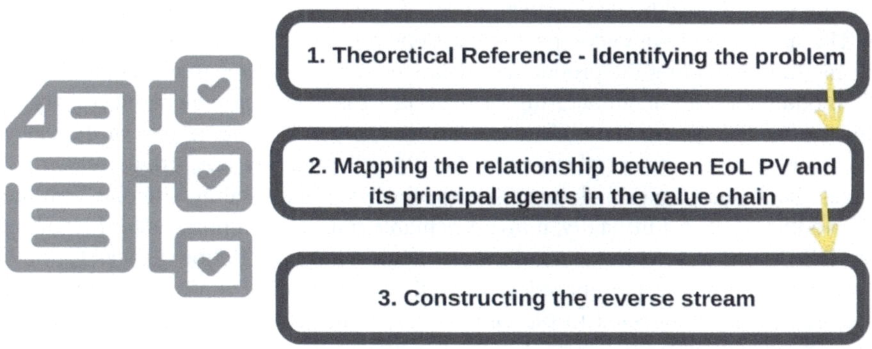

Fig. 1 Research steps undertaken in this study

210 articles, from which duplicates were removed. Subsequently, we applied a filter involving the review of titles and abstracts to identify articles that could offer insights into the EoL PV phase from the perspective of mapping the relationship between its main value chain stakeholders. This was done by answering the following questions: "Does the article address renewable/sustainable/clean energy or, more specifically, PV solar energy?" and "Does the article contribute to mapping the EoL PV value chain?" Articles that did not receive a "Yes" to these questions were excluded from the sample. This process resulted in 114 articles. A second filtering phase was then carried out, wherein articles from the initial filter were comprehensively read to trace the CE route in the EoL PV phase, ultimately resulting in 47 articles.

Afterward, we expanded our data collection to include gray literature to aid in constructing and proposing a reverse logistics stream in the PV segment. The research into gray literature involved sourcing information about the sector from studies by the research and consultancy firm Greener (2022), statistics from the Energy Research Company (EPE 2022), and data from the National Electric Energy Agency (ANEEL 2022b). Furthermore, examining online catalogs of PV installers and websites of PV recyclers' worldwide enabled an understanding of the current relationship dynamics between the EoL phase and its stakeholders in the value chain.

4 Results

Observing the interactions among various actors within the value chain is essential for determining how they generate and transfer knowledge and conduct business to enhance the adoption and dissemination of a segment (Juntunen and Hyysalo 2015). Mapping the value chain helps identify aspects that represent competitive advantages for the involved actors. In the study by Garlet et al. (2020), the PV value chain is broadly categorized into upstream, midstream, downstream, and auxiliary sections, encompassing all activities performed by different actors, from the production of materials required for the installation of the PV system to its delivery to the end consumers, followed by its deactivation and disposal at the end of its lifespan.

The upstream chain encompasses the steps preceding the manufacturing of the PV module, including the production of raw materials for the structural part of the PV module and the manufacturing of materials found in components necessary to balance the PV system. The midstream sector represents the core part of the value chain and includes the production of the PV cell, junction box, frame, electrical cables, back sheet, encapsulating film, special PV glass, sealing silicone, and the PV module itself. Additionally, it involves manufacturing components for balancing the PV system, such as inverters, fixing structures, and installation devices for both direct and alternating current sides, as well as the meter. The downstream sector is more directly connected to the end consumer and covers activities such as the integration of the PV system, which comprises project development, PV system installation, and grid connection. It also involves the operation and maintenance of the PV system, system deactivation, waste disposal, and material recycling. The auxiliary value chain identifies the essential activities that support the main structure and enable the interaction of companies, academic institutions, and the government to exchange technical knowledge and market information, thus allowing the PV sector to position itself competitively in the marketplace.

In this context, the EoL of PV is situated within the downstream sector. However, all preceding activities, such as those in the upstream and midstream sectors and subsequent activities in the auxiliary value chain, are related to the EoL of PV, as the dependency relationship between them mutually influences the interaction among the parties and actors. Figure 2 conceptualizes the chain as a cycle, where the EoL serves to benefit the upstream or other chains once again.

Importantly, the theoretical classification outlined above delineates a linear path within the PV segment. Conversely, the EoL of PV can alter this scenario by employing the concept of reverse logistics to close the current open- loop system and

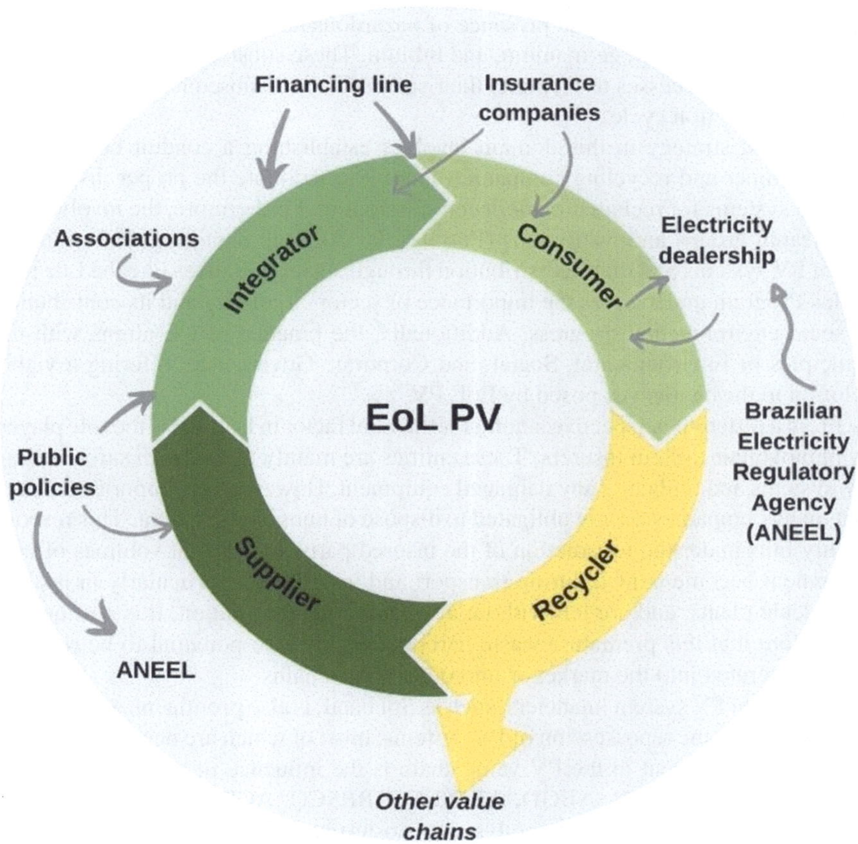

Fig. 2 Relationship between EoL PV and value chain stakeholders in a reverse stream

propose a circular approach for the sector. Figure 2 illustrates the concept of reverse logistics for the PV chain.

The stakeholders directly engaged with EoL PV systems include consumers and recycling companies responsible for repurposing the PV system. Consumers, positioned at the end of the value chain, play a pivotal role in determining the ultimate fate of PV systems. They maintain a direct relationship with integrating firms, which offer services such as installation and maintenance for PV systems. Recycling companies are the key connectors in ensuring the secondary use of PV equipment and elevating awareness among end-users. Conversely, integrator firms are intricately linked with the manufacturers and suppliers of the equipment constituting the PV system.

Photovoltaic recycling companies typically engage in recycling other forms of electronic waste or specialize exclusively in the PV sector. The recycling chain includes secondary components of the systems, such as inverters, cables, and mounting structures, due to their ease of reuse in electronic equipment and metal recycling. However, the primary components, like modules, necessitate distinct recycling methodologies because of the presence of hazardous and critical elements, such as lead, cadmium, silicon, germanium, and lithium. These substances demand specialized recycling processes to facilitate their separation and subsequent reintroduction into the production cycle.

A critical strategy in this domain involves establishing a conduit between the end consumer and recycling companies, aiming to facilitate the proper disposal of disused systems for reclamation or decommissioning. Furthermore, the involvement of research groups and institutions (Portal Solar 2022) in exploring the secondary life of PV systems and their redistribution through social initiatives like the Luz para Todos Program underscores the importance of sector circularity and its contribution to socio-environmental progress. Additionally, the practice of CE aligns with the principles of Environmental, Social, and Corporate Governance, offering a viable solution to the challenges posed by EoL PV.

From a different perspective, another influential factor in EoL PV is the role played by photovoltaic system insurers. These entities are mainly tasked with safeguarding PV systems and replacing any damaged equipment. However, it is important to note that many companies are not obligated to dispose of unused equipment. This responsibility falls under the jurisdiction of the insured party. Significant volumes of new equipment become tainted during transport and installation, particularly in insured large-scale plants, and are left without a suitable final destination. It is pertinent to underscore that this premature waste harbors considerable potential to be repaired and reintegrated into the market or into other value chains.

The role of PV system financiers, such as Sol Fácil, is also prominent, as consumer defaults lead to the repossession of PV systems, most of which are new and efficient. Another vital element in the PV value chain is the influence of solar energy associations like ABSOLAR, ABGD, ABINEE, ABESCO, AGSOLAR, and ABENS. These associations can act as catalysts for fostering forums and discussions with government bodies, particularly concerning the regulation of EoL PV.

In conclusion, the consolidation of a CE offers a regenerative system architecture that minimizes resource inputs and waste outputs by decelerating, narrowing, and closing material and energy cycles. This approach presents a solution capable of mitigating the environmental impacts associated with increasing PV waste. The proposition for a reverse stream could simplify this process, as enhancing information exchange among the diverse stakeholders in the PV value chain is essential for navigating the transition towards a CE industry.

5 Conclusions

In this study, the CE approach within the photovoltaic segment was investigated by mapping the relationship between end-of-life photovoltaic modules and their primary stakeholders in the value chain, contributing to the implementation of circular solutions in the market through the proposition of a reverse stream. Upon identifying the research problem, a systematic literature review was conducted to enhance theoretical knowledge about the PV value chain. Consequently, the study mapped the relationship between EoL PV modules and their principal stakeholders in the value chain, enabling an analysis of the connections between the parties involved in this segment. The theoretical classification of the PV value chain is broadly categorized into upstream, midstream, downstream, and auxiliary value chains. While this classification traditionally follows a linear path, end-of-life photovoltaic modules can alter this landscape, introducing circularity to the sector through the concept of reverse stream proposed in this study.

The study reaches its completion with the discussion on the mapping of actors involved in the photovoltaic value chain, their primary activities, and the demonstration of each actor's involvement in the reverse stream, proposing a shift from the current linear model of the photovoltaic sector to a circular approach. This proposed model could offer benefits to all stakeholders within the chain. Notably, end consumers are primarily responsible for initiating the reverse flow movement to ensure the sector's circularity, as they are identified as the initial link in contact with photovoltaic equipment at the end of its life cycle. Thus, their crucial role in fostering a CE within the PV market segment cannot be overstated. For consumers to take an active part and show interest in initiating the reverse route in the photovoltaic market, raising their awareness is undeniably essential.

It is important to emphasize that this study proposes a solution to the current linear production model in which the PV segment operates, enabling the repurposing of photovoltaic equipment within its value chain or across other value chains. However, it is crucial to acknowledge that the transition of the photovoltaic market to a circular production model should be undertaken with a comprehensive restructuring of the entire lifecycle of this production chain, from the conscious sourcing of raw materials to the design of circular products, incorporating the principles of the CE.

As a direction for future research, the goal is to conduct an opinion survey with the key stakeholders identified in the photovoltaic value chain, including consumers with operational generation systems, industry entrepreneurs, and researchers, using a data collection tool to identify potential solutions and understand their views on photovoltaic waste and their perspectives on implementing a reverse stream in the photovoltaic market.

References

Amato DD, Korhonen J (2021) Integrating the green economy, circular economy and bioeconomy in a strategic sustainability framework. Ecol Econ 188:107143. https://doi.org/10.1016/j.ecolecon.2021.107143

ANEEL (2022a) Capacidade de Geração Do Brasil. SIGA- Sistema de informações de Geração da Agência Nacional de Energia Elétrica. https://app.powerbi.com/view?r=eyJrIjoiNjc4OGYyYjQtYWM2ZC00YjllLWJlYmEtYzdkNTQ1MTc1NjM2IiwidCI6IjQwZDZmOWI4LWVjYTctNDZhMi05MmQ0LWVhNGU5YzAxNzBlMSIsImMiOjR9. 1 March 2022

ANEEL (2022b) Geração Distribuida. Agência Nacional de Energia Elétrica. https://app.powerbi.com/view?r=eyJrIjoiZjM4NjM0OWYtN2IwZS00YjViLTllMjItN2E5MzBkN2ZlMzVkIiwidCI6IjQwZDZmOWI4LWVjYTctNDZhMi05MmQ0LWVhNGU5YzAxNzBlMSIsImMiOjR9. 1 March 2022

Bao Q et al (2020) A human-centered design approach to evaluating factors in residential solar PV adoption: a survey of homeowners in California and Massachusetts. Renew Energy 151:503–513. https://www.scopus.com/inward/record.uri?eid=2-s2.0-85076215462&doi=10.1016%2Fj.renene.2019.11.047&partnerID=40&md5=03d3bc46a9d5aac1945f01de5caab59f

Chowdhury MdS et al (2020) An overview of solar photovoltaic panels' end-of-life material recycling. Energy Strateg Rev 27:100431

Cucchiella F, D'Adamo I, Rosa P (2015) End-of-life of used photovoltaic modules: a financial analysis. Renew Sustain Energy Rev 47:552–561. https://doi.org/10.1016/j.rser.2015.03.076

Cunha D (2023) Social sciences & humanities open photovoltaic electronic waste in Brazil: circular economy challenges, potential and obstacles, vol 7

Di A, Sohail V, Rosa H, Rohail P (2023) The transition towards circular economy and waste within accounting and accountability models: a systematic literature review and conceptual framework. Environ Dev Sustain 25(1):734–810. https://doi.org/10.1007/s10668-021-02078-5

EPE (2022) Empresa de Pesquisa Energética. https://www.epe.gov.br/pt

Farrell CC et al (2020) Technical challenges and opportunities in realising a circular economy for waste photovoltaic modules. Renew Sustain Energy Rev 128:109911

Garlet TB, Ribeiro JLD, de Souza Savian F, Siluk JCM (2020) Value chain in distributed generation of photovoltaic energy and factors for competitiveness: a systematic review. Solar Energy 211:396–411

Gautam A, Shankar R, Vrat P (2021) End-of-life solar photovoltaic e-waste assessment in India: a step towards a circular economy. Sustain Prod Consump 26:65–77. https://www.scopus.com/inward/record.uri?eid=2-s2.0-85091761849&doi=10.1016%2Fj.spc.2020.09.011&partnerID=40&md5=305e58cc576ef113c9620b262ca0729d

Greener (2022) Estudos. https://www.greener.com.br/estudos/. 24 June 2022

Hazen BT, Russo I, Confente I, Pellathy D (2020) Supply chain management for circular economy : conceptual framework and research agenda

INPE (2017)Atlas Brasileiro de Energia Solar 2 Ed., 80. http://urlib.net/rep/8JMKD3MGP3W34P/3PERDJE

IRENA (2016) End-of-life management: solar photovoltaic panels. International Renewable Energy Agency and the International Energy Agency Photovoltaic Power Systems

Jain S, Sharma T, Gupta AK (2022) End-of-life management of solar PV waste inIndia: situation analysis and proposed policy framework. Renew Sustain Energy Rev 153. https://www.scopus.com/inward/record.uri?eid=2-s2.0-85117687386&doi=10.1016%2Fj.rser.2021.111774&partnerID=40&md5=dc0a02b52d4a2695612cbcedd38a1af3

Juntunen JK, Hyysalo S (2015) Renewable micro-generation of heat and electricity—review on common and missing socio-technical configurations. Renew Sustain Energy Rev 49:857–870
Klepa RB et al (2020) Potencial De Resíduos Fotovoltaicos Reutilizados Através Da Logistica Reversa Verde (Lrv), p 84891
Krishnan N, Ravi Kumar K, Singh C (2023) How solar radiation forecasting impacts the utilization of solar energy : a critical review. J Clean Prod 388:135860. https://doi.org/10.1016/j.jclepro.2023.135860
Mahmoudi S et al (2019) End-of-life photovoltaic modules: a systematic quantitative literature review. Resour Conserv Recycl 146:1–16
Majewski P et al (2021) Recycling of solar PV panels-product stewardship and regulatory approaches. Energy Policy 149. https://www.scopus.com/inward/record.uri?eid=2-s2.0-85097458099&doi=10.1016%2Fj.enpol.2020.112062&partnerID=40&md5=506fc0c9ead1a300b9ebf5cc896e7ec3
Maka AOM (2022) Solar energy technology and its roles in sustainable development, pp 476–483
Martins FF, Castro H (2020) Raw material depletion and scenario assessment in European Union—a circular economy approach. Energy Rep 6:417–422. https://doi.org/10.1016/j.egyr.2019.08.082
Ndzibah E, Pinilla-De La Cruz GA, Shamsuzzoha A (2022) End of life analysis of solar photovoltaic panel: roadmap for developing economies. Int J Energy Sector Manage 16(1):112–128. https://www.scopus.com/inward/record.uri?eid=2-s2.0-85109808147&doi=10.1108%2FIJESM-11-2020-0005&partnerID=40&md5=83b55d38c6ce48de05b7e9eddcbd68ac
Okoroigwe FC et al (2020) Photovoltaic modules waste management: ethical issues for developing nations. Energy Technol 8(11):2000543
Portal Solar, Rüther R, Pivatto M (2022) Qual será o destino dos módulos fotovoltaicos após a repotenciação dos sistemas? Portal Solar. https://www.portalsolar.com.br/noticias/opiniao/opiniao-tecnologia/qual-sera-o-destino-dos-modulos-fotovoltaicos-apos-a-repotenciacao-dos-sistemas
Rigo PD, Siluk JCM, Lacerda DP, Spellmeier JP (2022) Competitive business model of photovoltaic solar energy installers in Brazil. Renew Energy 181:39–50
Saidani M et al (2019) A taxonomy of circular economy indicators. J Clean Prod 207:542–559. https://doi.org/10.1016/j.jclepro.2018.10.014
Salvi A et al (2023) Considering the environmental impact of circular strategies: a dynamiccombination of material efficiency and LCA. J Clean Prod 387. https://www.scopus.com/inward/record.uri?eid=2-s2.0-85145980994&doi=10.1016%2Fj.jclepro.2023.135850&partnerID=40&md5=b8b45c552aea7e8713ca13bba408584a.
Santos JD, Alonso-García MC (2018) Projection of the photovoltaic waste in Spain until 2050. J Clean Prod 196:1613–1628. https://www.scopus.com/inward/record.uri?eid=2-s2.0-85049345853&doi=10.1016%2Fj.jclepro.2018.05.252&partnerID=40&md5=0ee10cbdd85ef64c0b5afc8183444e60
Sharma HB et al (2021) Evaluation of heavy metal leaching under simulated disposal conditions and formulation of strategies for handling solar panel waste. Sci Total Environ 780:146645
Sica D et al (2018) Management of end-of-life photovoltaic panels as a step towards a circular economy. Renew Sustain Energy Rev 82:2934–2945
Simpson G, Clifton J (2017) Testing diffusion of innovations theory with data: financial incentives,early adopters, and distributed solar energy in Australia. Energy Res Soc Sci 29:12–22. https://www.scopus.com/inward/record.uri?eid=2-s2.0-85018431111&doi=10.1016%2Fj.erss.2017.04.005&partnerID=40&md5=34f76785a0447a249688e30be657f03d
Tognato C, De Oliveira GG, Oliveira A (2023) An overview of circular economy principles and sustainable development goals. Resourc Conserv Recycl 190:106850. https://doi.org/10.1016/j.resconrec.2022.106850
Tsanakas JA et al (2020) Towards a circular supply chain for PV modules: review of today's challenges in PV recycling, refurbishment and re-certification. Prog Photovolt 28(6 SI):454–464
Urbinati A, Chiaroni D, Chiesa V (2017) Towards a new taxonomy of circular economy business models. J Clean Prod 168:487–498. https://doi.org/10.1016/j.jclepro.2017.09.047

Van Opstal W, Smeets A (2023) Circular economy strategies as enablers for solar PV adoption in organizational market segments. Sustain Prod Consump 35:40–54. https://doi.org/10.1016/j.spc.2022.10.019

Integrated Environmental and Economic Assessment for the Replacement of the Flooring System in a Public Educational Building

Bárbara Pretto Biasi, Constance Manfredini, Ana Carolina Passuello, and Robert J. Ries

Abstract The sustainability of society is interrelated with improving the sustainability of the built environment. In this scenario, public buildings have a lasting influence on society throughout their life cycle, with significant environmental and economic impacts, particularly during the usage, operation, and maintenance phases. The objective of this study is to evaluate the potential environmental and economic impacts on the flooring system over the reference period of 30 years, taking into account three alternatives (wood, ceramic, and vinyl) applied in the replacement phase of a public educational building located in Porto Alegre, Brazil. In addition to encompassing the Sustainable Development Goals (SDGs): (SDG 11) Sustainable cities and communities, (SDG 12) Responsible consumption and production; (SDG 13) Action against global climate change—referring to the United Nations (UN) 2030 Agenda, the research method includes applying the integration of Life Cycle Assessment (LCA) and Life Cycle Cost (LCC) tools following national and international regulations. Tools are used with the assistance of OpenLCA and Excel software. The results demonstrate disparities in the integrated LCA-LCC analysis among the different alternatives. Ceramic flooring achieves low costs but has the most significant environmental impact. While wooden flooring has the best environmental results, vinyl flooring has the highest economic costs. Thus, the discussion

B. P. Biasi (✉) · A. C. Passuello
B. Eng., Universidade Federal do Rio Grande do Sul, Porto Alegre, Brazil
e-mail: barbarapbiasi@gmail.com

A. C. Passuello
e-mail: ana.passuello@ufrgs.br

C. Manfredini
Architect and Urban Planner, Universidade Federal do Rio Grande do Sul, Porto Alegre, Brazil
e-mail: arqconstance@gmail.com

R. J. Ries
Architect and Urban Planner, University of Florida, Gainesville, FL, USA
e-mail: rries@ufl.edu

L. Brandli et al. (eds.), *The Contribution of Life Cycle Analyses and Circular Economy to the Sustainable Development Goals*, World Sustainability Series,
https://doi.org/10.1007/978-3-031-85300-5_9

seeks to provide conscious and balanced choices in flooring alternatives for technical managers in the public sector to make decisions.

Keywords Life Cycle Assessment (LCA) · Life Cycle Cost (LCC) · Public educational buildings · Flooring systems

1 Introduction

The sustainability of society is interconnected with improving the sustainability of the built environment. The construction sector is responsible for the consumption of resources and energy, substantial emissions into the environment, and the generation of waste. On the other hand, civil construction also has great potential for mitigating these impacts on buildings (Alshamrani 2021). In this context, public buildings are necessary since all members of society use them from time to time and, therefore, influence society throughout their entire life cycle. In this context, significant environmental and economic impacts may occur during these buildings' use, operation, and maintenance phases.

Creating sustainable public buildings is challenging and a concern for institutions due to cultural barriers, control methods, and system complexity (Salicath et al. 2016; Alshamrani and Alshibani 2020). In recent years, many countries have implemented mandatory legislative measures to incorporate sustainability strategies into existing and future buildings. Implementing public procurement consumption efficiency requirements in using, mitigating, and treating possible environmental impacts are examples of sustainable public initiatives (Salicath et al. 2016).

One way to influence the impacts generated when maintaining existing buildings is to select construction materials with reduced environmental impacts and economic viability. According to Tarantini et al. (2011), green public procurement is a policy tool that reduces the environmental impacts of services and products throughout their life cycle. The market influence fostered by green public procurement has the potential to motivate stakeholders such as suppliers, manufacturers, private companies, and consumers to adopt more sustainable production and consumption practices.

Furthermore, the United Nations (UN) Agenda 2030 Sustainable Development Goals (SDGs) encompass the sustainability of civil construction, including (SDG 11) Sustainable cities and communities, ensuring everyone's access to safe housing, adequate and affordable prices (SDG 12) Responsible consumption and production, with the promotion of sustainable public purchasing practices; and also (SDG 13) Action against global climate change, to define mechanisms to increase planning and management capacity (Iberdrola 2024). These objectives can be addressed based on an LCA-LCC assessment to understand and estimate potential environmental and economic impacts generated by buildings and ways to mitigate and compensate for these impacts.

The research reported in this chapter considers the integrated assessment of the environmental and economic impacts associated with replacing floor coverings throughout the entire life cycle of a building. The assessment can be applied to different types of buildings, and the choice of the case study considers the context of the study, the poor conditions of public buildings, and the availability of data.

The purpose of this study is to evaluate the potential environmental and economic impacts of three alternative flooring systems (wood, ceramic, and vinyl) over the reference period of 30 years in the renovation of a public educational building located in Porto Alegre, Brazil. The main focus is to compare and evaluate the different options for building materials used for the renovation, considering aspects of profitability and sustainability to support decision-making.

2 Literature Review

The implementation of the SDGs starts with their theoretical understanding, followed by the establishment of priorities and impacts (scope), the definition of monitoring and evaluation processes, and, finally, the communication of these results (SEBRAE 2023). In this context, the application of impact assessment is a systemic and direct tool for improving the sustainability of building construction systems.

Life Cycle Assessment (LCA) is a tool for assessing potential environmental impacts, such as climate change and acidification, among others, and is regulated by ABNT NBR 14040 (2009), ABNT NBR 14044 (2009), and BS EN 15978 (2011). Life Cycle Costing (LCC) is a tool for assessing economic impacts regulated by European regulations (BS EN 16627 2015), where all relevant costs for the building are estimated over its lifetime (Islam et al. 2015).

The European standard (BS EN 16627 2015) defines six groups for the study of the life cycle of a building, including pre-construction, product, construction, use, end-of-life, and improvement. The building use stage includes the maintenance substage, defined as the combination of technical and administrative actions associated with the useful life of maintaining a building or system in a functional state, including cleaning, maintenance, repainting, repair, and replacement of parts (Enshassi et al. 2015).

The integration of Life Cycle Assessment (LCA) and Life Cycle Costing (LCC) is frequently noted in the literature, as it provides greater decision-making support for life cycle considerations using multiple indicators (Alshamrani 2021). Among the articles on construction components, Ros-Dosdá et al. (2019) bring an environmental comparison of internal floor coverings to the use stage of a building. The following coatings are evaluated: inorganic (natural stone and ceramic tiles), polymers (carpet and PVC), and wood derivatives (laminate and parquet)—considering data available in EPDs. In a simplified comparison, it is noted that inorganic floors present better environmental performance, followed by laminate and PVC floors (more significant impacts around 25%), parquet, which presents promising results for low traffic areas, and finally, the carpet, which has the worst environmental impacts for all traffic scenarios.

Vilches et al. (2017) published a systematic review in Denmark; the article discusses the application of a framework on CE strategies for retrofitting buildings over a reference period of 50 years. Existing buildings are part of the city's heritage and should continue to be used. However, there is a tendency for remodeling and demolition resulting from non-compliance with current building standards and codes. Other research by Guardigli et al. (2019), Medgyasszay (2019), Trovato et al. (2020), and Užšilaityte and Martinaitis (2010) also noted the integrated study of environmental and economic assessments in public buildings.

3 Research Approach and Methods

The appropriate research strategy identified for this project is Design Science Research (DSR), characterized by the search for innovative, practical solutions to real-world problems (Vaishnavi and Kuechler 2007). According to March and Smith (1995), DSR can be divided into four types: constructs, models, methods, and implementation. The DSR method has a prescriptive character; it aims to define a set of steps to execute a task, representing a path toward solving the problem.

This chapter presents the application approach and methods used to conduct the research. Figure 1 demonstrates the illustrative strategy and method design discussed below.

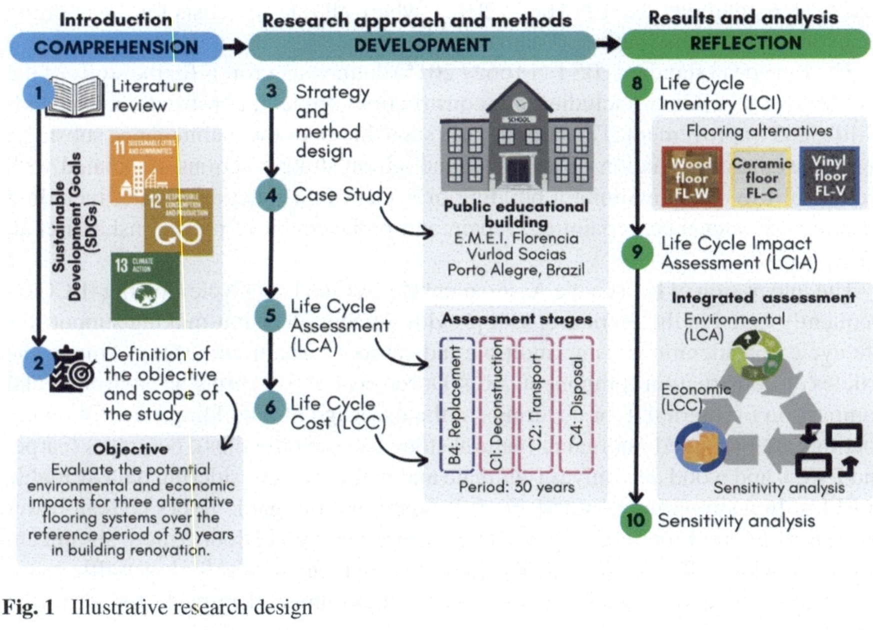

Fig. 1 Illustrative research design

The comprehension stage (Sect. 1 of the Introduction) serves to understand and contextualize the topic based on a literature review. Based on the analysis of the discoveries and challenges found in the literature, the objective and scope of the research are defined.

The development stage (Sect. 2 of Research Approach and Methods) refers to the definition and application of the method. It starts with the strategy and design of the method, which conditions the format of the method and its stages. This is followed by applying the method in a representative case study with environmental (LCA) and economic (LCC) assessment tools.

Finally, the Reflection stage (Sect. 3 of Results and analysis) comprises the analysis and discussion of the results. The Inventory Life Cycle Analysis (LCI) involves the compilation and quantification of inputs and outputs of a building, that is, the determination of information about evaluated construction systems, specification of materials, and definition of the useful life and processes for each of the constructive alternatives. The Life Cycle Impact Assessment (LCIA) aims to understand and evaluate the magnitude and importance of potential environmental and economic impacts on a system throughout the life cycle; this section analyzes the results of impacts for flooring options. Sensitivity Analysis investigates the estimated influence of methodological choices on final results for scenarios with variable useful life values.

3.1 Case Study

The case study is the Florencia Vurlod Socias Municipal Early Childhood School in Porto Alegre, Brazil. The school project is of educational buildings requiring numerous renovations to extend their useful life. In cost analysis, the floor covering system is highly relevant, and it was chosen for the evaluation of three commonly used alternative materials: ceramic floors, wooden floors, and vinyl floors.

The selected case study is the Florencia Vurlod Socias Municipal Early Childhood School, which represents a standard building project for early childhood education built by the City Hall of Porto Alegre/RS. This model has been replicated in the southern region of Brazil since the 1970s; the building under study was built in 1978. The building consists of six blocks interconnected by a central circulation on the ground floor and has 43 rooms (including classrooms, storage rooms, and service areas) with a total built area of 1,008.75 m^2.

The case study project is available in *AutoCAD* format. Therefore, the quantities and inputs are surveyed using budget spreadsheets, project information, and onsite measurements. Figure 2 shows the simplified floor plan of the building.

In 2023, Porto Alegre City Hall implemented an administrative process to maintain municipal public schools. Reporting information and maintenance budgets were made available with the support and authorization of the City Hall.

Table 1 provides information on the representativeness of the construction systems included in the budget for maintenance services for the complete building.

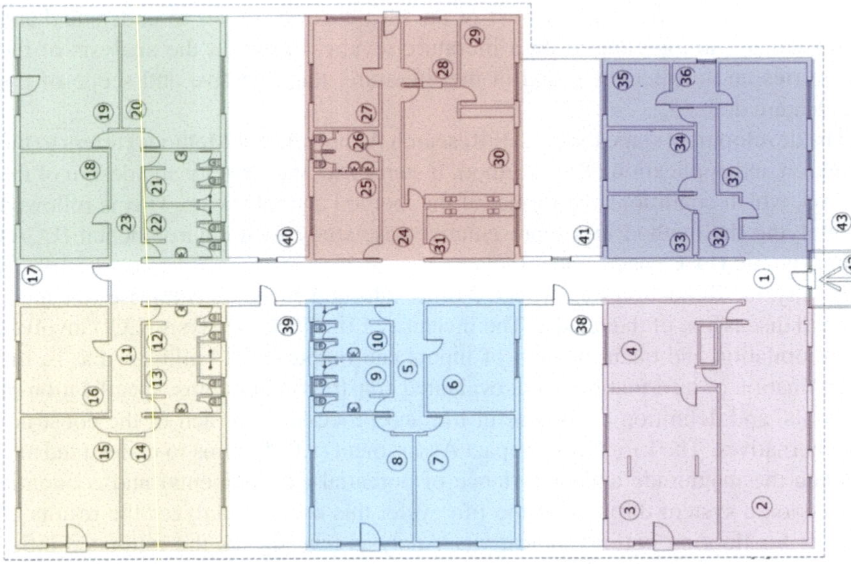

Fig. 2 Simplified Florencia Vurlod Socias Municipal Early Childhood School floor plan with different colors for each block

Table 1 Complete building maintenance budget per system

Item	System	Value (R$)	Representative (%)
1	Work management	R$ 52.44,552	6.18
2	Preliminary services	R$ 26.24,041	3.09
3	Repair of structures	R$ 14.38,854	1.70
4	Roofs	R$ 31.79,866	3.75
5	Frames	R$ 146.08,866	17.21
6	Floor coatings	R$ 286.81,303	33.79
7	Paintings coatings	R$ 229.38,206	27.03
8	Ceiling coatings	R$ 24.62,896	2.90
9	Hydrosanitary installations	R$ 31.30,902	3.69
10	Electrical installations	R$ 5.62,745	0.66
	Total	R$ 848.72,231	100.00

The school was built in 1978. The maintenance budget for the complete building was R$848,722.31 in 2023. Table 1 highlights the predominant representation of floor coverings with 33.79% (R$ 286,813.03), selected as the delimitation of the construction system for environmental and economic analyses of the present research. The photographic records are presented in Fig. 3 and represent the current situation of the building's floors.

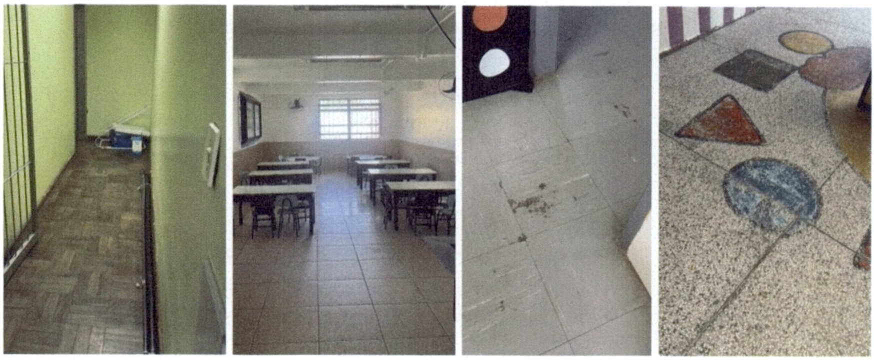

Fig. 3 Photographic records of the flooring system (wooden, ceramic, vinyl, and granite) in different rooms at Florencia Vurlod Socias School

As seen in Fig. 3, the building has floor coverings that vary depending on the room, including parts made of wooden blocks, ceramic, vinyl, and granite. The flooring system is in a poor state of conservation, showing specific pathologies such as fissures, cracks, lack of grout, detachment, dirt, and mold.

Floor coverings represent 0.84% of the total weight of the building's construction materials (7048.06 kg of the total 840,100.14 kg). In the research, areas of the land without floors, external floors, and wettable areas (such as kitchens, dining rooms, and bathrooms) that must receive ceramic floors for technical reasons are disregarded. For removal, the different types of floors verified in the inspection are taken into account; they are granite (119.76 m^2), ceramic (37.06 m^2), vinyl (302.44 m^2), and wooden planks (127.67 m^2)—totaling an area of 586.93 m^2 that is estimated for replacements. The floor mortar that is previously quoted in volume (in m^3) is transformed into the area (in m^2), maintaining the same thickness—it appears as a "subfloor" in the alternative ceramic floor (FL-C) and vinyl floor (FL-V) because a new layer is necessary with each replacement.

3.2 Aspects of Environmental and Economic Life Cycle Assessments

The method of this research includes the integrated application of environmental and economic tools: Life Cycle Assessment (LCA) by national regulations with international standards (ABNT NBR ISO 14040 2009; ABNT NBR ISO 14044 2009; BS EN 15804 2019); and Life Cycle Cost (LCC) following (BS EN 16627 2015). The LCA-LCC scope, together with the modeling aspects of the research, are presented in Table 2.

Table 2 Aspects of the objective and scope of environmental and economic assessment

Aspect	Description
Objective	Analyze and compare the environmental and economic impacts of three floor-covering alternatives
Characterization of the case study	Public educational building
Functional unit	Complete building (58,693 m^2)
Construction systems evaluated	Floor coverings (wood, ceramic, and vinyl)
Scope (system boundary)	Replacement stage (B4); End-of-life stage (C1, C2 and C4)
Geographical boundary	Porto Alegre, Rio Grande do Sul, Brazil
The reference period	30 years
LCA database	Ecoinvent version 3.6/EPDs
LCA software	OpenLCA version 1.11.0
LCA method	EPD 2013
LCA indicators	Abiotic depletion (ADP-N); Abiotic depletion, fossil fuels (ADP-F); Acidification (AP); Eutrophication (EP); Global warming (GWP); Ozone layer depletion (ODP); Photochemical oxidation (PO-NOX)
LCC database	SINAPI (February 2023)/SBC (March 2023)
LCC software	Excel
LCC indicators	Present value (PV); Initial purchase and installation costs, replacement costs for each construction material, demolition costs, and transportation costs for final disposal

In the case of LCA, two inventory database options (Ecoinvent version 3.6 and EPD System) are defined based on the availability of construction elements compatible with the building alternatives. The LCA impact category methodology (EPD 2013) is used by the alternatives selected from the Environmental Product Declaration (EPD) System. Ecoinvent alternatives are calculated using the same methodology, delimiting seven environmental impact indicators.

LCA modeling is conducted with OpenLCA software (version 1.11.0). The processes are preferably selected from the Ecoinvent database (version 3.6), available in OpenLCA software and of the Rest of the World (RoW) type. In unavailable elements, processes taken from the EPD System were used. For transportation, the Ecoinvent database (version 3.6) process is considered.

In the case of LCC, the inventory database considers SINAPI and SBC using Excel software and indicators to calculate the present value (PV). The costs are obtained from the SINAPI (CAIXA 2023) and SBC databases, commonly used for budgeting public works, with regionalized data reflecting Porto Alegre's context. In the case of SINAPI, the data are from the state of Rio Grande do Sul, while in the case of SBC, the data is specific to Porto Alegre. The unit costs reported by the databases include 22.67%, representing the BDI (Benefits and Indirect Expenses). BDI includes indirect costs not directly incorporated into the final product but are part of the total cost, such as taxes, interest, and profits.

The methodology used is Present Value (PV), which includes the initial purchase and installation costs, replacement costs for each construction material, demolition costs, and transportation costs for final disposal. Economic impacts are calculated in Excel spreadsheets using national indices, INCC-M to project future values, and SELIC to return to present values.

The cost of construction alternatives over the 30-year reference period is calculated using equations adapted from the European standard BS EN 16627 (2015). Equation 1 of Future Costs (Ct) is used to calculate one of the variables in Eq. 2 of Present Value (PV) of the Life Cycle Cost, expressed as follows:

$$C_t = C_i \times (1 + INCC)^t \quad (1)$$

where: C_t are future costs in period t; C_i are the initial costs of the gift; INCC is the future profitability rate; t is the number of years.

$$\mathrm{PV} = \sum_{t=0}^{N} \frac{C_t}{(1 + SELIC)^t} \quad (2)$$

where: PV is the present value of the Life Cycle Cost; C_t is the future costs in period t; N is the reference period; SELIC is the present discount rate; and t is the number of years.

4 Results and Analysis

The results and analyses are discussed in the Life Cycle Inventory (LCI), Life Cycle Impact Assessment (LCIA), and Sensitivity analysis.

4.1 Life Cycle Inventory (LCI)

Life Cycle Inventory (LCI) is the assessment phase of the life cycle, which involves compiling and quantifying a product's inputs and outputs throughout its life cycle following regulations (BS EN 15978 2011; BS EN 16627 2015). The research LCI is structured based on project data and access to information from onsite inspections.

The selection of the analyzed construction system considers the greatest representation (%) of the total budget value (Table 1). Thus, when evaluating the alternatives for the flooring system, there is more significant influence and economic optimization. The assessment of flooring systems considers materials verified onsite and commonly utilized in public buildings, with three alternatives being considered: wood, ceramic, and vinyl, detailed below.

- Ceramic floor (FL-C): ceramic floor covering, type plates straight edge porcelain tile, extra, smooth, mono-color, satin or polished, dimensions 60 × 60 × 2 cm (width, length, thickness), cementitious grout, fixed with type AC III adhesive mortar, applied on mortar subfloor, 1:4 ratio (volume of cement and medium wet sand), mechanical preparation with a 600 L concrete mixer;
- Wooden floor (FL-W): wooden flooring of the ipê type (heartwood) or equivalent from the region, dimensions of 7 × 21 × 2 cm (width, length, thickness), fixed with PVA-based glue, placed on the existing subfloor;
- Vinyl flooring (FL-V): semi-flexible vinyl flooring in sheets, smooth pattern, dimensions 30 × 30 cm (width, length) and thickness 3.2 mm, fixed with contact glue or water-based acrylic adhesive, applied to the subfloor of mortar, mixed 1:4 (volume of cement and medium wet sand), mechanically prepared with a 600 L concrete mixer.

Replacements are documented over a 30-year reference period, starting with an initial replacement defined at year zero due to the status of the building at that time. Throughout this period, complete material replacements occur after their useful life. Ultimately, by year 30, the possibility of demolishing the building is contemplated. The values of the replacement periods, as determined, are rounded to the highest integer value, following the requirements of BS EN 16627 (2015). The estimated Service Life (SL) and Number of Replacements (NR) values used in the research are (i) reference scenarios of own estimates and (ii) ABNT NBR 15575-1 (2021) estimation test scenario—presented in Table 3.

According to Table 3, considering the reference scenario, each alternative has a different SL according to the manufacturers of the flooring types. For the wooden floor alternative (FL-W), the SL is determined at 25 years (two replacements), assuming that it is an existing building and the reference period is 30 years. It is worth mentioning that periodic maintenance—such as cleaning and waxing wooden floors—is not included in this research.

Table 3 Service life (SL) values for base scenario (own estimate) and test scenario (NBR 15575) with a reference period of 30 years for the number of replacements (NR)

Floor coverings		Reference scenario: own estimate		Test scenario: NBR 15575 standard	
Alternative	Material	SL (years)	NR	SL (years)	NR
FL-W (wooden floor)	Wood	25	1	17	1
FL-C (ceramic floor)	Cement mortar	13	2	17	1
	Ceramic	13	2	17	1
FL-V (vinyl floor)	Cement mortar	10	2	17	1
	Vinyl	10	2	17	1

Table 4 Processes selected for environmental (LCA) and economic (LCC) assessment

Flooring alternatives		LCA processes		LCC processes	
		Database	Description	Database	Description
FL-W	Wood	EPD System	Wood flooring—TARKETT—al thickness of 22 mm from HSG plant	SINAPI 101751	Wooden parquet floor 7 × 21 cm, fixed with PVA-based glue
FL-C	Ceramic	Ecoinvent	Ceramic tile {RoW}|production|Cut-off, U	SINAPI 87263	Ceramic floor covering with porcelain tiles measuring 60 × 60 cm applied in rooms with an area larger than ten m^2
FL-V	Vinyl	EPD system	Natural homogeneous vinyl flooring—TARKETT	SINAPI 101727	Semi-flexible vinyl flooring in sheets, smooth pattern, thickness 3.2 mm, fixed with glue
FL-C FL-V	Cement mortar	Ecoinvent	Cement mortar production | cement mortar | Cutoff, U—RoW	SINAPI 87302	Cement mortar 1:4 (in the volume of cement and medium wet sand) for the subfloor, mechanical preparation with 600 l concrete mixer

For the ceramic floor alternative (FL-C), the SL considered is aligned to the ABNT NBR 1555-1 (2021) performance standard of 13 years and three replacements in total. For vinyl flooring (FL-V), an SL of 10 years with three replacements is standard, according to research by Libraga (2022) and data from manufacturers that define a useful life period of between 8 and 15 years.

The test scenario selects intermediate values for the estimated SL of the research systems, a conservative criterion for defining the Brazilian standard's Project Useful Life (VUP) periods (ABNT NBR 15575-1 2021). It is noted that the values are standardized by the floor coverings regardless of the type of material, causing more significant discrepancies with the reality of the systems in use.

The assessment processes are defined in separate databases for the LCA-LCC tools. The processes for assessing environmental impacts (LCA) are taken from the Ecoinvent version 3.6 and EPD System databases. The processes for assessing economic impacts (LCC) are determined based on compositions from the SINAPI and SBC bases. Table 4 presents the processes selected to represent each alternative.

4.2 Life Cycle Impact Assessment (LCIA)

Life Cycle Impact Assessment (LCIA) is the phase of Life Cycle Assessment that aims to understand and evaluate the magnitude and importance of potential environmental and economic impacts for a product system throughout its life cycle, according to regulations (BS EN 15978 2011; BS EN 16627 2015).

The integration of environmental (LCA) and economic (LCC) impacts of the research seeks to implement a comparative analysis of alternative floor systems to define the best options for decision-making in the maintenance process of public schools. The total impacts found are presented in Table 5 and visually in Fig. 4.

As seen in Table 5 for the flooring system, the wooden floor (FL-W) obtained reduced results in the environmental part and the lowest present value (Life Cycle Cost) among the alternatives. Therefore, the results position wooden floors as a priority solution for both analyses. Ceramic flooring (FL-C) has costs (PV) close to wooden flooring (FL-W)—a difference of R$ 522.69—however, the worst environmental performance, except the abiotic resource depletion category non-fossil (ADP-N) and ozone depletion (ODP). Vinyl flooring (FL-V) presents intermediate values for environmental impacts but is the most expensive alternative in terms of present value (PV), 42.25% higher than the best economic option (FL-W).

Figure 4 presents the results obtained for the flooring system's impacts. Ceramic flooring (FL-C) has the most significant impacts on five environmental indicators—except for abiotic depletion (ADN-P) and ozone layer depletion (ODP). However, for the cost category (economic impacts), the alternative shares the lowest costs with wooden floors (FL-W).

Table 5 Total environmental and economic impacts of flooring system alternatives for stages B4, C1, C2, and C4 of the complete building life cycle (586.93 m^2) over the 30-year reference period

Impact categories		FL-W (wooden floor)	FL-C (ceramic floor)	FL-V (vinyl floor)
ADP-N (kg SB eq)	Abiotic depletion	2148E−02	6137E−04	1060E+00
ADP-F (MJ)	Abiotic depletion, fossil fuels	1077E+05	8427E+05	3795E+05
AP (kg SO_2 eq)	Acidification	3772E+01	3953E+02	1841E+02
EP (kg PO_4 eq)	Eutrophication	1012E+01	1273E+02	6141E+01
GWP100 (kg CO_2 eq)	Global warming	9441E+03	8985E+04	4939E+04
ODP (kg CFC11 eq)	Ozone layer depletion	1030E−03	5618E−03	6422E−03
PO-NO_X (kg C_2H_4 eq)	Photochemical oxidation	5923E+00	2267E+01	8825E+00
Custo (R$)	Present value—PV	284.66,296	285.18,565	404.92,632

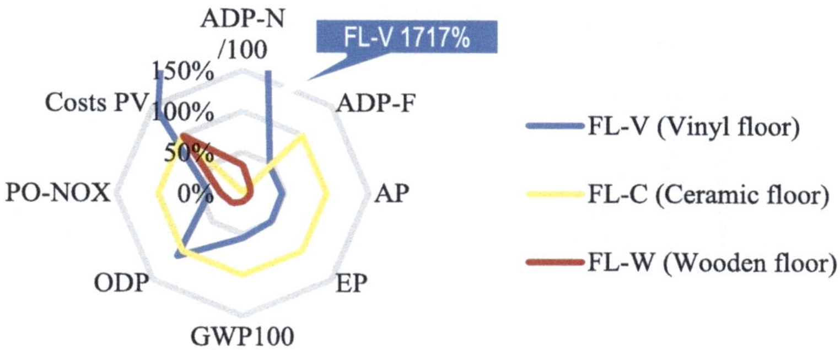

Fig. 4 Comparison of the results of total environmental and economic impacts of the alternative scenarios of ceramic (FL-C), wood (FL-W), and vinyl (FL-V) floors about the reference base scenario (FL-C) for stages B4, C1, C2, and C4 of the life cycle of the complete building (586.93 m^2) in the reference period of 30 years. *Note* The abiotic depletion value (ADP-N) is divided by 100 for graphical representation

Wooden floors (FL-W) obtained the best results in all categories, both environmental and economic, as they are natural materials and have a low level of industrial treatment compared to the components of ceramic and vinyl flooring options. The wooden floor solution also had similar environmental results in other studies (Libraga 2022).

Wood also had similar environmental results in other studies (Libraga 2022).

Vinyl floors (FL-V) have considerable environmental impacts in two categories—abiotic depletion (ADN-P) and ozone depletion (ODP)—but the greatest cost from an economic point of view. FL-V has the highest number of replacements throughout the useful life of the building, with increased impacts. It is worth mentioning that the vinyl floor is glued, making it necessary to change the subfloor each time the floor is replaced.

To deepen the analysis of flooring alternatives, the impacts are analyzed by the life cycle stage (Fig. 5) and materials (Fig. 6), which are discussed below.

Figure 5 presents the results found for the impacts by the stage of the flooring system's life cycle. In both assessments (environmental and economic), the most significant impacts occur during the replacement stage (B4); at this stage, the different inputs and materials required for the new construction of the system are accounted for, including expenses related to production and transportation.

There are high impacts in the replacement stage (B4) of the vinyl floor (FL-V) for abiotic depletion (ADP-N). For wooden floors (FL-W), a carbon balance is required to balance the global warming category (GWP), where negative CO_2 emissions occur in stage B4 (− 12.44%) and positive ones (20.31%) in stages C1, C2, and C4. The economic evaluation noted that the costs of demolition, transportation, and final disposal (C1, C2, and C4) present the most significant values among the indicators, reaching around 20% in the three alternatives—justified by the floor removal service. Among the systems analyzed, floors are the biggest waste generators (primary material and subfloor).

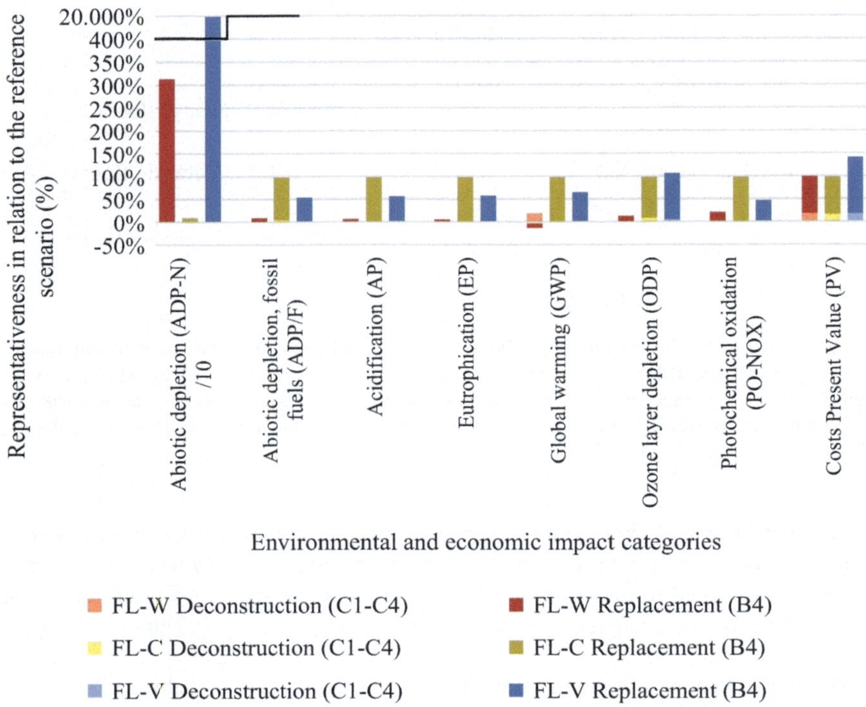

Fig. 5 Comparison of the results of total environmental and economic impacts by life cycle stage of the alternative scenarios of ceramic (FL-C), wood (FL-W), and vinyl (FL-V) floors about the reference base scenario (FL-C) for stages B4, C1, C2 and C4 of the life cycle of the complete building (586.93 m^2) in the reference period of 30 years. *Note* The abiotic depletion value (ADP-N) is divided by 10 for graphical representation

The maintenance stages are the most subjective and influential components regarding the impacts of alternatives. It is essential to consider that the present research only considers replacements, not periodic maintenance, such as cleaning or waxing floors.

The research by Minne and Crittenden (2015) investigates the environmental and economic impacts of different types of flooring in residential areas. Their findings reveal that maintenance procedures can significantly contribute to flooding impacts. When considering environmental scores, including high-maintenance activities can lead to score increases ranging between 31% (for hardwood) and 114% (for carpets). Additionally, maintenance costs significantly increase total costs—accounting for 49% for wood and approximately 30% for linoleum, vinyl, and carpet costs. The expected lifespan of the building plays a crucial role in determining the ideal floor, as costs and environmental impacts change dynamically over time (Minne and Crittenden 2015).

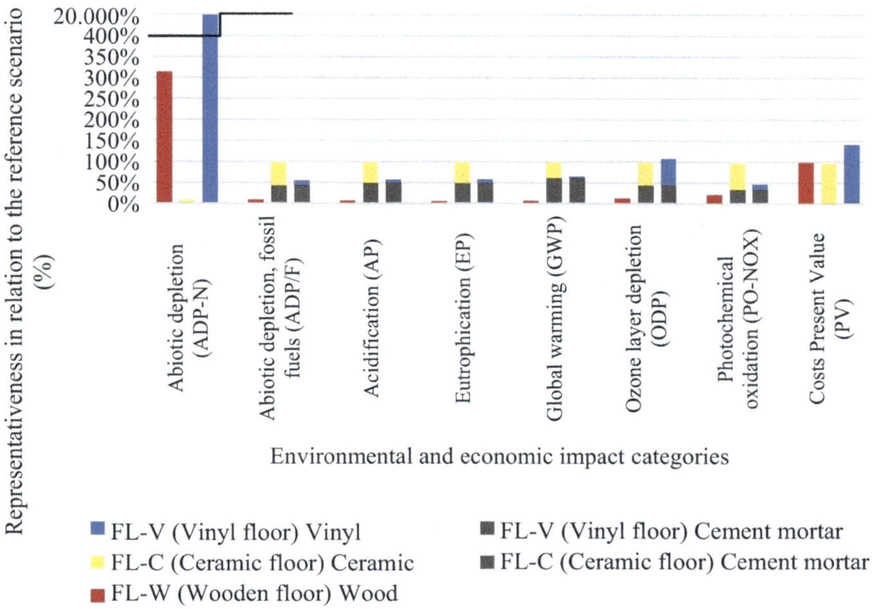

Fig. 6 Comparison of the results of total environmental and economic impacts by materials of the alternative scenarios of ceramic (FL-C), wood (FL-W), and vinyl (FL-V) floors concerning the reference base scenario (FL-C) for stages B4, C1, C2 and C4 of the life cycle of the complete building (586.93 m^2) in the reference period of 30 years

Figure 6 presents the results of the impacts of the materials used for each flooring system alternative. For the ceramic (FL-C) and vinyl (FL-V) floor compositions, subfloor layers of the flooring system are made with mortar for regularization. Vinyl material (FL-V) significantly impacts abiotic depletion (ADP-N). The wooden floor (FL-W) does not require a subfloor layer, achieving low environmental impacts in most indicators. For the economic assessment (costs, in R$), the biggest impacts are for the primary materials: wood, ceramic, and vinyl coverings.

The environmental impacts attributed to the subfloor are significant in six environmental indicators, with representation (%) approximately 44.08% for abiotic, non-fossil depletion of fossil fuels (ADP-F), 50.18% for acidification (AP), 49.72% for eutrophication potential (EP), 62.90% for global warming (GWP), 45.46% for ozone depletion (ODP) and 34.59% for Photochemical Oxidation (PO-NO_X)—the impacts of the subfloor are the same for ceramic (FL-C) and vinyl (FL-V) floors, considering the same amount of material (volume of subfloor) and several replacements. In other words, using cementitious materials to replace floors generates high environmental impacts, even if economically, it has no noticeable influence.

In summary, Fig. 7 shows the classification of the analyzed alternatives based on environmental (LCA) and economic (LCC).

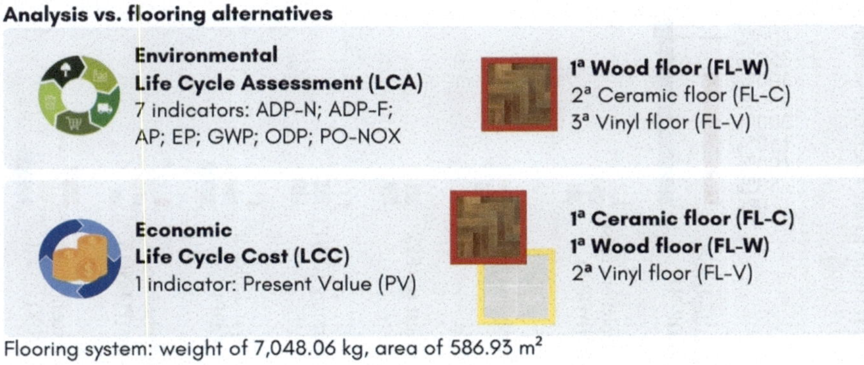

Fig. 7 Illustrative summary of total environmental and economic impacts of alternative flooring system scenarios for stages B4, C1, C2, and C4 of the complete building life cycle (586.93 m^2) in the 30-year reference period

Figure 7 illustrates that using wooden flooring (FL-W) is the best option, providing the best environmental results with the lowest costs. Ceramic flooring (FL-C) reaches a cost very close to the most economical option but has significant environmental impacts. At the same time, vinyl flooring (FL-V) presents the worst environmental scenario and the highest economic cost. The results demonstrate that the number of replacements of alternatives over the 30 years is relevant to prioritizing data closer to reality.

4.3 *Sensitivity Analysis*

Sensitivity analysis estimates the influence of methodological choices on final results, one of the main definitions being the service life of flooring alternatives. Uncertainties regarding predicting the useful life of different materials are relevant as they directly influence the impact assessment. In this way, a comparison is made between two scenarios with variable useful life values: (i) base scenario of own estimate and (ii) NBR 15575-1 (2021) estimation test scenario. The results of this analysis are presented in Fig. 8.

Figure 8 demonstrates the impact of different useful life values on the flooring system. In this comparison, the useful life of the floors varies between the Base Scenario: ceramic floor (FL-C) with 25 years, wood (FL-W) with 13 years, and vinyl (FL-V) with 10 years. For the NBR 15575-1 test scenario, the useful life of the flooring alternatives is 17 years (Table 3).

When analyzing the graph, there is a trend in environmental and economic impacts between the alternatives in both scenarios, even with different useful life values for the alternatives. The test scenario (NBR 15575) presents lower environmental impacts in the seven indicators for the ceramic (FL-C) and vinyl (FL-V) flooring alternatives

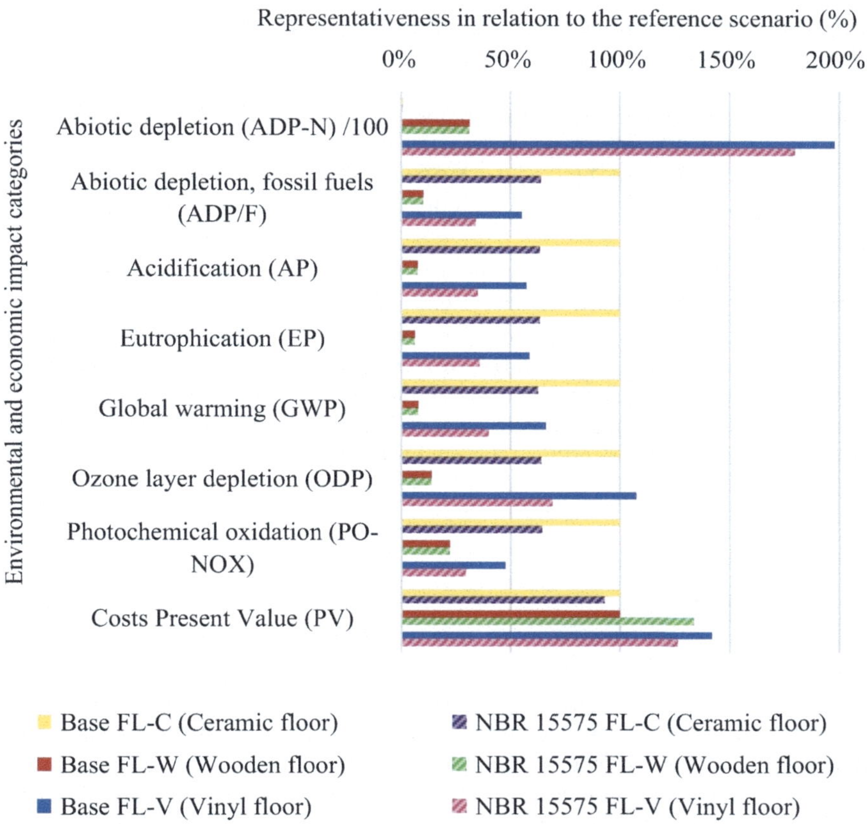

Fig. 8 Comparison of the results of environmental and economic impacts due to variation in the useful life (base and NBR 15575) of alternative scenarios for ceramic (FL-C), wood (FL-W), and vinyl (FL-V) floors about the base scenario of reference (FL-C) for stages B4, C1, C2 and C4 of the complete building life cycle (586.93 m^2) in the 30-year reference period. *Note* The abiotic depletion values (ADP-N) are divided by 100 for graphical representation

due to the longer useful life (17 years) and fewer replacements—two substitutions in the base Scenario and a substitution in the NBR 15575 Scenario. The idea is the same for the economic indicator (PV), in which there is a reduction in costs throughout the life cycle of the two alternatives (FL-C and FL-V).

The alternatives for ceramic flooring (FL-C) reveal a 7% difference in costs throughout the life cycle (PV) between the base scenario (R$ 285,185.65) and NBR 15575 (R$ 265,211.09). Vinyl floor (FL-V) has more significant impacts in the Abiotic Depletion, non-fossil (ADP-N) category, reaching values of 155.07% about FL-C for the Base Scenario and 103.38% for the Scenario NBR 15575—the economic indicator points to a difference of 16% for vinyl flooring (FL-V), with costs for the base (R$ 404,926.32) and NBR 15575 (R$ 360,427.67). In other words, the longer the useful life and durability of the system, the lower the economic costs (PV) for replacement.

The wooden floor (FL-W) maintains the same results for the seven environmental indicators, influenced by the number of replacements maintained with one replacement over 30 years. For the present value (PV) indicator, wooden flooring costs increase by around 34% in the NBR 15575 Scenario (R$ 381,381.36) and Base Scenario (R$ 284,662.96) because the year of replacement influences the PV calculation.

For the NBR 15575 Scenario, the priority alternatives for generating lower environmental impacts for the flooring system are wooden (FL-W), ceramic (FL-C), and vinyl (FL-V) floors. The priority for cost reduction throughout the life cycle is flooring: ceramic (FL-C), wood (FL-W), and vinyl (FL-V). The main difference between the base scenario and NBR 15575 is the prioritization of ceramic and wooden floors: the costs of ceramic floors (base with R$ 285,185.65 and NBR 15575 with R$ 265,211.09) and wooden floors, base with R$ 284,662.96 and NBR 15575 with R$ 381,381.36.

5 Conclusion

Building sustainability is a developing topic in the construction sector, in which decision-making for construction systems and materials is focused on an economical and immediate approach, commonly omitting environmental factors and the building's life cycle. Another consideration is the lack of planning for periodic preventive maintenance of buildings, which leads to technical limitations and functional problems of use by users, in addition to promoting the deterioration of systems and increased costs in global renovations.

The built environment is clearly related to the Sustainable Development Goals (SDGs), bringing to light discussions about sustainable cities (SDG 11), responsible consumption (SDG 12), and action against global climate change (SDG 13). Among the tools for analyzing and improving building sustainability, the LCA-LCC integrated assessment aims to understand and evaluate the magnitude and importance of potential environmental and economic impacts throughout the life cycle. This way, mitigating impacts, selecting construction alternatives, and optimizing processes is possible.

In this context, this research evaluates the impacts of three alternative coatings (ceramic, wood, and vinyl) for the case study of a public school building in Porto Alegre, Brazil. It is worth highlighting that the results found are exclusive to the evaluated case study, the environmental processes, the economic compositions, and other specifications within the scope of the impact assessment. Therefore, the parameters must influence the results and trends if the method is replicated in other contexts. Reconciling construction elements' environmental and economic compositions is challenging, as the information originates from different databases. Another obstacle encountered is the need for LCA databases related to the local context, with limited availability in Ecoinvent processes and Environmental Product Declarations (EPDs).

Possible solutions include compiling regional processes and product data according to their manufacturers and creating a specific database for the maintenance stage to assist in developing precise guidelines and recommendations for professionals involved in construction decisions. Designing accurate and clear policies must encompass all stages of the building's life cycle, where monitoring the deadlines set out in the Building Use, Operation, and Maintenance Manual and maintenance history are crucial.

LCA-LCC integration is a complex procedure that allows for the evaluation of the interdependence of various factors. Such a study is relevant to establishing building environmental and economic assessment criteria, mainly for selection and preference for specific materials. Studies aimed at the maintenance stage seek to improve the comfort conditions of occupants, the performance of the building, and the optimization of preservation processes. Furthermore, it supports technical teams' decision-making and serves as a reference for discussions and advances on the topic internationally.

References

ABNT NBR 15575-1 (2021) ABNT NBR 15575-1: Residential buildings—performance part 1: general requirements. ABNT—Brazilian Association of Norms Techniques. www.abnt.org.br

ABNT NBR ISO 14040 (2009) ABNT NBR ISO 14040: environmental management—Life Cycle Assessment—principles and framework. ABNT—Brazilian Association of Norms Techniques

ABNT NBR ISO 14044 (2009) ABNT NBR ISO 14044: environmental management—Life Cycle Assessment—requirements and guidelines. ABNT—Brazilian Association of Norms Techniques

Alshamrani OS (2021) Integrated LCA-LCC assessment model of offsite, onsite, and conventional construction systems. J Asian Arch Build Eng. https://doi.org/10.1080/13467581.2021.1942001

Alshamrani OS, Alshibani A (2020) Automated decision support system for selecting the envelope and structural systems for educational facilities. Building and environment, p 181. https://doi.org/10.1016/j.buildenv.2020.106993

BS EN 15804 (2019) Product Category Rules for Type III environmental product declaration of construction products to

BS EN 15978 (2011) BS EN 15978: sustainability of construction works: assessment of environmental performance of buildings: calculation method. British Standards Institution—European Committee for Standardization

BS EN 16627 (2015) BS EN 16627: sustainability of construction works: assessment of economic performance of buildings: calculation methods. British Standards Institution—European Committee for Standardization

CAIXA CEF (2023) SINAPI: methodologies and concepts—national civil construction cost and indices research system, 9th edn. http://www.caixa.gov.br

ENVIRONMENTAL PRODUCT DECLARATION (EPD) methodology (2013). International EPD® System. Available at: https://environdec.com/. Accessed on March 15.

Enshassi A, El Shorafa F, Alkilani S (2015) Assessment of operational maintenance in public hospitals buildings in the Gaza strip. https://core.ac.uk/download/478582415.pdf

Guardigli L, Ferrer C, Peters C, Fotopoulou A, Bragadin MA, Ferrante A (2019) Rehabilitation of public housing buildings in a life cycle perspective. IOP Conf Ser Earth Environ Sci 296(1). https://doi.org/10.1088/1755-1315/296/1/012038

Iberdrola (2024). We lead the way to a zero-emissions future. https://www.iberdrola.com/

Islam H, Jollands M, Setunge S, Haque N, Bhuiyan MA (2015) Life Cycle Assessment and Life Cycle Cost implications for roofing and floor designs in residential buildings. Energy Build 104:250–263. https://doi.org/10.1016/j.enbuild.2015.07.017

Libraga J (2022) Method for evaluating environmental impacts in the design process of public buildings. Federal University of Rio Grande do Sul

March ST, Smith GF (1995) Design and natural science research on information technology, vol 15, 4th edn. Decision Support Systems

Medgyasszay P (2019) Comparative analysis of an existing public building made from natural materials and reference buildings designed from common materials. IOP Conf Ser Earth Environ Sci 323(1). https://doi.org/10.1088/1755-1315/323/1/012140

Minne E, Crittenden JC (2015) Impact of maintenance on life cycle impact and cost assessment for residential flooring options. Int J Life Cycle Assess 20(1):36–45. https://doi.org/10.1007/s11367-014-0809-z

Ros-Dosdá T, Celades I, Vilalta L, Fullana-i-Palmer P, Monfort E (2019) Environmental comparison of indoor floor coverings. Sci Total Environ 693. https://doi.org/10.1016/j.scitotenv.2019.07.325

Salicath E, Liyanage P, Fladberg D (2016) Activity-based life-cycle costing of public assets: a case study of schools in Norway. In: Proceedings of the 10th world congress on engineering asset management

SEBRAE (2023) Brazilian support service for micro and small businesses. https://sebrae.com.br/

Tarantini M, Loprieno AD, Porta PL (2011) A life cycle approach to green public procurement of building materials and elements: a case study on windows. Energy 36(5):2473–2482. https://doi.org/10.1016/j.energy.2011.01.039

Trovato MR, Nocera F, Giuffrida S (2020) Life-cycle assessment and monetary measurements for the carbon footprint reduction of public buildings. Sustainability (Switzerland) 12(8). https://doi.org/10.3390/SU12083460

Užšilaityte L, Martinaitis V (2010) Search for the optimal solutions for public building renovation regarding the life cycle. J Environ Eng Landsc Manag

Vaishnavi VK, Kuechler W (2007) Design science research methods and patterns: innovating information and communication technology. Taylor & Francis

Vilches A, Garcia-Martinez A, Sanchez-Montañes B (2017) Life Cycle Assessment (LCA) of building refurbishment: a literature review. In: Energy and buildings. Elsevier Ltd, vol 135, pp 286–301. https://doi.org/10.1016/j.enbuild.2016.11.042

Wittocx L, Buyle M, Audenaert A, Seuntjens O, Renne N, Craeye B (2022) Revamping corrosion damaged reinforced concrete balconies: Life cycle assessment and life cycle cost of life-extending repair methods. J Build Eng 52. https://doi.org/10.1016/j.jobe.2022.104436

Tackling Waste, Enhancing Wealth: Circular Economy Solutions for Fulfilling the Green Development Agenda and Sustainable Development Goals: An Indian Perspective

Rishav Ray and Dr. Deeksha Dave

Abstract The pursuit of sustainable development has become critical in tackling the urgent environmental concerns that governments confront across the world. In this aspect, the circular economy has emerged as a possible paradigm for achieving environmental sustainability while also stimulating economic growth. India, with its growing population and rapid economic expansion faces tremendous challenges with regards to waste management and sustainable resource usage. Traditional linear economic models which follow take-make-dispose policy have resulted in widespread resource depletion, environmental damage, and social inequality. However, the circular economy represents a paradigm change by giving emphasis on reduction, reuse, recycling, and regeneration of materials and products. The objective of the paper is to investigates the relevance of circular economy solutions in the context of India, emphasizing on how they could assist in achieving Sustainable Development Goals (SDGs), Such as SDG 9 (Industry, Innovation, and Infrastructure), SDG 12 (Responsible Consumption and Production), SDG 13 (Climate Action) and the green development agenda set during its G20 presidency. India, with a population of over 1.43 billion, faces climate change challenges and must achieve net-zero carbon emissions by 2070. Climate technology firms could play a crucial role in this transformation, but a significant investment of over US$1 trillion is required by 2030. India's diversified business environment and rapidly developing industrial sector make it an attractive case study for sustainability transition. India can lead the global Circular Economy movement by focusing on product design, new materials, digital technologies, industrial symbiosis, recycling, and circular business models. The study explores the potential of a circular economy model

R. Ray (✉)
Ph.D. Scholar-Environmental Science, School of Interdisciplinary and Trans-Disciplinary Studies (SOITS), IGNOU, New Delhi, India
e-mail: rishavrey@gmail.com

Dr. D. Dave
Associate Professor-Environmental Studies, School of Interdisciplinary and Trans-Disciplinary Studies (SOITS), IGNOU, New Delhi, India

L. Brandli et al. (eds.), *The Contribution of Life Cycle Analyses and Circular Economy to the Sustainable Development Goals*, World Sustainability Series,
https://doi.org/10.1007/978-3-031-85300-5_10

in India's green development agenda and sustainable development goals. It reveals that transitioning to this model offers benefits such as increased resource efficiency, reduced environmental harm, and economic growth. India is at a critical moment to adopt circular economy concepts, which could address environmental issues and promote fair and sustainable development. Efforts like waste-to-energy projects, eco-design practices, producer responsibility initiatives, and remanufacturing and recycling sectors could shift India's economy towards a more sustainable future. Using case studies and best practices from India, it will further analyze multiple aspects of circular economy implementation in India, including regulatory frameworks, technology developments, institutional processes and waste management practices to improve resource efficiency, minimize waste output, creating new economic opportunities and encourage sustainable consumption habits. In conclusion, this study argues that embracing the principles of the circular economy is critical for India to accomplish its green development agenda, SDGs commitments and promoting inclusive growth.

Keywords Circular economy · Responsible consumption · Green development agenda · Inclusive growth · Sustainable development goals (SDG's)

1 Introduction

The global emphasis on sustainable development has significantly increased due to the pressing environmental issues faced by governments worldwide (Smith et al. 2020). One promising approach to address these challenges is the adoption of the circular economy concept, which not only promotes environmental sustainability but also drives economic prosperity (Ellen MacArthur Foundation 2019). In recent years, it has gained increased global attention due to, among other things, the awareness that resource security and efficiency are critical for the growth of countries and organizations. Several governments and businesses around the world have adopted the circular economy concept as a solution for reconciling what appear to be conflicting goals of economic growth and environmental sustainability (Ghisellini et al. 2016). Recently, it has become very popular as a reaction to the unsustainable methods of the traditional linear economy, which is known for its 'take, make, dispose' approach. Its goal is to change the definition of growth, with a focus on positive impacts that benefit society as a whole. It involves slowly separating economic activity from the use of limited resources, and eliminating waste from the system through design. Based on a shift to sustainable energy sources, the circular economy strengthens economic, environmental, and social resources. The growth of circular economy practices in various sectors is motivated by several related factors that all play a role in turning waste into a valuable resource. The increasing global focus on environmental sustainability is a key factor driving both consumers and businesses to prioritize eco-friendly initiatives. Strict rules and policies are helping to facilitate this change by reducing carbon footprints and improving resource efficiency. Governments and international organizations, such as the European Union, have introduced legal frameworks to promote the adoption of circular economy principles.

Pearce and Turner were the pioneers in incorporating the term circular economy into an economic framework. By building on the concept that "everything is connected," the writers analyzed the old linear economic system and created a fresh economic model called the circular economy, which is based on the principles of thermodynamics. This model highlights the connection between the economy and the environment by incorporating three economic functions of the environment: providing resources, assimilating waste, and serving as a source of utility. Additionally, the economic benefits of a circular economy are becoming increasingly clear. Businesses are recognizing the financial advantages of using fewer raw materials and the possibility of earning profits through recycling products. According to the take-make-dispose concept, conventional linear economic models have led to notable resource depletion, environmental harm, and social inequality (United Nations Environment Programme 2018). Instead, the circular economy focuses on reducing, reusing, recycling, and regenerating resources and products, marking a groundbreaking shift (European Commission 2020a, b). Nations can strive to minimize their impact on the environment, improve resource usage, and promote inclusive growth by embracing this novel economic approach (World Economic Forum 2021).

India, the world's biggest democracy and a worldwide mindscape influencer, has coexisted with a variety of social and environmental variables such as population growth, political turmoil, food and water scarcity, fast urbanization, environmental degradation, and climate change. However, India is making progress toward the United Nations (UN) Sustainable Development Goals (SDGs) "Agenda 2030" target, with waste processing increasing from 18% in 2014 to 70% by 2021. The present disruptive developments necessitate an urgent call to action to plan development and stimulate economic expansion while combating climate change and developing future waste management and resource preservation strategies. The circular economy promotes a shift from linear 'take-make-waste' to multi-life cycle circular value chains in business models, using a design-thinking approach for more effective and prudent resource usage. Currently, about 377 million people live in cities, creating 55 million tons of municipal solid waste (MSW) per year. Furthermore, this figure is expected to skyrocket, reaching 125 million tons yearly by 2031. Despite the circular economy's great importance, the sector currently has a varying understanding of the idea, posing a substantial hurdle to its widespread acceptance in India. It is anticipated that by 2050, India would receive yearly benefits of US$ 624 billion (Rs 40 lakh crore) from reduced negative externalities. By 2030, the transformation to a circular economy may boost global economic output by $4.5 trillion. Furthermore, in contrast to the existing growth climate, India's circular economy development path may yield an annual value of $218 billion (Rs 14 lakh crores) by 2030 and $624 billion (Rs 40 lakh crores) by 2050. The implementation of a circular economy in India will necessitate the creation of an enabling environment that promotes the discovery and acceptance of innovative business models. Currently, 377 million people live in cities, producing around 55 million tonnes of Municipal Solid garbage (MSW) (such as organic garbage and recyclables such as paper, plastic, wood, glass, etc.) each year, with this figure predicted to rise to 125 million MT per year by 2031. Furthermore, approximately 75–80% of MSW is collected; only 22–28% is processed, with the remainder disposed in dump yards. MSW output is expected to reach 165 million tons by 2031, rising to 436 million tons by 2050. By 2030, India is predicted to be the world's third-largest economy, accounting for around 8.5% of global GDP. The circular economy has the potential to boost India's growth while simultaneously bringing major environmental advantages, resulting in a more sustainable and resilient framework. According to the National Chemical Laboratory (NCL) and the PET Packaging Association for Clean Environment (PACE), India's recycled Polyethylene Terephthalate (PET) plastic sector is worth between $400 and $550 million. PET is recycled at a rate of 90% in India, surpassing Japan (72%), Europe (48%), and the United States (31%). As a result, India has huge prospects for implementing a circular economy. India has the potential to emerge as the fastest growing major economy in the world, as it is one of the fastest rising economies globally. The rapid economic growth and increasing household incomes have led to an increase in consumer spending, expected to reach USD 4 trillion by 2025. However, the country is encountering obstacles in waste management and resource sustainability due to its fast-growing population and economy (Dwivedi et al. 2021). Throughout history, the Indian culture has placed importance on preserving resources. Religious activities, folklore, art, and culture have always been ingrained in the Indian mindset, permeating all aspects of daily life. The origins of sustainable development and climate awareness can be linked to our practice of viewing "waste as wealth," where goods are reused and repurposed in various ways through a circular economy model. The importance of living harmoniously with the environment and using resources wisely is emphasized in ancient texts like the Atharva Veda and teachings of Mahatma Gandhi. India, with 1.3 billion inhabitants making up 18% of the world's population, resides on a mere 2.4% of the planet's land area, making it vulnerable to significant resource limitations. In order to accommodate its growing population, attain the targeted economic growth rate, and confront the challenge of dwindling resources, the nation needs to embrace a positive, all-encompassing, and environmentally conscious development plan.

A structured methodology was adopted to collect, analyze, classify, and categorize data on waste management practices and circular economy initiatives in India. Primary data sources include surveys, interviews, and field observations with stakeholders involved in waste management and circular economy practices. Secondary data includes existing literature, reports, policy documents, and academic articles related to circular economy initiatives, sustainable development goals (SDGs), and waste management in India. Data collection techniques include structured questionnaires to gather quantitative data on current waste management practices, challenges, and perceptions towards circular economy solutions. Interviews provide qualitative insights on policy perspectives, industry practices, and community involvement. Document analysis reviews and analyses policy documents, reports, and academic literature to understand the regulatory framework, best practices, and case studies. The purpose of this paper is to investigate the relevance of circular economy solutions in the context of India, with a focus on how they can help achieve Sustainable Development Goals (SDGs) such as SDG 9 (Industry, Innovation, and Infrastructure), SDG 12 (Responsible Consumption and Production), SDG 13 (Climate Action), and the green development agenda set up during India's G20 presidency. Using case studies and best practices, it will examine various aspects of circular economy implementation in India, such as regulatory frameworks, technological developments, institutional processes, and waste management practices, in order to improve resource efficiency, reduce waste output, create new economic opportunities, and promote sustainable consumption habits.

2 Circular Economy: Concept and Relevance

2.1 The Concept of Circular Economy

The Circular Economy is a sustainable method of managing waste that aims to prevent waste by focusing on product design and extending product life, as well as reusing, repairing, refurbishing, and remanufacturing. This involves maximizing circularity through the design of products that have a longer lifespan, are simple to upkeep, fix, enhance, renovate, manufacture again, or recycle, and incorporate materials like bio-based, lower in resource usage, or completely recyclable. Designing products and extending their lifespan are vital aspects of implementing the Circular Economy. Manufacturers could incentivize eco-friendly design practices like modularity, standardized components, and durability to produce items that are easier to repair, reuse, and recycle. A circular economy revolves around three principles: eliminating waste and pollution, maintaining the use of products and materials, and renewing natural systems (Kirchherr et al. 2017) (Fig. 1).

This method tackles environmental degradation while also generating economic opportunities and social benefits. An example is how circular approaches can lower greenhouse gas emissions, improve resource utilization, and generate employment in industries like manufacturing, waste management, and recycling as outlined in Geissdoerfer et al. It is crucial to reducing climate change. By reducing waste and encouraging the continuous utilization of resources, it effectively decreases greenhouse gas emissions related to material production and waste handling. For instance, research conducted by the suggests that implementing circular economy approaches in five crucial sectors—cement, aluminum, steel, plastics, and food—could reduce emissions by 9.3 billion tonnes of CO2e by 2050, which is the same as removing current emissions from all transportation worldwide. Efficient use of resources is a fundamental aspect of the circular economy. With the ongoing increase in global material demand, the strain on limited resources becomes more evident. The goal of the circular economy is to optimize the use of resources by prolonging the lifespan of products through activities such as reuse, refurbishment, remanufacturing, and recycling (Ghisellini et al. 2016). An example is the European Union's Circular

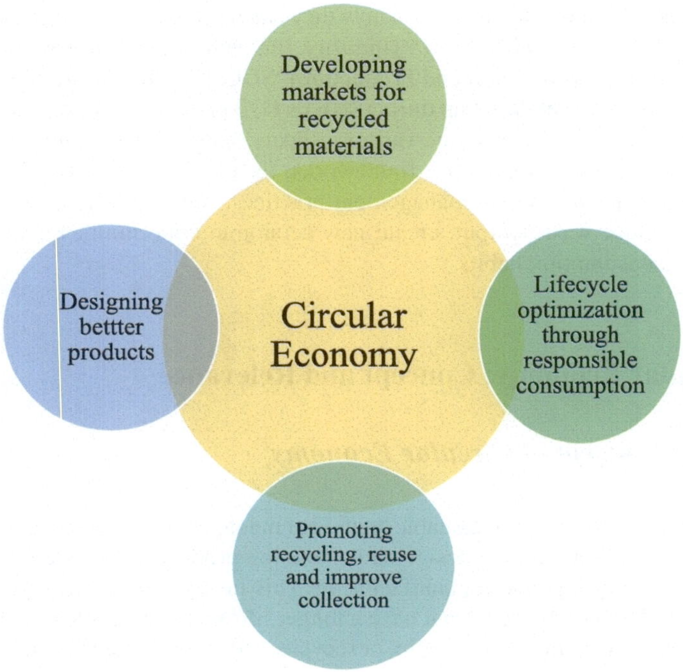

Fig. 1 Objectives of circular economy

Economy Action Plan, which is a key component of the European Green Deal. It aims to boost recycling rates and promote sustainable product design, with the objective of doubling the rate of circular material use within the next ten years (European Commission 2020a, b).

Corporates, governments, non-governmental organizations, and academics have all begun to study the circular economy. The following '5R' concepts are at the core of establishing circularity in any product, process, or service:

- **Reduce**: The emphasis is on increasing resource efficiency by prioritizing the use of regenerative and restorative resources. The underlying manufacturing process must guarantee that renewable, reusable, and non-toxic resources are efficiently used as sources of materials and energy.
- **Reuse**: This includes two aspects: the first is to reuse the valuable elements/components of a product wherever feasible, and the second is to encourage broader usage of product-as-a-service through sharing platforms. The goal is to encourage sharing of underutilized assets or developing a network of shared services, whereby the service provider normally has ownership of the product throughout its lifespan, and the consumer pays for the time or usage of it, guided by either a short or long-term contract.

- **Recycle**: The focus is on developing a closed loop system to use wasted material as a source of secondary resources, through extensive recycling.
- **Re-manufacture**: The aim behind this technique is to develop new goods by using waste streams through cooperation and collaboration between multi-sector industry players or a symbiotic connection between a network of enterprises in a specific value chain.
- **Repair/Refurbishment**: The goal is to maintain and prolong the life of an already manufactured object by planning for the future. It is critical to ensure that the product is developed for an adequate lifetime and long-term use from the start. While the product is in use, maintain, repair, refurbish, and enhance it to extend its life and provide it a second life through take-back techniques, where appropriate (Table 1).

Reconsidering standard linear methods of production and consumption, the circular economy economic model strives for a more regenerative and sustainable approach. In a circular economy, techniques like reuse, recycling, and remanufacturing are employed to prolong the lifespan of items, reducing both waste production

Table 1 Linear versus circular economy

Aspects	Linear economy	Circular economy
Resource use	Extracts raw materials from nature	Utilizes waste as a resource, focuses on recycling and reusing materials
Product lifecycle	Products are designed for a short lifespan and are often disposed of after use	Products are designed for longevity, repairability, and recyclability
Waste management	Generates significant waste that is usually disposed of in landfills or incinerated	Aims to minimize waste through recycling, repurposing, and biodegradation
Environmental impact	Higher carbon footprint and environmental degradation due to continuous resource extraction and waste generation	Lower carbon footprint and environmental impact by reducing resource extraction and maximizing resource efficiency
Economic model	Linear, one-way model of "take, make, dispose"	Circular, regenerative model of "reduce, reuse, recycle"
Value creation	Value is created through the production and sale of new products	Value is created through maintaining, reusing, and recycling existing products
Business approach	Focus on volume and sales growth	Focus on sustainability, efficiency, and resource conservation
Consumer role	Consumers are end-users who dispose of products after use	Consumers are participants in a loop who return products for reuse, repair, or recycling l
Innovation	Innovation is driven by the creation of new products and technologies	Innovation is driven by sustainable design, materials science, and waste reduction technologies

and its environmental impact. This method differs from the standard linear economy that involves a "take-make-dispose" cycle and leads to pollution and depletion of resources. Governments, companies, customers, and researchers are collaborating to transform goods, procedures, and structures in the shift towards a circular economy to optimize resource use and establish a closed-loop model. Creating plans for waste minimization, fostering collaboration across industries to optimize resource utilization and create synergies, and producing durable and recyclable products are crucial aspects of the circular economy (Kirchherr et al. 2017).

2.2 *Relevance Towards Environmental Sustainability*

The primary strategy being utilized to achieve the Sustainable Development Goals (SDGs) by 2030 is based on the circular economy. This idea not just addresses environmental issues, but also provides considerable economic advantages. Research by the Ellen MacArthur Foundation (2015) suggests that the circular economy can greatly cut down on carbon emissions and sustain resources. The rise of digital platforms enables enhanced sharing of resources and management of product lifecycles, thus integrating circular principles into business models (World Economic Forum 2022). Reducing waste is one of the key advantages of implementing a circular economy. By creating goods that have a longer lifespan and can be easily fixed, reused, or recycled, we can significantly decrease the amount of waste produced. In Kamikatsu, Japan, a zero-waste program has been put into action with an 80% recycling rate, with a goal of reaching zero waste by 2030. The program includes a high level of community involvement and a strict sorting process, demonstrating how circular principles can be effectively applied within the community. Switching to a circular economy presents enormous economic potentials as well. A report from Accenture (2020) suggests that the circular economy has the potential to create an extra $4.5 trillion in economic output by 2030. The increase is fueled by innovative business models like product-as-a-service, sharing platforms, and advanced recycling technologies, that extract value from waste and prolong the lifespan of products.

IKEA and other companies have adopted circular strategies by introducing furniture leasing services and providing repair and recycling choices to customers, which helps promote a more sustainable way of consuming. There are many advantages in terms of society from a circular economy. Through the promotion of sustainable practices, healthier living environments, reduced pollution, and increased community engagement can be achieved. Circular economy efforts in developing nations can help reduce poverty and promote social equality by generating employment opportunities in recycling, repair, and remanufacturing industries. In India, the informal sector is important for waste management and recycling. Including these workers in formal circular economy frameworks can enhance their livelihoods and working conditions. Firms are adopting circular strategies to meet consumer demand for sustainable products and boost market competitiveness (McKinsey and Company 2021). In addition, collaborative actions involving governments, businesses, and NGOs are fostering an environment conducive to circular economy practices. Establishing strong relationships is crucial for expanding sustainable practices and bringing about systemic transformation (OECD 2021). Together, these factors are causing a shift towards a circular economy, turning waste into economic opportunities and supporting a greener future.

3 India and Circular Economy: Embarking on a New Growth

India's vast and rapidly expanding population can take advantage of the Circular Economy idea by incorporating traditional practices of reusing, reducing, and repairing into contemporary sustainability efforts. Historically, Indian culture has focused on valuing and respecting nature. This is shown through traditions like worshipping trees and rivers, and celebrating festivals like Vasant Panchami and Chhath Puja, which have a strong emphasis on nature worship (Rao 2018). The respect for nature in this culture leads to a dedication to protect and preserve natural resources, which is in line with the core concepts of a Circular Economy (Gupta and Sharma 2019). Traditional agricultural practices in India have historically utilized sustainable methods like crop rotation and utilizing cow dung as fertilizer to improve soil fertility, as well as promote water conservation and reduce waste (Jain and Singh 2020). In addition, Indian cooking tradition shows cleverness by using leftover items effectively such as turning leftover rice and lentils into healthy meals like khichdi and upma, reducing food wastage (Kumari and Pandey 2017). Furthermore, excess food is frequently reused as compost or animal fodder, showcasing the circular nature of Indian traditions (Dhaliwal and Singh 2019). In the field of textiles, Indians have traditionally favoured clothing made from natural materials like cotton and wool, known for their durability and potential for reuse or recycling. This custom is in line with the concepts of circular fashion, where garments are created to last long and be easily recycled, lessening the environmental effects of the textile sector (Sengupta and Roy 2021). In addition, traditional Indian craftsmanship often includes the utilization of reused and repurposed materials, as artisans commonly use leftover cloth, paper, and metal to create new items like bags and home decor. Moreover, it is common in India to reuse glass bottles and plastic containers for different uses until they are no longer usable, which helps conserve resources and reduce waste (Sharma and Kumar 2019). By incorporating and expanding on these customary methods, India can take significant steps towards creating a strong Circular Economy model that reflects its cultural values and tackles present sustainability issues.

The G20 declaration in September 2023 in New Delhi highlighted the crucial role of the Circular Economy, emphasizing Extended Producer Responsibility (EPR) and resource efficiency for sustainable development goals. To maintain this progress, a strong framework is essential, aligning with the vision of Prime Minister Narendra Modi and the G20 declaration, and encouraging stakeholders like industries and investors. This framework should support the shift towards a Circular Economy by rewarding sustainable practices and innovation, while also guaranteeing economic viability and profitability for all stakeholders (Jain and Sharma 2023). This significant announcement led to the creation of the Resource Efficiency and Circular Economy Industry Coalition (RECEIC), demonstrating a shared dedication to improving eco-friendly waste management and significantly decreasing waste production by 2030 (MoEFCC 2023). The G20 declaration also highlighted the significance of zero-waste projects in furthering the Circular Economy agenda. Moving towards a Circular Economy framework is in line with India's obligations to the Paris Agreement, the UNSDGs, specifically Goal 12 on responsible consumption and production, and initiatives like LiFE (Ravindranath et al. 2023).

3.1 India's Transition to a Circular Economy

The economy of India at present operates mainly in a linear fashion, involving the extraction, processing, and transformation of resources into products for sale to consumers. After they have been used, these items are thrown away, resulting in a substantial amount of waste being produced. India produces more than 62 million tonnes of waste annually, as reported by the Central Pollution Control Board (CPCB). Approximately 43 million tonnes, which accounts for 70%, are gathered, with about 12 million tonnes receiving treatment, while 31 million tonnes are disposed of in landfills. It is projected to reach 165 million tonnes by 2030. Most of this garbage is thrown away in landfills or abandoned in open areas, causing harm to the environment and public health. According to the SPCBs/PCCs, there are 4437 ULBs, 3043 towns and cities, and 968 Class I & II towns and cities across the country. In the year 2016, India contributed 31 million tonnes of trash to the existing waste in its landfills. According to the Annual Report 2018–19 on solid waste management by the Central Pollution Control Board (CPCB), India disposed of 23.35 million tonnes of waste in 3159 dumpsites in 2019, despite four years of the Swachh Bharat Mission (SBM) (Table 2).

A recent Accenture analysis suggests that implementing circular economy business models in India could unlock nearly $500 billion in economic value by 2030. The substantial amount of waste generated in India underscores the necessity of establishing circular economy practices. Each year, approximately 377 million urban residents produce around 55 million tons of municipal solid waste. India is expected to have an urban population of 600 million by 2030 and 814 million by 2050, leading to a projected increase in waste generation to 165 million tonnes by 2030 and 436 million tonnes by 2050 (UN DESA 2019). By 2030, it is estimated that annual greenhouse gas emissions from municipal solid waste will increase to 41.09 million tons. The 'waste to wealth' approach and the concept of circular economy are being more and more embraced in sustainable resource management. The idea of turning waste into wealth tackles environmental and economic issues by converting discarded materials into valuable assets. It fits nicely with the concepts of a circular economy that

Table 2 Summarizing circular economy practices in waste management and waste-to-wealth initiatives worldwide

Region	Circular economy practices in waste management	Waste-to-wealth initiatives	References
Europe	Implementing aggressive waste reduction objectives, expanding recycling infrastructure, and encouraging circular product design through initiatives like the EU circular economy action plan	Innovative methods are being developed to transform waste into useful resources, such as converting organic garbage into biogas and creating recycled materials for manufacturing	Circular economy action plan progress report
United States	Adoption of sophisticated recycling methods, such as chemical recycling and pyrolysis, to turn plastic waste into feedstock for new goods, as well as campaigns promoting the use of recycled materials in production	Implementation of waste-to-energy projects to produce renewable energy from waste, as well as the use of landfill gas for power generation and heating	Advancing the circular economy: national strategy for sustainable materials management. (U.S. environmental protection agency 2023)
China	To reduce waste creation, strict waste management legislation, investments in waste-to-energy projects, and the promotion of circular economy concepts in industrial parks and urban design should be implemented	The emphasis is on transforming waste into riches through efforts like converting agricultural waste into biofuels and using industrial by products as raw materials in other sectors	Ministry of ecology and environment of the People's Republic of China. (Economy Development Report 2023)
India	Adoption of technology-driven waste collection systems, construction of material recovery facilities, and promotion of decentralized waste management efforts in both urban and rural regions	Introduction of waste-to-wealth initiatives aiming at turning organic trash into compost and biogas, as well as schemes to recycle plastic garbage into valuable items	Annual report on circular economy initiatives, Government of India (2022)
Australia	Implementing container deposit programs to encourage recycling, expanding organic waste processing facilities, and promoting circular procurement methods in government procurement regulations	Implementing waste-to-wealth programs, such as converting organic waste into renewable energy and creating recycled materials for infrastructure projects	National waste policy: circular economy action plan department of agriculture, water and the environment, Australia (2023)

support reusing, recycling, and natural system rejuvenation through the development of closed-loop systems.

3.2 Steps Taken Towards Creating a Circular Economy

Adopting Circular Economy principles offers the potential to separate economic expansion from resource exhaustion and environmental harm, leading to a more prosperous and fair society while protecting natural ecosystems for future generations (Chaturvedi and Dhar 2020). Circularity in this situation involves the effective movement and use of resources, materials, and products, going beyond just resource efficiency and recycling. Government initiatives like NMCG and Swachh Bharat Abhiyan stress the significance of circularity in water resource and waste management. Additionally, private sector engagement in circular economy practices is increasing, motivated by economic and environmental factors. India's waste management challenges have worsened due to rapid population growth and industrialization, requiring the need for creative solutions. In reaction, the government has initiated the ambitious Waste to Wealth Mission, indicating a significant change in waste management approaches. The magnitude of the challenge is immense: India is on track to become the most populous nation by 2027, and urban areas are projected to almost double in population to 814 million by 2050. Given these forecasts, circular economy efforts are becoming increasingly important as essential routes to reaching sustainable development in India.

India is set to become the world's third-largest economy, with Prime Minister Narendra Modi stressing the importance of a circular economy for fair and sustainable development. The goal of the circular economy is to reduce waste and pollution, promote the reutilization of products and resources at optimal value, and restore the environment. India needs to harmonize its economic growth with circular principles due to an increasing population, urbanization, environmental issues, and global obligations. The focus of the circular economy is on reducing waste and maximizing the efficient use or reuse of resources. India should adopt circular economy practices to enhance industrial productivity, competitiveness, and resource efficiency. The country's strategy for promoting the circular economy involves using various methods such as regulations, financial incentives, raising awareness, and developing skills. Moving towards a circular economy is essential to achieve India's NDC goals of reducing greenhouse gas emissions and reaching the Sustainable Development Goals for sustainable consumption and production.

3.3 Swacch Bharat Mission

The initiation of the Swachh Bharat Mission in 2014 aimed at enhancing cleanliness, hygiene, and waste management practices. The goal involves efforts to encourage waste separation, recycling, and composting, with the objective of transforming India into a "zero-waste" nation. Decreasing the amount of waste produced is key in achieving a circular economy (https://swachhbharatmission.gov.in/).

3.4 Atal Innovation Mission

The Atal Innovation Mission was initiated in 2016 with the aim of fostering innovation and entrepreneurship in India. The objective involves promoting the growth of circular business models and fostering the uptake of sustainable technologies (https://aim.gov.in/).

3.5 National Resource Efficiency Policy (NREP)

The National Resource Efficiency Policy (NREP) is a crucial policy in reaching the target of a circular economy. The NREP started in 2019 to encourage sustainable production and consumption, improve resource efficiency, and lessen the environmental impact of economic activities. The policy promotes the adoption of circular business models like product-as-a-service, leasing, and sharing, while also encouraging the use of recycled materials. Alongside the NREP, Extended Producer Responsibility (EPR) is also part of the strategies aimed at promoting the development of a circular economy. EPR holds manufacturers and producers accountable for managing the post-consumer waste created by their products under a regulatory framework. The EPR framework promotes sustainable product design practices, encourages increased use of recycled materials, and supports waste management and recycling efforts. (https://moef.gov.in/wp-content/uploads/2019/07/Draft-National-Resourc.pdf).

3.6 Waste to Wealth Mission

The Waste to Wealth Mission has been initiated by the Prime Minister's Science, Technology, and Innovation Advisory Council (PMSTIAC). Its goal is to use technology to bring social and economic benefits to 1.3 billion people in the country. This will also allow us to address issues such as garbage management, inadequate air quality, and rising water source pollution. The Chief Scientific Advisor to the Government is advancing this mission. Furthermore, in order to meet India's pledge to achieve the UN Sustainable Development Goals, the 'Project Management Unit' (PMU) has been set up by this department. Invest India, along with India's National Investment Promotion and Facilitation Agency, also provides support. The goal of the W2W Mission is to enhance the waste management system in India by showcasing creative solutions and models to achieve a country with zero landfill and zero waste. It shows a collective attempt to utilize circular economy principles to tackle the nation's growing waste crisis and also create economic opportunities at the same time. The goal is to protect, responsibly utilize, and repair land, air, and water resources with the help of scientific and technological contributions. It aids Urban

Local Bodies (ULBs) in developing circular economy models for waste management and improves waste management processes in India. The program also aids in the advancement of contemporary technologies to promote a cleaner, more environmentally friendly atmosphere. It enhances the Swachh Bharat Mission and Smart Cities Project through the utilization of science, technology, and innovation. The goal is to develop sustainable waste management models and best practices for cities to adopt. The goal is to address India's waste management issues by discovering and confirming cost-effective technologies for waste treatment, including waste separation, waste processing, and resource reuse. (https://www.psa.gov.in/waste-to-wealth).

The goal of the mission is to create a circular economy for waste, focusing on recovering and reusing resources rather than disposing of them. This includes promoting the utilization of recycled materials in the production of goods, composting, and developing new markets for products crafted from recycled materials. Raising public awareness and developing skills are essential for effective waste management, such as teaching people about separating waste and proper disposal, providing training for communities to manage waste processing units, and enhancing the abilities of local authorities and partners to enforce waste management plans. It is crucial for India's waste management challenges to have collaboration among different entities like government agencies, research institutions, private sector companies, NGOs, and community organizations in order to develop and execute innovative solutions. The mission has the opportunity to provide various advantages to India such as decreased reliance on landfills, enhanced cleanliness and public health, addressing climate change, generating employment, recovering resources, and promoting a circular economy. Social benefits consist of increased awareness of public health, community growth, and chances for entrepreneurial endeavours. Educational programs linked to the mission can enable individuals to embrace responsible waste management behaviours, resulting in a more sustainable way of living. The mission's success depends on efficient execution, cooperation among stakeholders, and ongoing development in waste management technologies and practices. In general, the Waste to Wealth Mission provides a thorough strategy to address India's waste management issues, leading towards a cleaner environment, better economy, and a more sustainable future. India's Waste to Wealth Mission is a plan created to tackle waste management issues and support a sustainable future.

The goal is to turn trash into useful materials and promote a long-lasting future by supporting various United Nations Sustainable Development Goals (SDGs). It plays a key role in achieving SDG 12 through encouraging responsible consumption and production practices like separating waste, recovering resources, and utilizing recycled materials. It further supports SDG 11 through its emphasis on sustainable urban areas, promoting local accountability for waste disposal. The goal also supports SDG 13 by providing other renewable energy options, decreasing dependence on fossil fuels, and lessening greenhouse gas emissions. It has the possibility of generating new job openings in waste management, recycling, and associated fields, which aligns with SDG 8 focusing on decent work and economic growth. The mission

Table 3 Circular economy and its relation with sustainable development goals (SDG's)-2030

Resource efficiency	Human well-being	Environmental impacts
SDG-7 (Affordable and clean energy	SDG-3 (Good health and well being	SDG-6 (Clean water and sanitation)
SDG-12 (Responsible consumption and production)	SDG-5 (Gender equality)	SDG-13 (Climate action)
	SDG-11 (Sustainable cities and communities)	SDG 15 (Life on land)
	SDG-8 (Decent work and economic growth)	

also helps advance SDG 3 by reducing improper waste disposal, preventing environmental pollution, diseases, and promoting better living conditions for good health and well-being. In addition to the mentioned SDGs, the Waste to Wealth Mission also contributes to achieving other goals like SDG 6 (Clean Water and Sanitation) and SDG 15 (Life on Land) through the promotion of sustainable practices that safeguard ecosystems and biodiversity. India's goal with this project is to convert waste into useful materials, which will help reduce environmental damage, boost the economy, and improve overall societal welfare. The Waste to Wealth Mission aims to establish a sustainable and prosperous cycle by embracing a holistic strategy that includes waste reduction, reuse, recycling, and recovery. This transformative plan is supported by a dedication to creativity, teamwork, and development that includes everyone. Transitioning to a circular economy involves moving away from the linear 'take-make-dispose' model towards a regenerative approach that focuses on resource efficiency, waste reduction, and value retention (Table 3).

3.7 Mission Lifestyle for Environment (LiFE)

The Prime Minister of India, Shri Narendra Modi, introduced the concept of LiFE at COP26 in Glasgow on 1 November 2021. Mission Lifestyle for Environment (LiFE), a worldwide movement initiated by Prime Minister Narendra Modi, seeks to encourage both individuals and communities to change their behavior in order to address the climate emergency and preserve the environment. This mission draws inspiration from India's age-old customs and principles, focusing on promoting sustainability and living in harmony with the environment. The goal of the mission is to promote awareness among individuals regarding their consumption habits and to make intentional use of resources. Achieving this goal is possible by implementing basic steps like cutting down on waste, reusing materials, and utilizing energy-saving devices. In addition, Mission LiFE aims to establish a group of individuals known as "pro-planet people" who are dedicated to practicing sustainable living. The community has the ability to collaborate in order to increase understanding of environmental concerns and endorse sustainable behaviors. It focuses not just on personal

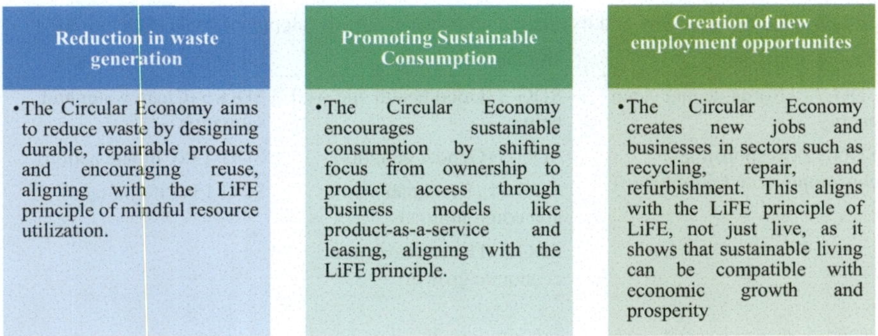

Fig. 2 Circular economy objectives in alignment with mission LiFE (Lifestyle for the environment)

choices, but also on impacting larger socio-economic systems. The goal is to initiate sustainable actions at the local level in order to generate a ripple effect that influences broader systemic shifts. For example, when more people and communities start using sustainable practices, it can impact market demand, prompting industries to transition to more sustainable ways of production. This change has the potential to greatly decrease the environmental footprint and resource consumption in different industries. Moreover, the mission's ultimate goal involves impacting government policies to endorse sustainable consumption and production. Mission LiFE's goal is to establish conditions that support a sustainable future by promoting green technologies and sustainable practices through regulatory frameworks and incentives. This comprehensive strategy understands that addressing the climate emergency necessitates collective actions on small and large scales. Mission LiFE's principles are in line with international sustainability targets, specifically the United Nations' Sustainable Development Goals (SDGs). The mission helps to reach goals for responsible consumption and production (SDG 12), climate action (SDG 13), and sustainable communities (SDG 11) by highlighting the link between personal actions and larger systemic changes. Blending traditional knowledge with sustainable techniques creates a culturally relevant model that can motivate other countries to embrace comparable methods. (Ministry of Environment, Forests, and Climate Change 2023) (Fig. 2).

4 Innovation Ecosystem in India: Corporates Institutions and Businesses in the Circular Economy Space

India, with a population of over 1.43 billion, is highly susceptible to the challenges of climate change, making it a key player on the global climate risk index (Global Climate Risk Index 2020). In order to reduce these risks and guarantee a stable future, it is crucial to reach net-zero carbon emissions by 2070 (Government of India

2020). The potential to drive this transition lies significantly with the private sector and climate technology startups. As per (Unitus and Climate India 2022), the climate tech market in India needs a massive investment of more than US$1 trillion by 2030 to meet sustainability goals effectively (Unitus Ventures 2022). Although there has been significant improvement, the climate technology industry in India is still in its early stages compared to more developed sectors. Yet, there are encouraging indications of expansion, as shown by the increase in investments in climate technology in 2021, exceeding the combined amount of the previous three years. From 2014 to 2022, around US$2 billion has been invested in the industry, showing an increasing dedication and focus (Ministry of Environment, Forest and Climate Change 2022). Nevertheless, there is ample opportunity for growth and progress in the field of climate technology in India. India is actively participating in global climate efforts with programs like Mission LiFE and the Green Credit Initiative. Prime Minister Narendra Modi's urging for practical steps showcases India's dedication to addressing environmental concerns. Moreover, India's dedication to sustainable development is evident through its leadership in initiatives like the Green Development Pact and its prompt fulfillment of NDC commitments. On a global scale, India has played a role in initiatives such as the International Solar Alliance and the Global Biofuel Alliance. The formation of the Global River Cities Alliance, led by India, demonstrates the country's commitment to proper river management and urban regeneration. Throughout COP28, India stressed the significance of the Global Stock take results in improving climate action. India envisions a sustainable future governed by the S.E.E.D.S framework, which emphasizes global collaboration, adaptation, community empowerment, innovation, and climate justice.

In recent years, there has been a noticeable increase in the global conversation about sustainability, driven by worries related to climate change, resource scarcity, and social inequity. Businesses are facing more scrutiny for their environmental and social impacts, which is pushing them to reassess their ways of operating. Due to its diverse corporate environment and quickly expanding industrial sector, India presents an intriguing case to study the transition towards sustainability. Amid mounting worries related to global warming, pollution, and the dwindling of resources, companies in various sectors are realizing the importance of adopting eco-conscious methods and integrating sustainability into their fundamental activities. India has witnessed a significant shift in business attitudes towards environmental sustainability, driven by the acknowledgment that sustainability is not only a moral obligation but also a wise financial choice. Indian startups and corporations, recognized for their flexibility and entrepreneurial ethos, have grasped that embracing sustainable practices is not only ethically sound but also strategically advantageous. These companies are shaping a brighter tomorrow, advocating for beneficial changes, and reducing their environmental impact through the use of sustainable practices. The Indian startup ecosystem is flourishing due to innovation and energy, propelled by an increasing consciousness of environmental accountability, thus encouraging green entrepreneurship. These new businesses are developing items for sustainable farming, environmentally-friendly transit, waste disposal, and renewable power. They aim to secure funding from entities that place importance on environmental, social, and governance standards, like impact investors, sustainable investment funds, and green bonds. This access to capital is crucial for accelerating the growth of the green economy. With severe global concerns such as climate change, pollution, and resource depletion, there has never been a greater need for new solutions. Fortunately, a new generation of entrepreneurs is rising to the challenge, leveraging innovation and technology to build firms that value not only profit but also environmental responsibility. To solve environmental issues, a number of Indian corporations are utilizing creativity and technology. Tata Group is an excellent illustration of a business leader in environmental sustainability. To lessen their carbon footprint, Indian corporations are putting different energy-efficiency initiatives into practice. This entails putting energy-efficient technology into use, streamlining procedures, and updating infrastructure. Some of the top Indian Corporates who are incorporating environmentally friendly measures have been discussed below.

4.1 Tata Group

The Tata Group, which operates companies in a number of industries including steel, automotive, energy, and hospitality, has made significant strides in integrating sustainability into its operations. Tata Steel, for example, has undertaken a number of measures to lessen its carbon footprint, such as increasing its energy efficiency, investing in renewable energy, and implementing greener production techniques (https://www.tatasteel.com).

4.2 Infosys

One of India's top providers of IT services, Infosys, is another noteworthy example. Infosys has integrated energy efficiency, waste minimization, and the use of renewable energy sources as pillars of its sustainability approach. In order to become carbon neutral and lower its per-person energy usage, the corporation has set high goals. Sustainable business practises used by Infosys have not only reduced their negative effects on the environment but also significantly reduced costs and improved operational efficiency. Businesses such as ITC Limited in the retail industry have shown a strong dedication to environmental sustainability.

4.3 ITC

Under its "Sustainability 2.0" strategy, ITC has put in place a number of programs, including as carbon-positive operations, sustainable agriculture methods, and water management. ITC works to increase agricultural productivity while reducing environmental effect in rural communities by promoting sustainable farming techniques through its e-Choupal initiative. Likewise, Wipro, a founding partner of Transform to Net Zero, is emphasizing sustainability via waste management, biodiversity conservation, water efficiency, and energy efficiency methods. Using procurement, supply chain management, employee engagement, customer stewardship, and trade associations, it involves incorporating sustainability into business operations. With a 55% reduction by the year 2030, Wipro Infotech hopes to achieve zero emissions by 2040 (https://www.itcportal.com/sustainability/sustainability-integrated-report-2023/ITC-Sustainability-Integrated-Report-2023.pdf).

4.4 Zomato

The popular food delivery app i.e. Zomato, initiated green delivery project, launched by, is to reduce carbon emissions associated with meal deliveries. With the use of electric cars, reforestation initiatives to offset emissions, and route optimization, the company hopes to make a portion of its delivery fleet carbon neutral. Through partnerships with restaurants to offer recyclable or biodegradable options, Zomato also promotes the use of eco-friendly packaging materials. Customers are informed about the advantages eco-friendly delivery has for the environment and are encouraged to choose green delivery options by the initiative. Through the measurement and disclosure of important data, such as packaging waste, carbon emissions, and other sustainability indicators, Zomato encourages transparency.

4.5 Mother Dairy Fruit and Vegetable Pvt. Ltd

Mother Dairy Fruit & Vegetable Pvt. Ltd. (MDFVPL) is the first dairy cooperative in India to register with the Central Pollution Control Board and implement EPR. The company has collected and processed 7284 MT of post-consumer used plastic trash, which is made up of Single Layered Plastic (SLP) and Multi-Layered Plastic (MLP) waste. MDFVPL has also switched to eco-friendly packaging, used fewer plastic bags, and substituted paper trays for plastic ones in addition to wooden spoons.

4.6 Godrej

The biggest furniture brand in India, Godrej Consumer Products, is now included in the 2019 Dow Jones Sustainability Emerging Markets Index. The largest furniture brand in India, Godrej Interio, wants to achieve carbon neutrality, zero waste, and a 25% decrease in energy use. The business regularly produces environmentally friendly goods, such as the first line of refrigerators in the nation to be entirely green (https://www.godrej.com/Resources/sustainability-full-report.pdf).

Apart from the above-mentioned corporates, Many Indian institutions and organizations are playing a pivotal role in promoting a circular economy, a system aimed at eliminating waste and resource consumption. They formulate policies, conduct research, and innovate in areas like waste management and recycling technologies. Indian businesses are recognizing the importance of circular economy principles for long-term sustainability and competitiveness. The country has a thriving start-up ecosystem focused on sustainability and circular economy solutions. Non-governmental organizations and civil society groups raise awareness about the circular economy and advocate for sustainable practices. Indian institutions collaborate with international organizations to align their efforts with global sustainability goals. They also invest in skill development programs to build a workforce capable of implementing circular economy practices. Some of the institutions who are promoting circular economy measures have been discussed below:

4.7 Confederation of Indian Industry (CII)

Confederation of Indian Industry (CII) has been actively promoting the circular economy in India through various initiatives. These include policy advocacy, awareness and education, business engagement, research and innovation, capacity building, international collaboration, and recognition and awards. CII has worked with government bodies to develop frameworks that encourage resource efficiency, waste reduction, and sustainable production practices. They have organized workshops, seminars, and conferences to raise awareness about the benefits and principles of the circular economy among businesses, policymakers, and the public. CII also works closely with businesses across sectors to promote circular business models, innovation, and best practices. CII supports projects aimed at developing sustainable technologies, processes, and products that minimize waste generation, maximize resource efficiency, and promote recycling and reuse. CII offers capacity-building programs and training sessions to equip businesses with the knowledge and skills needed to implement circular economy practices effectively. It also collaborates with international organizations and industry associations to exchange best practices and insights on the circular economy. CII recognizes and rewards businesses that demonstrate exemplary performance in adopting circular economy principles through initiatives like the CII-ITC Sustainability Awards (https://www.cii.in/).

4.8 The Energy and Resource Institute (TERI)

The Energy and Resources Institute (TERI) is a prominent player in India's promotion of the circular economy. They conduct extensive research on resource efficiency, waste management, sustainable production, and consumption patterns, providing evidence-based insights to policymakers. TERI organizes capacity-building programs, workshops, and training sessions to enhance awareness and understanding of the circular economy among various stakeholders. Demonstration projects showcase innovative circular economy solutions in practice, illustrating the feasibility and benefits of adopting circular practices across sectors like waste management, renewable energy, water conservation, and sustainable agriculture. TERI collaborates with government agencies, industry associations, research institutions, and international organizations to drive collective action towards advancing the circular economy agenda. They engage in advocacy efforts to raise awareness about the importance of the circular economy and promote policy reforms that facilitate its mainstreaming. They also promote innovation and technology transfer to accelerate the transition towards a circular economy, supporting the development and commercialization of sustainable technologies, processes, and products. TERI also collaborates with educational institutions and research organizations to integrate circular economy principles into academic curricula and research programs, nurturing talent and fostering interdisciplinary collaboration. This contributes to building a skilled workforce capable of driving sustainable development in line with circular economy principles. (https://www.teriin.org/).

An increasing number of creative businesses are entering India's with the goal of solving environmental issues, especially in the area of waste management, plastic alternatives. Moreover, Customer demand for eco-friendly substitutes for traditional items is rising as environmental challenges become more widely known. Sustainability and biodegradability are being introduced in a variety of areas by green entrepreneurs looking to capitalize on this trend. Some of these green businesses along with their initiatives have been discussed below:

4.9 Karo Sambhav

Karo Sambhav is a leading e-waste management startup in India that focuses on creating sustainable solutions for electronic waste management. They collaborate with various stakeholders including manufacturers, recyclers, and consumers to establish responsible recycling practices. Karo Sambhav operates collection centers across the country and organizes awareness campaigns to promote e-waste recycling. Their initiatives have helped divert significant amounts of electronic waste from landfills. (https://www.karosambhav.com/).

4.10 Saahas Zero Waste

Saahas Zero waste is a social company that specializes in resource recovery, recycling, and trash segregation while providing all-inclusive waste management solutions. They collaborate with communities, businesses, and municipalities to put into effect sustainable waste management techniques. The "Zero Waste Program," one of its noteworthy projects, seeks to divert 90% of garbage from landfills by encouraging source segregation and composting. Several Indian towns have seen the effective implementation of trash management programs by Saahas Zero trash (https://saahaszerowaste.com/waste-management-services-bangalore/).

4.11 Waste Ventures India

Improved waste management techniques in urban and peri-urban regions are the main goal of Waste Ventures India, a social company. Via their inclusive business approach, they give garbage pickers employment possibilities in addition to delivering waste collection and recycling services to homes, companies, and institutions. Garbage Ventures India has material recovery facilities that select, process, and sell recyclable garbage to businesses. Through their efforts, waste's negative effects on the environment have been lessened, and underprivileged populations have benefited socioeconomically (https://www.business.repurpose.global/).

4.12 Blue Planet Environmental Solutions

Blue Planet Environmental Solutions is a company that focuses on sustainable waste management and recycling solutions. Their mission is to create a zero-waste world by transforming waste into valuable resources. They offer core services such as waste collection and segregation, recycling and resource recovery, and energy from waste. They also provide consultancy and advisory services, such as sustainability consulting and compliance with environmental regulations. Innovative technologies include smart bins and IoT solutions, AI and machine learning, and blockchain for waste management. These technologies optimize waste collection routes and reduce operational costs. They also use blockchain technology to ensure transparency and traceability in the waste management process. Blue Planet Environmental Solutions has a significant environmental impact, reducing landfill use, reducing carbon footprint, and conserving natural resources. They also implement public awareness campaigns and school programs to educate children about sustainability and waste management. They also partner with local and national governments to implement waste management solutions and with businesses to develop customized waste management strategies. Blue Planet Environmental Solutions is at the forefront of the waste management industry, driving innovation and promoting sustainable practices that benefit both the environment and society (https://blueplanet.asia/).

Apart from the above-mentioned green Startups, Sustainable alternatives to single-use plastics are offered by creative enterprises like Bakeys Edible Cutlery and Pappco Greenware. Some examples of these alternatives are disposable dinnerware made of biodegradable materials, such as bamboo, and edible cutlery made from millet flour. By providing sustainable solutions to their customers, these entrepreneurs are contributing to the circular economy and reducing the amount of plastic pollution. India's fast urbanization has resulted in worsening traffic congestion and air pollution in major cities. Green business owners are tackling these issues by endorsing ecologically friendly and sustainable modes of transportation. Companies leading the way in electric vehicle (EV) technology, such as Ampere Electric and Ather Energy, provide electric bikes and scooters as green substitutes for conventional gasoline-powered autos. Such as Recykal, Plastics for Change, and TerraBlue XT are all focusing on waste management and circular economy principles. Recykal offers technology solutions for optimizing resource use, Plastics for Change connects waste collectors, TerraBlue XT develops IoT-based waste management solutions (Table 4).

4.13 Circular Economy Action Plan for Managing Wastes in the Electrical and Electronic Goods and Plastic Sectors: The Road Ahead

Implementing a circular economy for electrical and electronic products is essential for sustainable resource management and minimizing environmental harm. Important tactics for advancing a circular economy involve designing products to last longer and be easily taken apart, implementing extended producer responsibility (EPR), establishing take-back and recycling initiatives, focusing on material efficiency and recycling, promoting remanufacturing and refurbishment, introducing digital product passports and tracking systems, and redefining criteria for bulk consumers. The existing definition of bulk consumers lacks any reference to their requirement for yearly submissions. The informal sector should play a significant role in enhancing a strategic collection system for e-waste instead of being completely excluded from the industry. A system for evaluating recyclers could be implemented, with the government designating companies as Grade 1, 2, or 3 recyclers and mandating that other dismantlers and recyclers must send their materials to those unable to fully process them. Bulk consumers who fail to responsibly dispose of their e-waste should also be included in environmental compensation measures. It is important to thoroughly research current recycling companies before making a decision, and all government entities, schools, and universities should send their materials to approved recyclers. Additionally, motivators for circular design may consist of tax refunds, eco-friendly certification, acknowledgment from the public, financial support for

Table 4 Future action plan for achieving the goals of circular economy in various cities of India

Regions	Proposed measures	References
Delhi	Implementation of innovative waste sorting techniques based on AI and machine learning, growth of decentralized composting facilities, and integration of waste-to-electricity projects into the city's energy grid	Sharma, A., and Gupta, S. (2023). Waste Management and Research, 41(9), 904–916
Mumbai	Innovative plastic recycling activities, such as transforming plastic trash into construction materials and expanding waste-to-energy projects, are being implemented to minimize landfill dependency and enhance resource recovery	Patel, R., and Desai, M. (2023). Journal of Environmental Management, 302, 114,031
Bengaluru	Adoption of technology-driven garbage collection solutions, such as IoT-enabled smart bins and route optimization algorithms, as well as the construction of community recycling centers for electronic waste management	Kumar, S., and Rao, P. (2022). Waste Management, 133(Part A), 220–229
Kolkata	Implementation of community-based waste segregation initiatives, backed by mobile applications for citizen involvement, as well as the growth of composting facilities to convert organic waste into nutrient-rich soil conditioners	Chatterjee, S., and Das, A. (2022). Resources, Conservation and Recycling, 184, 105467
Chennai	To reduce waste creation, enterprises should implement required extended producer responsibility (EPR) legislation, as well as incentives for eco-friendly product design and circular buying practices	Annual report on circular economy initiatives, Government of India (2022)
Pune	Zero waste programs will be implemented in conjunction with local communities, with an emphasis on trash reduction and reuse, as well as the creation of material recovery facilities to recover valuable materials from waste streams	Jain, R., and Shah, K. (2023). International Journal of Environmental Studies, 80(6), 897–910

research and innovation, specialized contests for innovation, partnerships, and platforms for sharing resources. Tax refunds or lower tax rates may incentivize manufacturers to integrate sustainable methods. Green certification incentivizes companies following circular design principles by providing them with government contracts or exclusive market access. (https://www.meity.gov.in/writereaddata/files/Circular_Economy_EEE-MeitY-May2021-ver7.pdf).

Recognition awards or programs in the public eye highlighting companies excelling in circular design, improving their reputation and encouraging competition in the sector. Funding for research and innovation should be dedicated to advancing eco-friendly technologies and processes in the electronics industry, with a focus on promoting circularity through innovative methods. This could involve funding for sustainable technologies, research centers working together, innovation competitions, and partnerships between industry players, government organizations, and NGOs. Fairphone, Apple, Dell, Renault, IKEA, Sony, and the Netherlands are all incorporating circular practices into their products. Fairphone, a company based in the Netherlands, creates smartphones focusing on ethical sourcing, durability, and the ability to be easily repaired. Their products are designed with modularity in mind, enabling the simple replacement of individual components such as batteries and cameras. Apple's Liam Robot was created to take apart old iPhones, which enables the recycling and reusing of parts, ultimately decreasing the demand for new raw materials. Dell has set up a closed-loop recycling system where they collect used electronics and reuse the materials to make new products. Renault is currently developing initiatives to repurpose or recycle old EV batteries, such as exploring ways to utilize retired EV batteries for energy storage across different uses. IKEA introduced a program called Buy Back and Resell, where customers can bring back used furniture, as well as electronics such as lamps and chargers, in return for store credit. Sony has implemented programs for the return and recycling of its products to promote the responsible disposal and recycling of electronic waste. The Netherlands is working on a national program to encourage circular practices in the electronics industry, with cooperation among government bodies, manufacturers, and stakeholders to create and execute plans for eco-friendly design, recycling, and reusing electronic goods (Table 5).

The National Circular Economy Framework, which was originally released in November 2023 (CII 2023), presents essential tactics for encouraging a circular

Table 5 Circular economy action plan for managing wastes in the electrical and electronic goods

Key strategies	Outcomes
Product design for durability and disassembly	Designing products for long-term use and easy disassembly to facilitate repair, reuse, and recycling
Extended producer responsibility (EPR)	Holding manufacturers responsible for the entire lifecycle of their products, including collection, recycling, and disposal
Take-back and recycling programs	Establishing programs for consumers to return used products for recycling or proper disposal
Material efficiency and recycling	Maximizing resource efficiency and recycling to reduce waste and conserve resources
Remanufacturing and refurbishment	Reusing components or products through refurbishment or remanufacturing to extend their lifespan
Digital product passports and tracking	Implementing digital tracking systems to trace a product's lifecycle and facilitate easier recycling or refurbishment
Utilization of informal sector	Incorporating the informal sector into e-waste management for strategic collection mechanisms
Rating mechanism for recyclers	Introducing a rating mechanism for recyclers and recognizing them as Grade 1, 2, 3 recyclers
Environmental compensation for bulk consumers	Extending environmental compensation to bulk consumers for responsible e-waste disposal

economy specifically in the plastics sector. Some measures to enhance recycling efforts are creating designs with recycling in mind, enforcing Extended Producer Responsibility (EPR), improving recycling facilities, encouraging bottle-to-bottle recycling, educating consumers on recycling and disposal, backing new recycling technologies, enforcing plastic ban and rules, urging businesses to incorporate recycled plastics, adopting circular business practices, facilitating cooperation among key players, promoting consumer attitude adjustments, and supporting research for alternative materials to plastics. India is at the forefront of adopting EPR for plastic waste, but it is important to closely oversee its implementation. The existing GST system on plastic items discourages the utilization of recyclable materials, resulting in a scenario where both new plastic products and recycled granules are taxed twice. It is crucial to invest in and expand recycling infrastructure to efficiently process collected plastics and transform them into new products. Promoting bottle-to-bottle recycling encourages the utilization of recycled plastics in making new goods, establishing a closed-loop process that minimizes the requirement for producing new plastic. Consumer behaviour change promotes changes in the way consumers reduce plastic consumption, choose reusable options, and make sustainable decisions. Investigation and creativity are crucial for discovering substitute materials for conventional plastics and investigating waste-to-energy techniques that reduce the environmental consequences of burning plastics. Establishing monitoring and reporting systems is important to monitor the progress of efforts to recycle and reduce plastic. Coca-Cola India introduced the initial 100% recycled PET water bottle, and Refillable delivers detergent and cleaning liquids directly to customers' homes. Infinity Box provides food containers that can be used multiple times and partners with food delivery applications. Banyan Nation repurposes High Density polymers into new shampoo and lubricant bottles to address India's plastic issue (Table 6).

5 Circular Economy and Fulfilling the Green Development Agenda and SDG's

5.1 Policy and Regulations

Policy interventions are crucial in assisting the shift towards a Circular Economy. Regulatory standards, design requirements, tax reforms, extended producer responsibility, and public awareness interventions are key elements in implementing a national Circular Economy framework. Regulations play a crucial role in this process. Guidelines provided by regulations help industries, businesses, and individuals in implementing Circular Economy practices. This offers a shared structure for sustainable business practices. The National Resource Efficiency Policy (NREP) that was introduced in 2019 is an extensive policy that seeks to encourage sustainable production and consumption practices, improve resource efficiency, and decrease the environmental effects of economic activities. Current regulations consist of Extended

Table 6 Circular economy action plan for managing wastes in the plastic sector

Key strategies	Outcomes
Designing for recycling	Designing products with recyclability in mind to facilitate easier recycling and reprocessing
Promoting bottle-to-bottle recycling	Encouraging the use of recycled plastics in manufacturing new products to create a closed-loop system
Raising consumer education	Educating consumers about recycling practices, proper disposal, and the importance of reducing plastic use
Supporting innovative recycling technologies	Investing in and promoting technologies that enhance plastic recycling efficiency and sustainability
Enacting plastic bans and regulations	Implementing regulations to restrict the use of certain plastic products and promote alternatives
Encouraging businesses to use recycled plastics	Incentivizing businesses to incorporate recycled plastics into their products and packaging
Promoting consumer behaviour change	Encouraging consumers to reduce plastic consumption, opt for reusable alternatives, and make sustainable choices
Establishing monitoring and reporting mechanisms	Setting up transparent mechanisms to track progress in plastic recycling and reduction efforts
Working with international partners	Collaborating with international partners to develop consistent standards and guidelines for a circular economy in plastics

Producer Responsibility (EPR) programs, green credits, and prohibiting single-use items. EPR programs necessitate producers to handle post-consumption products, promoting longevity, repairability, and recyclability. Circular Economy initiatives could be incorporated into green credits programs. Rules for waste management could potentially include a wider range of waste types, such as tires and agricultural waste. India's public procurement policies may be changed to prioritize circular products and services. Requirements for using recycled materials could lead to a higher need for recycled content. Rules may stop companies from greenwashing by mandating that they support their statements with real actions. Product design standards should be enhanced to guarantee longevity, repairability, and recyclability. Businesses may have an obligation to disclose their performance in the circular economy. Rules could also stop planned obsolescence, which causes products to fail early.

5.2 Incentives to Businesses

Governments can provide different incentives to businesses that encourage the adoption of Circular Economy practices. Some examples are tax incentives, subsidies, funding, and guaranteed loans. Non-monetary rewards comprise of favouritism in public procurement, accolades for top companies, and support and education to aid businesses in adopting these procedures. These rewards mitigate the dangers of borrowing and encourage practices that are sustainable. (OECD 2021).

5.3 Public Awareness on Circular Economy

Public awareness is essential for a successful shift towards a circular economy. Governments, businesses, and non-profit organizations have the ability to initiate public education campaigns, community events, social media usage, and partnerships in order to enhance awareness. Various channels, including television, radio, social media, and print media, can be utilized for these campaigns. Events within the community, like festivals and fairs, have the potential to boost circular economy enterprises and their products. Collaborations have the ability to expand the reach of messaging and target a larger audience effectively. The purpose of the "A New Plastics Economy" initiative by the Ellen MacArthur Foundation is to advocate for a Circular Economy specifically for plastics. The "Circular Economy Action Plan" of the European Commission and the "Circular Economy Strategy" of the UK both focus on increasing public awareness by utilizing educational resources, social media promotions, and partnerships with companies. The goal of these initiatives is to educate and involve the public in order to transition successfully to a Circular Economy. (CII 2021).

5.4 Investment in Research and Development for Circular Economy

Investing in Research and Development (R&D) is essential for moving towards a sustainable future. It assists in creating technologies and methods that back the Circular Economy, decrease waste and pollution, enhance resource efficiency, and generate fresh economic possibilities. India can lead the global Circular Economy movement by focusing on research and development in product design, new materials and technologies, digital technologies, industrial symbiosis opportunities, expanding recycling materials, supporting circular business models, and creating markets for recycled products. India can lead the global Circular Economy movement by emphasizing modular components, diagnostic tools, eco-friendly raw materials, 3D printing, and digital technologies.

5.5 *Promotion of Public–Private Partnerships for Circular Economy*

Public–Private Partnerships (PPPs) are partnerships between public and private sectors that aim to achieve common goals. They can support the Circular Economy by developing infrastructure, promoting research, providing financial support, and raising awareness. PPPs can increase investment in Circular Economy projects, accelerate innovation, and increase public awareness of the benefits of the Circular Economy. They involve a wide range of stakeholders, including businesses, governments, and non-governmental organizations, fostering a more sustainable future. The Circular Economy Covenant in the Netherlands, a PPP between the government and businesses, has reduced waste by 20%. In the UK, the Ellen MacArthur Foundation's Circular Economy 100 network has developed projects like Circular Cities and Regions Initiative. In China, the government has launched PPP initiatives to support the Circular Economy, such as developing industrial parks and promoting recycled materials in the construction industry (https://www3.weforum.org/docs/White_paper_Circular_Economy_in_Cities_report_2018.pdf).

5.6 *Circular Economy Knowledge Network*

The Indian government has the potential to create a Circular Economy Knowledge Network to support sharing of information and cooperation across different sectors, such as government, business, academia, and civil society. This system has the potential to expedite progress in the Circular Economy by enabling the exchange of ideas and technologies, offering training opportunities for policymakers and businesses, and cultivating a skilled workforce. It may also establish fresh markets for recycled materials and circular products, increase understanding of their advantages, and collaborate with other knowledge networks. The network might arrange frequent events and workshops to exchange knowledge and work together on Circular Economy initiatives. It could also increase knowledge through workshops, educational materials, and social media. The network might also support policies and initiatives to speed up the shift to a Circular Economy.

5.7 *Circular Economy Park*

Circular Economy parks, also referred to as eco-industrial parks or resource recovery parks, are designated regions that aim to enhance resource efficiency and minimize waste by fostering cooperation between businesses and stakeholders. These parks offer a physical structure and encouraging atmosphere for businesses to embrace Circular Economy principles, like industrial symbiosis, circular supply chain management, and waste valuation. Businesses can enhance their profitability and sustainability by decreasing resource consumption, waste disposal costs, and energy expenses. Public parks frequently provide common facilities like sewage treatment facilities, recycling stations, and sustainable energy sources, thereby lowering the need for individual expenditure and ongoing expenses. Sharing knowledge and working together in a park can help create new and creative solutions. Regulatory assistance and benefits, like tax incentives or subsidies, motivate companies to implement environmentally friendly practices. Circular Economy parks around the world include Kalundborg Symbiosis in Denmark, Eco-Industrial Park of As Pontes in Spain, and Circular Economy Park of the Eastern Metropolitan Region in Thailand. In India, new circular parks are being established like the Industrial Symbiosis Hub by the National Industrial Corridor Development Corporation (NICDC) in Tuticorin, Tamil Nadu, and The Circular Economy Park by the Maharashtra Industrial Development Corporation (MIDC) in Talegaon, Maharashtra.

5.8 *Institutional Frameworks*

The implementation of a thorough institutional framework is essential for India's impactful Circular Economy initiative. This involves creating a clear plan and direction for the circular economy with input from various stakeholders such as government, businesses, academia, and civil society. It is important to fill institutional gaps by creating new institutions or revamping existing ones in order to enhance coordination among various agencies. To guarantee successful implementation of the circular economy, it is essential to invest in capacity building so that all stakeholders possess the required knowledge and skills. Fostering innovation can be achieved by investing in research and development and offering tax incentives to businesses that innovate with new circular technologies and business models. In order to establish a more effective institutional structure, the creation of the National Circular Economy Authority (NCEA) is necessary. This entity will be in charge of formulating and executing the national Circular Economy plan, as well as overseeing the efforts of various government departments and organizations, while also offering assistance to businesses and other interested parties. The present regulations do not support the circular economy, requiring collaboration between the NCEA and other government bodies to reform them. It is crucial to invest in the research and development of innovative circular technologies and business models as well. This could be accomplished by forming public–private partnerships or providing grants to academia and businesses.

In addition to the steps mentioned earlier, the government needs to set up a fund for Circular Economy to offer financial assistance to companies adopting circular strategies. Creating a Circular Economy curriculum for both schools and universities at a national level will guarantee that every student is educated on the circular economy and its significance. A potential layout for a National Circular Economy Authority in India may consist of a governing body, an executive committee, technical committees, and a secretariat. The NCEA is responsible for creating and executing the national Circular Economy strategy, overseeing ministries and agencies, aiding businesses and stakeholders, advancing circular economy research, and increasing awareness of the circular economy. Besides these roles, the NCEA may create and enforce criteria for recycled materials and products, offer monetary rewards to companies embracing circular practices, support the establishment of circular economy infrastructure, and oversee resource usage and waste management. Ensuring an inclusive and fair institutional framework is vital to guarantee all stakeholders, including marginalized communities, have a say in the circular economy strategy's development and execution. The Circular Economy has the potential to thrive in India if a more inclusive and equitable institutional framework is established.

6 Conclusion

The global focus on sustainable development has grown due to pressing environmental issues faced by governments worldwide. The adoption of the circular economy concept, which promotes environmental sustainability and drives economic prosperity, has gained increased attention due to the importance of resource security and efficiency for country and organizational growth. The circular economy aims to change the definition of growth by focusing on positive impacts that benefit society as a whole, separating economic activity from limited resource use and eliminating waste through design. The growth of circular economy practices is motivated by factors such as the increasing global focus on environmental sustainability, strict rules and policies, and legal frameworks introduced by governments and international organizations like the European Union. Pearson and Turner were pioneers in incorporating the term circular economy into an economic framework, building on the concept that "everything is connected." They created a new economic model called the circular economy, based on thermodynamic principles, which highlights the connection between the economy and the environment by incorporating three economic functions: providing resources, assimilating waste, and serving as a source of utility. The economic benefits of a circular economy are becoming increasingly clear, with businesses recognizing the financial advantages of using fewer raw materials and the possibility of earning profits through recycling products. Conventional linear economic models have led to significant resource depletion, environmental harm, and social inequality. The circular economy focuses on reducing, reusing, recycling, and regenerating resources and products, marking a groundbreaking shift. Nations can strive to minimize their impact on the environment, improve resource usage, and promote inclusive growth by embracing this novel economic approach. The global conversation about sustainability has increased in recent years due to concerns about climate change, resource scarcity, and social inequity. Businesses are increasingly recognizing the importance of adopting eco-conscious methods and integrating sustainability into their operations. India, with its diverse corporate environment and rapidly expanding industrial sector, presents an intriguing case to study the transition towards sustainability. Indian startups and corporations, known for their flexibility and entrepreneurial ethos, are embracing sustainable practices not only ethically sound but also strategically advantageous. The Indian startup ecosystem is flourishing due to innovation and energy, driven by an increasing consciousness of environmental accountability. These new businesses are developing sustainable farming, environmentally-friendly transit, waste disposal, and renewable power. They aim to secure funding from entities that place importance on environmental, social, and governance standards, such as impact investors, sustainable investment funds, and green bonds. With severe global concerns such as climate change, pollution, and resource depletion, there is a greater need for new solutions. A new generation of entrepreneurs is leveraging innovation and technology to build firms that value not only profit but also environmental responsibility. Indian corporations are using creativity and technology to solve environmental issues, such as Tata Group, which is implementing energy-efficiency initiatives to reduce their carbon footprint. Circular Economy principles aim to separate economic expansion from resource exhaustion and environmental harm, leading to a prosperous and fair society while protecting natural ecosystems for future generations. Government initiatives like NMCG and Swachh Bharat Abhiyan emphasize the importance of circularity in water resource and waste management. Private sector engagement in circular economy practices is increasing due to economic and environmental factors. India's waste management challenges have worsened due to rapid population growth and industrialization, necessitating creative solutions. The government has initiated the Waste to Wealth Mission, indicating a significant change in waste management approaches. India's rapid population growth and urbanization are causing a need for circular economy efforts to reach sustainable development. Prime Minister Narendra Modi emphasizes the importance of a circular economy for fair and sustainable development. The goal of the circular economy is to reduce waste and pollution, promote resource reutilization, and restore the environment. India should adopt circular economy practices to enhance industrial productivity, competitiveness, and resource efficiency. The country's strategy for promoting the circular economy involves regulations, financial incentives, raising awareness, and skill development. Moving towards a circular economy is essential to achieve India's National Development Goals of reducing greenhouse gas emissions and reaching the Sustainable Development Goals for sustainable consumption and production. India's growing population can benefit from the Circular Economy concept by integrating traditional practices of reusing, reducing, and repairing into contemporary sustainability efforts. The country's cultural values, such as respecting nature through festivals like Vasant Panchami and Chhath Puja, promote the protection and preservation of natural resources. Traditional agricultural practices, such as crop rotation and cow dung use, improve soil fertility and reduce waste. Indian cooking traditions also showcase the circular nature of food, with leftovers being used to create healthy meals like khichdi and upma. Excess food is often reused as compost or animal fodder, showcasing the circular nature of Indian traditions. In textiles, Indians prefer clothing made from natural materials like cotton and wool, known for their durability and potential for reuse or recycling. This aligns with the concepts of circular fashion, where garments are designed to last long and be easily recycled. Traditional Indian craftsmanship often uses reused materials, such as cloth, paper, and metal, to create new items like bags and home decor. Glass bottles and plastic containers are also reused until they are no longer usable, conserving resources and reducing waste. The G20 declaration in September 2023 highlighted the importance of the Circular Economy, emphasizing Extended Producer Responsibility (EPR) and resource efficiency for sustainable development goals. To maintain progress, a strong framework is essential, aligned with Prime Minister Narendra Modi's vision and the G20 declaration, encouraging stakeholders like industries and investors. This framework should support sustainable practices, innovation, and economic viability while ensuring profitability for all stakeholders. Policy and regulations are crucial for a circular economy transition, including the National Resource Efficiency Policy, incentives for businesses, public awareness campaigns, investment in research and development, public–private partnerships, and a Circular Economy Knowledge Network. These measures aim to promote sustainable practices, reduce waste, and enhance resource efficiency. Public–private partnerships can support the circular economy by developing infrastructure, promoting research, and providing financial support. Institutional frameworks are also essential for successful implementation, fostering innovation through research and development and tax incentives. To sum up, looking into circular economy options in the context of India's green development agenda and sustainable development goals highlights the importance of using inventive approaches to address waste and increase wealth at the same time. The research shows that transitioning to a circular economy model offers various advantages such as enhanced resource efficiency, minimized environmental harm, and promotion of economic growth. Analyzing the perspective of India shows that the nation is at a pivotal moment where adopting circular economy concepts could help tackle pressing environmental issues and create opportunities for fair and sustainable development. Efforts like waste-to-energy projects, eco-design practices, expanded producer responsibility initiatives, and the growth of remanufacturing and recycling sectors have the potential to shift India's economy towards a more sustainable and resilient future. Nonetheless, achieving the complete benefits of circular economy solutions necessitates cooperation among diverse stakeholders, such as governments, businesses, educational institutions, and the general public. Policy frameworks need to be created and put into action to promote circular behaviours, drive innovation, and guarantee fair distribution of benefits across different sectors of society. In addition, it is crucial to educate and raise awareness in order to promote a sustainable mindset and induce changes in behaviour for both consumers and producers. Working together with partners and sharing knowledge can facilitate the exchange of best practices and establish a supportive environment for the advancement of circular economy initiatives. This study has important practical implications for policy formation, commercial innovation, environmental and social benefits, education, and international collaboration. It can inform policymakers about effective strategies for incorporating circular economy principles into national and regional development plans, make evidence-based policy design and implementation recommendations, create supportive regulatory frameworks, and strengthen institutional capacity and governance structures. This research highlights the significance and feasibility of transitioning to a circular economy model in India. By adopting the concepts of waste reduction, resource efficiency, and value retention, the country may pave the way for greener and more prosperous growth, while also contributing to the global effort to accomplish the Sustainable growth Goals. As we move forward, we must remain dedicated to using the transformational power of circular economy solutions to build a more sustainable, fair, and resilient future for future generations.

7 Future Recommendation

Future research on circular economy solutions in India should include longitudinal studies, impact assessment frameworks, behavioural studies, technological innovations, policy analysis, sectoral case studies, community engagement and inclusion, education and awareness campaigns, innovative business models, and policy recommendations. These recommendations seek to enhance knowledge and implementation of circular economy solutions, as well as to contribute to the Green Development Agenda and the Sustainable Development Goals by giving practical insights for policymakers, businesses, and stakeholders from civil society.

References

Chaturvedi A, Dhar A (2020) Circular economy in India: challenges and opportunities. Curr Sci 118(4):491–497. https://www.jstor.org/stable/26915718

Dhaliwal S, Singh S (2019) Circular economy in India: opportunities and challenges. Sustain Prod Consumption 20:28–35

Dwivedi S, Khan MH, Pandey KK (2021) Sustainable development goals and circular economy: an Indian perspective. In: Sustainable development goals (SDGs) and climate justice. Springer, Singapore, pp 287–298

Ellen MacArthur Foundation (2015) Towards the circular economy: accelerating the scale-up across global supply chains. Retrieved from https://www.ellenmacarthurfoundation.org/publications/towards-the-circular-economy-vol-3-accelerating-the-scale-up-across-global-supply-chains

Ellen MacArthur Foundation (2019) Circular economy: the concept. Ellen MacArthur Foundation. https://www.ellenmacarthurfoundation.org/circular-economy/concept

European Commission (2020a) Circular economy action plan. Retrieved from https://ec.europa.eu/environment/circular-economy/index_en.htm

European Commission (2020b) Circular economy action plan. European commission. https://ec.europa.eu/environment/circular-economy/index_en.htm

Ghisellini P, Cialani C, Ulgiati S (2016) A review on circular economy: the expected transition to a balanced interplay of environmental and economic systems. J Clean Prod 114:11–32

Gupta R, Sharma A (2019) Transition to a circular economy in India: a systematic literature review. Resour Conserv Recycl 141:127–139

http://envigreen.in/. Accessed 17 March 2024

https://business.outlookindia.com/news/mother-dairy-aims-to-turn-plastic-waste-neutral-by-fy-2023-24--news-200010. Press trust of India; 2nd Jan 2022. Accessed 12 March 2023

https://greenfeels.in/blogs/sustainability-basics/eco-friendly-alternatives-to-plasticn. Jain P; 18 eco friendly alternatives to plastic in your life, 07 August 2022. Accessed 22 March 2024

https://planet.outlookindia.com/news/ficci-and-hul-to-set-up-centre-for-sustainability-leadership-news-415498. Outlook planet; 15th June 2023. Accessed 17 March 2024

https://saahaszerowaste.com/waste-management-services-bangalore/. Accessed 17 March 2024

https://saathipads.com/. Accessed 21 March 2024

https://www.business.repurpose.global/amp/waste-ventures-india-a-pioneer-in-urban-solid-waste-management. Accessed 17 March 2024

https://www.godrej.com/Resources/sustainability-full-report.pdf. Setting standards. Accessed 14 March 2024

https://www.itcportal.com/sustainability/sustainability-integrated-report-2023/ITC-Sustainability-Integrated-Report-2023.pdf Shaping the future. Accessed 4 March 2024

https://www.karosambhav.com/. Accessed 17 March 2024

https://www.linkedin.com/pulse/bioplastics-booming-sustainable-packaging-solution-santosh-g-e3inc/. Santosh bioplastics booming: are they the sustainable packaging solution? February 20, 2024. Accessed 20 March 2024

https://www.motherdairy.com/pdf/PR/2021/2021-06-04.pdf. Mother dairy reaffirms its commitment towards environment safety (2021) Accessed 10 March 2024

https://www.tatasteel.com/media/18368/tata-steel-brsr.pdf 2022–23. Accessed 28 Feb 2024

Jain P, Sharma A (2023) Circular economy and sustainable development in India. In: Jain P, Sharma A (eds) Sustainable development in India: concepts, theories, practices, and challenges. Springer, pp 109–121. https://doi.org/10.1007/978-981-16-8893-4_9

Jain R, Singh SP (2020) Sustainability of Indian agriculture: an assessment of its performance and determinants. Ecol Ind 117:106638

Kirchherr J, Reike D, Hekkert M (2017) Conceptualizing the circular economy: an analysis of 114 definitions. Resour Conserv Recycl 127:221–232

Kumari S, Pandey S (2017) Food loss and waste in India: how much of food is lost and wasted and where? Sustain Prod Consumption 12:1–10

McKinsey and Company (2021) The circular economy: moving from theory to practice. Retrieved from https://www.mckinsey.com/business-functions/sustainability/our-insights/the-circular-economy-moving-from-theory-to-practice

Ministry of Environment, Forest and Climate Change [MoEFCC] (2023). India takes lead in circular economy at G20 summit. Press Information Bureau. https://pib.gov.in/PressReleasePage.aspx?PRID=1784816

OECD (2021) Business models for the circular economy: opportunities and challenges for policy. Retrieved from https://www.oecd.org/environment/business-models-for-the-circular-economy-opportunities-and-challenges-for-policy.htm

Rao P (2018) Traditional ecological knowledge and cultural practices for sustainable development: a case study of South Asian countries. Sustain Dev 26(2):182–194

Ravindranath NH, Sathaye J, Sharma A, Murthy IK (2023) Sustainable development in India: issues and policies. In: Ravindranath NH et al (eds) India in a warming world: integrating climate change and development. Springer, pp 345–360. https://doi.org/10.1007/978-981-16-8830-9_17

Sengupta S, Roy S (2021) Circular economy in Indian fashion industry: opportunities and challenges. J Text Sci Fashion Technol 7(3):84–88

Sharma S, Kumar V (2019) Barriers to circular economy adoption in Indian manufacturing firms: an interpretive structural modeling approach. J Clean Prod 224:932–947

Smith A, Stirling A, Berkhout F (2020) The governance of sustainable socio-technical transitions. Res Policy 49(8):104084

United Nations Environment Programme (2018) Global resources outlook 2019: natural resources for the future we want. United Nations Environment Programme. https://www.resourcepanel.org/reports/global-resources-outlook-2019
World Economic Forum (2021) The circular economy imperative: going beyond waste management. World Economic Forum. https://www.weforum.org/reports/the-circular-economy-imperative-going-beyond-waste-management
World Economic Forum (2022) How the circular economy can help us achieve net zero. Retrieved from https://www.weforum.org/agenda/2022/01/circular-economy-net-zero/

Life Cycle Approaches

A Methodological Approach to Integrate Circularity, Ecosystem Services and the SDGs Under the Life Cycle Assessment Framework

Phoebe Koundouri, Stathis Devves, Conrad Landis, and Lydia Papadaki

Abstract This study outlines a methodological framework for incorporating the Sustainable Development Goals (SDGs), circular indicators, ecosystem services into the Life Cycle Assessment (LCA) framework. LCA provides a comprehensive assessment of environmental effects, but it does not guarantee a shift from linear to circular systems. Therefore, by incorporating the tenets of circular economy, the approach considers the movement of resources, the life cycle of materials, and their overall durability. Also, by integrating ecosystem services, the assessment of the effects on both ecosystems and human welfare is conducted across the entire life cycle. In addition, these assessments are mapped to particular SDGs, thereby quantifying contributions to sustainable development objectives. In the past, research has frequently focused on the analysis of these frameworks in isolation, treating them as distinct components of sustainability. The integrated methodology outlined herein offers a holistic structure for evaluating sustainability throughout the life cycles of products, thereby fostering circularity, conserving ecosystems, and advancing the Sustainable Development Goals.

Keywords Circular economy · Sustainable development goals · Ecosystem services · Life cycle assessment

---

P. Koundouri (✉) · S. Devves · C. Landis · L. Papadaki
DIEES, School of Economics, ReSEES, Athens University of Economics and Business, Athens, Greece
e-mail: pkoundouri@aueb.gr

P. Koundouri
Department TME, Technical University of Denmark, Copenhagen, Denmark

P. Koundouri · L. Papadaki
SDU, ATHENA Information Technology, Research Center, Marousi, Greece

P. Koundouri
AE4RIA, Athens, Greece

L. Brandli et al. (eds.), *The Contribution of Life Cycle Analyses and Circular Economy to the Sustainable Development Goals*, World Sustainability Series,
https://doi.org/10.1007/978-3-031-85300-5_11

1 Introduction

Maintaining goods, materials, and resources inside the economy for as long as feasible helps to minimise waste and resource usage, hence defining circularity—a fundamental idea of the circular economy (CE) (Kirchherr et al. 2017). From linear (take-make- disposal) to circular (reuse-recycle) systems, circular economy measurements—including circularity indices—evaluate the degree to which economic activity lower resource intake and waste output. Although these measures provide insightful analysis of material efficiency and waste minimisation, they are sometimes inadequate to assess the more general environmental effects linked with these developments.

By means of impact analysis spanning the whole lifetime of a product, from raw material extraction to manufacture, use, and disposal, Life Cycle Assessment (LCA) offers a complete knowledge of environmental performance in contrast (Bulle et al. 2019; Hauschild et al. 2018; Itsubo and Inaba 2014). This whole strategy guarantees capture of upstream and downstream processes including emissions, resource depletion, and ecological repercussions. LCA by itself, therefore, does not always lead to a circularity. Here the combination of LCA with circularity indices becomes absolutely vital. When paired with circularity indicators (Rigamonti and Mancini 2021), LCA's results can more precisely match the Sustainable Development those (SDGs) (UN Life Cycle Initiative 2024; United Nations 2022), especially those linked to responsible consumption and production (SDG 12), climate action (SDG 13), and life on land and water (SDGs 14 and 15). For instance, circularity in product design—that is, maximising recyclability or prolonging product lifespan—may lower environmental consequences, as judged by LCA, while simultaneously helping to furthermore general sustainability goals.

Achieving several SDGs and maintaining life depend on the basic ecosystem services—benefits that humans get from ecosystems. Practices of the circular economy, like resource conservation and pollution reduction, directly affect ecosystem services by so limiting environmental damage and so maintaining biodiversity (Geissdoerfer et al. 2017). For example, linked to circularity, better water usage efficiency in industrial operations can increase ecosystem services connected to flood control and provision of clean water. But using models like LCA helps one to better understand the relationships between circularity projects and ecosystem services, hence enabling their full advantages. Linking the two helps companies to better grasp how circularity initiatives affect ecosystem health, therefore matching economic operations with environmental preservation, as stressed in SDGs 6 (clean water) and 15 (life on land).

Combining LCA with circularity indices will help companies evaluate their environmental effect and advance towards sustainable growth from a more all-around standpoint. This strategy not only emphasises material and energy economy but also helps to clarify how circularity helps to meet the SDGs. Although circularity indices are less exact in effect assessment, they are a useful tool for presenting sustainability performance to stakeholders even as LCA studies offer thorough environmental data.

Crucially for obtaining support and financing for circular projects, this communication may boost public confidence, increase customer interactions, and help supply chains to establish their image.

This study is motivated by the growing demand for dependable, integrated systems able to monitor and advance sustainability. As a transforming concept, the circular economy offers to lower environmental impact and boost social and economic advantages. Still, given the different scopes and approaches of these systems, matching circularity with LCA, ecosystem services, and SDGs remains challenging. Thus, this study aims to develop a comprehensive framework for evaluating and advancing sustainability, focusing on the integration of circularity indices, ecosystem services, and the SDGs into the LCA. The circular economy, a transformative concept, offers a reduction in environmental impact and social and economic benefits. Circularity indices are useful for informing stakeholders about sustainability developments and aligning business plans with SDG metrics. However, matching circularity with LCA, ecosystem services, and SDGs remains challenging due to their different scopes and approaches. This research aims to meet global sustainability targets and foster the development of stronger, circular corporate models.

2 State-of-Play for LCA

A full set of the environmental outcomes, associated with a certain process followed by a business model or workflow, is perfectly described, modelized and quantified by the Life Cycle Analysis (LCA) approach. The ISO 14040 states that the LCA definition describes the evaluation of the inputs, outputs and the potential environmental impacts of a process/service workflow system throughout its entire life cycle (ISO 2006). The first step in the LCA is to define the objectives of the LCA: the manufacturing or service streamline, the origin of the raw materials, upstream flow, regional characteristics, and the lifetime for the actual functionality of the product or service itself (Fig. 1).

Next, it is required to gather the resources for input raw materials and to monitor or compute the environmental emissions for every step of the workflow. Following data collection, a set of environmental impact categories are created from the materials, trash (toxic or not), and emissions to the air, ground, and water (terrestrial or marine areas). This transformation, from material use and emissions to the environmental impacts, as grouped into Human Health (HH) Resource (RE) and Ecosystems (EC)

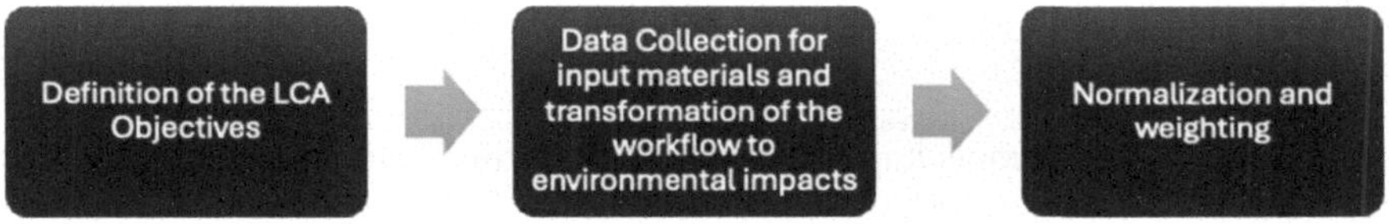

Fig. 1 The Life Cycle Assessment approach

categories, is taking placing throughout the core of the LCA, the Life Cycle Impact Assessment (LCIA).

There are several methods of LCIA throughout impact categories, where midpoint and endpoint indicators are derived. The midpoint indicators are expressed in units of the corresponding impact (i.e., toxicity or eutrophication of the soil, the fresh or marine water, or the air). The endpoints group the individual midpoint indicators of each main group (HH, RE, EC), and then endpoints are measured either on DALY-Disability Adjusted Life Year (HH), species year- species loss over a given time period (EC), or $ (for Resources Cost). It depends on each individual method to determine which impact categories are transformed into midpoint indicators, since each method focuses on certain impact categories and derives the corresponding indicators. Some categories are given special attention by certain methods.

The primary LCA methodologies are listed here in a review list together with the indicators for the most significant impact categories. When it comes to midpoint-end-point indications, several of these areas are often addressed by practically all approaches. For certain approaches, the linkage modelling between the midpoint and endpoint indicators is still waiting because the links between them have not yet been quantified. Therefore, the endpoints of these procedures have not been formally stated by us.

2.1 Distinctive Characteristics of the Primary LCIA Methodologies

ReCiPe

The ReCiPe method is a comprehensive set of midpoint indicators that are closely linked to impact categories, providing a low uncertainty for their value. This method transforms environmental data into an environmental scorecard, focusing on the consequences or endpoint indicators and results. The impact category set is a concentrated result of a wider list recorded through LC inventory analysis, allowing for comparative views of environmental consequences across various supply chains and raw material inputs. ReCiPe presents 18 midpoint indicators corresponding to specific impact pathways, such as climate change, water depletion, terrestrial ecotoxicity, and land use (Table 1). It also presents three endpoint indicators focusing on the final effects on human health, ecosystem biodiversity, and resource consumption. The method can be applied at a global level and is a data-oriented approach (Tables 2 and 3).

LIME

LIME is a methodical modeling process that transforms midpoint effect categories into endpoint impact categories, such as the Climate Change (CC) Impact Category.

Table 1 Mapping of the main impact categories onto the midpoint indicators for each LCIA method

Impact Categories	Midpoints Nr of Recipe	Midpoints Nr of LIME	Midpoints Nr of EDIP	Midpoints Nr of Impact 2002+	Midpoints Nr of Eco-indicator 99	Midpoints Nr of CML	Midpoints Nr of TRACI	Midpoints Nr of MEEuP
EC-Agricultural land occupation								
EC-Climate Change								
EC-Consumption of biotic resource								
EC-Freshwater ecotoxicity								
EC-Freshwater eutrophication								
EC-Marine ecotoxicity								
EC-Natural land transformation								
EC-Terrestrial acidification								
EC-Terrestrial ecotoxicity								
EC-Urban air pollution								
EC-Urban land occupation								
EC-Waste								
HH-Climate Change								
HH-Climate Change (as life supporting function)								
HH-Human toxicity								
HH-Indoor air pollution								
HH-Ionizing radiation								
HH-noise								
HH-Ozone depletion								

(continued)

Table 1 (continued)

HH-Particulate matter formation								
HH-Photochemical oxidant formation								
RE-Energy								
RE-Fossil depletion								
RE-Metal depletion								
RE-Non Hazardous Waste								
RE-total (social welfare)								
RE-Water consumption								
WE-accidents								
WE-allergy and Neurotoxic damage due to occupational exposure to chemicals								
WE-Cancer								
WE-Monotonous repetitive work								
WE-noise								
WE-Reprotoxic damage								
Working Environment (WE)								
Grand Total	**18**	**18**	**16**	**13**	**10**	**13**	**11**	**14**

It uses the CC midpoint to compute specific damages and their effects, such as infectious diseases, heat stress, disasters, and harm to plants or crops. It establishes a connection between midpoint data and associated environmental effects, evaluating endpoint damages that propose safeguarding subjects (AoP) in terms of productivity, ecosystem biodiversity, or human health. LIME uses conjoint analysis to apply the weighting strategy, employing a survey-based methodology to assess consumers' willingness to pay and tolerance for goods and services. It also includes social and economic elements. However, when considering the subjectivity of weighting techniques, potential methodological flaws may arise. LIME is data-driven, but data collection may not be feasible for certain product production or service provider categories.

EDIP

Table 2 Mapping of the main impact categories, onto the endpoint indicators for each LCIA method

Impact Categories	Endpoints Nr of Recipe	Endpoints Nr of LIME	Endpoints Nr of EDIP	Endpoints Nr of Impact 2002+	Endpoints Nr of Eco-indicator 99	Endpoints Nr of CML	Endpoints Nr of TRACI	Endpoints Nr of MEEuP
EC-Climate Change								
EC-Consumption of biotic resource								
EC-Waste								
EC-Total								
HH-Climate Change (as life supporting function)								
HH-Indoor air pollution								
HH-noise								
HH-Total								
RE-Energy								
RE-Non Hazardous Waste								
RE-Total (primary production)								
RE-Total (social welfare)								
Grand Total	3	4	0	4	3	0	3	0

EDIP is another well-established methodology, with a comprehensive documentation, developed in Denmark (Hauschild and Potting 2000). EDIP follows a similar midpoint approach within LCIA. EDIP also presents an evaluation of the working environment; this is a unique feature of this method compared to others. Characterization factors are applied for more than 40 European regions, based on a weighted average across Europe. Another add-on of the EDIP method is regarding the ozone depletion characterization factors. They have been selected for upcoming periods of 5 and 20 years due to the possible abolition of CFCs and HCFCs, dealing with the ozone depletion effect.

Impact 2002

This method brought new concepts and approaches to compare and improve the evaluation of human toxicity and ecotoxicity consequences, not based on a conservative

Table 3 Definition of endpoint indicators

Endpoint indicator	Definition	References
EC-climate change	Since the 1800s, burning fossil fuels like coal, oil, and gas has been the primary cause of climate change, which has been mostly caused by human activity affecting the natural ecosystems	United Nations (2024)
EC-consumption of biotic resource	Biotic resources are materials or resources that come from the biosphere, such as live organisms, forests, and the materials made from them. This primarily refers to fossil fuels, such as coal, gas, oil, etc. Abiotic refers to everything other than living organisms, such as heavy metals, fresh air, and land	Allied Academies (2024)
EC-waste	Every human action generates garbage. Construction and demolition debris, mining and extraction waste, oil and gas production waste, fossil fuel waste, sewage sludge, hazardous, industrial non-hazardous, medical, radioactive, municipal solid waste, and animal and agricultural waste are all types of waste	US EPA (2023)
EC-total	The sum of idividual ecosystem-related impact categories	(Author)
HH-climate change (as life supporting function)	Climate Change can be interpreted as the damage on life support systems that deserve protection for their own sake	Jolliet et al. (2003)
HH-Indoor air pollution	Due to the variety of indoor air pollutants, exposure to these pollutants can result in a range of health problems, including allergies, respiratory disorders, and sick house syndrome	Itsubo and Inaba (2014)
HH-noise	Noise (e.g., caused by the transportation of goods) can be disturbing to humans	Nielsen and Laursen (2005)
HH-total	The sum of individual human health-related impact categories	(Author)
RE-energy	This impact category covers energy policy, statistics, trends, application use, power generation and distribution efficiency, energy supply security, and accounting units	JRC (2024)
RE-non-hazardous waste	Beneficial use is the process of replacing some or all of the virgin materials in a natural or commercial product with these materials, either as generated or after additional processing, in a way that satisfies product specifications, offers a functional benefit, and does not endanger human health or the environment	US EPA (2024)
RE-total (primary production)	This impact category includes the influence on production of phytoplankton (aquatic primary production) and production of terrestrial plants (terrestrial primary production)	Itsubo and Inaba (2014)
RE-total (social welfare)	Human health is largely affected at the raw material production stage, and it is thought that the main factors causing harm are the CO_2 and SOx emissions produced as a result of energy use throughout the raw material production process	Itsubo, (2011)

assumption. The method presents 14 midpoint categories, which are transformed into 4 endpoint indicators, or, in other words, into 4 damage categories: human health, ecosystem quality, climate change, and resources. The method presents a framework with a holistic approach for environmental impacts, providing an extended range of characterization factors (according to a similar size of substance range); therefore, the evaluation of manufacturing or services with increased complexity can be technically satisfied. Throughout the advanced incorporation of human toxicity and ecotoxicity aspects, the assessment of human health has greatly improved. Data requirements for a wider range of substances are something to take seriously into account. A specialised software for LCA is also necessary.

Eco-indicator 99

This is a simplified method with a single score. The Eco-Indicator 99 has been designed for a single-scored result. We measure the environmental impacts in millipoints (mPt). This signifies the possible harm that a manufacturing process or a service supply chain can inflict over the course of their entire life cycle. The method is focused on three endpoints: human health, ecosystem quality, and resource depletion. Due to the simple single-score result, it is easy to communicate the main environmental impact and its severity, which is reflected to someone who is not an expert.

CML 2002

CML 2002 is a methodology that focuses on 15 midpoint categories. Characterization factors transform the data of inventory analysis into a common-used measurement unit per individual midpoint category. Therefore, this method became a best practice guide regarding the main impact assessment phase with respect to the ISO 14040 LCA standardization. Another useful feature of this method is that it is usually integrated into software applications dealing with LCA. On the other hand, the method does not explicitly link the midpoint with the endpoint results.

TRACI

North America developed this valuable methodology specifically for its use. It focuses on midpoint impact categories and indicators, especially in 10 environmental consequences per impact category, called "stressors." Inventory analysis is providing the material units to be transformed into a midpoint indicator by means of characterization factors with high relevance to the North American region. The midpoint framework contains specific impact categories for human health. Therefore, the context of health risks is supported by useful insights with regards to specific manufacturing processes or service supply chains. TRACI doesn't establish a direct connection with the midpoint and endpoint outcomes. Similarly, to CML 2002, the weighting approach is up to the developer's choice and it covers the data collection needs, which are not such a large set but still necessary.

MEEuP

This method aspires to assess when the products with energy intensity fulfil the criteria for CE labelling, by the adoption of the Life Cycle approach. Inventory data is transformed to the midpoint impact categories. The method is in compliance to European agreements, international treaties and as such the indicators and methods. There is no reference to endpoint impact categories or indicators. There are 14 midpoint impact indicators, relating to Human Health, Ecosystems and Resources depletion.

2.2 LCA Normalisation and Weighting

In any event, the key findings and conclusions of the LCA study depend critically on the scaling of the environmental outcomes with respect to the various effect categories to a common unit. As a result, the impact's magnitude can be determined regardless of the group it belongs to. This phase is referred to as LCA normalisation. If more action is required, think about whether one category is more significant than the other(s) with an A or B grade. We refer to this as a weighting step. Typically, this final stage provides a single score that highlights the total environmental score. However, the subjectivity of the weighting coefficient values is related to this one score, thus the weighting process—which includes providing the necessary technical and informational background about these values—should be meticulously followed.

When normalisation incorporates weighting as well, it becomes possible to compare the overall environmental impact, particularly across various product or service scenarios. Normalisation without weighting provides sufficient comparability to draw conclusions about the major effects produced by different environmental impact categories. Naturally, the most environmentally significant potentials across many impact categories are easier to evaluate if normalisation excludes weighing; however, when weighting is included, the benefit of the total footprint must be acknowledged (Venditti and Biomaterials 2016). As far as normalization is concerned, the definition itself recommends the calculation of the magnitude of the impact category indicator, according to a reference number. There are two alternative calculations.

The first one is called internal normalization, by comparing 2 or more values of the same impact indicator and setting one of them as a reference, for instance setting it as the value "1". Therefore, if the other value(s) to be compared, perform a double (or 3-times) score in comparison to the reference, then these indicators are (internally normalized to the value 2 (or 3), and this a dimensionless number. The second (external) normalization method is to compare any of the impact indicators values with a worl wide reference, a value relevant to a global value of the specific issue. For instance, if the GWB has a value of 1,4 *10^(13) Kg eq CO_2, with regard to the global annual value per capita in 1995, therefore, if the value of GWB of Product A is around 6.70 kg eq CO_2, then the (external) normalization value of Product A is 1,6 * 10^(−13) (dimensionless number). Regarding the weighting method, (Gloria et al. 2007) incorporates some stakeholder perceptions about the relative importance of different impact categories (Table 4).

3 State-of-Play for Circularity Metrics

Pearce and Turner (1990) introduced the notion of circular economy (CE) in the early 1990s. This model aims to convert the old open-ended economy into a closed-loop system, focusing on the material aspect. Subsequently, numerous scholars and practitioners have embraced various definitions for circularity (Winans et al. 2016). Kirchherr et al. (2017) define a circular economy as an economic system that replaces the disposal of products at the end of their life with a series of interventions aimed at promoting circularity.

Displays the R-strategies in a sequential manner, beginning with three strategies (R0-R2) that aim to avoid the wasteful use of resources. This is followed by five strategies (R3-R7) that aim to extend the lifespan of products from a technical and economic perspective. Finally, at the bottom of the hierarchy, the more commonly known R-strategies of recycling and energy recovery (R8-9) are included, which involve the retrieval of materials or their energy value from discarded products. The three tactics mentioned are commonly referred to as narrowing, slowing down, and closing the resource loop. Narrowing involves utilising fewer resources per product, slowing down involves extending a product's useful life, and closing the resource loop involves generating a circular flow of resources (PACE 2021).

To identify the most suitable CE strategies for a specific case study, the "Big Five Structural Wastes" suggests asking whether materials or products can be used for longer or more intensively and if more of the resource used than necessary or is its use causing harm. As displayed in Fig. 2, each of these questions can be remedied by a diverse strategy. Indicatively, if we notice "Premature End-of-Use", we can extend the life of the product using long life loop strategies, such as R4-R6, as described in Fig. 2.

In 2021, OECD aimed at compiling input, process, and output indicators used by governments at various levels, specifically to track and assess the advancement of current circular economy plans. The inventory offers a comprehensive summary of circular economy measurement frameworks. It assists in identifying gaps in measurements and can serve as a source of inspiration for governments seeking to establish or utilise indicators to enhance policies related to the circular economy (OECD 2021). Circular economy indicators are categorised into five primary categories by the inventory, namely: Environment, Governance, Economic and Business, Infrastructure and Technology, and Jobs. To further expand this initiative, we mapped them onto 10 R-Strategies displayed in Fig. 2. Table 5 provides a list of examples of cross-sectorial circular indicators along with their units mapped onto the R-Strategies category by category. They provide measurable ways to assess the effectiveness of each R-strategy within the circular economy framework. This mapping provides a

Table 4 Importance of the Environmental impact related to voting preferences and temporal perspective

	All time horizons (100%)				Short-term time horizon (24%)				Medium-term time horizon (31%)				Long-term time horizon (45%)			
	All	Producer	User	LCA Expert	All	Producer	User	LCA Expert	All	Producer	User	LCA Expert	All	Producer	User	LCA Expert
Global warming	29	16	30	50	7	5	9	7	43	26	43	60	52	30	57	68
Fossil fuel depletion	10	12	7	10	15	13	12	15	7	13	3	13	4	10	1	5
Criteria air pollutants	9	7	6	13	18	11	11	48	2	3	3	1	1	2	0	1
Water intake	8	7	10	5	7	7	8	3	10	8	14	6	8	8	9	6
Cancerous	8	8	6	6	8	11	6	5	6	4	6	4	9	9	6	7
Ecological toxicity	8	8	11	3	6	5	9	2	9	12	11	3	9	9	13	5
Eutrophication	6	8	6	3	8	8	9	4	6	10	5	5	3	4	2	2
Land use	6	6	9	3	7	7	11	3	6	6	8	3	5	6	6	3
Noncancerous	5	11	4	2	6	12	5	3	4	6	2	2	6	17	2	2
Smog formation	4	4	3	2	7	6	6	4	1	3	1	1	0	1	0	1
Indoor air quality	3	5	3	1	7	9	6	4	1	1	1	1	0	0	0	0
Acidification	3	4	4	1	4	6	6	1	2	4	2	1	2	2	2	1
Ozone depletion	2	3	2	1	2	3	3	1	2	4	2	1	2	3	1	1
Inconsistency	0.03	0.05	0.03	0.05	0.04	0.06	0.06	0.05	0.02	0.04	0.02	0.06	0.06	0.11	0.08	0.13

Source Gloria et al. (2007)

Fig. 2 10 R strategies for circular economy

tool set for each R-Strategy, enabling the circularity assessment in a more holistic way.

3.1 Connection and Alignment Between LCA and Circularity Indexes

LCA methodology is not capable of automatically providing circularity pathways and assessments due to the lack of a "cradle to grave" analysis in linear models. To improve its role in circularity assessments, modern circularity-oriented models must be incorporated, and phase changes in material use must be accounted for using specific technical guidance. LCA is necessary to measure the circularity index due to its ability to evaluate environmental, economic, and social results from certain circular economy (CE) applications. The magnitude of change produced by CE interventions and recommendations can be defined by exact LCA paths, allowing for improvements in existing CE strategies. To define the correlation between LCA and circular indices, it is crucial to link the degree of environmental options to circularity recommendations. Packaging of goods can improve the circularity index but must also improve the environmental impact of the supply chain, as assessed by

Table 5 Circular indicators mapped onto the 10 R-strategies: examples

R-strategies within production chain		Examples of cross-sector circular economy indicators (Unit)				
		Environmental	Governance	Economy	Technology	Societal
Smarter product use and manufacture	R0-refuse	Reduction in resource extraction (tons of raw material saved per year)	Policies or regulations limiting single-use products (number of single-use product bans or policies enacted)	Savings on raw material procurement (% reduction in procurement costs or monetary value saved (USD/year)	Technological adoption of alternative products (% of market share of alternative products)	Consumer behavior changes towards minimalism and product refusal (% reduction in single-use product consumption or survey-based metrics on consumer preferences)
	R1-rethink	Carbon footprint reduction (tons of CO_2-equivalent reduced)	Innovation incentives (number of government incentive programs or % increase in R&D spending)	Revenue from new business models (e.g.), product-as-a-service, sharing platforms etc.) (USD/year)	Multi-functional product adoption (% increase in multi-functional product sales)	Demand for sustainable products (% increase in consumer demand for versatile products (based on sales or surveys))
	R2-reduce	Waste reduction per output (kg of waste per unit of production or % waste reduction)	Efficiency regulations (number of regulations promoting resource efficiency)	Savings on materials and energy (USD saved per unit of output or % reduction in production costs)	Resource-efficient technologies (energy consumption per unit produced (kWh/unit) or material intensity (kg/unit))	Consumer awareness (survey metrics on consumer awareness or % increase in sales of resource-efficient products)
Extend lifespan of product and its parts	R3-reuse	Reduction in resource use (tons of resources saved or kg of reused products per capita)	Reduction in resource use (tons of resources saved or kg of reused products per capita)	Growth in second-hand markets (% increase in second-hand market revenue (USD/year))	Development of reusable designs (% of products designed for reuse or modular product market share)	Acceptance of used products (% of population purchasing reused products or number of reuse transactions)

(continued)

Table 5 (continued)

R-strategies within production chain		Examples of cross-sector circular economy indicators (Unit)				
		Environmental	Governance	Economy	Technology	Societal
	R4-repair	Waste reduction via repair (kg of waste diverted from landfills)	Right-to-repair laws (number of right-to-repair laws or policies enacted)	Growth in repair services (% increase in repair industry revenue or jobs created in repair sector)	Repairability improvements (repairability score or % of products designed for easy repair)	Preference for repair over replacement (% of population repairing products instead of replacing)
	R5-refurbish	E-waste reduction (tons of e-waste avoided or kg of refurbished electronics sold)	Certification for refurbished products (number of certified refurbishment programs)	Refurbishment market growth (% increase in refurbished product market value (USD/year))	Diagnostic and refurbishment tools (number of diagnostic tools developed or processing time per unit (hours))	Consumer trust in refurbished goods (% increase in refurbished product sales or survey-based trust scores)
	R6-remanufacture	Reduction in resource extraction (tons of raw materials saved)	Remanufacturing standards (number of remanufacturing certifications or standards developed)	Cost savings from remanufacturing (cost savings per unit (USD/unit) or reduction in material procurement costs)	Remanufacturing technology advancements (% of parts or components remanufactured or energy intensity per remanufactured unit (kWh/unit))	Adoption of remanufactured products (% increase in remanufactured product sales or consumer acceptance ratings)

(continued)

Table 5 (continued)

R-strategies within production chain		Examples of cross-sector circular economy indicators (Unit)				
		Environmental	Governance	Economy	Technology	Societal
	R7-repurpose	Waste diversion via repurposing (kg of waste repurposed or % reduction in waste-to-landfill)	Local or regional repurposing policies (number of policies promoting repurposing)	New markets created by repurposing (USD/year revenue from repurposed products or jobs created in repurposing sector)	Repurposing innovation (number of technologies developed for repurposing)	Community participation in repurposing (number of repurposing events or initiatives or % of population involved in upcycling or repurposing)
Useful application of materials	R8-recycle	Material recovery (recycling rate (% of material recovered) or tons of material recycled per year)	Recycling mandates and targets (% compliance with recycling targets or number of recycling policies enacted)	Revenue from recycled materials (USD/year from sale of recycled materials)	Advanced recycling technologies (recycling efficiency (% of materials recovered) or energy consumption per ton of material recycled (kWh/ton))	Public participation in recycling programs (% of population participating in recycling or kg of recycled material per capita)
	R9-recover	Energy recovered from waste (MWh of energy recovered per ton of waste or % of waste converted to energy)	Waste-to-energy policies (number of waste-to-energy projects supported by legislation)	Economic value of energy recovered (USD/year from energy recovery)	Waste-to-energy technology improvements (energy conversion efficiency (%) or reduction in emissions per MWh of energy recovered)	Public perception of waste-to-energy (survey-based metrics on public acceptance or % of waste-to-energy plant utilization)

an LCA method. Implementing a Multi-Criteria Decision Analysis (MCDA) can help integrate sustainability and circularity assessment into single indicators (Zanghelini et al. 2018).

The integration of Life Cycle Assessment (LCA) and Circularity indicators can help define the best-fit overall result, transforming linear models into circular economies. A valid 2-dimentional approach can reduce environmental impact caused by wasted resources by circulating processes in water, biological substances, and energy across streams and equipment hierarchy. The best circularity-oriented solution can be assessed based on environmental improvements compared to existing linear model conditions. However, externalities like resource use and environmental negative unbalance may need to be addressed through an LCA approach. Gap analysis, life cycle inventory analysis, life cycle impact assessment, and ISO 14040 compliance orientation can measure the impact and describe necessary interventions for environmental and circularity improvements. Improved LCA methodologies include CE-oriented alignment and circularity indicators for incorporating material and environmental circularity performance. Different levels of process can be defined according to upstream, intermediate, and downstream stages of processing, including environmental performance of each stage.

Rigamonti and Mancini (2021) propose a method to integrate LCA and circularity indices, focusing on the integration of environmental metrics into circularity topics. They highlight the need to measure the synergies and trade-offs between LCA and circular indicators to examine how circularity efficiency impacts human health, ecosystems, biodiversity, and the correlation between material flow and renewable energy use. The shortlist review supports the researcher's understanding of LCA methods applied to each case study and the definition of the circularity field (and Table 6).

This list contains a range of activities, including manufacturing (washing machines, alkaline batteries, dairy processing industry, and upstream farming), repair services (tyres), consumers popular interaction with the supply chain (food and beverage packaging), bottle waste management, and construction support services (asphalt). Therefore, there is a wide field of topics and case studies to create an SDG-oriented analysis, taking into consideration the most comprehensive LCA method in combination with the most consistent circularity index approach, as follows in the next section, defining the state-of-play in SDGs.

4 State-of-Play for SDGs

4.1 The Linkage Between LCA and SDGs

The Sustainable Development Goals (SDGs) are closely linked to the products and services provided by companies and institutions. Metrics presenting the contribution to the SDG's targets are necessary, and evaluating environmental and social impacts

Table 6 An indicative list of recent studies that include both LCA evaluation and circularity assessment

References	Application	Methodologies when applied separately
Bracquené et al. (2020)	Washing machines	LCA, product circularity indicator (developed in the study) and MCI
Glogic (2020)	Alkaline batteries	LCA and MCI
Lonca et al. (2018)	Tires	LCA and MCI (adapted)
Pauer et al. (2019)	Food packaging	LCA and circularity indicators (input related: recycled content, reuse rate, renewable content; output related: recyclability, recycling rate, recycling output rate, downcycling factor, reuse rate, compostability; energy: share of renewable energy)
Schmidt et al. (2020)	PET bottle waste management	LCA and six material efficiency measures including the circularity potential developed by Eriksen et al. (2019)
Stanchev et al. (2020)	Anaerobic treatment of dairy processing effluents	Agudelo-Vera et al. (2012) proposed two new circularity metrics: the Material circularity performance indicator, which is based on the Demand Minimisation Index, and the Environmental circularity performance indicator, which is based on the ratio of the total environmental benefits and costs. These metrics are based on material flow analysis and LCA
References	*Application*	*Methodologies when integrated*
Mantalovas and di Mino (2020)	Asphalt mixtures	A composite indicator of environmental sustainability and circularity assessment, where environmental sustainability is quantified by LCA and circularity by MCI (adapted)
Niero and Kalbar (2019)	Beer packaging	A MCDA is used to connect two categories of indicators: (i) material circularity-based indicators, including the material reutilization score and MCI, and (ii) a number of life cycle-based indicators that are pertinent to the beer industry, including climate change, abiotic resource depletion, acidification, particulate matter, and water consumption

Source Rigamonti and Mancini (2021)

through LCA methods is applicable for many case studies. However, connecting SDG indicators with impact category indicators is challenging. It is important to consider the world-wide oriented spectrum of the SDGs, as LCA methodologies have been developed with reference to manufacturing, specific services, or process flow scales.

The UN Life Cycle Initiative launched the project "Linking the UN Sustainable Development Goals to the Life Cycle Impact Pathway frameworks" to build a robust structure in the linkage between LCA and SDGs (UN Life Cycle Initiative 2024; UNEP 2020). This will support companies in their SDG and ESG-oriented reports.

Business units, companies, or institutions can link the impact categories derived from inventory analysis to relevant SDG targets, allowing quantification of LCA impact pathways to specific SDG-related metrics. This allows businesses to measure their contribution to SDG achievement throughout the entire supply chain, identify and monitor challenges throughout the life cycle of one product, and produce strategies for sustainability improvements.

This linkage can support business products or services in focusing on specific SDGs for which a certain set of achievements should be implemented. SDG-oriented measurement can be monitored through a time series set, and the quantity of products related to SDG improvement can be measured. Information and monitoring can support this procedure, and the environmental, social, or resource depletion results of LCA can be used to determine which supports or not the enhancement of an SDG or individual's targets.

Goedkoop et al. (2009) developed a methodology for three areas of protection: human health, ecosystems, and resources, with an endpoint indicator for each area. For that purpose, a well-established LCA method, the ReCiPe (see above) was used. The goal was to align the impact categories at the endpoint level with areas of protection in policy and sustainable development, such as human health, ecosystem quality, resource availability, and man-made environment. This approach ensures that endpoint indicators can be calculated using midpoint calculations, as a PEC/PNEC-based toxicity midpoint cannot be used with a PDF-based ecosystem quality endpoint. An overview of the midpoint categories and characterization factors is presented in Table 7, while the quantitative connection between midpoint and endpoint categories is presented in Table 8.

The specific results for each impact category can be transferred into a 5-point scale, as illustrated in Table 9, following the identification of the impact categories. The scale's specifics are further defined based on the nature of impact (e.g., social or environmental). The quantitative results of each LCA impact category need to be compared with a benchmark value, or a reference value, thanks to a baseline scenario. Therefore, a differentiation, a figure of variation per impact category, towards an (environmentally) improved or getting worse, due to an intervention solution, such as a reuse, saving, or material recycling method, is applied. This variation defines the performance with regards to the 5-point scale. Another possible approach is to compare the impact category results with an already recorded impact result due to the application of the same or similar intervention recorded in the specific industry, manufacturing process, or service value chain. The conceptual configuration of the 5-point reference scale is shown in Table 10.

A comprehensive toolkit for completing the process consists of the mapping of certain SDG targets to LCA impact categories. An illustration of how they relate to each other is displayed in Tables 11 and 12. If an impact category makes a positive (or negative) contribution to a single SDG target, and the remaining targets make a neutral contribution derived from the corresponding impact category, the overall contribution to the SDG is positive (Case 12). If all contributions are neutral in case 2 and all contributions are positive (or negative) in case 4, the same characterisation

Table 7 Midpoint categories and characterization factors

Impact category name	Abbr.	Indicator name	Unit*
Climate change	CC	Infra-red radiative forcing	$W \times yr/m^2$
Ozone depletion	OD	Stratospheric ozone concentration	$ppt^{\dagger} xyr$
Terrestrial acidification	TA	Base saturation	$yr \times m^2$
Freshwater eutrophication	FE	Phosphorus concentration	$yr \times kg/m^3$
Marine eutrophication	ME	Nitrogen concentration	$yr \times kg/m^3$
Human toxicity	HT	Hazard-weighted dose	–
Photochemical oxidant formation	POF	Photochemical ozone concentration	kg
Particulate matter formation	PMF	PM_{10} intake	kg
Terrestrial ecotoxicity	TET	Hazard-weighted concentration	$m^2 \times yr$
Freshwater ecotoxicity	FET	Hazard-weighted concentration	$m^2 \times yr$
Marine ecotoxicity	MET	Hazard-weighted concentration	$m^2 \times yr$
Ionising radiation	IR	Absorbed dose	$man \times Sv$
Agricultural land occupation	ALO	Occupation	$m^2 \times yr$
Urban land occupation	ULO	Occupation	$m^2 \times yr$
Natural land transformation	NLT	Transformation	m^2
Water depletion	WD	Amount of water	m^3
Mineral resource depletion	MRD	Grade of decrease	kg^{-1}
Fossil resource depletion	FD	Upper heating value	MJ

Source Goedkoop et al. (2009)

applies to the overall score for each SDG. A "++" or "—" symbol is used to denote a stronger contribution, whether positive or negative.

5 State-of-Play for Ecosystem Services

Nevertheless, circularity serves an additional function by facilitating the regeneration process of natural systems, thereby promoting plantation, animal life, and biodiversity. Ecosystem services (ES) can be classified into four categories: (1) provisioning, (2) regulatory, (3) cultural, and (4) supporting (Fig. 3). Provisioning services are natural benefits extracted from nature, such as food, water, timber, wood fuel, and plant materials. Regulating services are provided by ecosystem processes that moderate natural phenomena, such as pollution and decomposition. Cultural services contribute to cultural advancement and include knowledge building, creativity, and recreation. Ecosystems are sustained by underlying natural processes like photosynthesis, nutrient cycling, and soil creation, which are essential for their existence.

Table 8 Quantitative connection between midpoint and endpoint categories

Impact category abbreviation	Unit	Characterisation factor name	Abbreviation
CC	kg (CO_2 to air)	Global Warming Potential	GWP
OD	kg (CFC-115 to air)	Ozone Depletion Potential	ODP
TA	kg (SO_2 to air)	Terrestrial Acidification Potential	TAP
FE	kg (P to freshwater)	Freshwater Eutrophication Potential	FEP
ME	kg (N to freshwater)	Marine Eutrophication Potential	MEP
HT	kg (14DCB to urban air)	Human Toxicity Potential	HTP
POF	kg (NMVOC6 to air)	Photochemical Oxidant Formation Potential	POFP
PMF	kg (PM10 to air)	Particulate Matter Formation Potential	PMFP
TET	kg (14DCB to industrial soil)	Terrestrial Ecotoxicity Potential	TETP
FET	kg (14DCB to freshwater)	Freshwater Ecotoxicity Potential	FETP
MET	kg (14-DCB7 to marine water)	Marine Ecotoxicity Potential	METP
IR	kg (U235 to air)	Ionising Radiation Potential	IRP
ALO	$m^2 \times yr$ (agricultural land)	Agricultural Land Occupation Potential	ALOP
ULO	$m^2 \times yr$ (urban land)	Urban Land Occupation Potential	ULOP
NLT	m^2 (natural land)	natural Land Transformation Potential	NLTP
WD	m^3 (water)	Water Depletion Potential	WDP
MRD	kg (Fe)	Mineral Depletion Potential	MDP
FD	kg (oil†)	Fossil Depletion Potential	FDP

Source Goedkoop et al. (2009)

Table 9 5-point scale

2	Performance is a lot better than the benchmark
1	Performance is better than the benchmark
0	Performance is equal to the benchmark
−1	Performance is worse than the benchmark
−2	Performance is a lot worse than the benchmark

Source UNEP (2020)

Table 10 The conceptual configuration of the 5-point reference scale

2	The environmental impact is a lot lower than the reference product (>10%)
1	The environmental impact is significantly lower than the reference product (5 to 10%)
0	There is no significant difference in environmental impact
−1	The environmental impact is significantly higher than the reference product (5 to 10%)
−2	The environmental impact is a lot higher than the reference product (> 10%)

Source UNEP (2020)

Table 11 LCA impact categories and specific SDG targets, case study No. 1

LCA impact category				(Positive/negative/neutral) contribution to SDGs
Ecosystems	EC-freshwater eutrophication	SDG 14	Life below water	+
		SDG 14.1	Mean area that is protected in marine sites important to biodiversity (%)	+
		SDG 14.3	Fish caught from overexploited or collapsed stocks (% of total catch)	n
		SDG 14.7	Marine biodiversity threats embodied in imports (per million population)	n

Source Based on UNEP (2020)

Table 12 LCA impact categories and specific SDG targets, case study No. 2

LCA impact category				(Positive/negative/neutral) contribution to SDGs
Ecosystems	EC-freshwater eutrophication	SDG 2	Zero hunger	+
		2.5	Human trophic level (best 2–3 worst)	+
		2.6	Cereal yield (tonnes per hectare of harvested land)	+
		2.7	Sustainable nitrogen management index (best 0–1.41 worst)	+
		2.8	Yield gap closure (% of potential yield)	+

Source Based on UNEP (2020)

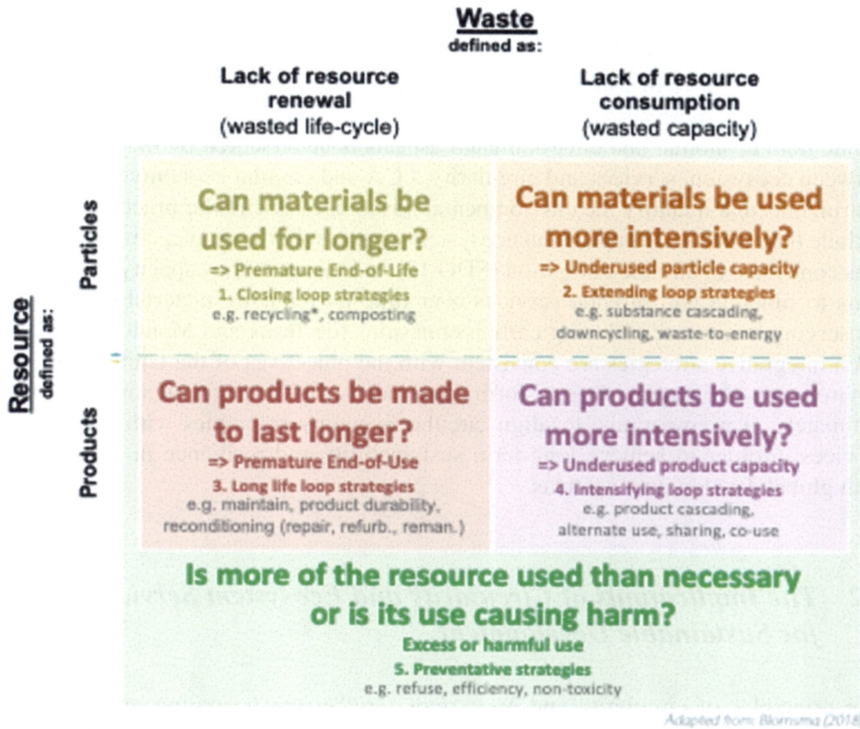

Fig. 3 The big five structural wastes (Blomsma and Brennan 2017)

5.1 *The Interplay Between Circularity, Ecosystem Services and LCA*

We have already underscored the importance of circularity in the reduction of pollution and waste by adhering to the most efficient design for products or services that have the least environmental impact, as well as the importance of material savings, the extension of product life, and the utilisation of repair or refurbishment methods for spare parts and critical in order to reduce the demand for new products or spares and to reduce the cost of the entire supply chain.

Circularity is crucial for the conservation and rehabilitation of ecosystems, as they provide essential services for the supply chain and the circular economy. Implementing resource-efficient management is essential for conserving materials and promoting recycling and reuse. Refuse minimization and pollution reduction can also protect ecosystems, while biodiversity and soil/water quality can improve communities, leading to high-quality products, food, cultivation, and healthy human activities. Table 13 presents several strategies that incorporate ES and circularity within various human activities, displaying environmental alternatives that enhance sustainability, and that can be supported by a framework delineating the challenges and opportunities in certain sectors.

Additionally, the necessity of comprehensive sustainability evaluations that incorporate both economic and environmental aspects is underscored by the correlation between ecosystem services and circularity. LCA and circular economy metrics are instruments that quantify the environmental advantages of circular practices, which include their beneficial impacts on ecosystem services. For example, circular strategies contribute to climate mitigation (SDG 13) and maintain the capacity of ecosystems to function and provide services over time by reducing material extraction, conserving water, and reducing carbon emissions (de Jesus and Mendonça 2018). Such integrative strategies are consistent with the objectives of the United Nations Environment Programme to transform economies in order to attain sustainability. Ultimately, it is imperative to align circular economy principles with ecosystem services in order to achieve long-term sustainability and resilience in accordance with global development agendas.

5.2 The Implications of Circularity and Ecosystem Services for Sustainable Development

The principles of circularity and ecosystem services are becoming more closely linked with sustainable development frameworks, providing a means to establish a more environmentally restorative and resource-efficient economy. This transition seeks to disentangle economic growth from environmental damage and resource extraction. The SDGs, particularly SDG 12 (Responsible Consumption and Production), are supported by circular strategies, such as product design optimisation and closed-loop systems, which mitigate negative externalities such as pollution and resource depletion (de Jesus and Mendonça 2018; Geissdoerfer et al. 2017).

The maintenance of biodiversity and human well-being is contingent upon ecosystem services, which are the advantages that humans derive from ecosystems, including pure air, water, and sustenance. By reducing the pressure on natural systems through resource conservation and pollution control, circularity is essential for the maintenance of these ecosystem services. For instance, the capacity of ecosystems to regenerate and provide essential services is improved by strategies such as sustainable waste management, soil restoration, and water recycling. Costanza et al. (1997) contend that ecosystem services are the bedrock of human existence. Consequently, safeguarding these services is consistent with the overarching objectives of sustainable development. By incorporating circularity into economic practices, the negative effects on ecosystem services, including habitat loss and biodiversity decline, are mitigated, thereby promoting a more harmonious relationship between economic activities and environmental health. Consequently, circular economy strategies can facilitate the implementation of critical SDGs, such as SDG 6 (Clean Water and

Table 13 Ecosystem services and circularity challenges and opportunities mapped onto six economic sectors

Sectors	Challenges	Opportunities
Agriculture	Water extreme consumption	Water savings using automations, AI and weather forecasting technologies
	Soil health and overuse	Diversification of cultivation paths
	Fertilizers use	Soil health practices and methods
	Exposure to climate change phenomena	Deployment of Automation, forecasting techinques
Energy	Fossil fuel deployment and GHG emmissions	Renewables uptake in farming, urban, manufacturing and tourism activities
	Extreme energy intensity	Energy efficiency techniques
	Exposure to climate change phenomena, such as drought or water level reduction in renewable energy reservoirs	Renewable storage systems, weather precaution projects (for floods etc.). Recycling batteries and other energy equipment, spares etc.
Industrial manufacturing	Raw material consuption	Reuse, refurbishment
	Waste productions	Circular manufacturing and sharing waste among industrial symbiosis streams
	GHG emmissions	Activation of air quality regulations
	Other emmissions	Activiation of water purification methods
Urban areas	GHG emmissions	Reneable infrastructures and renewable deployment, green mobilibty
	Heat waves and phenomena exposed to climate change	Green parks, urban gardens etc.
	Inequal access to energy, water and other crucial resources	Communities providing affordable access to crucial resources
		People involved in recycling, refurbishment
Tourism	Natural attractions could be deteriorated, due to the excessive use	Promoting responsible consumption
	Overall infrastructure cannot correspond to tourism demand and the infrastructure for local jobs and services (i.e. housing for health/education/goverental proffesionals)	Design and reliable estimation for the appropriate infrastructure, resilience

(continued)

Table 13 (continued)

Sectors	Challenges	Opportunities
	Extreme consumption and waste production	Promotion of local products
		Promotion of local environmental regenerating services and building potential for eco-tourism, as a result

Sanitation) and SDG 15 (Life on Land), by advocating for environmental practices that safeguard ecosystems from unsustainable exploitation.

6 Combining All Elements. The Alkaline Batteries Case Study

To display how all abovementioned approaches can come together, the current example builds on previous work developed by Glogic et al. (2021), who combined the material circularity indicator (MCI) and LCA methods and applied them to several scenarios for the circularity improvement of alkaline batteries. We build on this study by integrating the environmental indicators that are derived from the reuse and resource conservation to the SDGs. Alkaline batteries are attractive for the circular economy due to their widespread use, single-use life cycle, and small size, with disposal being discouraged due to potential toxicity, higher environmental impacts, and their indirect role in the recycling rates of other battery types. To improve circularity, strategies may vary from recycling to the use of secondary materials in manufacturing, and the improvement of battery durability. These strategies serve as the foundation for six alternative scenarios (Fig. 4). The common industrial practice is represented by the baseline scenario. We examine three complementary improvement scenarios (S1, S2, and S3) and two competing (or substitute) scenarios (S1x and S2x) among the improvement scenarios. The baseline scenario and a single circular strategy are combined in each improvement scenario (Fig. 5).

These strategies also capture all four ecosystem services. The ES impacted by alkaline batteries include pollution and toxicity control and climate regulation (Regulation Services), habitat and biodiversity support and nutrient cycling (Supporting Services), raw material supply (Provisioning Services), and aesthetic and recreational value (Cultural Services). Pollution and toxicity control involve the release of harmful materials into the environment, affecting water purification, soil formation, and nutrient cycling. Climate regulation services, such as greenhouse gas emissions, can be mitigated by promoting recycling and reducing waste disposal. Habitat and biodiversity support involves preserving natural habitats and nutrient cycling, which can be impacted by improper disposal. Provisioning services involve the sustainable use of finite resources like metals and minerals, which can be supported by

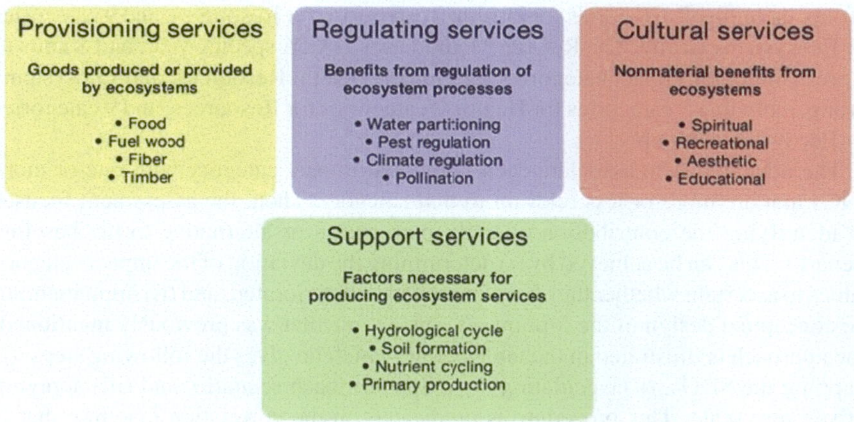

Fig. 4 Types of ecosystem services. *Source* Falkenmark et al. (2007)

(S0): Baseline- No Improvement, the current industrial practice.

(S1) Use of recycled content: secondary zinc and steel from galvanized steel production (open-loop recycling)

(S1x) Use of recycled content: secondary zinc and steel in a closed loop

(S2) Recycling considering route 1 with the assumption of secondary application of metallurgical slag

(S2x) Recycling considering route 2

(S3) Improved utility

Fig. 5 A list of different recycling strategies for zinc and steel production. *Source* Glogic et al. (2021)

CE strategies that promote recycling and reuse. Cultural services involve preserving the aesthetic and recreational value of natural ecosystems, such as natural parks, forests, and clean waterways. By promoting circularity, CE strategies can indirectly contribute to the preservation of these services.

The next steps involve the population of the LCA and MCI models to measure how the increase in the circularity corresponds to the reduction or increase in impacts on endpoint categories of the ReCiPe impact assessment method in the context of the six scenarios. MCI scores are normalised to the net circularity value, and the endpoint category values of each improvement scenario are calculated relative to the baseline scenario. Table 14 displays the absolute values of each scenario for endpoint categories and MCI. The ReCiPe Method uses DALY, species-year, and $ units to express midpoint impact categories for Human Health, Resources, and Ecosystems

Group, including 5 categories for Health, 2 categories for Resources, and 9 categories for Ecosystems Group. The ReCiPe Method uses DALY, species-year, and $ units to express midpoint impact categories for Human Health, Resources, and Ecosystems Group, including 5 categories for Health, 2 categories for Resources, and 9 categories for Ecosystems Group.

The next step is to associate each individual impact category with one or more SDGs that are more or less relevant to that category. Then, the assessment focuses on identifying the contribution of each intervention or alternative to the baseline scenario. This can be achieved by (i) determining the deviation of the impact category values to ascertain whether they have improved or deteriorated, and (ii) implementing the conceptual design of the 5-point reference scale that was previously mentioned. The approach is illustrated in the table below, which involves the following steps: (i) mapping the SDGs, (ii) calculating deviations for each scenario, and (iii) applying a five-point scale. This procedure is predicated on the correlation labelling that is provided in Tables 11 and 12.

When we examine a specific LCA impact category, such as Freshwater Eutrophication, we can understand its impact on one or more SDGs, as well as individual targets or indicators within these SDGs. SDG Target 6.2, which pertains to the freshwater withdrawal performance (the percentage of available freshwater resources), bears a specific impact on the overall SDG 6, also known as the Clean Water and Sanitation issue. Freshwater Eutrophication, an LCA impact category, links to the SDG 6.2 indicator (freshwater withdrawal) by contributing to water turbidity, which poses a health risk to the population due to the presence of harmful microorganisms such as viruses, bacteria, and parasites.

Table 15, presents different alternative circularity scenarios, from S0 to S2x, in which every single intervention can improve the effect on a specific SDG. For example, the S3 circularity scenario can reduce the LCA impact "Freshwater Eutrophication" from $9.1 * 10^{-13}$ to $7.6 * 10^{-13}$ (species.yr), based on Table 14. This results in an improvement of 16.5%, which in turn contributes to the improvement or reduction of freshwater withdrawal (SDG 6.2). According to the 5-point scale (Table 10) this improvement is translated into the +2 scale, since the environmental impact (changed by 16.5%) is a lot lowered than the reference product (>10%). We emphasize that S0 represents the baseline scenario, devoid of any improvement or intervention, in line with current industrial practice. On the contrary, Scenario S1 produces an increase in Freshwater Eutrophication" from $9.1 * 10^{-13}$ to $9.7 * 10^{-13}$ (species.yr), based on Table 14. As a result, this becomes a 6.6% worsening, which contributes (negatively) to the worsening (increasing) of freshwater withdrawal (SDG 6.2). Then, according to the 5-point scale (Table 10) this worsening is translated into the −1 scale, since the environmental impact (changed by 6.6%) is slightly higher than the reference product (<10%). We employ the same methodology as previously described to compute and evaluate the residual values pertinent to the contribution of LCA impact to the indicators of the SDGs.

Table 14 The values of the midpoint impact categories according ReCiPe method

HH-human toxicity	DALY	3.7 × 10 − 8	3.6 × 10 − 8	3.7 × 10 − 8	3.1 × 10 − 8	3.6 × 10 − 8	3.5 × 10 − 8
HH-ozone depletion	DALY	7.2 × 10 − 12	6.9 × 10 − 12	7.1 × 10 − 12	6.0 × 10 − 12	7.4 × 10 − 12	6.0 × 10 − 12
HH-ionizing radiation	DALY	6.6 × 10 − 11	6.5 × 10 − 11	6.6 × 10 − 11	5.5 × 10 − 11	5.1 × 10 − 11	5.5 × 10 − 1
HH-particulate matter formation	DALY	2.6 × 10 − 8	2.5 × 10 − 8	2.5 × 10 − 8	2.1 × 10 − 8	2.3 × 10 − 8	2.5 × 10 − 8
HH-photochemical oxidant formation	DALY	6.4 × 10 − 12	6.5 × 10 − 12	6.3 × 10 − 12	5.3 × 10 − 12	6.6 × 10 − 12	4.6 × 10 − 12
HH-climate change	DALY	4.4 × 10 − 8	4.7 × 10 − 8	4.3 × 10 − 8	3.7 × 10 − 8	4.1 × 10 − 8	3.6 × 10 − 8
HH-total	DALY	1.1 × 10 − 7	1.1 × 10 − 7	1.1 × 10 − 7	8.9 × 10 − 8	1.0 × 10 − 7	9.6 × 10 − 8
RE-metal depletion	$	1.6 × 10 − 2	1.6 × 10 − 2	1.6 × 10 − 2	1.4 × 10 − 2	1.6 × 10 − 2	1.0 × 10 − 2
RE-fossil depletion	$	1.4 × 10 − 3	1.5 × 10 − 3	1.4 × 10 − 3	1.2 × 10 − 3	1.3 × 10 − 3	1.1 × 10 − 3
RE-total	$	1.8 × 10 − 2	1.8 × 10 − 2	1.8 × 10 − 2	1.5 × 10 − 2	1.7 × 10 − 2	1.2 × 10 − 2
EC-climate change	Species.yr	2.5 × 10 − 10	2.7 × 10 − 10	2.4 × 10 − 10	2.1 × 10 − 10	2.3 × 10 − 10	2.1 × 10 − 10
EC-freshwater eutrophication	Species.yr	9.1 × 10 − 13	9.7 × 10 − 13	9.1 × 10 − 13	7.6 × 10 − 13	8.7 × 10 − 13	1.0 × 10 − 12
EC-marine ecotoxicity	Species.yr	3.7 × 10 − 13	3.2 × 10 − 13	3.7 × 10 − 13	3.0 × 10 − 13	3.2 × 10 − 13	3.4 × 10 − 13
EC-natural land transformation	Species.yr	9.1 × 10 − 11	8.4 × 10 − 11	9.1 × 10 − 11	7.6 × 10 − 11	7.9 × 10 − 11	− 7.3 × 10 − 11
EC-agricultural land occupation	Species.yr	2.7 × 10 − 11	2.5 × 10 − 11	2.7 × 10 − 11	2.2 × 10 − 11	1.3 × 10 − 11	2.2 × 10 − 11
EC-freshwater ecotoxicity	Species.yr	1.6 × 10 − 12	1.4 × 10 − 12	1.6 × 10 − 12	1.3 × 10 − 12	1.3 × 10 − 12	1.7 × 10 − 12
EC-terrestrial ecotoxicity	Species.yr	3.7 × 10 − 12	3.6 × 10 − 12	3.7 × 10 − 12	3.1 × 10 − 12	4.1 × 10 − 12	− 1.6 × 10 − 12
EC-urban land occupation	Species.yr	1.4 × 10 − 11	1.4 × 10 − 11	1.4 × 10 − 11	1.2 × 10 − 11	1.4 × 10 − 11	1.2 × 10 − 11
EC-terrestrial acidification	Species.yr	1.3 × 10 − 12	1.4 × 10 − 12	1.3 × 10 − 12	1.1 × 10 − 12	1.3 × 10 − 12	1.3 × 10 − 12

(continued)

Table 14 (continued)

EC-total	Species.yr	3.9 × 10 − 10	4.0 × 10 − 10	3.8 × 10 − 10	3.3 × 10 − 10	4.6 × 10 − 10	1.7 × 10 − 10
MCI	–	1.9 × 10 − 1	2.4 × 10 − 1	2.5 × 10 − 1	3.3 × 10 − 1	2.3 × 10 − 1	2.5 × 10 − 1

Abbreviations: Human Health—HH, Resources—RE, and Ecosystems—EC
Source Glogic et al. (2021)

7 Discussion and Implications of the Results for Future Studies

This methodological study provides a unique perspective on the LCA and its relationship to sustainability frameworks by incorporating CE strategies and the Ecosystem Services framework, in contrast to previous research on the aforementioned. In the past, research has frequently concentrated on the analysis of circularity and LCA in isolation, considering them as distinct components of sustainability. Nevertheless, this study underscores the combined value of these two approaches in facilitating a more sophisticated comprehension of environmental consequences, thereby contributing to the broader conversation regarding the optimisation of circularity and sustainability for particular industries. The CE's potential to reduce waste, minimise resource consumption, and close material cycles has been the primary focus of previous research. For example, Kirchherr et al. (2017) have conducted research that has emphasised the macro-level advantages of CE, with a particular emphasis on overarching themes such as resource conservation, economic savings, and systemic changes in industrial processes. Nevertheless, a significant number of these studies fail to establish a direct correlation between CE practices and quantifiable environmental outcomes, particularly in the context of specific industrial sectors such as alkaline battery manufacturing.

In terms of the case study that we previously discussed, the environmental impacts of battery manufacturing are influenced by specific interventions and scenarios, which can be used to mitigate them. These impacts are categorized into climate change, resource depletion, and ecotoxicity, and are associated with specific SDG targets. The study distinguishes itself by linking LCA impacts with SDG indicators and measuring changes based on each intervention concept. This approach allows for the identification of trends, discussion of additional concerns, and evaluation of sustainability initiatives. Linking LCA impact categories to SDG indicators can guide alternative pathways, such as correlating environmental impacts to environmental or social reflections on SDG targets. Additionally, environmental or resource depletion impacts may be associated with economic performance, such as establishing a new innovative and profitable branch in sustainable manufacturing by conserving material and resources through re-use and refurbishment.

Future research can contribute to the creation of new policy packages. The differentiation in priority significance of LCA impact categories can offer guidelines for

Table 15 Example: (i) Impact categories mapping onto the SDGs, (ii) Deviations for each scenario and (iii) Implementation of the 5-point scale

LCA impact category	Circularity scenario (Baseline = S0)	SDG 6		Clean water and sanitation	Rational about effect of LCA impact to SDG target	% change to baseline scenario (S0)	5-point scale contribution to SDGs	(Positive/negative/neutral) contribution to SDGs
EC-freshwater eutrophication	S0	SDG6	6.2	Freshwater withdrawal (% of available freshwater resources)	Increased turbidity, or cloudiness, of the water (potential health risk due to the existence of harmful microorganisms like bacteria, viruses, and parasites)		0	n
EC-freshwater eutrophication	S1	SDG6	6.2	Freshwater withdrawal (% of available freshwater resources)	Increased turbidity, or cloudiness, of the water (potential health risk due to the existence of harmful microorganisms like bacteria, viruses, and parasites)	−6.6	−1	–
EC-freshwater eutrophication	S2	SDG6	6,2	Freshwater withdrawal (% of available freshwater resources)	Increased turbidity, or cloudiness, of the water (potential health risk due to the existence of harmful microorganisms like bacteria, viruses, and parasites)	0	0	n

(continued)

Table 15 (continued)

LCA impact category	Circularity scenario (Baseline = S0)	SDG 6		Clean water and sanitation	Rational about effect of LCA impact to SDG target	% change to baseline scenario (S0)	5-point scale contribution to SDGs	(Positive/negative/ neutral) contribution to SDGs
EC-freshwater eutrophication	S3	SDG6	6.2	Freshwater withdrawal (% of available freshwater resources)	Increased turbidity, or cloudiness, of the water (potential health risk due to the existence of harmful microorganisms like bacteria, viruses, and parasites)	16.5	2	++
EC-freshwater eutrophication	S1x	SDG6	6.2	Freshwater withdrawal (% of available freshwater resources)	Increased turbidity, or cloudiness, of the water (potential health risk due to the existence of harmful microorganisms like bacteria, viruses, and parasites)	4.4	0	n+
EC-freshwater eutrophication	S2x	SDG6	6.2	Freshwater withdrawal (% of available freshwater resources)	Increased turbidity, or cloudiness, of the water (potential health risk due to the existence of harmful microorganisms like bacteria, viruses, and parasites)	−9.9	−1	–

(continued)

Table 15 (continued)

LCA impact category	Circularity scenario (Baseline = S0)	SDG 6		Clean water and sanitation	Rational about effect of LCA impact to SDG target	% change to baseline scenario (S0)	5-point scale contribution to SDGs	(Positive/negative/neutral) contribution to SDGs
EC-freshwater eutrophication	S0	SDG6	6.5	Population using safely managed water services (%)	Increased turbidity, or cloudiness, of the water (potential health risk due to the existence of harmful microorganisms like bacteria, viruses, and parasites)		0	n
EC-freshwater eutrophication	S1	SDG6	6.5	Population using safely managed water services (%)	Increased turbidity, or cloudiness, of the water (potential health risk due to the existence of harmful microorganisms like bacteria, viruses, and parasites)	−6.6	−1	–
EC-freshwater eutrophication	S2	SDG6	6.5	Population using safely managed water services (%)	Increased turbidity, or cloudiness, of the water (potential health risk due to the existence of harmful microorganisms like bacteria, viruses, and parasites)	0	0	n

(continued)

Table 15 (continued)

LCA impact category	Circularity scenario (Baseline = S0)	SDG 6		Clean water and sanitation	Rational about effect of LCA impact to SDG target	% change to baseline scenario (S0)	5-point scale contribution to SDGs	(Positive/negative/neutral) contribution to SDGs
EC-freshwater eutrophication	S3	SDG6	6.5	Population using safely managed water services (%)	Increased turbidity, or cloudiness, of the water (potential health risk due to the existence of harmful microorganisms like bacteria, viruses, and parasites)	16.5	2	++
EC-freshwater eutrophication	S1x	SDG6	6.5	Population using safely managed water services (%)	Increased turbidity, or cloudiness, of the water (potential health risk due to the existence of harmful microorganisms like bacteria, viruses, and parasites)	4.4	0	n+
EC-freshwater eutrophication	S2x	SDG6	6.5	Population using safely managed water services (%)	Increased turbidity, or cloudiness, of the water (potential health risk due to the existence of harmful microorganisms like bacteria, viruses, and parasites)	−9.9	−1	–

From the Author

relocating the resource in a more efficient way to get more sustainability performance results. By measuring and comparing the as-is and to-be situation of LCA impacts figures, which are based on specific sustainability interventions, we can monitor the performance of these interventions or policies in relation to the achievements of SDG indicators. In addition, circularity indicators can be included in LCA software or digital platforms, while the calculation of impact category results can be based on one or more alternative circular solution scenarios and linked to the SDGs. Specific sectors and industries can adopt a variety of sustainable options due to the impact of sustainable manufacturing on the supply chain and end users.

If integrated into the whole information chain, scientific digital libraries linking the impact category fields, SDG targets, and circular innovation solutions could expedite the process. Research communities comprising a variety of academic institutions, business people, developers, experts, and other potential stakeholders could work together to produce, develop, expand, and carry out similar projects in the near future, taking into account the demands of sustainable development goals with an eye towards an updated and extended agenda beyond 2030, as well as the requirements for the efficient use of raw materials and their effects on the environment, society, and economy. To conclude, the monitoring of actual trends, the knowledge of potential gaps, and the assessment of sustainability applications and innovations can equip business professionals, policymakers, and end users with the necessary tools to increase the adoption of sustainable solutions in the economy and society.

8 Conclusions

It is reasonable to affirm that circular strategies' environmental performance cannot be evaluated solely by the circularity indicators. Consequently, an LCA study must be conducted to confirm that the circularity of the system also results in an improvement in environmental performance, in addition to a study based on circularity indicators. This research provides a distinctive and substantial contribution to the comprehension of the circular economy's application in specific industrial contexts. It establishes a new standard for the assessment of the real-world effects of CE practices by combining LCA with circularity indicators and connecting these to SDGs and to the ecosystem services. The study's methodological innovations and focused application, which build upon prior research, encourage further investigation into the broader societal benefits and optimisation of circularity across various industries.

Future studies could apply this methodology in various case studies, from examining the extent to which CE practices can further diminish dependence on non-renewable energy sources during manufacturing processes to investigating the long-term ecological and biodiversity implications of circular economy practices. Future research could also further explore the social and economic implications of circular practices, despite the fact that the environmental advantages of CE are evident. This could entail an evaluation of the impact of CE strategies on community health, economic resilience, and job creation in various industries.

Furthermore, comprehensive training programmes, skill- and CE Staregiy-specific toolboxes, and other pertinent multidisciplinary knowledge-based assemblies, along with the training of company personnel at all levels, can foster new initiatives, contemporary projects, and inventive methodologies pertaining to the implementation of circularity, the improvement of Areas of Protection (LCA), and the routes towards the Sustainable Development Agenda. This might be shared with the industrial manufacturing network, the provision of sustainable goods and services, and the supply chain's stakeholders, communities, associations, facilitators, and consumers in terms of behavioural awareness.

References

Academies A (2024) Biotic resources. Sci J. https://www.alliedacademies.org/scientific-journals/biotic-resources-2901.html

Agudelo-Vera CM, Mels A, Keesman K, Rijnaarts H (2012) The urban harvest approach as an aid for sustainable urban resource planning. J Ind Ecol 16(6):839–850. https://doi.org/10.1111/J.1530-9290.2012.00561.X

Blomsma F, Brennan G (2017) The emergence of circular economy: a new framing around prolonging resource productivity. J Ind Ecol 21(3):603–614. https://doi.org/10.1111/JIEC.12603

Bracquené E, Dewulf W, Duflou JR (2020) Measuring the performance of more circular complex product supply chains. Resour Conserv Recycl 154:104608. https://doi.org/10.1016/J.RESCONREC.2019.104608

Bulle C, Margni M, Patouillard L, Boulay AM, Bourgault G, De Bruille V, Cao V, Hauschild M, Henderson A, Humbert S, Kashef-Haghighi S, Kounina A, Laurent A, Levasseur A, Liard G, Rosenbaum RK, Roy PO, Shaked S, Fantke P, Jolliet O (2019) IMPACT World+: a globally regionalized life cycle impact assessment method. Int J Life Cycle Assess 24(9):1653–1674. https://doi.org/10.1007/S11367-019-01583-0

Circularise (2023) R-strategies for a circular economy. https://www.circularise.com/blogs/r-strategies-for-a-circular-economy

Costanza R, D'Arge R, De Groot R, Farber S, Grasso M, Hannon B, Limburg K, Naeem S, O'Neill RV, Paruelo J, Raskin RG, Sutton P, Van Den Belt M (1997) The value of the world's ecosystem services and natural capital. Nature 387(6630):253–260. https://doi.org/10.1038/387253a0

de Jesus A, Mendonça S (2018) Lost in transition? Drivers and barriers in the eco-innovation road to the circular economy. Ecol Econ 145:75–89. https://doi.org/10.1016/J.ECOLECON.2017.08.001

Eriksen MK, Damgaard A, Boldrin A, Astrup TF (2019) Quality assessment and circularity potential of recovery systems for household plastic waste. J Ind Ecol 23(1):156–168. https://doi.org/10.1111/JIEC.12822

Falkenmark M, Finlayson M, Gordon LJ, Bennett EM, Chiuta M, Coates D, Ghosh N, Gopalakrishnan M, De Groot RS, Jacks G, Kendy E, Oyebande L, Moore M, Peterson GD, Mora Portuguez J, Seesink K, Tharme R, Wasson R (2007) Agriculture, water, and ecosystems: avoiding the costs of going too far

Geissdoerfer M, Savaget P, Bocken NMP, Hultink EJ (2017) The circular economy—a new sustainability paradigm? J Clean Prod 143:757–768. https://doi.org/10.1016/J.JCLEPRO.2016.12.048

Glogic E, Sonnemann G, Young SB (2021) Environmental trade-offs of downcycling in circular economy: combining life cycle assessment and material circularity indicator to inform circularity strategies for alkaline batteries. Sustainability 13(3):1040. https://doi.org/10.3390/SU13031040

Glogic EYSSG (2020) Confronting challenges of combining and comparing material circularity indicator with life cycle assessment indicators: a case of alkaline batteries. SETAC Europe 30th annual meeting

Gloria T, Lippiatt B, Cooper J (2007) Life cycle impact assessment weights to support environmentally preferable purchasing in the United States. https://doi.org/10.1021/es070750

Goedkoop M, Heijungs R, Huijbregts M, De Schryver A, Struijs J, Van Zelm R (2009) ReCiPE 2008: a life cycle impact assessment method which comprises harmonised category indicators at the midpoint and the endpoint level. https://www.researchgate.net/publication/302559709_ReCiPE_2008_A_life_cycle_impact_assessment_method_which_comprises_harmonised_category_indicators_at_the_midpoint_and_the_endpoint_level

Hauschild M, Potting J (2000) Guidelines from the danish environmental protection agency spatial differentiation in life cycle impact assessment. The EDIP2003 methodology

Hauschild MZ, Rosenbaum RK, Olsen SI (2018) Life cycle assessment, vol 2018. Springer International Publishing

ISO (2006) ISO 14040:2006—environmental management—life cycle assessment—principles and framework. https://www.iso.org/standard/37456.html

Itsubo N, Inaba A (2014) LIME 2 life-cycle impact assessment method based on endpoint modeling

Itsubo N (2011) For comprehensive environmental assessment including assessment of environmental impacts on biodiversity and water. https://lca-forum.org/english/pdf/No10-11.pdf

Jolliet O, Margni M, Charles R, Humbert S, Payet J, Rebitzer G, Rosenbaum R (2003) IMPACT 2002+: a new life cycle impact assessment methodology. Int J Life Cycle Assess 8(6):324–330. https://doi.org/10.1007/BF02978505

JRC (2024) Review of the MEErP. Publications office of the EU. https://op.europa.eu/en/publication-detail/-/publication/03ac5f5a-eb3b-11ee-bf53-01aa75ed71a1

Kirchherr J, Reike D, Hekkert M (2017) Conceptualizing the circular economy: an analysis of 114 definitions. Resour Conserv Recycl 127:221–232. https://doi.org/10.1016/J.RESCONREC.2017.09.005

Lonca G, Muggéo R, Imbeault-Tétreault H, Bernard S, Margni M (2018) Does material circularity rhyme with environmental efficiency? Case studies on used tires. J Clean Prod 183:424–435. https://doi.org/10.1016/J.JCLEPRO.2018.02.108

Mantalovas K, Di Mino G (2020) Integrating circularity in the sustainability assessment of asphalt mixtures. Sustainability 12(2):594. https://doi.org/10.3390/SU12020594

Nielsen P, Laursen J (2005) Integration of external noise nuisance from road and rail transportation in lifecycle assessment. https://www2.mst.dk/udgiv/publications/2005/87-7614-581-6/html/kap09_eng.htm

Niero M, Kalbar PP (2019) Coupling material circularity indicators and life cycle based indicators: a proposal to advance the assessment of circular economy strategies at the product level. Resour Conserv Recycl 140:305–312. https://doi.org/10.1016/J.RESCONREC.2018.10.002

OECD (2021) The circular economy in cities and regions: synthesis report. https://doi.org/10.1787/10ac6ae4-en

PACE (2021) Circular indicators for governments accelerating action in the circular economy

Pauer E, Wohner B, Heinrich V, Tacker M (2019) Assessing the environmental sustainability of food packaging: an extended life cycle assessment including packaging-related food losses and waste and circularity assessment. Sustainability 11(3):925. https://doi.org/10.3390/SU11030925

Pearce DW, Turner RK (1990) Economics of natural resources and the environment, p 378. https://books.google.com/books/about/Economics_of_Natural_Resources_and_the_E.html?hl=el&id=ex8vaG6m4RMC

Rigamonti L, Mancini E (2021) Life cycle assessment and circularity indicators. Int J Life Cycle Assess 26(10):1937–1942. https://doi.org/10.1007/S11367-021-01966-2/TABLES/1

Schmidt S, Laner D, Van Eygen E, Stanisavljevic N (2020) Material efficiency to measure the environmental performance of waste management systems: a case study on PET bottle recycling in Austria, Germany and Serbia. Waste Manage 110:74–86. https://doi.org/10.1016/J.WASMAN.2020.05.011

Stanchev P, Vasilaki V, Egas D, Colon J, Ponsá S, Katsou E (2020) Multilevel environmental assessment of the anaerobic treatment of dairy processing effluents in the context of circular economy. J Clean Prod 261:121139. https://doi.org/10.1016/J.JCLEPRO.2020.121139

UN Life Cycle Initiative (2024) Linking the UN SDGs to life cycle impact pathway frameworks. https://lca-net.com/projects/show/linking-the-un-sdgs-to-life-cycle-impact-pathway-frameworks/

UNEP (2020) LCA-based assessment of the sustainable development goals

United Nations (2022) Transforming our world: the 2030 agenda for sustainable development. Department of economic and social affairs. https://sdgs.un.org/2030agenda

United Nations (2024) What is climate change? United Nations. https://www.un.org/en/climatechange/what-is-climate-change

US EPA (2023) Wastes. https://www.epa.gov/report-environment/wastes

US EPA (2024) Sustainable management of industrial non-hazardous secondary materials. https://www.epa.gov/smm/sustainable-management-industrial-non-hazardous-secondary-materials

Venditti RA, Biomaterials F (2016) Environmental life cycle assessment PSE 476/FB 576 lecture 14: normalization, weighting and aggregation to single score

Winans K, Kendall A, Deng H (2016) The history and current applications of the circular economy concept. https://doi.org/10.1016/j.rser.2016.09.123

Zanghelini GM, Cherubini E, Soares SR (2018) How multi-criteria decision analysis (MCDA) is aiding life cycle assessment (LCA) in results interpretation. J Clean Prod 172:609–622. https://doi.org/10.1016/J.JCLEPRO.2017.10.230

An Assessment of the Nexus Between the Ricardian Thoughts on Comparative Advantage and Circular Economy in Developing Economies

Oluwabunmi O. Adejumo, Joshua F. Obisanya, and Akintoye V. Adejumo

Abstract Economics for sustainable development enables us to appreciate the factors that promote and enhance the quality of life. The nascent drive towards sustainability requires national economies to identify and evolve their ingenious resources for sustainability that are consistent with their capacity. Identifying strength and ingenuities can birth effective global interdependencies and possibly lead to trade where comparative advantages are effectively harnessed for sustainable development. The Ricardian theory of comparative advantage postulates that "nations can gain an international trade advantage when they focus on producing goods with the lowest opportunity costs compared to other nations". The drive towards a circular economy can serve as part of the fulfillment of this mandate through production and utilization of environmentally compliant products. However, the comparative advantage can be larger when Indigenous technology of production embraces waste conversion into other products of environmental and economic value. Therefore, based on the proposition of David Ricardo, this study assesses the plausibility of negotiating economic activities from the area of strength for the emancipation of developing economies, such that they can benefit from trading activities by concentrating their efforts on innovatively harnessing and developing sustainably the things they are best at.

Keywords Circular economy · Sustainable development · Comparative advantage · Opportunity cost · Life cycle

---

The original version of the chapter has been revised. A correction to this chapter can be found at https://doi.org/10.1007/978-3-031-85300-5_27

O. O. Adejumo (✉) · J. F. Obisanya
Institute for Entrepreneurship and Development Studies, Obafemi Awolowo University, Ile-Ife, Nigeria
e-mail: jumobum@gmail.com

A. V. Adejumo
Department of Economics, Obafemi Awolowo University, Ile-Ife, Nigeria

L. Brandli et al. (eds.), *The Contribution of Life Cycle Analyses and Circular Economy to the Sustainable Development Goals*, World Sustainability Series,
https://doi.org/10.1007/978-3-031-85300-5_12

1 Introduction

The economics for sustainable development centers on keeping and possibly improving the current financial capacity of countries while keeping in perspective the need to preserve an economy's environmental, social, and cultural heritage (Daly 2014; Jeronen 2020). This objective bothers on the need to harmonize short-term economic pursuits with the long-term goal of environmental preservation and financial stability (Ozili and Iorember 2023). The principle of sustainability requires that development spans through to the future with the least minimal effects. The clamor for sustainable development has been described as a heart cry to global economic activities that engage the use, exploitation, and overexploitation of environmental resources and has been acclaimed to be a public good (Giddings et al. 2002; Nar 2014). Thus, a sustainable economy relies on an efficient utilization and reutilization of resources that culminate into societal transformations and portends high prospects to achieve the United Nations' Sustainable Development Goals (Mikhno et al. 2021).

Most developing economies are still in the blind drive for the growth of national income without critical reflections on growth contents and components that could have severe implications for future generations (Cobbinah et al. 2015). The unsustainable ideology and practices that make up the gross domestic product of production to consumption have continuously engineered linear economic tendencies that perpetuate unsustainable practices and frustrate the efforts to achieve sustainable development (Tien et al. 2019). For instance, a cursory look at some of the components of economic activities that make up the Gross Domestic Product (GDP) still includes deforestation, fossil fuel generation, and combustion that causes pollution, as well as the generation and proliferation of waste materials from overconsumption, plastics, steel and iron, manufacturing, construction, and recycling via poor or polluting technology, among others.[1] Some of the identified challenges that fester environmental polluting economic activities include the need for industrialization and economic growth, taste and luxury, urbanization, and exploration of natural resources for further production, exportation, and income (Ukaogo et al. 2020).

Unlike the linear economy, the circular economy (CE) has been largely passed as a type of response to the challenge of sustainable development which has been beclouded by linear economic tendencies (Marino and Pariso 2016; Afteni et al. 2021). In the CE model, the nascent economic challenge is to elongate the life span of resources, materials, and products possibly (Niero and Kalbar 2019). One method to actualizing this within the CE model is to activate and operationalize the 6Rs of sustainability- rethink, refuse, reduce, reuse, repair, and recycle- across the life span of resource utilization and re-utilization (Hapuwatte and Jawahir 2021; Manavalan and Jayakrishna 2019). The direction of a circular economy is a form of development that is people-centric and concerns the preservation of the earth (Afteni et al. 2021). Thus, this wave of thought takes us beyond the traditional economic desire of meeting individual want to ensuring posterity of the earth and its resources. Therefore, this

[1] https://www.un.org/en/climatechange/science/causes-effects-climate-change.

study expands on the construct of economics for sustainable development by hinging on the concept of circularity and downplay linear economic tendencies.

2 The New Economic Model and the Life Cycle Assessment Thought

In the nascent literature, is the *New Economic Model* which postulates that the existence of resources, products, and materials can be extended with relation to their intrinsic value, process, re-utilization, and maintenance (Chang et al. 2014). Undoubtedly, at the heart of resource and economic efficiency (that is, productivity) is the economics of circularity (Figge et al. 2018). The circular economy postulates a repetitive process where the goods, resources, and materials enter a closed-circuit system, and where used products are converted from waste into reusable materials in subsequent production cycles (Winans et al. 2017). Thus, the main goal of the New Economic Model is to reconcile or harmonize the goals of environmental and business sustainability.

An instance is drawn from the need for a sustainable economy through the replacement of fossil-fuel systems with a bio-based driven economy (Khan et al. 2021). The biological-based economy is largely typified within the tenets of a circular system which features efficient use of resources, as well as sustainable production and consumption practices (Lewandowski 2018). Thus, a sustainable bioeconomy relies on efficient utilization of biological resources that culminate into societal transformations and portends high prospects to achieve the global goals of sustainability (Eversberg and Fritz 2022).

Figure 1 is a hypothesized framework that explains the sustainable use of materials within economic activities. The cycle moves from resource generation to processing, utilization, waste production, and material reuse. However, within this framework, the obvious phenomenon for sustainability is not just material reuse/recycling in itself; rather, a critical issue is the innovative process for recycling, and even more critical is the environmental friendliness of the production process (Lopes et al. 2022). These issues of innovation technology in the process of actualizing sustainability are the crux of this write-up. Therefore, in the process of material re-utilization, decision-makers need to identify the most significant areas of impact and implement targeted strategies for sustainable production and consumption.

A major assessment tool for the plausibility of product and production circularity is the Life Cycle Assessment (LCA) tool (Tóth Szita 2017; Dahiya et al. 2020). Life Cycle Assessment (LCA) is a systematic methodology used to evaluate the environmental impacts of a product, process, or activity throughout its entire life cycle, from raw material extraction to end-of-life disposal (Clabeaux et al. 2020). LCA considers all stages, including manufacturing and process technology, transportation, use, and disposal (Walker and Rothman 2020). It considers various environmental impact categories such as energy consumption, resource depletion, emissions, and waste

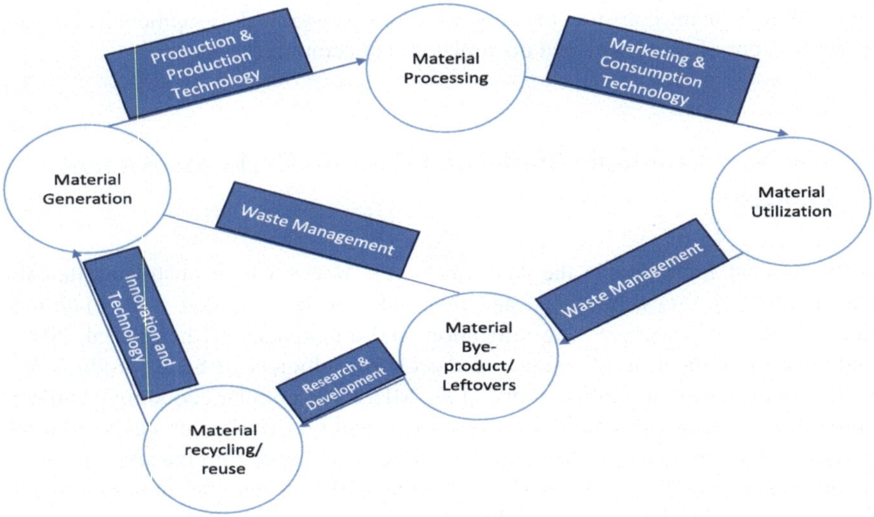

Fig. 1 Circular transmission of materials for sustainability

generation (Hauschild et al. 2018). LCA involves collecting and analyzing data on inputs and outputs at each stage of the life cycle to assess the environmental performance of the evaluated system. The goal is to provide a holistic and quantitative analysis that enables informed decision-making and the identification of opportunities for improving the environmental sustainability of products and activities (Curran 2017). Thus, the LCA supports informed decision-making by quantifying and assessing impacts of material utilization on issues of resource depletion, energy consumption, emissions, and waste generation.

Thus, to fully actualize the goal of a sustainable dimension between ecological and business economic activities, adaptable technology via innovation has been identified as a critical go-between for environmental preservation and raw materials use/re-use (Jeronen 2020; Zhong et al. 2021). Therefore, hinging on the thoughts of the LCA and with particular emphasis on process technology, this study examines the prospect for circularity in material utilization with emphasis on clean production techniques for environmental sustainability.

3 The Ricardian Thought of Comparative Advantage

To understand the Ricardian model, a distinction is made between the theory of Absolute Advantage versus Comparative Advantage. The theory of Absolute Advantage (propounded by Adam Smith) emphasizes the provision and trading of goods or services in greater quantity for the same cost or the same quantity at a lower cost than its competitors (Uysal 2019). The theory of Comparative advantage (propounded by David Ricardo) is an economy's ability to produce a particular good or service at a lower opportunity cost than its trading partners.[2]

The theory of comparative advantage emphasizes lower opportunity costs, high productivity, and specialization (Siddiqui 2018). Indeed, the principle of comparative advantage offers some of the most compelling reasons supporting international trade. Some basic facts of the Ricardian model relevant to the development of low-income economies:

- Less productive nations can benefit from free trade with their more productive neighbors.
- Low-wage countries are not likely to use their production cost advantage in many circumstances.

Arguing from the perspective of natural, developed, or imitated strength, different factors have been observed to motivate trade. These factors include differences in technology, differences in resource endowments, differences in demand, the presence of economies of scale, and the presence of government policies (White 2014; Algieri et al. 2018). For example, the Ricardian model of trade notes that the differences in technologies between countries lead to trade that everyone could benefit from. Meanwhile, the Heckscher-Ohlin model emphasizes that endowment differences are a central motivation for trade; and as a result, it concludes that there will be winners and losers from trade.[3] Thus, in the parlance of sustainability and based on the Ricardian thoughts, when countries begin to tilt their production towards their ingenious advantage, they can change the basis for trade and possibly begin to influence the outcomes from trade and set the pace for economic sustainability vis-à-vis environmental concerns and consciousness.

4 The Relevance of the Ricardian Thoughts of Comparative Advantage and the Circular Economy

Comparative economics for sustainable development enables us to appreciate the factors that enhance the quality of life on issues of environmental, economic, and social well-being (Naeem et al. 2016). This can be achieved by studying and understanding the different resources, norms, political and economic ideologies, systems, and realities embedded in each eco-economy. The nascent drive towards sustainability requires national economies to identify and evolve their ingenious resources for sustainability that are consistent with their capacity. Identifying individual strengths and ingenuities can birth effective global interdependencies and possibly lead to trade where comparative advantage is effectively harnessed for sustainable development (Mushanyuri and Mzumara 2013).

[2] https://www.elgaronline.com/edcollchap/book/9781784715489/C11.xml.

[3] https://saylordotorg.github.io/text_international-trade-theory-and-policy/s05-the-ricardian-theory-of-compar.html.

The Ricardian theory of comparative advantage postulates that "nations can gain an international trade advantage when they focus on producing goods that produce the lowest opportunity costs as compared to other nations" (Faccarello 2015). The drive towards a circular or bio-economy requires producing and utilizing environmentally friendly products with economic value. However, these resources will possess greater environmental and economic value when they are homegrown and can culminate into greater economic returns for an economy. Therefore, negotiating from the area of strength is the proposition of David Ricardo, such that countries can benefit from trading with each other by concentrating their efforts on the creation of the things they are best at while trading for the things they cannot manufacture or use unfriendly technologies to produce. Thus, concerning sustainability, clean processes or production technologies can form the new basis for trade, measures, and development.

Developing the technology that will engineer a circular economy requires innovation that evolve through novel or cumulative investments (Sehnem et al. 2022). While innovation remains a product of investment in Research and Development (R&D), the priority placed on developing technology for commercial purposes requires institutional coordination and intentional capital investments (Lehmann et al. 2022). Also, cumulative investments entail individual and institutional identification, harnessing and development of localized technologies to meet up with modern technological demand, and mass production strategies that will incorporate international trade.

5 Case Studies

This section presents a case study of two recycling industries in Nigeria with a primary emphasis on their production technology. The first case assesses the activities of an iron and steel rolling company, while the second examines the activities of a weaving firm that recycles textile and plastic wastes.

Case Study 1: A Case Study of Ife Iron and Steel Nigeria Limited, (IISNL) Osun State, Nigeria Description of the Plant Location: IISNL is a privately owned (by Indian nationals) recycling business enterprise that fabricates iron rods from metal scraps and wastes. The plant is located in a community, Ile-Ife, Osun State, South West Nigeria, along Ile-Ife/Ibadan Federal Road. Ile-Ife, an ancient Yoruba town, is about 218 km Northeast of Lagos. The inhabitants of the community are majorly low-income earners, who engage in farming, trading and other commercial activities (Saka and Busari 2020).

Production Technology: The input material is typically 100% metal scraps and the production technology is not purely clean. According to the study by Saka and Busari (2020), the inhabitants of the community complained about the consequences of emitted fumes from the factory on their health and well-being, which indicates that the production technology is not environmentally friendly. The health issues experienced by the dwellers of the community include but are not limited to skin rashes, catarrh, low sperm count, and sudden death. However, the company (IISNL) strongly declared that it has invested resources into ensuring that the emission to the atmosphere arising from its operational activities is minimal through capturing. Also, upholding this position, the company's Personnel Manager/spokesperson reiterated that the company is always capturing far above the World Health Organization regulated standard, which is 70–80%. According to the spokesman, as reported in Saka and Busari (2020), the factory is capturing not less than 80% of its emission and it is only 15% coming out of the factory because it is visible for anybody to see. He stated further that for every advantage, there is always a disadvantage. That is, industrialization comes with its peculiar challenges in turn.

Case Study 1 highlights the activities and impacts of a foundry firm that recycles metal scraps and waste into other iron and steel products for reuse. As laudable as the recycling initiative appears, there are still certain underlying issues that question the activities of the foundry. Specifically, the foundry becomes appalling when the process or production technology leads to unsustainable practices such as environmental pollution that are hazardous to human health. Thus, it is not sufficient to advance material longevity through recycling or reuse for business and trading purposes, especially when waste processing is involved. Rather, a careful assessment of the processing methods is critical to circular economy and should be carefully considered in extending the life cycle of a material or product compared to human life.

Case Study 2: Planet 3R Planet 3R is a social enterprise located in the ancient city of Ibadan, Oyo State, Nigeria. Planet 3R converts textile and plastic wastes into affordable eco-friendly products for low and middle-income earners using the 3R principle, (Reduce, Reuse, Recycle), as well as to save the planet Earth by weaving them into innovative items (https://www.planet3r.com/). The main goal of Planet 3R is to create a sustainable world with minimal t textile and plastic waste. Therefore, Planet 3R recycles clothes and nylons through indigenous weaving skills and technologies to create new products for the art savvy. The fascinating nature of recycled products include bags, shoes, travelling and lunch boxes as well as clothing fabrics. Planet 3R creates a huge brand appeal that attracts people to patronize products made from recycled materials. They also engage in grassroots and creative awareness programs geared towards achieving a clean environment.[4]

[4] https://afri-plastics.challenges.org/planet-3r/.

Practically, Planet 3R aims to mediate in the issues of unhealthy economic activities of dumping and burning, as well as empowerment and awareness creation on the need for environmental sustainability. For instance, a common behavioural pattern among the people in Nigeria is the dumping of plastic waste (especially sachets of bottles and nylon water) by the roadsides, among other waste. Eventually, these plastics become appalling and cause environmental menace. Therefore, Planet 3R uses her innovative skills of converting plastic waste into strands for weaving and converting this waste into other useable products is a practical extension of promoting environmental sustainability through an elongated life cycle process of plastic and textile waste. According to the CEO of Planet 3R, she notes that plastics and nylon are much cheaper to produce than recycle. As a result, factories keep producing it without much concern for the environment. But Planet 3R recycles around 140 pieces of pure water nylon per product, which is far from the 60 million sachets of pure water that Nigerians consume daily, and 130,000 tons of plastic makes its way into Nigeria's water bodies every year[5]; thus, indicating the prospects and need for more investments to achieve environmental sustainability in this business.

Case Study 2 depicts the activities and impact of a locally home-grown weaving firm that recycles textile and plastic waste. The weaving firm utilizes a traditional method of weaving traditional attires to weaving plastics and cloth waste. It spins several strands from plastics and textile waste and converts them into reusable products like shoes and clothing. The point of emphasis in the weaving firm is the application of indigenous traditional weaving methods to recycling, which is not only environmentally friendly but also safe for human health at the level of production and consumption (Ojukwu and Tariq 2020).

5.1 *Case Study Assessment*

Comparisons can be made from the two cases highlighted regarding recycling and trade. Both cases studied engage in innovative technological approaches to recycling and material waste to create new products that they can trade or sell. The foundry firm utilizes a fire technology that smelts iron waste to reproduce iron rods. These iron rods which are produced are usually used in construction projects such as the building of houses, bridges, and road networks (Ohimain 2013). Similarly, the weaving firm employs the spinning and weaving method to recycle and recreate new products for clothing which can equally be traded to generate income.

Figure 2a, b summarize the process of production across both firms. The highlighted features depict the tenets applicable to a circular economy which include

[5] https://fortomorrow.org/explore-solutions/planet-3-R.

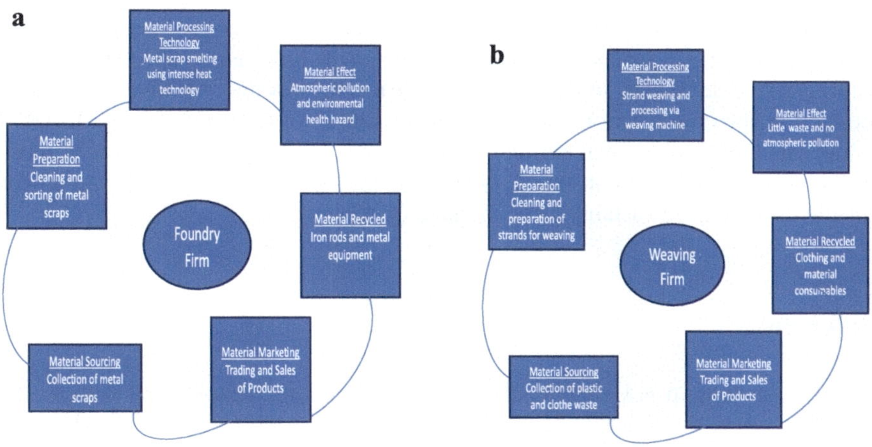

Fig. 2 **a** Material cycling in a foundry firm. **b** Material cycling in a weaving firm

material sourcing to preparation, processing, environmental pollution effects, products, and marketing; and the waste products can be recycled again across the product life cycle.

However, a closer look at Fig. 2a, b highlights the point of departure between the recycling firms. The major point of contrast between both firms is the eco-friendliness of the process technology for recycling by-products. While the foundry firm adopts a process technology that is environmentally unfriendly and in effect for trade, the weaving firm embraces a technology that is indigenous and eco-friendly. In essence, the indigenous weaving technology bears better sustainable prospects of being improved technologically and scaled up to engage mass production for local and international trading purposes.

6 Conclusion

This study delved into issues that described the connectedness of the Ricardian thoughts of comparative advantage with circular economy in developing economies. It subsumes the new economic model (circular and bio-based economy model) within the Ricardian thoughts of comparative advantage to drive sustainable development. This is based on Ricardo's assertion that countries can benefit from producing and trading in products that amplify areas of specialization while buying the things they are not as good at producing from other countries.

Judging from the assessment of Case Study 2, this study suggests that developing countries, such as Nigeria, should devote resources towards innovatively improving eco-friendly production technology. This is not only for local production but harnessing such technologies in a modern fashion to engage mass production that can spur international trade. On the other hand, trade connected with brown economy (non-green production technology), as in the first case study, does not offer a comparative advantage to developing countries. To this end, to drive sustainable production and consumption in developing economies, it is paramount for all stakeholders including policymakers, trade associations, and academics to focus more on the synergy of Ricardian thoughts of comparative advantage with development objectives to achieve sustainable economies.

References

Afteni C, Păunoiu V, Afteni M (2021) Study on the transition from the linear economy to the circular economy. Ann "Dunarea de Jos" Univ Galati, Fascicle V, Technol Mach Build 39(1):49–55

Algieri B, Aquino A, Succurro M (2018) International competitive advantages in tourism: an eclectic view. Tour Manage Perspect 25:41–52

Chang D, Lee CKM, Chen CH (2014) Review of life cycle assessment towards sustainable product development. J Clean Prod 83:48–60

Clabeaux R, Carbajales-Dale M, Ladner D, Walker T (2020) Assessing the carbon footprint of a university campus using a life cycle assessment approach. J Clean Prod 273:122600

Cobbinah PB, Erdiaw-Kwasie MO, Amoateng P (2015) Rethinking sustainable development within the framework of poverty and urbanisation in developing countries. Environ Dev 13:18–32

Curran MA (2017) Overview of goal and scope definition in life cycle assessment. Springer, Netherlands, pp 1–62

Dahiya S, Katakojwala R, Ramakrishna S, Mohan SV (2020) Biobased products and life cycle assessment in the context of circular economy and sustainability. Mater Circ Econ 2:1–28

Daly HE (2014) Beyond growth: the economics of sustainable development. Beacon Press

Eversberg D, Fritz M (2022) Bioeconomy as a societal transformation: mentalities, conflicts and social practices. Sustain Prod Consump 30:973–987

Faccarello G (2015) Comparative advantage. In: The Elgar companion to David Ricardo. Edward Elgar Publishing Limited

Figge F, Thorpe AS, Givry P, Canning L, Franklin-Johnson E (2018) Longevity and circularity as indicators of eco-efficient resource use in the circular economy. Ecol Econ 150:297–306

Giddings B, Hopwood B, O'brien G (2002) Environment, economy and society: fitting them together into sustainable development. Sustain Dev 10(4):187–196

Hapuwatte BM, Jawahir IS (2021) Closed-loop sustainable product design for circular economy. J Ind Ecol 25(6):1430–1446

Hauschild MZ, Rosenbaum RK, Olsen SI (2018) Life cycle assessment. Springer International Publishing, Cham. https://doi.org/10.1007/978-3-319-56475-3

Jeronen E (2020) Sustainable development. Encyclopedia of sustainable management. Springer International Publishing, Cham, pp 1–7

Khan N, Sudhakar K, Mamat R (2021) Role of biofuels in energy transition, green economy and carbon neutrality. Sustainability 13(22):12374

Lehmann C, Cruz-Jesus F, Oliveira T, Damásio B (2022) Leveraging the circular economy: investment and innovation as drivers. J Clean Prod 360:132146

Lewandowski I (2018) Bioeconomy: shaping the transition to a sustainable, biobased economy. Springer Nature, p 356

Lopes JM, Gomes S, Pacheco R, Monteiro E, Santos C (2022) Drivers of sustainable innovation strategies for increased competition among companies. Sustainability 14(9):5471

Manavalan E, Jayakrishna K (2019) An analysis on sustainable supply chain for circular economy. Procedia Manuf 33:477–484

Marino A, Pariso P (2016) From linear economy to circular economy: research agenda. Int J Res Econ Soc Sci 6225(6):2249–7382
Mikhno I, Koval V, Shvets G, Garmatiuk O, Tamošiūnienė R (2021) Green economy in sustainable development and improvement of resource efficiency
Mushanyuri BE, Mzumara M (2013) An assessment of comparative advantage of Mauritius. Eur J Sustain Dev 2(3):35–35
Naeem S, Chazdon R, Duffy JE, Prager C, Worm B (2016) Biodiversity and human well-being: an essential link for sustainable development. Proc Roy Soc B Biol Sci 283(1844):20162091
Nar M (2014) Global public goods and sustainable development. J Sustain Dev Stud 6(2)
Niero M, Kalbar PP (2019) Coupling material circularity indicators and life cycle based indicators: a proposal to advance the assessment of circular economy strategies at the product level. Resour Conserv Recycl 140:305–312
Ohimain EI (2013) Scrap iron and steel recycling in Nigeria. Greener J Environ Manage Public Saf 2(1):1–9
Ojukwu OA, Tariq T (2020) Rethinking sustainability in the fashion industry through indigenous culture
Ozili PK, Iorember PT (2023) Financial stability and sustainable development. Int J Financ Econ
Saka L, Busari D (2020) State, industries and regulation in Nigeria: Ife iron and steel company and environmental pollution. J Manage Soc Sci 9(1):750–768
Sehnem S, de Queiroz AAFS, Pereira SCF, dos Santos Correia G, Kuzma E (2022) Circular economy and innovation: a look from the perspective of organizational capabilities. Bus Strateg Environ 31(1):236–250
Siddiqui K (2018) David Ricardo's comparative advantage and developing countries: myth and reality. Int Crit Thought 8(3):426–452
Tien NH, Bien BX, Vu NT, Hung NT (2019) Risks of unsustainable economics development in Vietnam. Int J Manage Commer 1(4):4–9
Tóth Szita K (2017) The application of life cycle assessment in circular economy. Hungarian Agric Eng 31:5–9
Ukaogo PO, Ewuzie U, Onwuka CV (2020) Environmental pollution: causes, effects, and the remedies. In: Microorganisms for sustainable environment and health. Elsevier, pp 419–429
Uysal G (2019) Wealth of nations and intellectual capitals: Adam Smith approaches. J Mod Account Audit 15(8):411–414
Walker S, Rothman R (2020) Life cycle assessment of bio-based and fossil-based plastic: a review. J Clean Prod 261:121158
White R (2014) A shift in comparative advantage? Making sense of anti-trade sentiment: international trade and the American worker. Palgrave Macmillan US, New York, pp 37–57
Winans K, Kendall A, Deng H (2017) The history and current applications of the circular economy concept. Renew Sustain Energy Rev 68:825–833
Zhong R, Pei F, Yang K, Xia Y, Wang H, Yan G (2021) Coordinating socio-economic and environmental dimensions to evaluate regional sustainability—towards an integrative framework. Ecol Ind 130:108085

An Investigation on the Connections Between Circular Economy and Life Cycle Analysis in Relation to Sustainable Development Goals

Joshua Funminiyi Obisanya

Abstract Solutions to the all-encompassing sustainability phenomenon facing the world are known to be multifaceted. Notably among them are Circular Economy (CE) and Life Cycle Assessment (LCA), for achieving Sustainability Development Goals (SDGs). Also, integration of these two approaches offers a more comprehensive understanding that could help in making informed decisions towards a sustainable future. This suggests that there are connections between CE and LCA in relation to SDGs, However, the areas of their connections were not explicitly captured in the extant literature. Therefore, this study aims at identifying and elucidating more on the connections between the duos of CE and LCA vis-à-vis SDGs. The study employed a systematic literature review methodology to explore the research goal. As part of the findings from this study, a conceptual framework was developed, to show the flow of links among CE, LCA and SDGs as a basis for discussion on their connections and in-depth understanding of the phenomenon. Also, 10Rs strategies of CE that align with SDGs were identified. Besides, seven goals out of the seventeen goals in SDGs, were identified to connect with the goal of CE. More connections between CE and SDGs were also noted in the three key areas: implementation and application; alignment of SDGs targets with CE vision; and connection of indicators. Additionally, CE-LCA linkage was recognised as a comprehensive and robust tool for the advancement and assessment of resource efficiency of a product or process. This study has implications for various stakeholders, particularly in the sphere of sustainable development, to be more aware of connecting areas and factors of linkage between CE and LCA in relation to SDGs. The study then recommends that future studies on CE, LCA and SDGs, either as a standalone construct or in a combined form, leverage on this baseline information to promote sustainable development in all ecosystems.

Keywords Life cycle analysis · Circular economy · Resource efficiency · Sustainable development · Business enterprise · Framework

---

J. F. Obisanya (✉)
Institute for Entrepreneurship and Development Studies, Obafemi Awolowo University, Ile-Ife, Nigeria
e-mail: jobisanya@oauife.edu.ng

L. Brandli et al. (eds.), *The Contribution of Life Cycle Analyses and Circular Economy to the Sustainable Development Goals*, World Sustainability Series,
https://doi.org/10.1007/978-3-031-85300-5_13

1 Introduction

The presence of business enterprises—commercial ventures, undertakings, projects or corporations, of various sizes (nano, micro, small, medium or large enterprises), owners (public or private), and sectors (processing, production or manufacturing), is inevitable in all nations of the world. Business enterprises, as the bedrock of development and growth of the economy, both at micro and macro levels, are well-known to involve a series of socio-economic activities and interactions, which in turn could lead to at least a deleterious environmental issue, in many cases. Notably among the common environmental menaces associated with all kinds of business endeavours are generation and unhealthy handling of wastes, pollution of earth's key survival elements—air, water and soil, climate change, land degradation, loss of biodiversity and resource depletion (Akter et al., 2024). Over the years, awareness of the negative environmental implications and consequences of economic activities of business enterprises is on the increase and wide-reaching. In the same vein, concerted efforts of stakeholders, especially development practitioners, academics and researchers, industrialists, and government policymakers (Mensah, 2019), focusing on how to attain mutual healthy relationship cum interactions between business and its operating environmental components, have yielded the concept of sustainability. The concept is a mitigating and development strategy to resolve the negative environmental challenges in business operations. Sustainability aims majorly at preserving ecological balance, ensuring fair distribution of natural resources among generations and nurturing development (Aleksić et al., 2023). It is worth noting that sustainability originated in 1987, as it appeared in the Brundtland Report—Our Common Future, in World Commission on Environment and Development (WCED) in a quest for sustainable development (Imperatives, 1987; Rosa et al., 2023; Shi et al., 2019).

Fundamentally, according to Imperatives (1987) and Rosa et al. (2023), there are two main perspectives about sustainable development. Firstly, it emphasises the basic needs of the poor in the global village, which must be given principal priority. Secondly, it sheds light on the limitation of natural resources, which could hinder both current and future generations from meeting their needs. That is, achieving economic development and growth without depleting natural resources for future generations. Also, in their view—Chen et al. (2008), the authors affirmed that the three key constructs in the submission of Imperatives (1987): basic needs of the poor, limitation of natural resources and economic development, all converge towards the holistic description of sustainable development. In other words, the authors stated that sustainable development aims at balancing the integration of the three pillars (economy, society and environment) of sustainability, where the three pillars must coexist, interconnect and interact in a beneficial way with one another. This constitutes a three-pillar concept, which is predominantly known as the Triple Bottom Line (TBL)—the balance among the three pillars of sustainability (Akter et al., 2024).

Furthermore, in 2015, Sustainable Development Goals (SDGs) were launched by the United Nations to unveil the specific areas of development related to sustainability, which call for urgent attention and collaborative efforts of nations around the globe

(Allen et al., 2018). In this way, the SDGs give a comprehensive and elaborate understanding of the concept of sustainability, nowadays, as it gives an in-depth description of all the elements covered in the three pillars of sustainability (Purvis et al., 2019). Basically, the SDGs are also known as 2030 agenda or initiatives, which were formed by 193 countries under the auspices of the United Nations to holistically resolve the three-pillar issues in sustainable development around 17 focal areas with 169 goals that must be achieved by 2030. As a good framework for the realization of SDGs, circular economy (CE) emerged in the recent past. The concept of CE emphasises the transition to a new business principle that would concentrate not only on the economic aspect but also on the social and environmental impacts of such business, in a systematic manner (Aleksić et al., 2023). Nowadays, CE has gained increasing attention around the world among stakeholders in diverse disciplines, businesses and helm of governance, as a viable strategy for sustainable development and actualisation of SDGs (Rosa et al., 2023). Basically, CE is based on the principle of sustainable production and consumption with the prime objective of increasing the stock of materials. That is, more stock of materials is produced and consumed with less resource intake. The logic behind circularity in CE is to recover and re-inject wastes into the economic/business system (production) again. Thereby, it enables the resulting products and materials to be reused by the consumers, resold in the distribution channels, and recycled for further processing (production) of new products and materials. Moreover, Akter et al. (2024) and Aleksić et al. (2023) noted that measurement of the production and consumption pattern of an economy is indispensable for the implementation of CE. To achieve this, the authors declared that Life Cycle Assessment (LCA) is a proven method and an analytical framework to measure CE's impacts. Thus, according to Peña et al. (2021) LCA could be used effectively to measure the environmental consequences of a specific product or materials. Besides, most CE scholars agree that LCA is the most effective tool for evaluating options for CE solutions (Haupt & Zschokke, 2017). Although the use of LCA is well-noticeable at the product level, many of its applications in studies concerning CE and related subjects divulged the effective use of LCA at the societal level, which makes it a universal tool for measuring CE impacts (Roos Lindgreen et al., 2021). This affirms that connections exist between CE and LCA vis-à-vis SDGs. However, the connections were scattered in the extant literature on sustainable development and related domains, thus, this study sets to identify and elucidate more on the connections between the duos of CE and LCA in relation to SDGs.

2 Methodology

This is qualitative research which employed research methodology involving a systematic approach (Rossi et al., 2018) of reviewing relevant extant literature on a particular research issue, to generate information/findings that will resolve such issue. The stages in the systematic approach used for this study include:

Preliminary stage: This involves the identification of keywords and the database to be used. The keywords that are used revolve around sustainable development, sustainable development goals, circular economy, circular economy indicators, circularity, life cycle analysis, life cycle assessments, resource efficiency and waste management strategies. The keywords were searched via online databases, encompassing Scopus, Google Scholar, Web of Science, JSTOR and ScienceDirect.

Documents'/Materials' selection stage: This involves the type or nature of academic documents/materials to be selected and the selection criteria. Here, academic documents in the form of referred journal articles, books, conference presentations, which were published in the English Language and reputable outlets were selected. Additional selection criteria, generally based on the relevance, authority and currency of the documents, were used to include and exclude materials for further analysis.

Extraction of relevant information stage: This involves thorough reading and synthesis of full texts of selected materials. That is, the contents were summarised under themes without coding but with notes. Thereafter, the extracted materials were carefully paraphrased to make them more condensed, concise, coherent and manageable without changing the meaning of the contents. The result was a more concise and refined summary of the relevant literature regarding the focus of this study, as presented in the next section.

3 Findings and Discussion

Connections of the constructs—SDGs, CE and LCA to one another are depicted in the Fig. 1. The fundamental key elements of sustainability (FKES) in the figure, are referring to the economy, society and physical environment. Also, the arrows indicated a complementary system of the constructs. That is, understanding of each of the constructs is based on the interconnected links among the constructs, as shown in Fig. 1. The framework served as the basis for discussion of findings on the aim of this study.

4 Circular Economy (CE) in Sustainable Development Goals (SDGs)

This section presents connectedness of CE with SDGs. It starts with a synopsis of both SDGs and CE, in sequential order. This is followed by the identification of links between the duo construct—CE and SDGs. The section gives state-of-the-art information and an in-depth understanding of the relationship between the duos.

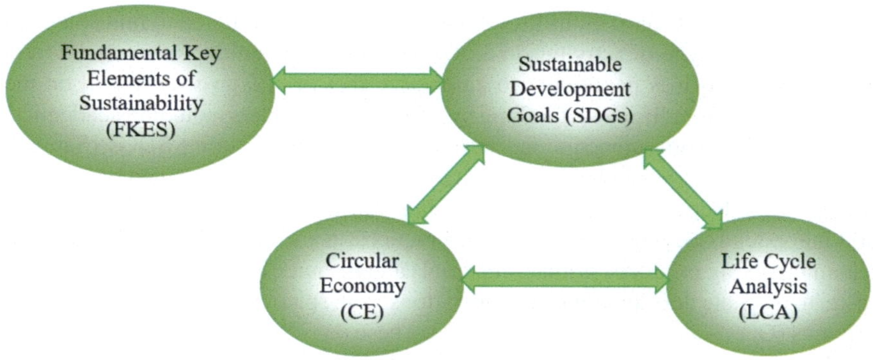

Fig. 1 A framework showing connections among SDGs, CE and LCA

5 A Synopsis of SDGs

The 2030 Agenda for SDGs has enriched and expanded the discourse on sustainable development in the contemporary era. It is no longer news that the agenda contain 17 goals, which aims at achieving simultaneously sustainable development in the areas of economy, society and environment. Principally, the agenda reveals a shift from concentrating largely on ecological issues towards more general societal challenges, encompassing zero hunger, good health, quality education, gender equality, sustainable cities, communities, and so on (Montag, 2023). The modern focus on sustainable development in relation to societal issues is to ensure that the well-being of humans in an environment is absolutely intact and sustainable. Thus, the concept of SDGs in this era emphasises more on poverty eradication, human-induced environmental damage prevention, and economic prosperity goals (Manioudis & Meramveliotakis, 2022).

By and large, the SDGs are aspirational goals of promoting healthy socio-economic relationships in an environmentally sustainable manner and unprecedented way. So far, 169 targets and 232 quantifiable indicators have been introduced to assess the realisation of SDGs by the year 2030. In this context, the goals refer to the expected changes or effects or the state of global affairs, the targets define how SDGs could be practically accomplished, and the indicators pinpoint how these targets could be measured. For example, Goal 1 of the SDGs states that poverty should end in all its forms worldwide. This goal sounds extremely ambitious and unreachable, for which the UN has set five targets. One of the targets is to eliminate severe poverty worldwide by 2030. This target is currently applied to individuals making less than USD 1.25 daily, and the indicator for this scenario is the percentage of people living below the poverty level internationally. Thus, goals, targets, and indicators in the sphere of SDGs seem complicated but logical. This is so because, without measurable indicators, it might be challenging to know the scope and type of activities needed to practically actualise those ambitious goals (Smith, 2020).

In recent times, several multidisciplinary studies have attempted to find solutions to the scope and type of activities in SDGs, CE stands out among others to address the phenomenon. It is paramount to emphasize that the CE evolved based on the lessons learned from life cycle thinking by various stakeholders including academics, political leaders, entrepreneurs and business owners, in pursuing SDGs (Samani, 2023). Also, it is crystal clear in the extant literature that CE emerged as a viable strategy for a deeper and better understanding of SDG scopes. With CE, life cycle thinking extends beyond environmental practices, it incorporates a more efficient system involving products, primary materials and resources that would generate less waste, to promote development and innovations in all spheres of economic and social activities (Gheewala & Silalertruksa, 2021). These days, it is widely recognized that the CE comprises numerous interconnected cycles of materials and products (Samani, 2023).

The idea of circularity, in the discourse of SDGs, is to recover wastes and inject them into the economic system again. Thereby, CE enables the products and materials to be reused by the consumers, resold in the distribution channels, and recycled for further processing (production) of new products and commodities. The prime objective in the circularity of resources, as evident in CE, is to increase the stock of materials, that is, more stock of materials produced and consumed with less resource intake. Therefore, the classical concept of circularity deals with maximizing the time of materials/products used in the economic system before it is subsequently passed down as a disposal in landfills. To this end, in the modern world, the idea of circularity, in the context of SDGs, is to facilitate economic development without compromising its environmental and societal consequences in which CE is regarded as one of the most effective tools to achieve it (Tan et al., 2022).

6 An Overview of CE

The concept of CE finds its root, according to Martins-Rodrigues et al. (2020), in the theory of thought of the Economics of Sustainable Development otherwise known as eco-development. The theory is based on the balance between the economy and the environment. That is, green economy, which is related to green businesses or jobs (Sulich & Rutkowska, 2020). Besides, Ryen et al. (2018) traced the outstanding existence of CE ideology to as far back as the 1970s. This is corroborated by Ekins et al. (2020), who declared that the idea of CE emanated from two long strands, that is, two conceptual streams, which could be traced to early days of the modern environmental movement in the 1960s and 1970s. While the first relating to the flow of materials through an economy, the second is concerned with thinking about the economic conditions that might bring about such a flow. The authors elucidated further that the two conceptual streams have a subsequent symbiotic relationship. Furthermore, several authors, such as Ghisellini et al. (2016) and Zink & Geyer (2017), pointed out that the first understanding of the construct (CE) was given by Pearce & Turner (1989). Their explanations hinged on interactions between natural resources and economic activities and their subsequent impacts on supplies for manufacturing as well as consumption. This concept (CE) has been explored in depth in recent years, according to Kirchherr et al. (2017). Nevertheless, the year 2015 marked fundamentally its peak by the action plan for the circular economy of the European Union, the SDGs and the Paris Agreement (Cifuentes-Faura, 2021).

To date, there is no universally agreed-upon description for the concept of CE. Hence, it has various interpretations from multiple sources. To this end, CE is considered as an overarching concept that consists of multiple strategies and approaches. For instance, in Rosa et al. (2023), CE was highlighted as a relevant change for companies, as it forces them to rethink the economic practices of today's society, to balance economic growth with sustainable development. In this situation, the mission of companies should include the guiding principles such as promoting research, sustainable development, innovation, and a collaborative strategy, which are the basis of CE. Moreover, del Mar Alonso-Almeida & Rodriguez-Anton (2019) regarded CE as possible solutions to multidimensional problems currently facing the world, which include but not limited to increase in global demand for resources, price volatility for raw materials, growing population and rise in consumption of resources and products around the world. Oriekhova (2019) viewed CE as main transformative and integrative philosophy, which is the basis of the ecological and economic policies of the public–private partnership. In their quest to synthesise a versatile definition for CE, Kirchherr et al. (2017) reviewed 114 definitions of CE. Then, the authors summed up the definitions of CE in the extant literature and described the concept as economic system that is based on business models, which replaces the concept of end-of-life (open linear system: take, make, use and dispose system). That is, CE encompasses strategies of reduction, reusing, recycling and recovery of materials in production/distribution and consumption processes that operate at micro level (products, companies, consumers), meso level (eco-industrial parks) and macro level (city, region, nation and beyond), according to the authors. They stated further that the main goal of CE is to achieve sustainable development, which implies creation of environmental quality, economic prosperity and social equity, for the benefit of current and futures generations (Rosa et al., 2023; Ekins et al. (2020); ; ; ;. Therefore, this study describes the concept of CE as a sustainable business model with multi-dimensional strategies that guarantee resource efficiency, no or little generation of waste, and exceptional societal well-being, during the processes of production and consumption in an economic system. With the description of CE for this study in mind, it is important to emphasize that, through CE, an economy could be reinvented by boosting its growth and making it more competitive and sustainable. Additionally, CE can stimulate innovation, reducing carbon and greenhouse gas emissions (Choudhary et al., 2015), improving the security of supply of raw materials, providing a reduction in energy use (Zhalechian et al., 2016), optimizing waste management, among other benefits.

7 CE Strategies to Achieve SDGs

The circularity strategies, as found in Ekins et al. (2020) and Samani (2023), are constantly employed in CE to achieve SDGs, could be denoted as R_0, R_1, R_2, R_3, R_4, R_5, R_6, R_7, R_8, and R_9. Altogether, this is known as 10Rs of resourceful management of product/output that might emanate from the circular production-consumption systems of CE and it is described as follows, according to the submissions of the authors on the strategies:

R_0 (Refuse strategy): This aims at making a product redundant by abandoning its function or by offering the same function with a radically different product. In other words, the goal of this strategy is to abandon the function, whether for the same or a new product, within the same life cycle.

R_1 (Rethink strategy): The strategy focuses on intensive product use, with either the existing or a new function, through sharing products, or by putting multifunctional products on the market.

R_2 (Reduce strategy): This helps to achieve efficient manufacturing and product use, in which resource and material consumption are reduced, while maintaining the same product and function within the same life cycle. That is, it increases efficiency in product manufacturing or usage by consuming fewer natural resources and materials.

R_3 (Re-use strategy): In this strategy, the product is reused with its original function after being discarded. It also involves reuse by another consumer of discarded product, which is still in good condition and fulfils its original function.

R_4 (Repair strategy): It focuses on the same product and function, but extends the product's lifespan. That is, repair and maintenance of defective product, so that it can be used with its original function.

R_5 (Refurbish strategy): This aims at prolonging the lifespan of the same product while maintaining its original function. That is, to restore an old product and bring it up-to-date.

R_6 (Remanufacture strategy): This seeks to prolong the product's lifespan while maintaining its original functionality. With this strategy, the original function of the product is preserved, and in some cases, even improved, but without introducing new functions. Specifically, the strategy uses parts of discarded product in a new product with the same function.

R_7 (Repurpose strategy): The strategy involves the usage of discarded products or their parts to create a new product with a new or different function.

R_8 (Recycle strategy): This strategy aims at producing a new lifecycle, where both the product and its function can undergo changes. While recycling activity towards reproducing the same product is referred to as a closed-loop recycling, recycling for a new product is known as an open-loop recycling. In closed-loop recycling, the same product with the same function is maintained, whereas in the case of open-loop recycling, both products and functions could change. Additionally, recycle strategy could depict processing of materials to obtain the same (high grade) or lower (low grade) quality product.

R_9 (Recover strategy): It is primarily focused on energy recovery. That is, incineration of materials with energy recovery.

The strategies affirmed CE as a regenerative system in which resource input and waste, emission, and energy leakage are minimized by slowing, closing, and narrowing material and energy loops (Schröder et al., 2019). Thus, CE sustainably turns a product at the end of its life cycle into resource(s) for other stakeholders via reusing and recycling what is still functional; remanufacturing, refurbishing, and repurposing the components that cannot be repaired; and regenerating and redistributing what is processed (Stahel, 2016). In this manner, CE as a replacement of traditional linear take, make, use, dispose production model, is potent in creating a positive value-driven closed loop system for an economy to develop sustainably (Hishammuddin et al., 2018).

8 Connections Between SDGs and CE

Suárez-Eiroa et al. (2019) attempted to establish the connection between SDGs and CE. The authors noted that CE is a prerequisite means and an indispensable strategy for realisation of goals in SDGs. They emphasised further that CE and SDGs have a compensatory relation. Besides, **a**mong the goals of sustainable development, according to the authors, seven goals are directly related to the application of the circular economy and the goals were declared as the primary goals of the concept of CE. The goals encompass Goal 7—affordable and clean energy, Goal 8–decent work and economic growth, Goal 11–sustainable cities and communities, Goal 12–responsible consumption and production, Goal 13–climate action, Goal 14–life below water, Goal 15–life on land (Aleksić et al., 2023).

In addition to the above fundamental information on the connections between the two constructs, this section also identified three more key connections on the duos, as they are jointly working towards sustainable development.

Implementation and Application: The concepts of SDGs and CE are inseparable, especially at the level of implementation, as all the goals, strategies and tools for their application are closely linked, to yield effective results (Mehta, 2023). It is the concern for ecological, social and economic dimensions of business operations in the circular economy that leads to the comparative concern for the trio principles of sustainable development. Also, through the creative usage of resources, including wastes (secondary raw materials) and renewable energy sources, CE acts in the direction of achieving sustainable development. This is so because it behaves responsibly and resourcefully towards the natural environment, its inhabitants and various economic interactions in the ecosystem. Besides, the model of circular economy is based on a production system respecting the principles of sustainable economic development, without any deleterious effect on the environmental elements – air, land and water (Marković et al., 2023).

Alignment of SDGs Targets with CE Vision: Two of the targets included in the UN Sustainable Development Goal 12 are ensuring sustainable consumption and production patterns with a focus on reducing waste generation (target 12.2), and the promotion of more efficient and sustainable resource management (target 12.5). Those targets are aligned with the CE vision of decoupling economic growth from resource constraints (Niero & Rivera, 2018).

Connection of Indicators: To measure and track progress towards SDGs and CE, there is a need for indicators. Generally, indicators have two principal purposes – forward looking to provide guidance and backward looking to provide feedback and review performance. Therefore, the choice of indicators is of paramount importance as it becomes a critical determinant of the behaviour of a system (Meadows, 1998). In other words, in-depth understanding of link of indicators between SDGs and CE are of prime importance in order to guide and also assist in reviewing policies, among other benefits, on sustainable development in developing and developed world, today. To this end, Meadows (1998) emphasised that indicators of sustainability, as found in SDGs, are also valid for those of CE. In the same vein, Ekins et al. (2020) reiterated that indicators of sustainable development could be found in the SDGs, which seek to integrate and achieve the key trio constituents of the concept of sustainability – environmental sustainability (the maintenance of important environmental functions), economic sustainability (maintenance of the capital stock, sound money, corporate profitability) and social sustainability (upkeep of social cohesion). According to the authors, the SDGs indicators form basis for CE indicators. Jerome et al., 2022 explained further that indicators of CE based on circularity strategies can be categorized as either single-focus (targeting one specific strategy) or multi-focus (considering multiple strategies). Remarkably, material recycling emerges as the primary focus for both single-focus and multi-focus circularity indicators. While single-focus indicators are easier to interpret, they might not encompass all life cycle phases and flows, making them unsuitable for capturing burden-shifting issues, that is, circularity phenomenon in SDGs. With this scenario, in order to provide an indication of what the indicators are measuring,conducted a wide-ranging literature review to identify 55 CE indicators (circularity indicator sets). The indicators were grouped into 10 categories based on a number of criteria—Levels (micro, meso, macro), Loops (maintain, reuse, recycle), Performance (intrinsic, impacts), Perspective (actual, potential), Usages (improvement, Benchmarking, communication), Transversally (generic, sector, specific), Dimension (single, multiple), Units (quantitative, qualitative), Format (web-based, Excel. formulas) and Sources (academics, companies, agencies). The authors noted that the only category that provides insight into what the indicators are measuring is one referring to loops. Albeit, the indicators of SDGs and CE are connected, there is no universally agreed specific number of indicators that can be used as standards for the duos.

The indicators and their connection between the constructs vary from one stage to another stage in the life cycle of a product among other factors, especially when evaluating the options that might reduce environmental impacts (Ekins et al., 2020). The authors pinpointed that a life cycle approach is therefore an indispensable and essential tool of eco-measurement to properly compare different options to explore within a circular economy approach. In this way, understanding of connection between life cycle assessment (LCA) and CE is germane.

9 A Background on CE-LCA Connections

In order to have a clearer understanding of LCA and its connection with CE vis-à-vis sustainable development, Mannan & Al-Ghamdi (2022) shed more light on the fundamental philosophy of (CE). The authors reiterated that CE is the concept of cradle-to-cradle in production and consumption systems, through the application of CE strategies. Additionally, the authors revealed that the concept is continually gaining popularity worldwide, since its inception, as it aims to increase resources efficiency through closed loop technologies. Nonetheless, the closed loop concept of CE does not always ensure environmental benefits. To this end, comprehensive environmental assessment tool or methodology such as life cycle assessment (LCA) is quite beneficial for robust environmental assessment of each strategy of CE. Therefore, this section presents a discourse on the findings about the connections between CE and LCA, as the combination of LCA and CE could result in more in-depth analysis and better understanding of economic, social, and environmental sustainability in SDGs according to Mannan & Al-Ghamdi (2022). This section starts with a description of LCA followed by findings on connectedness of LCA and CE.

10 A Description of LCA

By and large, a life cycle represents a scheme of identifying both opportunities and risks of a product or technology, all the way from raw materials to disposal. To achieve this, there is a continuum of life cycle approaches—from qualitative (life cycle thinking) to comprehensive quantitative approaches (life cycle assessment studies). Globally, these life cycle approaches have found application among various stakeholders, including people, researchers, companies and governments, working towards sustainable development. They have used the approaches in different areas: from day-to-day shopping, selecting office supplies for the workplace, engineering a new product design, or developing a new government policy (Ngan et al., 2021).

LCA is a comprehensive, robust scientific approach that addresses the environmental aspects and potential environmental impacts of any product or process throughout its life cycle, starting from raw material extraction to final disposal or recycling (Standardization, 2006; Finnveden et al., 2009). This cradle-to-grave approach refers to a holistic view of environmental interactions that compiles all input and output data associated with any product system or process. In that way, it analyzes environmental loadings over life cycle of any product or process by connecting to relevant inputs and outputs (Yoonus et al., 2020). These inputs could be listed as extracted raw materials from the earth that are required for any product stream or process, as well as the energy and water used, while outputs include emissions to land, air, and water, that is, solid wastes, by-products, and other emissions. At the end of the assessment, LCA summarizes the environmental impact of products or processes in the area of resources used, human health hazards, and ecosystem damage (Mannan & Al-Ghamdi, 2022). Assessment of all life cycle phases of a product or process enables LCA to estimate the cumulative impact results, which in turn, allows selection of the product or production route that is most environmentally friendly (Mannan & Al-Ghamdi, 2020). Therefore, LCA is apt to provide comparative conclusions, helping policymakers to find better alternatives by comparing all key environmental impacts (Seidel, 2016).

According to Ngan et al. (2021) LCA was first introduced as a quantitative environmental performance tool that evaluate the potential environmental impacts and resources used throughout a product's life cycle. Over time, the tool has been useful in quantifying environmental impacts of a product during its life cycle from the stage of raw materials acquisition to manufacturing of products to the disposal and waste management process. In their view, Rigamonti & Mancini (2021) LCA was described as a methodology developed for assessing and quantifying the latent environmental impacts associated with entire product life cycle and it is standardized in the ISO 14040 series. Samani (2023) provided all-inclusive description of LCA. The authors elucidated that the first LCA study was conducted in the 1960s in packaging studies, which primarily focused on energy use rather than emissions (Hauschild et al., 2018). Subsequently, it has been introduced as a methodology (Brändström & Saidani, 2022), a method (Bastianoni et al., 2023), or a tool (Corona et al., 2019). Also, Civancik-Uslu et al. (2018) highlighted that LCA was defined as a tool by the United Nations Environment Program (UNEP) and as a methodology by the ISO standard. In addition, Beltran, (2018) pinpointed that while LCA emerged as a method, the LCA tool has been developed over the past 30 years. As a standardized and well-known method, LCA is considered as the go-to approach for evaluating the environmental performance of products and services (van Stijn et al., 2021).

Noteworthy, when referring to LCA in general, attributional LCA is the focus. The attributional LCA aims at determining the environmental impacts that can be attributed to the studied product and it models the product system in isolation from the rest of the technosphere and economy, using average processes for the background systems. On the other hand, consequential LCA attempts to evaluate the environmental consequences of consuming the studied product by modelling the changes to the economy resulting from its introduction. Unlike attributional LCA, consequential LCA uses marginal processes for the background systems (Hauschild et al., 2018). Other types of LCA include prospective LCA and dynamic LCA. As LCA is continuously evolving nowadays, scholars have integrated LCA with other scientific domains to innovate different types of LCA. For instance, Blockchain LCA

(Shou & Domenech, (2022), transitional LCA (Ventura, 2022), and Life Cycle Gap Analysis-LCGA (Dieterle & Viere, 2022). Occasionally, the novelty lies not only in the methodology, but also in applying LCA. This is evident in a recent study on LCA conducted by (Ellsworth-Krebs et al., 2023) called Feminist LCA, in which a streamlined LCA compares three hair removal methods for women—shaving, waxing, or laser. To sum up, LCA, in any form, has been gaining wider acceptance and adoption in many business enterprises and industries, including but not limited to building and construction, manufacturing, oil and gas, life sciences and chemicals (Ngan et al., 2021).

11 Connectedness of LCA and CE

The extant literature on sustainable development emphasizes that more circularities does not necessarily lead to more sustainability as there are risks of problem shifting, rebound effect, and over-consumption. Therefore, there is a need for a tool or methodology to comprehensively evaluate circularity strategies in CE (Rigamonti & Mancini, 2021). In the same vein, Mannan & Al-Ghamdi (2022) expounded that implementation of a sustainable CE requires consideration of economic, social, and environmental circularity strategies (Niero & Rivera, 2018). This implies that for a successful implementation of the CE concept, material circularity and environmental performance should be assessed simultaneously, as circularity is not always equivalent to environmental sustainability (Haupt & Hellweg, 2019). In the quest for an assessment tool that will assist in choosing the best option or solution from circularity strategies in CE, LCA emerged as a comprehensive and perfect assessment tool. That is, a tool that is versatile in considering and assessing both material circularity and environmental impact of all strategies or options in CE, in a comprehensive manner. Thus, LCA is suitable for providing direct evaluation of strategies in CE regarding a reduction in input and use of natural resources, reduction in emissions, and a reduction in valuable material losses among others (Haupt & Hellweg, 2019; Mannan & Al-Ghamdi, 2022).

Moreover, according to Rigamonti & Mancini (2021), the Life Cycle Initiative proposed LCA for identification of the most promising CE strategies and options for improving the environmental performance of society's consumption and production patterns, as underscored in the recent position paper by the Life Cycle Initiative (Peña et al., 2021). Besides, LCA could be applied to build more consistent and robust CE strategies by considering potential upstream and downstream impacts and all relevant resources and impact categories. Corona et al. (2019), in their study, revealed that LCA is excellent for comparing and choosing the best among different strategies of CE, as its chief aim is to avoid shifting of burden. The authors adduced several reasons for their observation, which could also be tagged as benefits of LCA in relation to CE. According to the authors, LCA helps to understand and evaluate whether the claimed environmental benefits of CE solutions could be achieved and to what extent. Also, the methodology is perfect to identify the most critical aspect of strategies of CE that calls for urgent improvement (Peña et al., 2021). Therefore, as LCA is an approach to achieving sustainable development, which builds on the concept of life cycle thinking, it can help decision-makers to make more sustainable choices by providing a holistic and comprehensive perspective, offering insights not only on environmental aspects but also economic and social considerations of SDGs (Peña et al., 2021; Rigamonti & Mancini, 2021).

Largely, LCA is used, to focus only on a single activity (Helander et al., 2019). As a result, extant literature in this domain reveals that a lot of companies have started to analyse and reduce the environmental impacts of their processes, products and services using LCA. This has led to the development of various tools and instruments of LCA (Hervaa et al., 2011). These include but not limited to LCA for energy and carbon footprinting (Plassmann et al., 2010), water footprinting (Hoekstra et al., 2009), product environmental footprinting (Lehmann et al., 2015) and material footprinting (Lettenmeier et al., 2014).

As adoption of LCA in evaluation of strategies in CE is indispensable, Mannan & Al-Ghamdi (2022) reiterated that the CE concept aims at creating value for society and the economy while reducing environmental impacts, whereas LCA is a robust scientific tool for assessing the environmental load of any process or product. Also, the authors accentuated that LCA is a science-based environmental assessment tool with the potential to complement the strong vision of CE as well as improve its strategies in diverse ways. And so, the authors, highlighted the roles of LCA in CE as: testing CE models for products or processes, because in many cases, closed loops produce negative environmental impacts; discovering the limitations, impact hotspots, and inaccurate assumptions made during CE strategy, and thereby finding alternatives; and continuously improving CE strategies, to implement them in real business cases. Thus, while promoting circularity in all possible sectors, LCA continuously provides insight and assistance into the setting of key performance indicators. Likewise, it helps to define a baseline scenario, for evaluation of progress or any drawbacks in CE.

Consequently, a combination of CE principals and the robustness of LCA can be beneficial and potentially provide a holistic approach that could strengthen available CE alternatives. Mannan & Al-Ghamdi (2022) noted further that LCA could be utilized to compliment CE at three different stages. Firstly, for any hypothetical product or process (macro, micro, or regional level) the pros and cons of CE can be analysed through LCA. Secondly, possible better alternatives for the entire life cycle can be identified via LCA, once the limitations of a circular model are recognized during the first stage. Thirdly, LCA continuously sets goals and aims to improve circularity in real-life scenarios. Therefore, combining CE and LCA (CE-LCA) ensures mobilization of business with proven benefits for the environment and society. The CE-LCA provides an opportunity for product manufacturers to compare CE strategies along with complete environmental assessments. Thereby, it ensures a balanced environmental and ecological system for the new circular or looped product design method. Besides, CE-LCA is the preferred method for realizing the development of products and services for companies, product designers, and sustainability teams, as well as consumers (Valencia, 2017). Also, it results in a more in-depth analysis and understanding of economic, social, and environmental sustainability (Tóth Szita, 2017).

Ngan et al. (2021) added that CE-LCA offers an expansion to include not only the circular resources within the product system through reduction, reuse, recycled, repurposed, and remanufactured, but also the interconnections between different system boundary that leads to resourceful waste management, pollution reduction and regeneration of natural systems. Therefore, a CE-LCA framework could generate a concrete idea of how to improve the well-being of the ecosystem. In the same vein, it also includes the creation of a positive value-driven closed loop, where all the stakeholders in the economy cooperate in utilizing their resources efficiently and channel the unused or unutilized resources to the respective stakeholders (Ngan et al., 2021). According to Gervasio, (2019), resource efficiency (RE) is a reduction of the consumption of natural resources and the production of waste against current values, throughout the life cycle of a product. Therefore, RE is directly linked to the life cycle environmental performance of a product and by reducing this, it will ensure a better use of resources and a better management of waste. Also, the strategies (10Rs) of CE play critical roles towards the realisation of the efficiency of resources (Gervasio, 2019). Additionally, Ekins et al. (2020) stated that both CE and LCA could be used to evaluate or measure resource efficiency of a product or process.

12 Conclusions

This study examined the connections between Circular Economy (CE) and Life Cycle Analysis (LCA) regarding Sustainable Development Goals (SDGs). A conceptual framework that revealed the flow of connections/links among the key constructs – SDGs, CE and LCA of this study was developed, as a pathway for in-depth comprehension of the constructs and their connections. The study identified 10Rs strategies (Refuse, Rethink, Reduce, Re-use, Repair, Refurbish, Remanufacture, Repurpose, Recycle and Recover strategies) that are indispensable for resourceful management of production-consumption systems in CE vis-à-vis SDGs. Also, it revealed seven goals (Goal 7—affordable and clean energy, Goal 8–decent work and economic growth, Goal 11–sustainable cities and communities, Goal 12–responsible consumption and production, Goal 13–climate action, Goal 14–life below water, Goal 15–life on land) in SDGs that are directly related to the application of the circular economy. CE and SDGs were further found to link to each other in three major areas—implementation and application; alignment of SDGs targets with CE vision; and connection of indicators. Moreover, LCA was identified as a comprehensive, holistic and robust environmental assessment tool or methodology for evaluation of strategies in CE. Therefore, the study pinpointed CE-LCA linkage as a tool that could promote resource efficiency (RE) throughout the life cycle of a product, as both CE and LCA is versatile in evaluating or measuring RE of a product or process.

Furthermore, the implications for various stakeholders, such as government, business practitioners and scholars, in the domain of sustainable development and related issues, to be aware of connecting areas and factors of linkage between CE and LCA in relation to SDGs. This will serve as a guide and baseline information on the connectedness of the duos towards the seamless actualisation of the SDGs by 2030. The study then recommends that future studies on CE, LCA and SDGs, either as a standalone construct or in a combined form, should utilise this baseline information to promote sustainable development in all ecosystems of the globe.

References

Akter MMK, Haq UN, Uddin MA (2024) Sustainable development and circular economy. Waste Management in the circular economy. Springer International Publishing, Cham, pp 133–152
Aleksić A, Jovanović VM, Veselinović N (2023) Sustainability and product life cycle in circular economy. Econ Sustain Dev 7(1):29–38
Allen C, Metternicht G, Wiedmann T (2018) Initial progress in implementing the sustainable development goals (SDGs): a review of evidence from countries. Sustain Sci 13:1453–1467
Bastianoni S, Goffetti G, Neri E, Patrizi N, Ruini A, Sporchia F, Pulselli FM (2023) LCA based circularity indices of systems at different scales: a holistic approach. Sci Total Environ 897:165245
Beltran M (2018) Deepening the uncertainty dimension of environmental life cycle assessment: addressing choice, future and interpretation uncertainties. Leiden University
Brändström J, Saidani M (2022) Comparison between circularity metrics and LCA: a case study on circular economy strategies. J Clean Prod 371:133537
Chen AJ, Boudreau MC, Watson RT (2008) Information systems and ecological sustainability. J Syst Inf Technol 10(3):186–201
Choudhary A, Sarkar S, Settur S, Tiwari MK (2015) A carbon market sensitive optimization model for integrated forward–reverse logistics. Int J Prod Econ 164:433–444
Cifuentes-Faura J (2021) Environmental policies to combat climate change in europe and possible solutions: European green deal and circular economy. Int J Sustain Policy Pract 17(2):27
Civancik-Uslu D, Ferrer L, Puig R, Fullana-i-Palmer P (2018) Are functional fillers improving environmental behavior of plastics? A review on LCA studies. Sci Total Environ 626:927–940
Corona B, Shen L, Reike D, Carreón JR, Worrell E (2019) Towards sustainable development through the circular economy—A review and critical assessment on current circularity metrics. Resour Conserv Recycl 151:104498
del Mar Alonso-Almeida M, Rodriguez-Anton JM (2019) Circular supply chain and business model in apparel industry: an exploratory approach. In: The circular economy and Its implications on sustainability and the green supply chain. IGI Global, pp 66–83
Dieterle M, Viere T (2022) Life Cycle Gap Analysis for product circularity and sustainability—A case study with three different products. Mater Circ Econ 4(1):12
Ekins P, Domenech Aparisi T, Drummond P, Bleischwitz R, Hughes N, Lotti L (2020) The circular economy: what, why, how and where
Ellsworth-Krebs K, Niero M, Jack T (2023) Feminist LCAs: finding leverage points for wellbeing within planetary boundaries. Sustain Prod Consum 39:546–555
Finnveden G, Hauschild MZ, Ekvall T, Guinée J, Heijungs R, Hellweg S, Suh S (2009) Recent developments in life cycle assessment. J Environ Manage 91(1):1–21
Gervasio H (2019) The role of resource efficiency towards circular economy. In: IOP conference series: earth and environmental science. IOP Publishing 225(1):012057
Gheewala SH, Silalertruksa T (2021) Life cycle thinking in a circular economy. An introduction to circular economy, pp 35–53

Ghisellini P, Cialani C, Ulgiati S (2016) A review on circular economy: the expected transition to a balanced interplay of environmental and economic systems. J Clean Prod 114:11–32
Haupt M, Hellweg S (2019) Measuring the environmental sustainability of a circular economy. Environ Sustain Indic 1:100005
Haupt M, Zschokke M (2017) How can LCA support the circular economy?—63rd discussion forum on life cycle assessment, Zurich, Switzerland, November 30, 2016. Int J Life Cycle Assess 22:832–837
Hauschild MZ, Rosenbaum RK, Olsen SI (2018) Life cycle assessment, vol 2018. Springer International Publishing, Cham. https://doi.org/10.1007/978-3-319-56475-3
Helander H, Petit-Boix A, Leipold S, Bringezu S (2019) How to monitor environmental pressures of a circular economy: an assessment of indicators. J Ind Ecol 23(5):1278–1291
Herva M, Franco A, Carrasco EF, Roca E (2011) Review of corporate environmental indicators. J Clean Prod 19(15):1687–1699
Hishammuddin MAHB, Teck GLH, Chau LW, Ho CS, Ho WS, Idris AM (2018) Circular economy (CE): a framework towards sustainable low carbon development in Pengerang, Johor, Malaysia. Chem Eng Trans 63:481–486
Hoekstra AY, Chapagain AK, Aldaya MM, Mekonnen MM (2009) Water footprint manual: State of the art 2009. Water Footprint Network, Enschede
Imperatives S (1987) Report of the world commission on environment and development: our common future. Accessed Feb, 10(42,427)
Jerome A, Helander H, Ljunggren M, Janssen M (2022) Mapping and testing circular economy product-level indicators: a critical review. Resour Conserv Recycl 178:106080
Kirchherr J, Reike D, Hekkert M (2017) Conceptualizing the circular economy: an analysis of 114 definitions. Resour Conserv Recycl 127:221–232
Lehmann A, Bach V, Finkbeiner M (2015) Product environmental footprint in policy and market decisions: applicability and impact assessment. Integr Environ Assess Manag 11(3):417–424
Lettenmeier M, Liedtke C, Rohn H (2014) Eight tons of material footprint—suggestion for a resource cap for household consumption in Finland. Resources 3(3):488–515
Manioudis M, Meramveliotakis G (2022) Broad strokes towards a grand theory in the analysis of sustainable development: a return to the classical political economy. New Polit Econ 27(5):866–878
Mannan M, Al-Ghamdi SG (2020) Environmental impact of water-use in buildings: latest developments from a life-cycle assessment perspective. J Environ Manage 261:110198
Mannan M, Al-Ghamdi SG (2022) Complementing circular economy with life cycle assessment: deeper understanding of economic, social, and environmental sustainability. In: Circular economy and sustainability. Elsevier, pp 145–160
Marković M, Popović Z, Marjanović I (2023) Towards a circular economy: evaluation of waste management performance in European Union countries. SerbN J Manag 18(1):45–57
Martins-Rodrigues MC, Da Rosa LAB, Sousa MJ, Godoy TP (2020) Recent research topics in circular economy. Int J Econ Manag Syst 5
Meadows DH (1998) Indicators and information systems for sustainable development. The Sustainability Institute, Hartland Four Corners: US
Mehta S (2023) Circular economy and sustainable development goals. MET Manag Rev 10(1):24–29
Mensah J (2019) Sustainable development: Meaning, history, principles, pillars, and implications for human action: LITERATURE review. Cogent Soc Sci 5(1):1653531
Montag L (2023) Roadmap to a circular economy by 2030: a comparative review of circular business model visions in Germany and Japan. Sustainability 15(6):5374
Ngan SP, Ngan SL, Lam HL (2021) An overview of circular economy-life cycle assessment framework. Chem Eng Trans 88:1123–1128
Niero M, Rivera XCS (2018) The role of life cycle sustainability assessment in the implementation of circular economy principles in organizations. Procedia CIRP 69:793–798
Oriekhova T (2019) Circular economy as a global imperative. J Eur Econ 18(4), (71):415–424

Pearce DW, Turner RK (1989) Economics of natural resources and the environment. Johns Hopkins University Press
Peña C, Civit B, Gallego-Schmid A, Druckman A, Pires AC, Weidema B, Motta W (2021) Using life cycle assessment to achieve a circular economy. Int J Life Cycle Assess 26:215–220
Plassmann K, Edwards-Jones G (2009) Where does the carbon footprint fall?: Developing a carbon map of food production. IIED, no 4
Plassmann K, Norton A, Attarzadeh N, Jensen MP, Brenton P, Edwards-Jones G (2010) Methodological complexities of product carbon footprinting: a sensitivity analysis of key variables in a developing country context. Environ Sci Policy 13(5):393–404
Purvis B, Mao Y, Robinson D (2019) Three pillars of sustainability: in search of conceptual origins. Sustain Sci 14:681–695
Rigamonti L, Mancini E (2021) Life cycle assessment and circularity indicators. Int J Life Cycle Assess 26:1937–1942
Roos Lindgreen E, Mondello G, Salomone R, Lanuzza F, Saija G (2021) Exploring the effectiveness of grey literature indicators and life cycle assessment in assessing circular economy at the micro level: a comparative analysis. Int J Life Cycle Assess 26:2171–2191
Rosa LABD, Cohen M, Campos WYYZ, Ávila LV, Rodrigues MCM (2023) Circular economy and sustainable development goals: main research trends. Rev Adm Da UFSM 16(1):e9
Rossi PH, Lipsey MW, Henry GT (2018) Evaluation: a systematic approach. Sage publications
Ryen EG, Gaustad G, Babbitt CW, Babbitt G (2018) Ecological foraging models as inspiration for optimized recycling systems in the circular economy. Resour Conserv Recycl 135:48–57
Samani P (2023) Synergies and gaps between circularity assessment and life cycle assessment (LCA). Science of The Total Environment, 166611
Saidani M, Yannou B, Leroy Y, Cluzel F, Kendall A (2018) A taxonomy of circular economy indicators. arXiv preprint arXiv:1901.02709
Schröder P, Bengtsson M, Cohen M, Dewick P, Hofstetter J, Sarkis J (2019) Degrowth within–Aligning circular economy and strong sustainability narratives. Resour Conserv Recycl 146:190–191
Seidel C (2016) The application of life cycle assessment to public policy development. Int J Life Cycle Assess 21:337–348
Shi L, Han L, Yang F, Gao L (2019) The evolution of sustainable development theory: types, goals, and research prospects. Sustainability 11(24):7158
Shou M, Domenech T (2022) Integrating LCA and blockchain technology to promote circular fashion–A case study of leather handbags. J Clean Prod 373:133557
Smith MJ (2020) Sustainable development goals: genuine global change requires genuine measures of efficacy. J Maps 16(1):1–3
Stahel WR (2016) The circular economy. Nature 531(7595):435–438
Standardization IOF (2006) Environmental management: life cycle assessment: principles and framework. vol ISO, 14040
Suárez-Eiroa B, Fernández E, Méndez-Martínez G, Soto-Oñate D (2019) Operational principles of circular economy for sustainable development: linking theory and practice. J Clean Prod 214:952–961
Sulich A, Rutkowska M (2020) Green jobs, definitional issues, and the employment of young people: an analysis of three European Union countries. J Environ Manage 262:110314
Tan J, Tan FJ, Ramakrishna S (2022) Transitioning to a circular economy: a systematic review of its drivers and barriers. Sustainability 14(3):1757
Tóth Szita K (2017) The application of life cycle assessment in circular economy. Hung Agric Eng 31:5–9
Valencia E (2017) Why circular economy business models need LCA. PRe Sustainability
Ventura A (2022) Transition life cycle assessment: a new method to face ecological transition. Front Sustain 3:801668
van Stijn A, Eberhardt LM, Jansen BW, Meijer A (2021) A circular economy life cycle assessment (CE-LCA) model for building components. Resour Conserv Recycl 174:105683

Yoonus H, Mannan M, Al-Ghamdi SG (2020) Environmental performance of building integrated grey water reuse systems: life cycle assessment perspective. In: World environmental and water resources congress 2020, pp 1–7. Reston, VA: American Society of Civil Engineers
Zhalechian M, Tavakkoli-Moghaddam R, Zahiri B, Mohammadi M (2016) Sustainable design of a closed-loop location-routing-inventory supply chain network under mixed uncertainty. Transp Res Part e: Logist Transp Rev 89:182–214
Zink T, Geyer R (2017) Circular economy rebound. J Ind Ecol 21(3):593–602

Barriers to the Adoption of Circular Economy Principles in Nigeria

Sylvester Anaba

Abstract Nigeria, the largest economy in Africa, grapples with several challenges with the adoption of circular economy principles; resulting in a substantial portion of its resources remaining unrecycled. Consequently, the country faces sustainability and resource-efficiency dilemmas. Extant literature shows that the recycling of used resources may potentially reduce waste and foster optimum utilization of resources. However, only a few literatures have delved into the circular economy landscape in Nigeria. The study aimed to identify the challenges hindering the adoption of circular economy principles in Africa's largest economy, gauge the level of engagement in circular economy practices among individuals, evaluate governmental involvement in promoting circular economy initiatives, find out whether the adoption of circular economy is important to the sustainable development, evaluate the viability of circular economy to businesses, analyze the adequacy of infrastructure for the adoption of circular economy and find out whether education and awareness campaigns are required for the adoption of circular economy in Nigeria. The research was hinged on Robert Frosch and Nicholas E. Gallopoulos' Industrial Ecology Theory of Circular Economy. The study employed quantitative research methods. Questionnaires were disseminated utilizing online platforms such as Google forms, emails, WhatsApp, and various social media channels including LinkedIn, Facebook, and X (formerly known as Twitter). Descriptive statistics and factor analysis techniques were utilized for data analysis. Analyses were based on 95.0% confidence level. Findings from the study revealed that respondents identified lack of awareness and understanding (91.0%), technological limitations (91.0%), economic factors (90.0%), regulatory and policy barriers (80.0%), social and cultural factors (78.0%), and infrastructural limitations (64.0%) as challenges to the adoption of circular economy principles in Nigeria. Notably, level of individuals engagements in circular economy was low as only 13.8% respondents were involved in circular economy activities. Similarly, government support for circular economy was low as only 8.0% respondents believed that government is supportive. Moreover, all the respondents (except 13.0% that

S. Anaba (✉)
Center for Petroleum and Energy Economics (CPEEL), University of Ibadan, Ibadan, Nigeria
e-mail: sylvesteranaba@yahoo.com

L. Brandli et al. (eds.), *The Contribution of Life Cycle Analyses and Circular Economy to the Sustainable Development Goals*, World Sustainability Series,
https://doi.org/10.1007/978-3-031-85300-5_14

were neutral) believed that transitioning to a circular economy would be economically viable for businesses. Also, all respondents (except 11.0% that were neutral) indicated that adoption of circular economy is important for Nigeria's economic sustainability. Meanwhile, 61.0% respondents indicated that Nigeria does not have adequate infrastructures to adopt circular economy. Similarly, 78.0% respondents believed that education and awareness campaigns are required for the adoption of circular economy in Nigeria. These findings underscore the imperative for government to enhance educational efforts and raise awareness about circular economy concepts in Nigeria. Encouraging active participation in circular economy activities and providing requisite policies, infrastructure, and incentives are crucial steps toward fostering circular economy adoption in Nigeria's economic landscape.

1 Introduction

Nigeria lies roughly between latitudes 3° 15′–13° 30′ N and longitudes 2° 59′–15° 00′ E. It shares borders with the Republic of Benin to the west, Niger to the north, Chad to the northeast, and Cameroon to the east, while the Atlantic Ocean stretches along its southern coastline. The country has a land mass of about 923,768 km^2 remains one of the fastest growing populations in the globe (Department of Climate Change, Federal Ministry of Environment, Nigeria 2021). Nigeria, often referred to as Africa's powerhouse, is one of the most populated African countries and ranks number six in terms of an estimated overall population of 230 million people. According to the United Nations Population Fund (UNFPA), this number might increase to 400 million in 2050. Nigeria boasts the largest economy in Africa, with an annual Gross Domestic Product (GDP) growth rate of 3.46% in 2023 (Trading Economics 2024).

Annually, Nigeria generates 32 thousand tonnes of waste, making it the 7th largest waste polluter globally. Meanwhile, less than 30.0% of this waste undergoes recycling, whilst the remaining over 70.0% end up in sewers, landfills, water bodies and beaches (Ajibesin 2023). Research estimates show that Nigeria produces nearly 32 million tonnes of solid waste annually, making the country one of the biggest waste producers in Africa (PricewaterhouseCoopers 2021). Other studies' estimates indicate that, annual domestic solid waste generation rate in Nigeria is 63 million tonnes, equivalent to 0.45 kg per capita per annum. Accordingly, Lagos, Nigeria's most densely populated city with an estimated population of 24 million produces circa 13 thousand tonnes of waste daily (Ajibesin 2023).

Waste production is constantly increasing in Nigeria because of upsurge in population, lifestyle, and pattern of consumption. Consequently, the rate of waste generation in Nigeria has outweighed existing structures put in place to manage waste (Ajibesin 2023). Waste sector in Nigeria accounts for almost 4.0% of the total net national emission of 609,783Gg Co2-eq. Notably, this is anticipated to increase in the nearest future (Department of Climate Change, Federal Ministry of Environment, Nigeria 2021). Below pictures show typical waste sites in Nigeria.

Refuse dump in Lagos, Nigeria. *Sources* Adopted from Punch Newspaper (2022) and Business Day Newspapers (2022)

The surge in waste generation could be due to the adoption of a linear economy wherein resources are extracted, processed, distributed, consumed as illustrated in figure below.

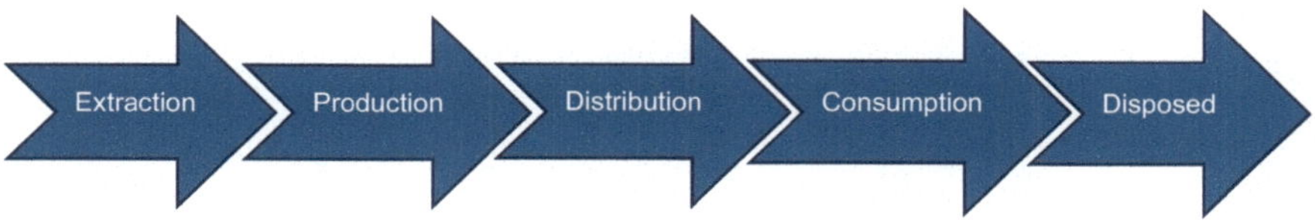

Linear economy. *Source* Adopted from Okafor et al. (2020)

These wastes could have been regenerated, reused, or recycled to minimize waste as well as engender economic sustainability as enshrined in circular economy illustrated in figure below.

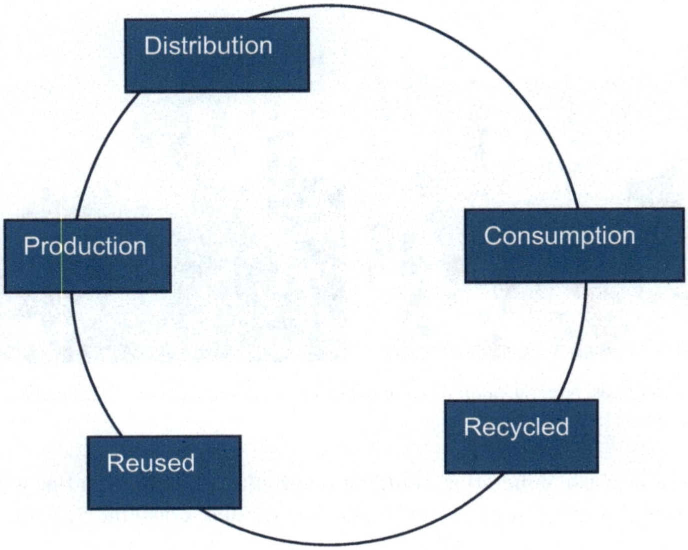

Circular economy. *Source* Adopted from Okafor et al. (2020)

It is against this backdrop that the study aimed:

1. To identify the barriers to the adoption of circular economy practices in Nigeria.
2. Find out the level of engagement in circular economy practices among individuals in Nigeria.
3. Ascertain government's level of support for circular economy practices in Nigeria.
4. Evaluate whether the adoption of circular economy is important to the sustainability development of Nigeria.
5. Ascertain the viability of the adoption of circular economy to businesses in Nigeria.
6. Find out whether infrastructures in Nigeria are adequate for the adoption of circular economy in Nigeria.
7. Analyze whether education and awareness campaigns are strongly required for the adoption of circular economy in Nigeria.

2 Theories of Circular Economy

The various theoretical lenses that can be adopted in analyzing circular economy are briefly discussed below:

1. **Reduce, Reuse, and Recycle (3Rs) Theory**: This is centered on the principle of reducing waste and increasing the use of resources. It involves giving a second life to a product by reusing them. It may entail repairing, refurbishing, and repurposing.

2. **Cradle to Cradle (C2C)**: This theory aims at eliminating waste and pollution. The idea is that resources used should not cause harm to the environment. It promotes products that have positive impacts, e.g. products that enhance biodiversity and reduce pollution.
3. **Performance Economy Theory of Circular Economy**: This theory emphasizes delivering the functionality that customers want, instead of ownership of the physical product. It encourages the designing of products for longevity, durability, and ease of maintenance to increase their service life and performance.
4. **Biomimicry Theory of Circular Economy**: This theory draws inspiration from nature to design more sustainable products and processes. It emulates biological principles found in nature to address human problems and improve the efficiency of resources.
5. **Regenerative Design Theory of Circular Economy**: This seeks to regenerate degraded ecosystems and landscapes via intentional design interventions. It supports co-creation with nature.
6. **Industrial Ecology Theory**: Here resources are circulated and reutilized rather than consumed and disposed of. It stipulates that industrial processes should mimic natural ecosystems, where waste from one process becomes a resource for another, creating a closed-loop system.

Amongst the several theories of Circular Economy, this study adopts the Industrial Ecology Theory. This is hinged on the fact that the theory focuses on the complete life cycle of products and processes. Thus, from the extraction of raw resources to the disposal (end-of-life) of used materials, opportunities for optimization and improvement are identified. Also, it encourages a holistic approach to industrial design and management, given the interconnections and feedback loops between diverse components of the system.

Nigeria needs to carefully consider Industrial Ecology Theory as a key initiative toward achieving a sustainable economy via the adoption of circular economy concept. Common amongst Nigerians is the culture of "consume and dispose". This culture deprives Nigerians from extracting value from disposed wastes since most of this waste end up in the refuse dump. Contrary to this, these wastes can be processed into another product that can be useful to people and can generate an additional revenue for the country.

2.1 *Hindrances to the Adoption of Circular Economy*

Nigeria has not fully articulated its circular economy road map (Rajput et al. 2020). This is a major challenge to the adoption of circular economy in the country. The inability to properly articulate plans for the adoption of circular economy in the country prevents the country from taking advantage of opportunities of transforming and contributing to food and resources security.

Furthermore, poor waste management is a serious threat to the adoption of circular economy in Sub-Saharan Africa (SSA), Nigeria inclusive (Debrah et al. 2022). The culture of waste management is still very poor as people dump their waste by the roadside, drainage, streams and rivers. Nigerians lack the culture of separating their waste into different types and converting same to useful products. This is due to the continuous adoption of linear economy which entails harvesting, using, and disposing. Consequently, the absence of enlightenment campaigns on the activities of waste management have immensely hindered the adoption of circular economy.

Considering these barriers from a different angle, firms that implement circular economy practices face several obstacles emanating from extant policies and regulations, economic issues, supply chains, adequate technology, consumer taste and preferences, as well as internal challenges peculiar to each of the firms (Rizos et al. 2021).

Meanwhile, barriers to the adoption of circular economy are interrelated. For example, firms that have a culture that does not accept recycling and reuse of waste may not develop designs that foster recycling. Subsequently, users of such products may not be aware that such waste can be recycled. This indicates that cultural barriers can facilitate technological barriers (Kirchherr et al. 2017).

We may further ask ourselves whether the barriers to the adoption of circular economy are the same in the developed and underdeveloped countries (Ting et al. 2023). Notably, some developed countries have extensively adopted the concept of circular economy compared to underdeveloped countries that have broadly not adopted this concept. Consequently, challenges in the adoption of circular economy may slightly differ between developed and undeveloped countries, especially when considering cultural, technological and awareness issues.

However, some barriers cut across both developed and undeveloped countries. Paramount amongst these barriers are technological, policy and regulatory, economic and cultural barriers (Galvao et al. 2018). Meanwhile, it is important to note that the severity level of these barriers across the developed and underdeveloped countries may differ.

2.2 *Need for a Circular Economy in Nigeria*

Nigeria's growing population and elevated waste generation point to the need to convert from linear economy to a circular economy. Circular economy provides alternative means of articulating economic activities which facilitate social and economic prosperity in the long term. A circular economy model originates its insight from nature's biological cycle and creates a closed loop system which reduces waste from the value chain. Beyond waste elimination and recycling, circular economy provides a reconsidering of the present economic model at a global level (Ajibesin 2023).

Similarly, a circular economy fosters enormous environmental and economic benefits (United Nations Environment Program 2019). This is true as it facilitates the reduction of harmful waste in the environment, thus preserving but the environment and lives. In another form, it promotes economic prosperity as waste can be repaired, recycled, and reused rather than been thrown away. Waste from one industrial process can become a valued input into the production of another product, hence increasing revenue to the country thereby contributing to the gross domestic product.

The circular economy is capable of decoupling resource use from economic development to help address major African challenges like food insecurity, scarcity of water, climate change, and management of natural resources. Out of 100 billion tonnes of virgin materials generated worldwide every year, only about 8.6% are recovered and reused, indicating a substantial circularity gap. A significant adoption of circular economy might create new opportunities (thereby contributing to the 17 sustainable development goals—SDGs) and generate new businesses. Circular economy could also facilitate more resilience to external shocks via the creation of regional value chains for self-sufficiency (African Development Bank 2021).

Circular economy could be a favorable substitute strategy of industrial development and job creation to the traditional manufacturing strategy. With the appropriate conditions, circular economy might generate new jobs for economic diversification, value creation and development of skills (Chatham House 2019).

Additionally, developing economies might struggle to achieve sustainable growth if there is no substantial investment in recycling, reusing, and repairing of used items. To achieve sustainable growth which can address the needs of future generations, radical shift from linear economy to circular economy is expedient. Obviously, it is becoming clearer that circular economy is becoming a new model for resilient economic growth (Chatham House 2019).

It is also important to note that creating and optimizing resource 'loops' along value chains could help meet the material needs of growing populations via significantly smaller rates of per capita primary resource use (Chatham House 2019).

2.3 Tradeoffs of Circular Economy

Transition from linear economy to circular economy requires diligence. Where coordinated and strategic efforts to the implementation of circular economy are lacking, there is a possibility that organizations may adopt tokenistic, harmful activities which preclude sustainability. Some may adopt substandard incineration practices which are harmful to humans and the environment (Chatham House 2019).

Similarly, the adoption of circular economy may lead to a substantial shift in industrial policy; resource intensive economies may support value addition but may also risk job losses among those that are employed in resource extraction and primary processing of resources (Chatham House 2019).

2.4 Initiatives of Circular Economy in Nigeria

Nigeria is incorporating circular economy concept to transition from a "take-make-dispose" (linear) economy to circular economy. Though the process is quite slow but pockets of activities that show that the country is interested in adopting circular economy have been observed over the years. These efforts include:

- **Extended Producer Responsibility (EPR) Programs**: This program places the responsibility for the collection and recycling of products on the producers of such products. Nigeria has adopted EPR with programs aiming specific sectors like electronic waste and batteries. The National Environmental Regulation was amended recently requiring producers to subscribe to EPR programs and to stop in appropriate waste treatment (Ikeah n/a).
- **Recycling Initiatives**: Several programs like the Lagos State Circular Economy program have been initiated to improve recycling activities in the country. This program is aimed at growing the recycling products market and encouraging private firms like Wecyclers and Recycle Points which focus on household waste collection for recycling (Lagos State Nigeria n/a).
- **Development of Circular Economy Frameworks**: The Nigeria Circular Economy Working Group (NCEWG) is a collaborative effort between the government, private sector, and civil society organizations to develop a roadmap and framework for a circular economy in Nigeria. This includes designing a National Circular Economy Program (NCEP) with a 2030 target date (African Development Bank Group n/a).
- **Financial Supports**: The Nigerian Government, the Global Environment Facility and United Nations Environment have announced a $15 million initiative to commence a circular e-waste system in Nigeria (United Nations Environment Programme 2019).

2.5 Circular Economy Policies and Regulation in Nigeria

Nigeria has developed some policies targeted at the adoption of circular economy. Howbeit, some of these policies lack will power for implementation due to over dependence on crude oil. However, the following are some of the policies initiated for the adoption of circular economy (Rajput et al. 2020):

- **Economic Recovery and Growth Plan (2017)**: This aims to diversify the economy by reducing dependence on crude oil. It intends to adopt circular economy principles into the mainstream economy.
- **National Policy on Plastic Waste Management (2020)**: This was drafted to set the foundation for the recycling of plastics and adoption of circular economy principles.

- **Extended Producer Responsibility (EPR)**: Nigeria is also considering an EPR due to international guidance. This may help the country to move to a closed-cycle manufacturing and effective recycling practices.
- Recently, Nigeria is planning to address e-waste problem, inclusive of a circular economy principle for electronics. If this plan is successful, it will bolster the circularity of electronic waste in the country.

2.6 Viable Sectors for Circular Economy in Nigeria

Though the activities of circular economy are yet to receive adequate attention in Nigeria, the country has some sectors that may benefit from the adoption of circular economy. These sectors include agri-food, plastics/packaging, and waste management (Rajput et al. 2020).

- **The Agriculture and Food Sector**: Adoption of circular economy principle in this sector may promote food productivity and help to minimize lack of food. The adoption of circular economy principles may reduce loss of food as harvesting, handling and storage practices improve.
- **Plastic and Packaging Sector**: Nigeria is one of the key importers of plastics, amounting to 70.0% of feedstock imported from Middle East, Europe and Asia, and 30.0% are produced locally (Rajput et al. 2020). This is a huge opportunity to apply circular economy practices in this space.
- **Automobile Sector**: Nigeria imports a lot of used cars. This presents an opportunity to develop an effective secondary resource market for this sector. This may require an end-to-end review starting from availability parts and restructuring of same for further use by motorists.
- **Waste Management Sector**: This industry plays an important role in the adoption of circular economy. It may help in the generation of employment opportunity for the teaming youths. It may also help to reduce pressure on the environment as collection and reprocessing of waste improves.

2.7 Major Drivers of Circular Economy in Nigeria

Geographical limitations and lack of resources have been major drivers for the implementation of circular economy. In 1990, Japan kickstarted efforts toward circular economy. Currently, the European Union (EU) and its member states are driving global efforts towards a circular economy. These efforts are focused on efficient minimization of the consumption of resources. Though waste management and recycling policies have been the pillars of circular economy regulatory framework, policies have gradually evolved toward extended producer responsibility, eco-modulation, and design (Weick et al. 2022).

3 Methodology

This study aims to identify the challenges hindering the adoption of circular economy principles in Africa's largest economy, gauge the level of engagement in circular economy practices among individuals, and evaluate governmental involvement in promoting circular economy initiatives. To achieve these objectives, the study employed quantitative research methods. Questionnaires were disseminated utilizing online platforms such as Google Forms, emails, WhatsApp, and various social media channels including LinkedIn, Facebook, and X (formerly known as Twitter). Descriptive statistics and factor analysis techniques were utilized for data analysis.

3.1 Research Design and Research Instruments

The study was hinged on Robert Frosch and Nicholas E. Gallopoulos' Industrial Ecology Theory of Circular Economy. It was majorly quantitative research with a mix of qualitative analysis of respondents' responses to a few open-ended questions. The study employed a mix of closed-ended, open-ended, and Likert-scale questions in the questionnaires. To analyze responses generated from the questionnaires, the study used descriptive statistics and factor analysis techniques. These data are described using tables, simple charts, graphs, and percentages. On the other hand, opened ended questions which allow respondents to express their thoughts were analyzed qualitatively, providing insights to how the government can improve circular economy in Nigeria.

3.2 Research Population and Sampling Design

The population for the study was made up of professionals across different sectors. Some of these sectors include the manufacturing sector, agriculture sector, construction sector, waste management sector, and other sectors like the finance sector and education sector etc. Responses to the questionnaires were 100.0% accepted as respondents showed full knowledge of the subject matter.

Questionnaires were distributed majorly through online platforms like Google Forms, emails, WhatsApp, and various social media channels including LinkedIn, Facebook, and twitter to target professionals. Consequently, the questionnaires were opened to targeted professionals of those platforms to respond. Notably, the study did not apply sampling design in the selection of respondents as it was opened to targeted professionals of some online platforms. The selection of online platforms wherein the questionnaires were distributed was hinged on no reason, rather than those platforms that the researcher felt will be easier to reach the target audience in. Hence, any online platform could have been used for the dissemination of the questionnaires.

Table 1 Description of data

Name of variable	Definition of variable	Description of variable	Type of variable
Factor 1	Lack of awareness and understanding	The variable assesses awareness and understanding of circular economy as one of the challenges to the adoption of circular economy in Nigeria	Qualitative variable
Factor 2	Economic factors	This variable evaluates economic factors like high upfront costs, lack of access to financing as one of the challenges to the adoption of circular economy in Nigeria	Qualitative variable
Factor 3	Technological limitations	Factor 3 measures technological limitations as one of the challenges to the adoption of circular economy in Nigeria	Qualitative variable
Factor 4	Regulatory and policy barriers	The variable gauges regulatory and policy barriers as one of the challenges to the adoption of circular economy in Nigeria	Qualitative variable
Factor 5	Social and cultural factors	This variable considers social and cultural factors (e.g. consumer preferences, traditional practices) as one of the challenges to the adoption of circular economy in Nigeria	Qualitative variable
Factor 6	Infrastructure limitations	Infrastructure limitations (e.g. poor waste collection system, lack of recycling of facilities) as one of the challenges to the adoption of circular economy in Nigeria	Qualitative variable

Source Questionnaire analysis

Meanwhile, these professionals are people with experience in their various fields, might have directly or indirectly involved in circular economy activities and have shown interest in driving economic sustainability.

3.3 Data Description

The data used in the study are briefly defined and explained in Table 1.

3.4 Computational Devices

Microsoft Excel and STATA were used as computational devices. The above-mentioned analytical software help to achieve a holistic and fulsome output from the analysis of data.

4 Presentation of Results and Interpretation

The results generated from the questionnaires were categorized based on demography, respondents' knowledge of the circular economy, barriers to the adoption of circular economy principles in Nigeria and transitioning to circular economy in Nigeria.

5 Demographic Results

Figure 1 provides a nuanced insight of the various sectors of operations that the respondents of the questionnaires operate in. The figure shows that 63.0% of the respondents work in the finance sector, followed by education (13.0%), manufacturing (6.0%), construction (5.0%), agriculture (3.0%) and others (11.0%), illustrating a sizable coverage of sectors. This indicates that majority of the respondents work in the finance sector and might have been familiar with the concept of sustainable economy.

Figure 2 shows the seniority levels of the respondents that took part in the survey. Figure 2 indicates 73.0% of respondents (mid-management—21.0%, senior management—31.0% and directors—31.0%) are in higher positions in their various sectors. However, 16.0% of the respondents are analysts and 11.0% are in lower positions. Generally, the graph indicates that the study is mostly made up of people in higher positions in their place of work.

Figure 3 indicates the various years of experiences of the respondents in their different sectors of operations. It signifies that 73.0% (23.0%, 21.0%, 19.0% and 10.0%) of the respondents have experience from 5 to 20 years and above. Nevertheless, 27.5% have experience between one to four years. This shows that respondents to the questionnaires are well knowledgeable in their fields.

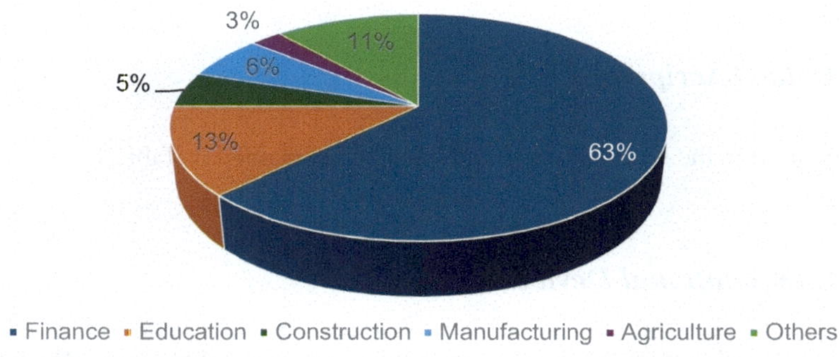

Fig.1 Respondents' sector. *Source* Questionnaire analysis

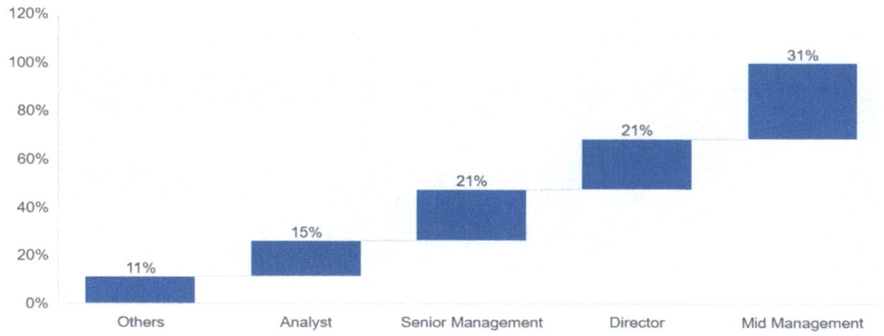

Fig. 2 Respondents' seniority level in their various sectors. *Source* Questionnaire analysis

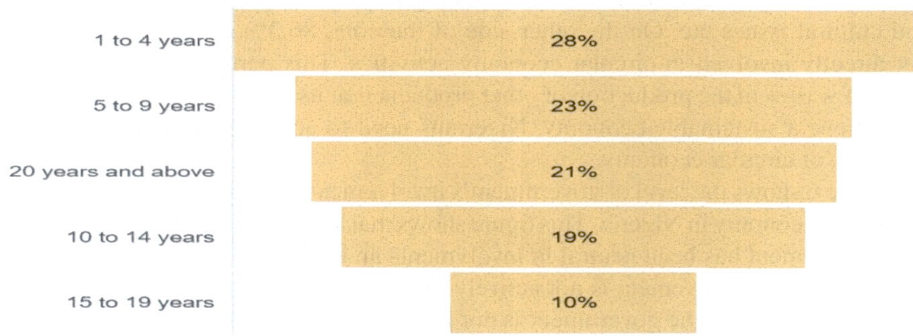

Fig. 3 Respondents' years of working in their various sectors. *Source* Questionnaire analysis

5.1 Results Based on Respondents' Knowledge of the Circular Economy Industry

Figure 4 shows the level of familiarity and knowledge of circular economy industry in Nigeria amongst the respondents of the questionnaire. Consequently, Fig. 4 shows that majority of respondents (90.0%) are familiar and have knowledge of the circular economy industry. This provides the comfort that respondents' responses are based on their wealth of knowledge of the circular economy industry in Nigeria. Based on this, it is believed that the questionnaire was able to generate quality and unbiased data from respondents. Meanwhile, the remaining 10.0% respondents indicate little knowledge of circular economy. This 10.0% with limited knowledge of circular economy might have some knowledge of the subject matter hence, their contribution to this study were believed to be relevant.

Figure 5 provides respondents' level of involvement in circular economy activities in Nigeria. The figure shows that 13.8% of respondents are directly involved in circular economy activities. This figure signifies that the level of direct involvement in circular economy amongst Nigerians is very low. This may partly be explained by lack

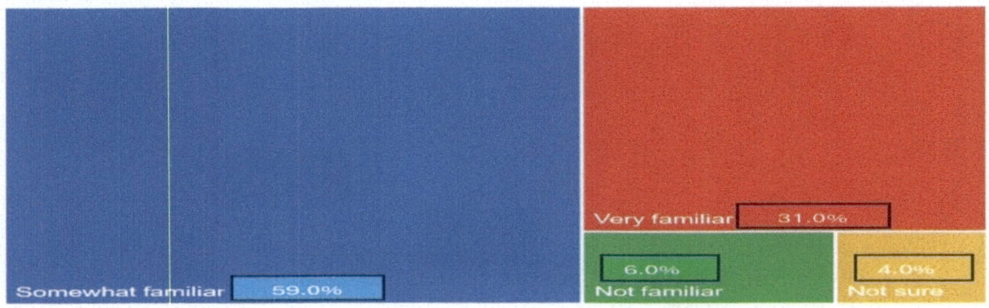

Fig. 4 Respondents' familiarity and knowledge of circular economy industry. *Source* Questionnaire analysis

of awareness, poor sensitization, inadequate government support, poor regulation, and cultural issues etc. On the other side of the coin, 86.3% of respondents were not directly involved in circular economy activities. This partly explains why the reuse of wastes in the production of other products is at its very minimum in Nigeria. To achieve a sustainable economy, Nigerians need to actively get involved in the activities of circular economy.

Figure 6 shows the level of government's involvement and support to the adoption of circular economy in Nigeria. This figure shows that half of the respondents (50.0%) said government has been neutral in involvments and support for circular economy. This signifies that govement is not actively involved in circular economy activities. It also indicates that the government is not fully supporting circular economy in the economy. Meanwhile, 34.0% said that government has not been supportive nor being involved in circular economy in Nigeria. Trailing is 8.0% participants that indicated that government has been absolutely not involved nor supportive. Surprisingly, only

Fig. 5 Respondents' involvement in circular economy in Nigeria. *Source* Questionnaire analysis

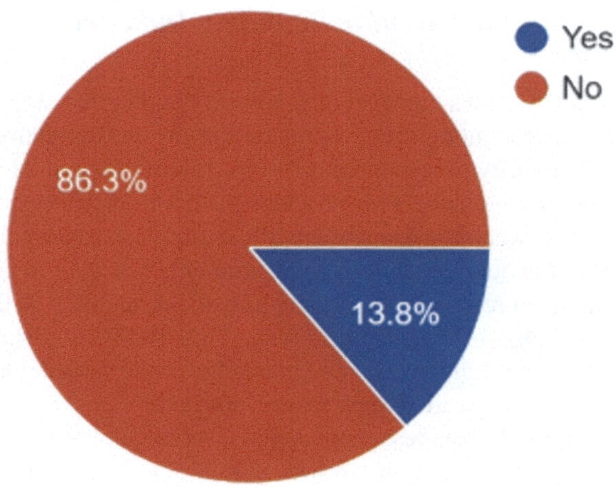

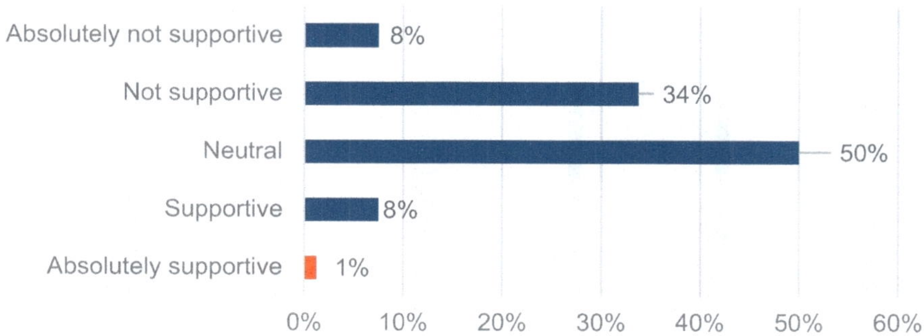

Fig. 6 Level of Government's involvement and support of circular economy. *Source* Questionnaire analysis

8.0% said that government has been supportive, and 1.0% said government has been very involved and supportive of circular economy in Nigeria.

The above statistics generally show that government support and involvement in circular economy is very low. This also underscores why economic sustainability through circular economy has been very low. Government's involvement and support for circular economy could come in form of regulations, and policies that encourage circular economy. The government can also be directly involved in procedures, systems and structures that facilitate the circular economy.

5.2 *Empirical Results Based on Barriers to the Adoption of Circular Economy*

Figure 7 provides some of the key barriers to the adoption of circular economy principles in Nigeria. Considering the empirical results, Fig. 7 identified lack of awareness and understanding (91.0%), technological limitations (91.0%), economic factors (90.0%), regulatory and policy barriers (80.0%), social and cultural factors (78.0%), and infrastructural limitations (64.0%) as challenges to the adoption of circular economy principles in Nigeria. In terms of ranking, lack of awareness and understanding, as well as technological limitations prove to be major challenges, followed by economic factors, regulatory and policy barriers, social & cultural factors, and infrastructural limitations.

Figure 7 clearly shows respondents knowledge of the various challenges facing the adoption of circular economy in the country. It will be expedient to address these challenges in order of their priority following the ranking of the challenges as revealed from the survey.

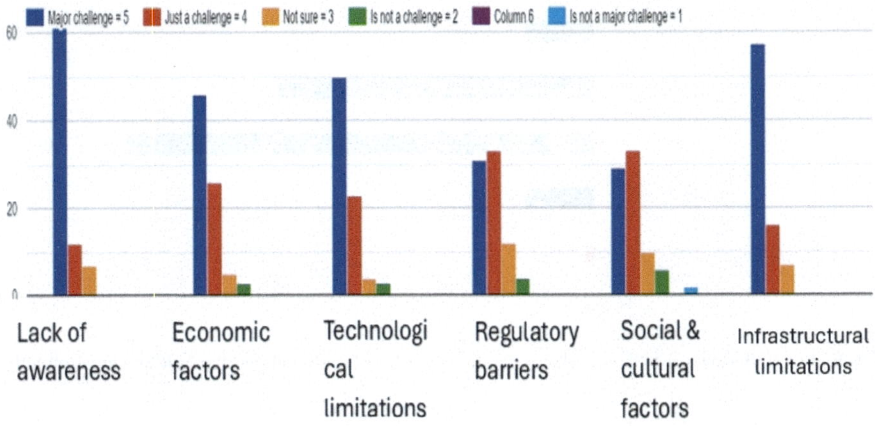

Fig. 7 Barriers to the adoption of circular economy principles in Nigeria. *Source* Questionnaire analysis

5.3 *Factor Analysis*

The study considered several factors affecting the adoption of circular economy principles in Nigeria. These factors are analyzed below.

Table 2 contains factor analysis of barriers to the adoption of circular economy in Nigeria, using the Principal Factors method, which is a common approach to factor analysis. Principal factors analysis aims to identify factors that account for the maximum variance in the data. The analysis identified four underlying factors that explain a significant portion of the variance in the data regarding barriers to adopting a circular economy.

Table 2 Factor analysis of barriers to the adoption of circular economy in Nigeria

Factor analysis/correlation number of obs. = 80				
Method: principal factors retained factors = 4				
Rotation: (unrotated) number of params = 15				
Factor	Eigenvalue	Difference	Proportion	Cumulative
Factor 1	1.498227	1.27508	1.0599	1.0599
Factor 2	0.22320	0.08824	0.1579	1.2178
Factor 3	0.13496	0.10812	0.0955	1.3133
Factor 4	0.02684	0.15362	0.0190	1.3323
Factor 5	– 0.12678	0.21612	– 0.0897	1.2426
Factor 6	– 0.34290	–	– 0.2426	1.0000

LR test independent versus *Saturated* chi2 (15) = 68.41 Prob > chi2 = 0.0000
Source Questionnaire analysis

Eigenvalues represent the variance explained by each factor. Factor 1 has the highest eigenvalue of 1.498, indicating that it explains the most variance in the data. This was followed by factor 2 (0.223) and factor 3 (0.134) and factor 4 (0.026). The column named "difference" shows the difference in eigenvalues between consecutive factors. This helps to evaluate how much more variance one factor explains compared to the next factor. It helps to determine the point at which adding more factors does not significantly increase the explanation of variance. The proportion column represents the proportion of total variance explained by each factor. Factor 1 explains 105.99% of the total variance, which is the cumulative percentage up to that factor. Conversely, factor 2 explains only 15.8%, factor explains 9.55% and factor 4 explains 1.90% of the total variance.

The Cumulative proportion column indicates the cumulative proportion of variance explained by all factors up to that point. It helps in understanding how much of the total variance is accounted for by the retained factors.

The Likelihood Ratio (LR) test compares the fitted model (in this case, the factor analysis model) to a saturated model where the data is perfectly reproduced. The chi-square value of 68.41 with a probability (Prob > chi2) close to 0 indicates that the fitted model significantly differs from the saturated model, suggesting that the factors are indeed explaining the variance in the data.

Also, the table mentions "Rotation: (unrotated)", indicating that the factors have not been rotated for better interpretability. Rotation methods like Varimax or Oblimin can be used to simplify the factor structure and make it easier to interpret the relationships between variables and factors. Overall, this table suggests that there are four underlying factors or dimensions influencing the barriers to adopting circular economy practices in Nigeria, with the first factor being the most significant. However, the unusual cumulative proportion value for Factor 1 (105.99%) may need further investigation to ensure the accuracy of the analysis.

Table 3 shows factor loading analysis of the unique variances of four of the factors that serve as barriers to the adoption of circular economy in Nigeria.

To fully understand the factors, Table 3 (factor loading table) shows the relationship between observed variables and latent factors in a factor analysis. The values in the table represent the factor loadings, which indicate how strongly each variable is related to each factor.

Table 3 Factor loading (pattern matrix) and unique variance

Variable	Factor 1	Factor 2	Factor 3	Factor 4	Uniqueness
Lack of awareness	0.4846	– 0.2620	– 0.0391	– 0.0933	0.6863
Economic factors	0.5552	– 0.1209	– 0.1919	0.0558	0.6372
Technological limitations	0.4433	0.2816	– 0.1787	0.0152	0.6920
Regulatory and policy barriers	0.6821	0.0684	0.1238	– 0.0409	0.5130
Social and cultural factors	0.3610	0.2014	0.1452	– 0.0266	0.8073
Infrastructure limitations	0.4040	– 0.1240	0.1680	0.1114	0.7808

Source Questionnaire analysis

The variable column lists the observed variables being analyzed, such as "Lack of awareness," "Economic factors," "Technological limitations," "Regulatory and policy barriers", "Social and cultural factors" and "Infrastructure limitations". The columns for factor 1, factor 2, factor 3, factor 4 represent the loadings of each variable on the respective factors extracted from the analysis. Loadings indicate the strength and direction of the relationship between the variable and the factor. These values range from − 1 to 1. These indicate a strong positive or negative correlation between the variable and the factor. Values closer to 0 indicate a weaker correlation.

From Table 3, a loading of 0.4846 in factor 1 for "Lack of awareness" suggests a moderately positive relationship between this variable and factor 1. Similarly, 0.5552 (economic factors), 0.4433 (technological limitations), 0.6821(regulatory and policy barriers) indicate a modest positive relationship with factor 2, 3 and 4 respectively.

The column for uniqueness shows the amount of variance in each variable that is not explained by the factors. It's essentially the unique contribution of each variable that is not accounted for by the shared factors. Higher values indicate more unique variance, suggesting that the variable is less well explained by the factors. It is ideal to have high factor loading and less uniqueness.

Factor 1: Variables such as "Economic factors," "Technological limitations," "Regulatory and policy barriers," and "Infrastructure limitations" have relatively high loadings on Factor 1. This suggests that these variables share a common variance explained by factor 1, which could represent an underlying construct related to economic and structural aspects.

Factor 2: "Social and cultural factors" and "Technological limitations" have moderate loadings on Factor 2. This factor may represent a combination of cultural and technological influences.

Factor 3: None of the variables have particularly strong loadings on Factor 3. The variables seem to be weakly related to this factor.

Factor 4: "Regulatory and policy barriers" has the highest loading on Factor 4, indicating that this factor might represent a distinct aspect related to regulations and policies.

Uniqueness: Most variables have relatively high uniqueness values, indicating that a substantial portion of their variance is not explained by the extracted factors. This suggests that there are other factors or unique characteristics influencing these variables that are not captured by the current analysis.

5.4 Factors Loading Analysis

Figure 8 provides the factor loading of the various unique variables that serve as barriers to the adoption of circular economy principles in Nigeria.

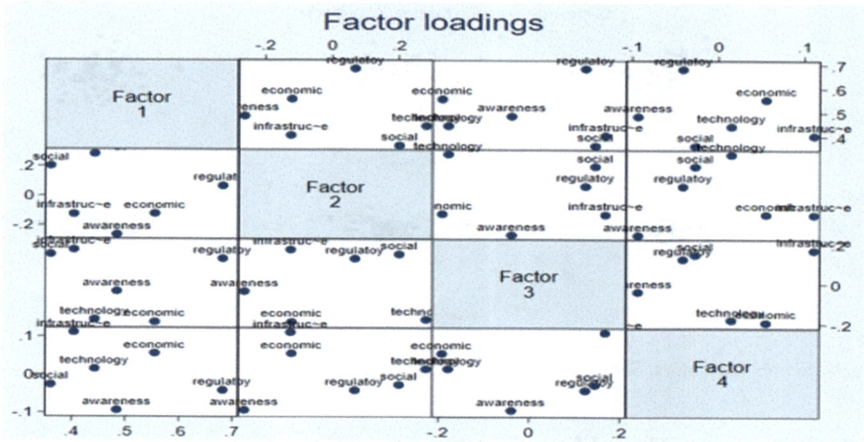

Fig. 8 Factor loading. *Source* Questionnaire analysis

Factor loadings range from − 1 to 1. A high positive loading (close to 1) indicates that the variable is strongly influenced by the factor.

A high negative loading (close to − 1) indicates that the variable has a strong negative influence from the factor. In other words, high values on the variable are associated with low scores on the factor and vice versa. A loading close to zero indicates that the variable is weakly influenced by the factor.

In Fig. 8, the text labels on the axes indicate factors (Factor 1, Factor 2, Factor 3, and Regulatory) while the data points represent variables. The distance of a data point from the origin reflects the magnitude of the factor loading, and the position of the point along the axis reflects the sign of the factor loading (positive or negative).

- Factor 1 seems to have strong positive loadings on variables economic awareness and technology awareness.
- Factor 2 appears to have strong positive loadings on variables infrastructure and social awareness.
- Factor 3 (technology) seems to have strong positive loadings on variables economic and social.
- Regulatory factors seem to have a strong negative loading on the variable economic awareness and a strong positive loading on the variable regulatory (redundant variable name).

Factor loading matrix shows the correlation between each observed variable and each factor. Loadings close to − 1 or 1 indicate that the factor strongly influences the variable. Loadings close to 0 indicate that the factor has a weak influence on the variable.

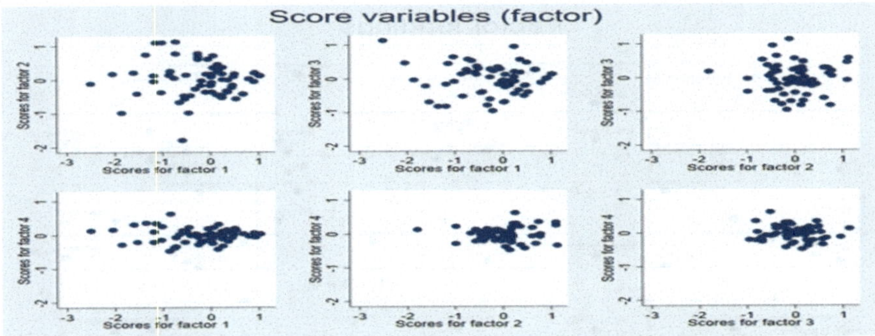

Fig. 9 Scores for factors. *Source* Questionnaire analysis

5.5 The Score Variables

Figure 9 provides the score variables of the factors that serve as barriers to the adoption of circular economy principles in Nigeria. The graph shows a series of six scatter plots, likely biplots, that visualize the relationship between multiple factors from a factor analysis. Each plot shows the scores for two factors on the x and y axes. For instance, the top left plot graphs the scores for factor 1 on the x-axis and factor 2 on the y-axis.

The top left plot shows a positive correlation between factor 1 and factor 2. This means that observations with higher scores on factor 1 also tend to have higher scores on factor 2.

The bottom left plot shows a negative correlation between Factor 4 and Factor 3. This means that observations with higher scores on Factor 4 tend to have lower scores on Factor 3.

5.6 The Scree Plot

Figure 10 is a scree plot of eigenvalues after factor analysis. The scree plot is a line graph that shows the eigenvalues on the y-axis and the number of factors on the x-axis. Eigenvalues are a measure of how much variance each factor explains in the data. The scree plot is used to determine the number of factors to retain in the analysis. The elbow in the curve is often used as a guide for the number of factors to retain.

In the scree plot, eigenvalues are typically plotted in descending order of magnitude. The point at which the plot levels off (forms an "elbow") indicate the number of factors that should be retained. Factors before this point contribute significantly to the explanation of the variance in the data, while those after it contribute relatively little.

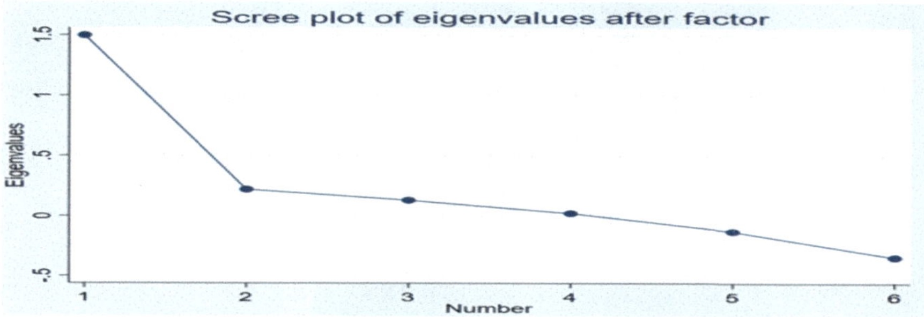

Fig. 10 Eigenvalues after factor. *Source* Questionnaire analysis

In a scree plot, you typically look for an "elbow" where the curve starts to flatten out. This elbow point often suggests the number of factors to retain because additional factors after this point explain much less variance in the data.

- The scree plot shows that the first four factors seem to explain a substantial amount of variance in the data.
- The eigenvalue for factor 5 and onwards appears to drop off significantly.

Based on the scree plot, it seems reasonable to retain the first four factors for further analysis. However, it's important to consider other factors in addition to the scree plot, such as the interpretability of the factors and the overall goals of the analysis.

6 Transitioning to Circular Economy in Nigeria

Figure 11 provides respondents' view of the importance of the adoption of circular economy to the sustainability development of Nigeria. The figure shows that the adoption of circular economy is very important to Nigeria's economic sustainability. For instance, all the respondents (except 11.0% that were neutral) indicated that the adoption of circular economy is important. The 11.0% respondents that were neutral may probably require further insights of circular economy and sustainable economic development.

At the same time, it is expedient to note that the importance of transiting to circular economy to achieving economic sustainability cannot be over emphasized. Consequently, Nigeria can emulate some advanced countries that have adopted circular economy to achieving sustainability. For instance, in Netherlands, the Dutch government has planned to become a fully based circular economy by 2050. Already, Netherlands has made massive progress in construction and agriculture. Also, Spain has a focus on sustainable development, ecological design, and recycling, with initiatives like banning single-use plastics and supporting textile recycling. Additionally, Sweden has been very effective in the management of waste and recycling systems.

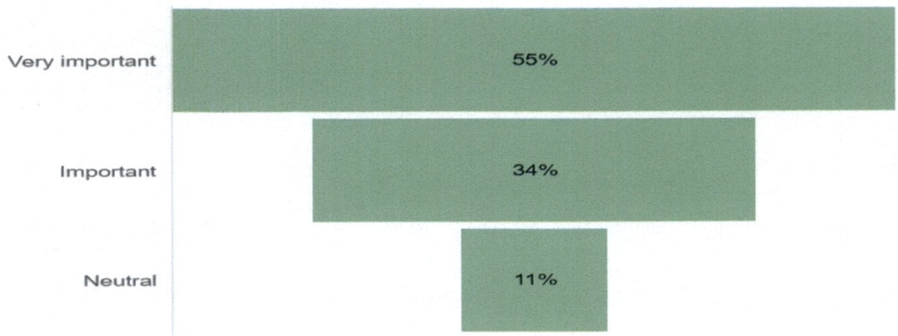

Fig. 11 Importance of circular economy to Nigeria's sustainability development. *Source* Questionnaire analysis

The country invested in advanced waste-to-energy technologies and has achieved a high rate of recycling.

Notably, the adoption of circular economy is perceived as an important source of achieving economic sustainability. With the enormous waste generated in Nigeria, the country is well position to convert its waste into useful item. However, technological backwardness, insufficient government support, cultural and structural issues have deprived the country of this opportunity.

Figure 12 indicates that the adoption of circular economy is viable to businesses in Nigeria. For example, all the respondents (except 13.0% that were neutral) believed that transitioning to a circular economy would be economically viable for businesses. Meanwhile, no respondent indicated that circular economy is inviable to businesses in the economy.

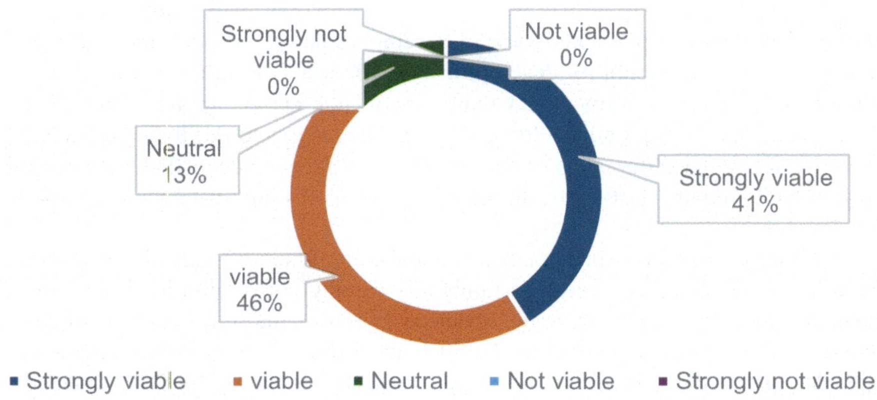

Fig. 12 Respondents' views of the viability of circular economy to businesses. *Source* Questionnaire analysis

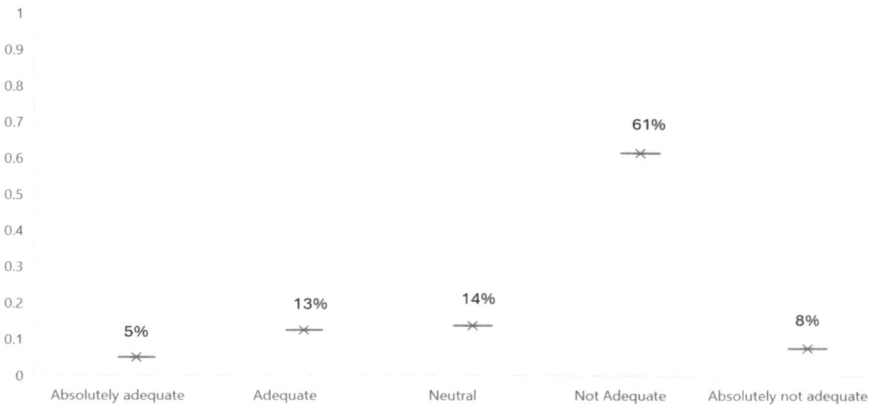

Fig. 13 Respondents' views on the adequacy of infrastructures. *Source* Questionnaire analysis

The adoption of circular economy may facilitate business activities in the economy. For instance, businesses involved in circular economy activities may consider the reuse of waste emanating from the use of their products. Similarly, these businesses may derive extra monetary value from reprocessing of waste. These businesses have a huge business opportunity in Nigeria due to the high volume of waste produce in the country. This waste may be transformed into new products lacking in the country. Also, circular economy provides a larger platform for ecofriendly firms that aim at addressing global warming issues in Africa (Nigeria to be precise). Awareness level of business opportunities in this space is still very low. Subsequently, businesses in this space stand a better chance of deriving value in circular economy adoption in Nigeria.

Furthermore, Fig. 13 shows that infrastructures in Nigeria are not adequate for the adoption of circular economy in Nigeria. The graph shows that 61.0% of the respondents believe that infrastructures are not adequate, whereas only 5.0% indicated that infrastructures are adequate for the adoption of circular economy in Nigeria.

It is simple to understand that Nigeria lacks basic technologies and infrastructures needed for the adoption of circular economy. The level of industrialization of the country is still low as manual production persist in the economy. Ultimately, the country exports its natural resources to the advanced countries at a cheap price and subsequently import the processed products at a higher price. This places the country at a disadvantage position. Moreover, the country can derive value from processing wastes into useful products.

The culture of disposing waste without reprocessing for reuse becomes strengthened due to lack of technology and infrastructures for reprocessing of waste. Ultimately, most Nigerians have no recycling equipment, hence, no value is derived from waste.

Figure 14 shows that education and awareness campaigns are strongly required for the adoption of circular economy in Nigeria. The graph indicates that 78.0%

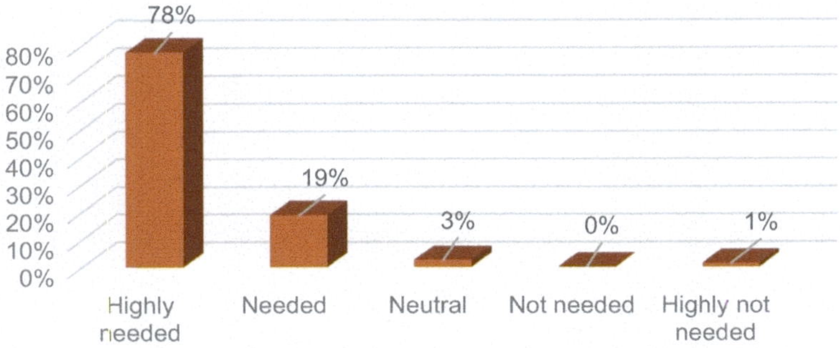

Fig. 14 Need for education and awareness campaigns. *Source* Questionnaire analysis

of the respondents believe that there is a high need for education and awareness campaigns for circular economy, but only 1.0% said there is no need for education and awareness campaigns.

Notably, lack of awareness and knowledge of circular economy seems to be one of the major challenges confronting the adoption of circular economy in Nigeria. To address this, the government and private sector must engage in continuous campaigns aim at educating Nigerians about the importance of the adoption of circular economy. These campaigns should also be targeted at increasing the level of participation of both the government and private entities in circular economy.

Ultimately, to achieve wider awareness, and participation level in circular economy, businesses must be incentivized. The government may encourage adoption of circular economy by giving tax holidays to businesses operating in this space. Also, firms operating in this space should further be tasked to encourage individuals in submitting their waste for a token. Gradually, this may change the narratives of Nigeria from a non-circular economy to a country driven by circular economy.

7 Discussion

The adoption of circular economy in Nigeria is still very low. This is also similar to other African countries. Looking deeply, the low adoption of circular economy is directly connected to the barriers identified in the study. Paramount amongst these barriers is lack of awareness and understanding, technological limitations, economic factors, regulatory and policy barriers, social and cultural factors, and infrastructural limitations. These barriers have different levels of Impact on the adoption of circular economy practices. For instance, lack of awareness and lack of technology may have a higher impact compared to economic barriers. On the same vein, some of these barriers are inter-related, as one barrier may lead to the crystallization of another barrier.

Meanwhile, the study pinpoints that the role of government in the adoption of circular economy cannot be overemphasized. Government's support for circular economy and the execution of relevant policies and regulation may facilitate the adoption of circular economy in the country. Government's policies have the capacity of compelling businesses and households to adopt circular economy practices.

Also, the private sector should consider the viability of circular economy in their various businesses. This may lead to the extension of their product/services lines as well as generate extra income. To achieve this, businesses might begin to manufacture products that can be recycled and reprocessed into new products.

8 Limitation of the Research

The study was conducted within a very short time. Hence, the Researcher worked tirelessly to meet up with the time scheduled for the completion of the study. It was also irksome to convince participants to fill and submit responses to the surveys in a timely manner.

9 Recommendations to Adopting Circular Economy

1. Government should sensitize the businesses to adopt appropriate technologies.
2. Government should invest in waste management, recycling, and resource recovery facilities to facilitate the adoption of circular economy.
3. The government should improve access to waste collection services and should invest in decentralized solutions for waste management.
4. Continuous education and awareness in all social media platforms.
5. The government should increase awareness campaigns on the benefits of circular economy practices among businesses, policymakers, and the public.
6. The government should set up a reward system for recycling.
7. Paying for nylon at the supermarket and getting paid for recycling plastic bottles are great incentives.

10 Conclusion

The study concludes that:

1. Lack of awareness, economic factors, technological factors, regulatory barriers, social and cultural factors, and infrastructural limitations were identified as challenges to the adoption of circular economy in Nigeria.
2. The level of engagement in circular economy practices among individuals in Nigeria is low.

3. Government shows very little support for circular economy practices.
4. Adoption of circular economy is very important to the sustainability development of Nigeria.
5. Adoption of circular economy is viable to businesses in Nigeria.
6. Infrastructures in Nigeria are not adequate for the adoption of circular economy in Nigeria.
7. Education and awareness campaigns are strongly required for the adoption of circular economy in Nigeria.

References

African Development Bank (2021) Circular economy. https://www.afdb.org/en/topics-and-sectors/topics/circular-economy. Retrieved on 06 Apr 2024

African Development Bank Group (n/a) Nigeria Circular Economy Working Group (NCEWG). https://www.afdb.org/en/topics-and-sectors/topics/circular-economy/nigeria-circular-economy-working-group-ncewg. Retrieved on 13 Apr 2024

Ajibesin O (2023) Unlocking Nigeria's future through the circular economy. https://www.circularinnovationlab.com/post/unlocking-nigeria-s-future-through-the-circular-economy. Retrieved on 06 Apr 2024

Business Day Newspapers (2022) Nigeria's 32m-ton annual solid waste ripe for investments. https://businessday.ng/big-read/article/nigerias-32m-ton-annual-solid-waste-ripe-for-investments/. Retrieved on 10 Apr 2024

Chatham House (2019) An inclusive circular economy—priorities for developing countries. https://www.chathamhouse.org/2019/05/inclusive-circular-economy. Retrieved on 06 Apr 2024

Debrah JK, Teye GK, Dinis MAP (2022) Barriers and challenges to waste management hindering the circular economy in Sub-Saharan Africa. Urban Sci 6:57. https://doi.org/10.3390/urbansci6030057. Retrieved on 30 July 2024

Department of Climate Change, Federal Ministry of Environment, Nigeria (2021) 2050 long-term vision for Nigeria (LTV-2050). //efaidnbmnnnibpcajpcglclefindmkaj/https://unfccc.int/sites/default/files/resource/Nigeria_LTS1.pdf. Retrieved on 06 Apr 2024

Galvao GDA, de Nadae J, Clemente DH, Chinen G, de Carvalho MM (2018) Circular economy: overview of barriers. CIRP J Manufact Sci Technol. https://doi.org/10.1016/j.procir.2018.04.011. https://www.sciencedirect.com/science/article/pii/S2212827118305262. Retrieved on 29 July 2024

Ikeah CK (n/a) Transition to circular economy in Nigeria: program and activities. https://www.unido.org/sites/default/files/files/202208/3_Transition%20to%20Circular%20Economy%20in%20Nigeria%20Programmes%20and%20Activities.pdf?. UNIDO. Retrieved on 13 Apr 2024

Kirchherr J, Hekkert M, Bour R, Huijbrechtse-Truijens A, Kostense-Smit E, Muller J (2017) Breaking the barriers to the circular economy. Chrome-extension: //efaidnbmnnnibpcajpcglclefindmkaj/https://circulareconomy.europa.eu/platform/sites/default/files/171106_white_paper_breaking_the_barriers_to_the_circular_economy_white_paper_vweb-14021.pdf. Retrieved on 30 July 2024

Lagos State Nigeria (n/a) Circular Lagos. https://circularlagos.com/. Retrieved on 13 Apr 2024

Okafor C, Ajaero C, Madu C, Agomuo K, Abu E (2020) Implementation of circular economy principles in management of endof-life tyres in a developing country (Nigeria). https://www.aimspress.com/article/doi/10.3934/environsci.2020027. Retrieved on 06 Apr 2024

PricewaterhouseCoopers (2021) 2021 Nigeria sustainability Outlook. Chrome-extension: //efaidnbmnnnibpcajpcglclefindmkaj/https://www.pwc.com/ng/en/assets/pdf/nigeria-sustainability-outlook-2021.pdf. Retrieved on 06 Apr 2024

Punch Newspaper (2022) Igando landfill: Lagos dump where food vendors, scavengers, others defy deadly stench to eke out a living. https://punchng.com/igando-landfill-lagos-dump-where-food-vendors-scavengers-others-defy-deadly-stench-to-eke-out-a-living/. Retrieved on 10 Apr 2024

Rajput J, Potgieter J, Aigbokhan G, Felgenhauer K, Smit TAB, Hemkhaus M, Ahlers J, Van Hummelen S, Chewpreecha U, Smith A, McGovern M (2020) Circular economy in the Africa-EU cooperation—Country report for Nigeria. Country report under EC Contract ENV.F.2./ETU/2018/004 Project: "Circular Economy in Africa-Eu cooperation", Trinomics B.V., Tomorrow Matters Now Ltd., Adelphi Consult GmbH and Cambridge Econometrics Ltd

Rizos V, Bryhn J, Alessi M, Righetti E, Fujiwara N, Stroia C (2021) Barriers and enablers for implementing circular economy business models: evidence from the electrical and electronic equipment and agri-food value chains. Adopted from https://circulareconomy.europa.eu/platform/en/knowledge/barriers-and-enablers-implementing-circular-economy-business-models. Retrieved on 29 July 2024

Ting LS, Zailani S, Sidek NZM, Shaharudin MR (2023) Motivators and barriers of circular economy business model adoption and its impact on sustainable production in Malaysia. Environ Dev Sustain 26:17551–17578. https://doi.org/10.1007/s10668-023-03350-6. Retrieved on 29 July 2024

Trading Economics (2024) Nigeria's indicators. https://tradingeconomics.com/nigeria/indicators. Retrieved on 10 Apr 2024

United Nations Environment Program (2019) UN report: time to seize opportunity, tackle challenge of e-waste. https://www.unep.org/news-and-stories/press-release/un-report-time-seize-opportunity-tackle-challenge-e-waste. Retrieved on 06 Apr 2024

United Nations Environment Programme (2019) UN report: time to seize opportunity, tackle challenge of e-waste. https://www.unep.org/news-and-stories/press-release/un-report-time-seize-opportunity-tackle-challenge-e-waste. Retrieved on 13 Apr 2024

Weick M, Rodysill J, Ray N (2022) Circular economy: navigating the evolving global policy landscape. https://www.ey.com/en_us/chemicals/circular-economy-navigating-the-evolving-global-policy-landscape. Retrieved on 13 Apr 2024

Integrating Life Cycle Analyses and Circular Economy: A Pathway Towards Sustainable Development in Bangladesh

Mahreen Mustafa

Abstract The textile sector of Bangladesh stands as a major economic pillar, yet it poses substantial environmental and social challenges in its journey towards sustainability. Integrating Life Cycle Analyses (LCA) and Circular Economy (CE) emerges as a crucial strategy to combat environmental degradation, resource scarcity, and socio-economic disparities. This paper underscores the significance of LCA and CE frameworks in enhancing resource efficiency, mitigating harmful environmental impacts, and fostering inclusive growth in Bangladesh's textile industry. Through applying these frameworks, the study identifies tangible benefits, such as reductions in water and energy consumption, lower carbon emissions, and advancements in waste recycling processes, all of which contribute to achieving Sustainable Development Goals (SDGs) in Bangladesh. By embracing a comprehensive and holistic approach that encompasses the entire product lifecycle and emphasizes circularity, Bangladesh can unlock pathways for sustainable development in the textile sector while confronting urgent environmental and social concerns. This study explores the interconnections among LCA, CE, and the Sustainable Development Goals (SDGs) in Bangladesh, synthesizing existing literature and empirical data to illuminate their implications. It delves into the synergies between LCA and CE, highlighting their complementary roles in reducing resource depletion, waste generation, and environmental harm across product lifecycles, including specific examples such as improved textile recycling initiatives and reductions in harmful chemical usage. Additionally, the paper examines diverse perspectives from key institutions in Bangladesh, assessing prospects, practices, and challenges of LCA and CE implementation. Identifying barriers hindering the adoption of LCA and CE practices, including policy, technical, and public participation obstacles, this paper offers insights for policymakers and lays the groundwork for future research endeavors. It emphasizes interdisciplinary approaches to strengthen the relationship between LCA and CE as a sustainable metric, advocating for policy interventions, capacity building, and stakeholder engagement. Ultimately, this research underscores the pivotal roles of LCA

M. Mustafa (✉)
University of Cambridge, Cambridge, England
e-mail: mm2462@cantab.ac.uk

L. Brandli et al. (eds.), *The Contribution of Life Cycle Analyses and Circular Economy to the Sustainable Development Goals*, World Sustainability Series,
https://doi.org/10.1007/978-3-031-85300-5_15

and CE, especially within the textile industry, in propelling Bangladesh towards a more resource-efficient, environmentally sustainable, and socially inclusive future, aligned with the SDGs.

Keywords Life Cycle Analyses (LCA) · Circular Economy (CE) · Sustainable Development Goals (SDGs) · Resource efficiency · Inclusive growth · Policy interventions · Green development · Textile industry · Bangladesh

1 Introduction

The phenomenal growth and transformation of Bangladesh can be attributed in large part to its textile sector. In the last few decades, Bangladesh's economy has grown significantly, moving from one based primarily on agriculture to one that is booming in the industrial and services sectors. The textile industry of Bangladesh is one of the biggest globally, supporting more than 80% of the nation's exports and providing jobs for over 4 million people (Bossavie et al. 2023; IFC 2019; Frayer 2022). This sector is significantly vital to Bangladesh's economy, employing millions and contributing substantially to the GDP and exports of the nation. Bangladesh's textile sector has become a fundamental pillar of the nation's economic growth and development. However, because the sector uses a lot of resources and pollutes the environment, questions have been raised about how sustainable it is for the environment.

The manufacture of textiles results in both financial losses and serious environmental problems. The clothing industry imposes significant detrimental impacts on the environment and public health, accounting for 20% (79 billion cubic meters) of the world's wastewater and more than 10% of global emissions (amounting 1.7 billion tonnes of CO_2) (Duhoux et al. 2022; Abbate et al. 2023; Nikolina 2019). Water scarcity in many countries is exacerbated by this exceptionally resource-demanding industry, which accounts for about four percent of the world's freshwater outflows and annual per capita water use of 23,900 L (UNCCD 2018; Scaling Circularity Report 2021). An even worse problem is that a major portion of the textile production becomes wastage. Every second, clothing amounting to one full garbage truck is delivered to a landfill to be either burned or buried (Ellen MacArthur Foundation n.d.). There is an annual loss of supplies valued at approximately USD 500 billion since only about one percent of the resources used to produce garments are recycled into new apparel (Abbate et al. 2023). In addition, concerns regarding health and safety have raised awareness of labor practices globally, particularly in the wake of the Rana Plaza calamity in Bangladesh (Saha et al. 2022). Nonetheless, not much lessons were learnt, nor many actions taken; the problems with labor welfare still remain. Three textile workers in Bangladesh lost their lives just over a salary disagreement that quickly turned into an aggressive outrage; it then resulted in a full-scale factory closure (Hossain 2023). As a direct result of continued protests, many garment workers lost their livelihoods. Not only does this significantly impact the mental health and financial security of the laborers and their households, but also has repercussions in a knock-on effect on the society as a whole. Moreover, from the standpoint of the firms, because of their importance to the socioeconomic structure of the industrial sites, the closing of these facilities has detrimental effects on nearby companies, suppliers of services, and the entire economy.

The existing linear system, mostly driven by rapid fashion, is harmful to the environment. The model's low rates of textile repair, reuse, and fiber-to-fiber recycling encourage customers to buy cheap, flimsy clothing, exacerbating the industry's negative environmental effects (Smith et al. 2017). Over 92 million tons of garbage are produced annually by the disposal of more than half of the clothing products (Abbate et al. 2023). Consequently, the Ellen MacArthur Foundation (2017a) predicts that, by 2050, there would be disastrous consequences that might have enormous effects on the environment and society. The adverse effects encompass surpassing carbon budgets, a substantial build-up of plastic microfibers in the seawater and marine ecosystem, and challenging control and management of garment waste. The foundation also predicts that the detrimental effects of the textile sector could result in a rise in reputational challenges, plummeting earnings, and tighter regulations. Microplastics are released at every stage of their lifecycle due to the wasteful use of non-renewable resources caused by the rising demand for textiles, including the creation of synthetic fibers from fossil fuels (Beylot et al. 2020). Moreover, Bangladesh, as well as China and India (two other major textile-producing countries), continue to remain significantly dependent on coal for their manufacturing processes (Konwar and Boruah 2020). Furthermore, when textile waste is improperly disposed of in landfills or incinerator facilities, dangerous chemicals and greenhouse gases (GHG) are released into the atmosphere. Besides, the procedures involved in textile dyeing and finishing lead to wastewater contamination, which poses high risks to human health, aquatic habitats, and waste and water treatment. (Panda et al. 2021). These environmental issues are made worse by energy-intensive daily clothing care procedures like washing, dry cleaning, drying, and pressing at home and facilities (Broega et al. 2017).

Understanding the environmental effects of textile manufacture, use, and disposal is essential to addressing the sustainability concerns that are inherent in this complicated sector. Accordingly, LCA has become a significantly essential method and instrument to mitigate these issues. With LCA, a comprehensive evaluation of a product's complete life cycle—which includes raw material extraction, usage, production, distribution, and disposal at the end of its useful life—can be conducted for various products of the textile industry (Salo et al. 2020; Wiedemann et al. 2022). By implementing the method of LCA, which assesses and quantifies environmental impacts at each stage, decision-makers can be better equipped to (i) identify crucial aspects of the garments production and management that require improvement and (ii) implement eco- design solutions as adequate. This scientifically grounded quantitative method addresses a wide range of environmental concerns from a life cycle viewpoint (Cicconi 2020).

Along with applying the holistic tool LCA, a transition to a CE is urgently needed to enhance garment utilization, reduce overconsumption, and boost recycling rates. According to Saha et al. (2022), business processes in the textile industry must shift from being linear (make, use, and discard) to circular (take, make, distribute, use, and recover). Transitioning to the circular economy, the textile industry will be adopting a sustainable and regenerative approach that in turn would mitigate waste and environmental damages (Ellen MacArthur Foundation 2021). The focus lies on the design, creation, manufacturing, and use of textiles in a way that encourages longevity, recycling, reuse, and conscientious disposal. Consequently, several countries are now adopting the CE approach within the textile industry due to its promising impacts on business models, policies of different institutions, and preferences of consumers and producers for sustainable and green products (Bailey et al. 2022; Blomsma et al. 2022; Rainville 2021; Khan et al. 2023; Colucci and Vecchi 2021; Jia et al. 2020).

Governments, NGOs and private sectors have started advancing sustainability policies, strategy plans and initiatives for the global apparel industry. The EU is spearheading initiatives to reuse and decrease the substantial amount of garbage produced. Environmental issues are given top priority in nations like the US and Canada, which place a strong emphasis on traceability and circularity in the value chain. Following sustainable standards is essential for Bangladesh, a significant exporter of RMGs, to keep conserving its competitive advantage in the global textile industry.

This chapter aims to emphasize that Bangladesh has the chance to set itself apart on the international arena by promptly embracing sustainability trends as the world moves toward clothing produced sustainably and ethically. The most effective strategy available to mitigate harmful environmental impacts, enhance resource efficiency, and foster inclusive growth in Bangladesh's textile industry—while also supporting the nation's pursuit of achieving its Sustainable Development Goals (SDGs)—is the application of Life Cycle Analysis (LCA) and the shift to a Circular Economy (CE). However, there is a notable gap in the existing literature that directly links the specific application of Life Cycle Analysis (LCA) and Circular Economy (CE) to the achievement of Sustainable Development Goals (SDGs) within Bangladesh's textile sector. This gap underscores the urgent need for further research in this area. Therefore, this chapter addresses this critical need by advocating for in-depth exploration of these topics in the context of Bangladesh.

Section 2 discusses the methodology used in this study. Section 3 provides the theoretical framework and literature review, exploring the concepts of LCA and CE and their connection to the SDGs. Section 4 emphasizes the urgency of integrating LCA and CE within Bangladesh's textile sector and provides an overview of international policies promoting these practices in major markets, and further discussing their implications for Bangladesh. Section 5 examines various institutional perspectives in Bangladesh regarding the implementation of LCA and CE. Section 6 identifies key barriers to the adoption of LCA and CE in the country. Section 7 draws on global experiences to offer potential policy recommendations for Bangladesh.

2 Methodology

This study primarily relies on secondary data to explore the topic comprehensively, gather relevant insights, and draw precise conclusions. Data sources include scholarly journal articles, academic research papers, databases, reports, and newsletters. A structured selection process was employed, focusing on literature with clear relevance to Life Cycle Assessment (LCA), Circular Economy (CE), and their contributions to Sustainable Development Goals (SDGs). However, additional criteria, such as the recency and geographical context of the data, were also considered to ensure the relevance and reliability of the findings.

To improve replicability, the study outlines a systematic approach to data categorization, grouping information based on key themes, such as institutional perspectives and sectoral initiatives in Bangladesh. The rationale behind these methodological choices is to ensure the clarity and alignment of the data with the study's three research questions:

1. What are the contributions of LCA and CE in attaining the SDGs?
2. What are the different institutional perspectives of Bangladesh on LCA and CE implementation, and what are the current roles and initiatives by different sectors?
3. What are the potential policy recommendations for Bangladesh, drawing from global achievements in the adoption of LCA and CE as sustainable metrics?

This structured methodology, combined with a transparent categorization process, allows for better replication of the study and strengthens its credibility for future research.

3 Theoretical Framework and Literature Review

3.1 Life Cycle Analysis (LCA)

Life Cycle Analysis (LCA) is a methodical quantitative approach that evaluates how products, services, and processes affect the environment. From the extraction of raw materials to disposal, it can be connected to every phase of the entire life cycle of products, services, and processes. Standardized by ISO 14040, 2006 and ISO 14044, 2006, the LCA is a four-phase process, encompassing: (i) goal and scope definition; (ii) inventory analysis; (iii) impact assessment; and (iv) interpretation (ISO 14040 2006; ISO 14044 2006; Guinée et al. 2011), as shown in Fig. 1. By assessing several environmental impact indicators factors like water footprint, carbon footprint, acidification, human toxicity, and eutrophication, LCA offers a holistic and LCA offers a thorough and all-encompassing understanding of the environmental burdens and benefits associated with products and services (Muthu 2017; Harsanto et al. 2023; Munasinghe et al. 2021). This evaluation assists in recognizing areas

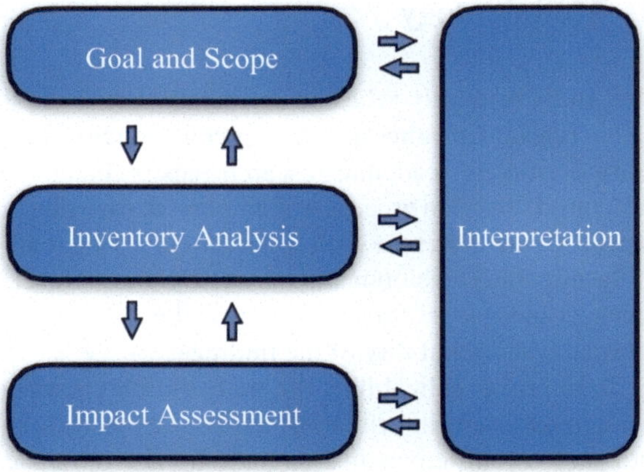

Fig. 1 Methodology of a Life Cycle Analysis (LCA)

for environmental enhancement across the entire life cycle of these products and services (Guinée 2002).

3.2 Circular Economy (CE)

The Circular Economy (CE) is an economic framework focused on prolonging the lifespan of textile products via repair and reuse and ensuring that end-of-life (EoL) products are recycled back into the economy (NIST 2022). The life cycle of textile products has historically been rather linear: when raw materials are extracted or mined, products are manufactured, distributed, consumed, and finally disposed of—typically in incinerators or landfills. Interest in and momentum for a shift to a CE are growing in order to ensure goods and resources cycling inside the economy and out of unintentional drains (such as the water, air, and land systems) (European Commission 2022; Piribauer and Bartl 2019; Ellen MacArthur Foundation 2019; Oregon State Legislature 2021). The CE paradigm, in accordance with Geissdoerfer et al. (2017), strongly emphasizes collaboration across value chains in order to create closed-loop systems that maximizes economic value while reducing environmental damages. Key elements of the CE include promoting repair and reuse, developing products for durability and recyclability, and transitioning from a linear "take, make, dispose" framework to a circular "reduce, reuse, recycle" paradigm (Stahel 2016). A circular economy, in contrast to a linear one, strives to eradicate waste via producing long-lasting, repairable, and reusable products from supplies that can be recycled and recovered at EoL (NIST 2022). Consequently, there is a push for increased circularity on a national and global level public and private institutions to mitigate climate change, avoid supply chain disruptions, preserve natural resources, and reduce pollution (European Commission 2018, 2022; Ellen MacArthur Foundation 2017a; United Nations 2022) (Fig. 2).

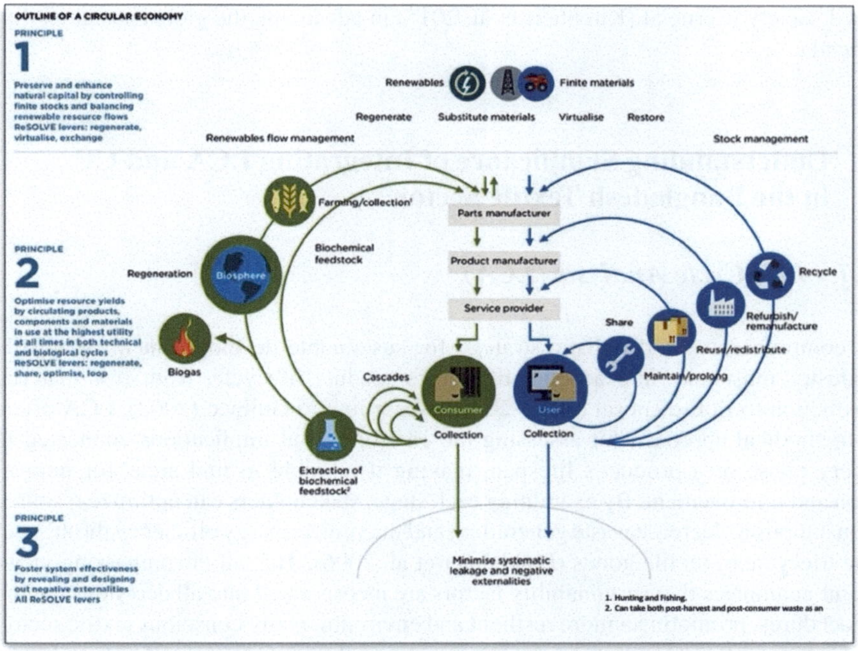

Fig. 2 Diagram of the circular economy system. *Source* Ellen MacArthur Foundation (2017b), edited by author

3.3 Contributions of LCA and CE in Achieving the SDGs

Sustainability in the textile sector refers to the idea of developing a system that is environmentally and socially responsible in addition to being economically profitable. It is vital for Bangladesh's textile industry, which mostly uses unregulated manufacturing techniques and is highly dependent on natural resources. Consequently, putting into practice sustainable practices has become more crucial as a need for preserving the environment and fostering economic growth. The LCA and CE approaches are closely tied to the UN's Sustainable Development Goals (SDGs), which tackle climate change (SDG 13) and offer frameworks promoting sustainable production and consumption (SDG 12). The LCA methodology makes it possible to quantify environmental consequences, which facilitates the making of well-informed decisions consistent with the goals of the SDGs (Tillman 2000). Similarly, the CE framework aligns with several SDG targets by encouraging sustainable economic growth, efficiency of resource management, and reduction of waste (Geissdoerfer et al. 2017). By finding synergies and trade-offs across environmental, economic, and social dimensions, the combination of LCA and CE can help achieve various SDGs (Blomsma and Brennan 2017). To fully utilize the potentials of LCA and CE principles, cooperative endeavors among governments, policymakers, enterprises, and civil society is crucial (Kirchherr et al. 2017) in advancing the global sustainability agenda.

4 Understanding Significance of Integrating LCA and CE in the Bangladesh Textile Sector

4.1 Life Cycle Analysis (LCA)

A comprehensive and holistic strategy for sustainable development in the textile industry must take into account the entire product lifecycle, from raw material extraction to EoL disposal (NIST 2022). According to Guinée (2002), LCA offers a methodical approach for assessing the environmental implications connected to every phase of a product's lifespan, making it possible to find areas for innovation and improvement. By examining each stage, stakeholders can optimize resource consumption, decrease waste generation, and maximize energy efficiency throughout the lifecycle of textile goods (Finkbeiner et al. 2006). This all-encompassing viewpoint guarantees that sustainability factors are incorporated into all decision-making procedures, promoting a more resilient and environmentally conscious textile sector. Therefore, different countries are now aware of the importance of LCA and have been applying the methodology in their textile industry (Fig. 3).

4.2 Circular Economy (CE)

The principles of the CE play a pivotal role in achieving sustainable development in the textile sector. CE aims to create a closed-loop mechanism in which resources are constantly exchanged throughout the economy and waste is minimized. It does so by promoting material recycling, reuse, and repurposing (Kirchherr et al. 2017). Within the textile industry, this refers to designing durable products, making it possible to recover and recycle textile fibers, and encouraging cutting-edge business models like leasing and sharing that increase the lifespan of products (Ghisellini et al. 2016). In addition to mitigating environmental effects, implementing CE techniques boosts economic activities and opportunities, improves resource efficiency, and stimulates competitiveness and innovation in the textile sector (Stahel 2016). In addition to the obvious environmental beneficial implications of this method, a CE also brings financial benefits. It is predicted that the transition to a CE will contribute USD 4.5 trillion into the world economy by 2030 (SGS 2023).

Many countries are transitioning to the circular economy paradigm from the linear economic model (Figs. 4 and 5). The environmental, economic, social and results have been extremely promising and beneficial for these countries. Bangladesh can learn lessons from these successful outcomes and work towards mainstreaming

	Title of Project	Purpose (and Method)	Main Conclusions
USA	Resource and Environmental Profile Analysis of a Man-Made Apparel Product: Woman's Knit Polyester Blouse	To assess the energy requirements and environmental emissions and solid waste for the life cycle of a polyester blouse.	The consumer use corresponds to 86% of the life cycle energy needs. Cold wash with line drying reduces laundering energy by more than 90%. Air emissions and solid waste are also very dominant in the consumer phase.
Finland	The Ecology of Hotel Textiles and Textile Services - an LCA Study on Best Available Applications and Technologies	To find the present best ecological solutions for hotel textiles and services, to develop an ecological index for products and processes with 'Best Available Technology'.	The use of polyester-cotton sheets has less environmental consequences than the use of 100% cotton sheets.
Switzerland	Comparative LCA of Three Textile Dyestuffs	Comparison between three red cotton reactive-dyestuffs. ECOSYS, a software from Ciba-Geigy, was used.	The fixation rates (low for the dyestuffs studied) is the most important factor, the dyeing process is the most emissive.
Denmark	Textile Products, Impact Assessment and Criteria for Eco-labelling	Impact assessment of cotton and polyester for T-shirts and bedlinen.	EU label criteria for T-shirts and bed linen were proposed.
Denmark	The Life Cycle of Cotton Rolls for Hand Drying	Simplified LCA on cotton-based towel rolls used for 10,000 hand-dryings.	The laundering process is the most important process related to environmental burdens.
Switzerland	Ökobilanz eines Baumwoll-T-Shirts mit Schwerpunkt auf den verwendeten Chemikalien	Comparison between the production of textiles (not fibers) and the use phase. To develop a method to study how persistent chemicals are degraded.	The use phase creates by far the highest environmental burdens. The energy used has the highest environmental impact. The method for fate developed gave e.g. the conclusion that dyes are very persistent.
Germany	Produktlinienanalyse Waschen und Waschmittel, Umweltbundesamt	Comparative product system assessment (LCA plus social, health, safety and economical aspects) of washing agents to optimize the use.	The three component system (washing agent, softener and bleaching agent separated) is the best. There is much to save in both environmental impacts and money if a household becomes conscientious.
Netherlands	De milieugerichte levenscyclusanalyse van vier typen vloerbedekking	LCA comparison between a wool broadloom carpet and one made of polyamide.	For most impact categories wool has the lowest impact (except eutrophication and non chemical waste).
Denmark	Livseyklusvurdering of produktorienteret miljöledelse hos Gabriel A/S	LCA comparison between three different padding textiles: Wool, Wool/nylon, Polyester.	Due to lack of data, this study cannot be used for rules for EPD's. Only characterization, no weighting. Prod. of chemicals included. The inventory data was not reported.

Fig. 3 Different nations performing LCAs in their textile sector. *Source* Dahllöf (2003), edited by author

circular approaches in order to reduce pollution and waste, prolong the life of supplies and products within the economy, introduce environmentally friendly goods into the market, develop more green and sustainable value chains, and ultimately leading the transition towards a greener future for Bangladesh and the globe.

4.3 Unlocking Pathways for Sustainable Development

Enhancing resource efficiency: For Bangladesh's textile industry, which is a leading contributor to the nation's economy, enhancing resource efficiency through LCA and CE approaches is paramount. LCA enables the industry to identify inefficiencies

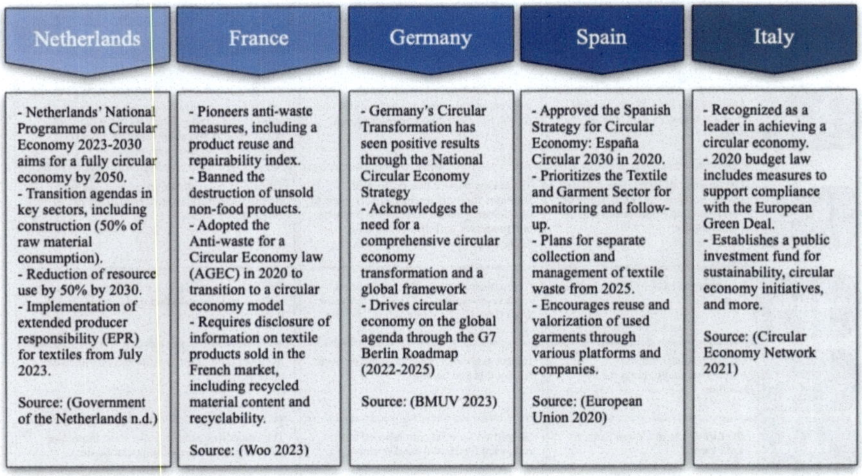

Fig. 4 Strategy plans and policies of European countries adopting circular economy

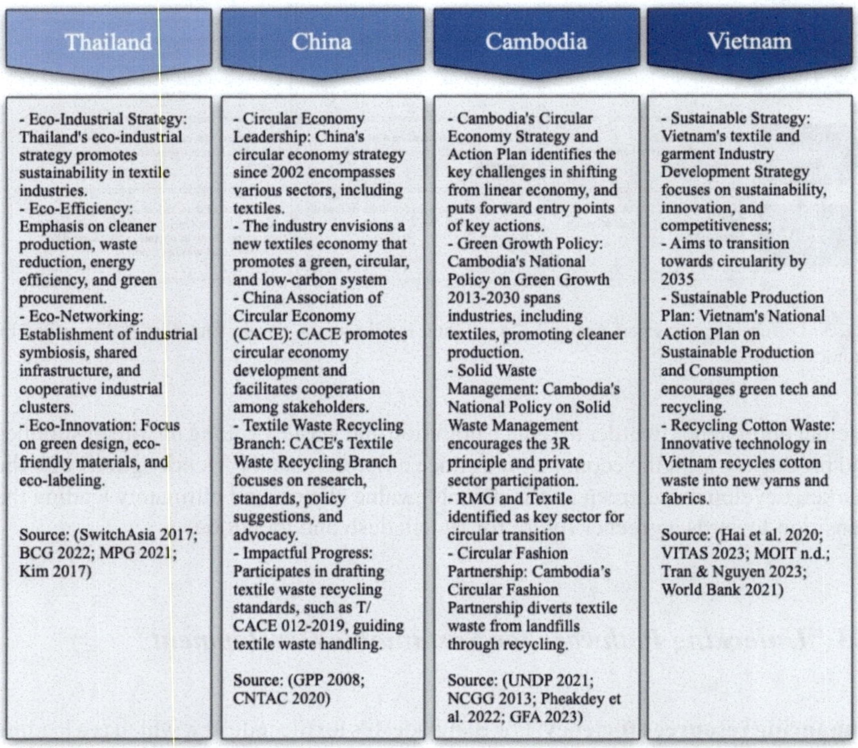

Fig. 5 Major textile nations advancing circular economy policy development

and waste across the entire product lifecycle (Azapagic 2004). Using CE principles can help the industry become less dependent on raw materials and produce less waste by facilitating the recycling and reuse of products (Geissdoerfer et al. 2017). Such measures conserve valuable resources as well as provide economic advantages by lowering production expenses and boosting competitiveness in global markets (Tukker 2015).

Mitigating environmental impacts: The textile sector is notable for having substantial adverse effects on the environment, especially when it comes to water use, chemical pollution, and greenhouse gas emissions. LCA provides a systematic framework to quantify and assess these environmental impacts, enabling targeted interventions to mitigate adverse effects (Clift et al. 2000). CE strategies, focusing on waste prevention, reuse, and recycling, can substantially mitigate the textile sector's environmental burden (Bocken et al. 2014). Implementing cleaner production techniques, optimizing water usage, and substituting hazardous chemicals with safer alternatives are essential steps towards achieving environmental sustainability in Bangladesh's textile sector (Subramanian and Kannan 2016).

Fostering inclusive growth: Inclusive growth is essential in order to ensure that the benefits of economic progress are shared fairly throughout society. LCA and CE approaches can contribute to fostering inclusive growth in Bangladesh's textile industry by offering SMEs with more opportunities, promoting green jobs, and enhancing workforce skills and capabilities (UNCTAD 2014). Moreover, CE emphasizes the importance of designing products which are accessible, affordable, and environmentally friendly, catering to the diverse needs and preferences of consumers (Rizos et al. 2016). Government, policymakers, industry stakeholders, and civil society must collaborate together to mainstream LCA and CE practices and ensuring that the textile industry's growth is sustainable, equitable, and beneficial for all.

Embracing an integrated and comprehensive approach to sustainable development in the textile industry unlocks pathways for transformative change. By synergistically applying LCA and CE principles, stakeholders can develop strategies that address environmental challenges, drive economic growth, and promote social equity (Lieder and Rashid 2016). Collaborative efforts involving policymakers, industry leaders, civil society, and consumers are essential to mainstreaming sustainable practices and fostering a culture of accountability and responsibility (Schmidt-Bleek 1998). By unlocking these pathways, the textile industry can move toward a more resilient and sustainable future, aligned with the underlying notions of sustainable development and the broader objectives of the SDGs.

5 Institutional Perspectives of Bangladesh on LCA and CE Implementation

5.1 Initiatives and Policies of Governmental Institutions of Bangladesh

The Government of Bangladesh has taken some policy and strategy initiatives to promote the LCA and CE mechanisms as a result of realizing the urgency of applying their principles in harnessing sustainable development in the nation's textile industry. The Ministry of Environment, Forest and Climate Change (MoEFCC) has actively participated in formulating and implementing policies with two goals: reducing environmental deterioration and improving resource efficiency (MoEFCC 2020). The Government of Bangladesh, in 2018, introduced the 'National Sustainable Development Strategy (NSDS) 2018–2030, emphasizing the need to adopt sustainable patterns of consumption and production, that incorporates LCA and CE principles into industrial practices to an extent (Bangladesh Planning Commission 2018). The government has also implemented several regulatory frameworks, such as the National 3R Strategy for Waste Management, the Environmental Conservation Act of 2021, and the Bangladesh Environment Conservation Rules to monitor and control industrial pollution and encourage environmentally friendly green practices (MoEFCC 2020; Van der ven 2022). The Environmental Conservation Act of 2021 included hazardous waste management regulations for e-waste, including collection obligations for producers (Huang 2021). There are no such laws pertaining to textile waste. Nonetheless, the Government of Bangladesh intends to create legislative frameworks to implement a CE in the textile sector (Hossain 2020). The 2010 National 3R Strategy for Waste Management gives industrial waste management top priority, although it lacks concrete targets (DoE 2010).

5.2 Roles and Contributions of Non-governmental Organizations of Bangladesh

The non-governmental organizations (NGOs) in Bangladesh have been greatly contributing to the advancement of sustainable practices and the incorporation of LCA and CE in the country's textile sector. To raise awareness of industry stakeholders of environmental sustainability and to promote the application of green practices, different institutions have been actively pursuing research, capacity-building, and advocacy initiatives. The Centre for Policy Dialogue (CPD) has been working through policy research and advocacy towards advancing sustainable development and promoting the inclusion of the private sector on adopting the LCA and CE frameworks in environmental conservation projects (CPD n.d.). Likewise, the Bangladesh Centre for Advanced Studies (BCAS) has also been conducting in-depth research on the impacts of the LCA and CE mechanisms to develop and implement sustainable practices tackling environmental issues, such as pollution prevention and resource conservation, in collaboration with government organizations, industry stakeholders, and other relevant parties (BCAS n.d.).

5.3 Bangladesh Private Sector Engagement and Innovations

In Bangladesh, the private sector has realized the potential advantages of embracing LCA and CE concepts in the the textile industry and has been introducing creative solutions to pressing social and environmental issues into practice more frequently. Investing in energy-efficient production methods, waste management systems, and green technologies has allowed major textile exporters and manufacturers like Recover Textile Fibers and Beximco Textiles to lessen their environmental impact and increase resource efficiency. Recover has partnered with Beximco to establish the biggest recycling facility in Bangladesh in an effort to scale the latter's production of recycled fiber (Husain 2023). The new factory provides a scalable method to convert textile waste into high-quality, sustainably sourced, low-impact recycled fiber (Husain 2023). Apart from implementing sustainable practices, the business sector has exhibited a dedication to advancing circularity by means of endeavors such product redesign, reuse and recycling activities, and the creation of environmentally friendly goods (Forbes 2021). Furthermore, trade groups such as the Bangladesh Garment Manufacturers and Exporters Association (BGMEA) have been actively pushing for sustainable practices and helping their members implement LCA and CE guidelines (BGMEA 2021).

In order to counteract waste and resource depletion resulting from textile manufacturing, BGMEA, Reverse Resources, Global Fashion Agenda (GFA), and P4G have designed the multi-sectoral project Circular Fashion Partnership (CFP), which aims to expand circularity in Bangladesh's textile sector (BGMEA 2021). Under the program, the GFA has designated 2 work streams: the first entails creating new clothing from recycled waste, and the second involves setting up a circular-fashion stock marketplace for extra apparel that has been collected over time from canceled orders (BGMEA 2021). In order to collect and reuse post-production waste, the collaboration seeks to initiate communication amongst key fashion companies, textile and garment manufacturers, and recyclers (BGMEA 2021). Additionally, the BGMEA has become part of the German Green Button initiative that offers certifications and recommendations to textile firms for environmentally friendly labor and production practices (GIZ 2023). Cradle to Cradle, a benchmark evaluating environmental and social sustainability, is another certification that a number of textile products from Bangladesh have obtained (EIG 2021). According to the UN Fashion Industry Charter for Climate Change, the BGMEA has set 2030 GHG emission reduction targets of 30% (BGMEA 2021). Furthermore, the BGMEA has been funding green- factories—that is, LEED-certified establishments—which reduce water and energy use by 30% and 40%, respectively (Finnwatch 2023). Both the building's overall energy efficiency and the industrial operations occurring within it are considered for LEED certification. Approximately 5% of Bangladesh's textile factories were LEED-certified as of December 2022, with 170 of them being located there (Finnwatch 2023).

Similar prospects to link stakeholders involved in the waste supply chain are provided by another platform created by Reverse Resources. For instance, manufacturers and recycling facilities in Bangladesh (as well as India) who wish to classify, market, and monitor textile wastes can do so through this platform (E-estonia 2021). Likewise, Simco Spinning & Textile has been taking the lead in gathering and reusing textile waste from the post-production process (Textile Today 2018). They reduce water usage, pollutants, and the need for dyes and chemicals by categorizing fabrics based on color and material using unique technology of mechanical recycling. To improve the traceability of their recovered products, they have started utilizing tracers (Textile Today 2018). The institution also made efforts to treat factory wastewater and reduce water use, which in turn also resulted in reduction of production costs. One of Bangladesh's top ten apparel companies accomplished a roughly 40% water usage reduction in 2017 and set further goals to reduce dyeing operations by 60% (Islam and Jabber n.d.). Moreover, initiatives such as rainwater harvesting, usually contingent on cost and capacity, became a standard LCA and CE practice in RMG enterprises through government requirements.

6 Key Barriers to LCA and CE Adoption for Bangladesh

6.1 Policy Barriers

Despite the increasing awareness and recognition of the importance of LCA and CE in Bangladesh, there are several policy barriers that hinder their widespread adoption in the textile industry. The existing regulatory framework often lacks comprehensive guidelines and incentives to encourage businesses to invest in sustainable practices and technologies (Ahmed et al. 2019). Moreover, the inconsistency and ambiguity in environmental policies and regulations create uncertainties for industry stakeholders, making it challenging for them to make informed decisions and long-term investments in sustainable technologies (Rahman and Khan 2020). Additionally, the lack of effective enforcement mechanisms and monitoring systems further undermines the implementation of sustainable practices in the textile sector (Hossain et al. 2021).

6.2 *Technical Challenges*

Technical challenges to adopting LCA and CE in Bangladesh's textile industry are multifaceted. Limited access to affordable green technologies and inadequate infrastructure, coupled with high initial implementation costs, are primary obstacles faced by industry stakeholders (Islam and Ali 2018). Moreover, a scarcity of technical expertise and a skilled workforce capable of managing sustainable initiatives hinder the integration of LCA and CE principles into textile production processes (Hasan and Hossain 2020). Additionally, the absence of standardized methodologies and tools for conducting LCA and designing circular business models tailored to the local context further complicates adoption (Ahmed et al. 2019). Transitioning to a CE model in Bangladesh's textile industry necessitates careful consideration of labor market implications. Reorienting garments as durable goods challenges the prevalent fast fashion paradigm, potentially leading to reduced working hours or job losses in the short term. Moreover, adopting circularity policies requires significant investments in digitalization and technical advancements, potentially altering the nation's labor market dynamics (Chauhan et al. 2022). Infrastructure remains a formidable barrier to embracing the circular economy in Bangladesh's textile industry. This encompasses both institutional and physical infrastructure, including manufacturing and waste management facilities, as well as educational and training programs. Increasing awareness about the circular economy's benefits and the necessary steps for its implementation is crucial to overcoming these hurdles (Kazancoglu et al. 2021). Collaboration among government, businesses, and civil society is imperative to provide essential infrastructure as well as education and training, requiring investment from both the public and corporate sectors (Saha et al. 2022). Implementing best practices, such as waste reduction and advanced recycling technologies, is essential to realizing Bangladesh's potential as a leader in the circular economy (Masum 2016).

6.3 *Public Participation Obstacles*

Although government regulations, industry practices, and research all play a major role in the implementation of a circular economy, consumers and conscientious consumerism take the lead in the process. Public participation and awareness play a crucial role in driving the implementation of LCA and CE in the textile industry. However, limited public awareness and understanding of sustainability issues, coupled with the lack of consumer demand for eco- friendly products, act as significant obstacles to the widespread adoption of sustainable practices (Rahman and Khan, 2020). Moreover, the absence of effective communication and engagement strategies by industry stakeholders and government agencies further limits public participation and involvement in sustainable initiatives (Hossain et al. 2021). Additionally, cultural and behavioral factors, such as the preference for fast fashion and low-cost products, contribute to the slow pace of adoption of LCA and CE principles among consumers in Bangladesh (Islam and Ali 2018).

7 Global Lessons and Potential Policy Recommendations for Bangladesh

Bangladesh can gain a lot of knowledge from the accomplishments of other nations that have successfully integrated the LCA and CE principles into their textile industries. By learning lessons from their policy approaches and customizing LCA and CE procedures to its local conditions, Bangladesh can effectively increase sustainability and resource efficiency. This may entail encouraging the use of recycled textiles, establishing collaborations between recycling businesses and textile makers, and establishing laws for ecologically friendly production practices. Such programs will not only reduce the detrimental impact that textile waste has on society and the environment, but they will also increase employment opportunities and economic growth in the recycling sector.

Immediate Policy Actions:

Adopting a Multi-Agency Cooperative Strategy among government agencies, the textile industry, and fashion brands is crucial for advancing textile circularity. Learning from Thailand's Sustainable Consumption and Production Roadmap 2017–2036, where various government bodies collaborate through the Steering Committee on Natural Resources and Environment (SwitchAsia 2017), Bangladesh can implement a similar approach. This involves coordinating efforts among relevant agencies to ensure alignment towards sustainable consumption and production goals. The Ministry of Commerce can lead this collaborative effort, facilitating data exchange, monitoring progress, and addressing challenges in the RMG sector. By fostering partnerships across sectors, including textile manufacturers and retailers, Bangladesh can promote sustainable practices and the use of recyclable materials. This multi- agency approach is vital for developing a comprehensive strategy to achieve circularity in Bangladesh.

Collaborating with other countries and national and international organizations that advocate for LCA and CE initiatives in the textile sector is crucial. Given the worldwide implications of textile recycling, international cooperation is essential for progress. Bangladesh can actively participate in international initiatives and agreements aimed at promoting textile recycling and reducing global textile waste. Being a member of the Regional 3R Forum in Asia and the Pacific can further strengthen Bangladesh to engage in efforts to promote 3R (Reduce, Reuse, and Recycle) strategies and policies in the region (Yousuf 2014). Through collaboration with other nations, Bangladesh can exchange knowledge, share best practices, and contribute to joint research and development efforts across the global textile value chain.

Community involvement and cooperation: Bangladesh can involve municipal communities in waste management by modeling itself after Thailand's Community-Based Integrated Solid Waste Management Projects (Mongkolnchaiarunya 2005; IUCN 2021). By encouraging methods like source separation, recycling, and composting, this strategy lessens the overall amount of waste produced. In this effort, effective cooperation between governmental and non-governmental institutions is essential.

Public awareness about recycling should be increased by educating the public about recycling, waste reduction, and the function of informal waste workers. This way the community would be encouraged to assist and cooperate in these initiatives. In partnership with regional authorities, the Ministry of Environment, Forest & Climate Change may launch public education campaigns to inform the public about the importance of textile recycling, the environmental effects of textile waste, and the benefits of recycling. A wide range of communication channels, such as advertisements, seminars, and educational materials distributed through educational institutions, community centers, and internet platforms, can be used for these awareness initiatives.

Medium-Term Policy Actions:

Private investors, who fund textile recycling companies, should be offered financial incentives including tax rebates, grants, loans, subsidies, or guarantees. Assistance of such kind can help develop cutting-edge and modern recycling technologies, acquire the adequate equipment and technology, and make it easier to build or expand recycling operations. More especially for situations when short-term business sustainability is questionable, cooperation and involvement from the government are even more crucial. China, for instance, lowered the value- added tax (VAT) on recycled plastics, promoting their use (GPP 2008; CNTAC 2020). Through initiatives like Horizon 2020, the LIFE program, and the European Structural and Investment Funds, the EU is offering a number of financial initiatives to aid with the shift to a CE (World Bank 2022). The European Investment Bank (EIB) is also participating in the financing of circular deals and investments under the 'EU Finance for Innovators' (InnovFin) Program and the European Fund for Strategic Investments (World Bank 2022). Furthermore, the Dutch bank ING is assisting clients establish circular propositions and solid business reasons for them by offering funding and advisory services for CE initiatives (World Bank 2022). Such efforts and assistance could be given to Bangladesh's textile recycling sector. The Bangladesh Ministry of Commerce can probably work with the National Board of Revenue (NBR) and the Bangladesh Bank (BB) to consider potential incentives that may be offered to encourage the textile sector.

Investment in recycling infrastructure is crucial as establishing adequate infrastructure is one of the most critical challenges facing the textile recycling approaches. Infrastructure for recycling, such as sorting premises, collection centers, and recycling facilities, should be developed by government agencies. For instance, the Dutch government has initiated pilot projects aimed at enhancing the current market dynamics that enable increased recycling (World Bank 2022). China has also taken the lead in the world in promoting the recycling of waste materials by establishing eco-industrial parks, in addition to setting goals to reduce resource consumption and eventually close industrial loops that convert outputs from one manufacturer into inputs for another (GPP 2008; CNTAC 2020; Mathews and Tan 2016). Means to fund such projects could be through encouraging public–private partnerships. Besides, recycled textiles require high-quality feedstock, which highlights the necessity of strategically located sorting facilities. The government may increase public access to recycling and encourage corporate involvement by putting in place a robust and effective recycling infrastructure.

Formalization and recognition are necessary to boost decent work and safety. Bangladesh can adopt Thailand's waste management strategy of formal recognition and integration of informal waste workers into the system (Asvanon 2022). This entails recognizing their vital role in waste reduction and recycling, while making sure they have legal protection and benefits. Similar to Thailand's Waste Picker Registration Project (Panitchart 2022), Bangladesh should place a high priority on the informal waste workers' health and safety, including women. Their safety and well-being can be improved by offering them health insurance, identification cards, safety gear, and training courses. Furthermore, promoting the establishment of waste picker cooperatives gives these laborers the collective capacity to bargain for reasonable prices for recyclables.

Strengthening the ties between the formal and informal sectors is vital to facilitate a fair and smooth transition to a circular economy. This can be done by promoting knowledge transfer, capacity building, information sharing, market access, technology adoption, quality enhancement, and social protection. Thailand has assisted the informal garbage collectors through the provision of identification cards, safety gear, health insurance, waste collection facilities, and training courses (Asvanon 2022).

Long-Term Policy Actions:

A National Strategy and Roadmap for Circular Economy transition should be developed for Bangladesh textile industry. Bangladesh developed the National 3R Strategy of Waste Management in 2010 with the goal of encouraging recycling of garbage by requiring waste to be separated at the source, building waste treatment plants, and creating a market for recyclable products (Yousuf 2014). In addition to that, Bangladesh can also gain from developing its own National Strategy and Roadmap for Circular Economy transition that places a clear emphasis on textile waste recycling. Thailand's Sustainable Consumption and Production Roadmap 2017–2036 and its creative Bio-Circular-Green model offer a concrete path for implementing a circular economy that promotes both environmental sustainability and economic growth (SwitchAsia 2017; BCG 2022; MPG 2021). Bangladesh can also learn from Cambodia's National Circular Economy Strategy and Action Plan 2022 (UNDP 2021), which serve as a roadmap for the country's circular economy transition. The Ministry of Commerce, in partnership with industry organizations and other pertinent stakeholders, can design a sector-specific roadmap for circular textiles that includes national indicators for efficient monitoring and assessment of advancement. The circular economy approach will need to be implemented across a number of sectors, and this will require a defined and coherent vision along with an exhaustive action plan.

Designing and implementing a Comprehensive Waste Management Policy Framework for Bangladesh textile sector is fundamental. It can be modeled following Cambodia's 'CE Strategy and Action Plan' or Thailand's 'National Solid Waste Management Master Plan' (UNDP 2021; Stratsea 2023), and then tailored according to the context of Bangladesh. Bangladesh should define clear targets for waste management, develop quantifiable goals for waste reduction and recycling, and highlight the importance of informal waste workers. While different institutions and policymakers adopt best practices from nations that have succeeded, it is crucial for Bangladesh to customize the policies and strategies in compliance with the nation's specific waste management barriers and challenges, waste composition, and demographic dynamics in its textile industry.

Industry specific regulations and guidelines are necessary for the application of LCA and transition to a CE in the textile sector of Bangladesh. Concrete recycling goals are outlined in Vietnam's 'Textile and Garment Industry Development Strategy to 2030, Vision to 2035' (VITAS 2023). Cambodia's 'Circular Economy Strategy and Action Plan' clearly outlines the main obstacles and procedures in switching from a linear economy model to a circular one, With an emphasis on consumption and production, eco-design and raw materials, collection, residual waste, and recycling (UNDP 2021). Thailand is one of the top nations in Asia for its robust Life Cycle Analysis (LCA) capabilities (STEAM Team 2021). Bangladesh can also benefit from building a sector-specific strategy for its own garment industry that focuses on improving its recycling processes and minimizing waste products. The Ministry of Commerce of Bangladesh can devise customized regulations and guidelines to tackle the distinct obstacles and prospects faced by the industry. By adhering to these rules, Bangladesh can improve resource efficiency, minimize waste, and establish sustainable practices.

Investments in Research and Development projects focused on textile products circularity is key for technological improvements in Bangladesh's textile industry. In addition to setting up research facilities and sponsoring research programs and projects, the government can promote cooperation between government agencies, businesses, and academics. More economical and effective techniques for recycling textile products may result from advancements in recycling technologies. Government assistance can be provided here in the form of loans and funding for academic institutions to create courses on innovative circular technologies and methodologies. Moreover, the government may also grant appealing incentives, that meet international standards, and enhance the investment climate, which would magnetize foreign investment into the textile recycling industry and further promote technological and expertise advancement.

8 Conclusion

The textile industry of Bangladesh, while a cornerstone of economic growth, faces critical environmental and social challenges on its path to sustainability. The integration of Life Cycle Analyses (LCA) and Circular Economy (CE) frameworks presents a viable strategy to address these issues, promoting resource efficiency and reducing environmental impact. This paper has highlighted the potential of LCA and CE to foster sustainable development by emphasizing the entire product lifecycle and circular practices. Through a comprehensive synthesis of literature and empirical data, the study underscores the alignment of LCA and CE with the Sustainable Development Goals (SDGs), illustrating their complementary roles in mitigating resource depletion, waste generation, and environmental degradation.

Furthermore, the paper has explored the perspectives of key institutions in Bangladesh, identifying the barriers to LCA and CE adoption, including policy constraints, technical challenges, and limited public participation. By offering insights for policymakers and emphasizing interdisciplinary approaches, the research advocates for policy interventions, capacity building, and stakeholder engagement to strengthen the implementation of LCA and CE.

Ultimately, this study reinforces the importance of LCA and CE in driving Bangladesh's textile industry towards a more sustainable and inclusive future. By aligning with the SDGs, the adoption of these frameworks can ensure a resource-efficient, environmentally sustainable, and socially equitable trajectory for the sector, contributing significantly to the nation's overall sustainable development goals.

Ethics Approval and Consent to Participate

Consent for Publication Not applicable.

Data Availability Statement Not applicable.

List of Abbreviations

ISO	International Organization for Standardization
EoL	End-of-life
LCA	Life Cycle Analysis
CE	Circular Economy
SDG	Sustainable Development Goal
UN	United Nations
SME	Small and medium-sized enterprise
NGO	Non-governmental organization

References

Abbate S, Centobelli P, Cerchione R (2023) From fast to slow: an exploratory analysis of circular business models in the Italian apparel industry. Int J Prod Econ 260:108824. https://doi.org/10.1016/j.ijpe.2023.108824

Ahmed Z, Mahmud S, Acet H (2019) Circular economy model and sustainable development nexus in Bangladesh. https://doi.org/10.1504/IJSSOC.2025.10053223

Asvanon R (2022) Nothing to waste here: Five lessons about informal waste workers to improve urban waste systems. SEI Publications. https://www.sei.org/perspectives/5-lessons-informal-waste-workers-improve-urban-waste-systems/

Azapagic A (2004) Developing a framework for sustainable development indicators for the mining and minerals industry. J Cleaner Prod 12(6):639–662. https://www.sciencedirect.com/science/article/abs/pii/S0959652603000751

Bailey K, Basu A, Sharma S (2022) The environmental impacts of fast fashion on water quality: a systematic review. Water 14(7):1073. https://www.mdpi.com/2073-4441/14/7/1073

Bangladesh Centre for Advanced Studies (BCAS) (n.d.) Sustainable development initiatives. https://www.bcas.net/

Bangladesh Garment Manufacturers and Exporters Association (BGMEA) (2021) Embracing circular economy in Bangladesh apparel industry. The apparel story. https://www.bgmea.com.bd/uploads/newsletters/apparel-story-october-november-2021.pdf

Bangladesh Planning Commission (2018) National sustainable development strategy (NSDS) 2018–2030. Government of Bangladesh

BCG (2022) Bio-circular-green (BCG) economy action plan 2021–2027. https://www.nstda.or.th/en/images/pdf/bcg_action_plan.pdf

Beylot A, Ardente F, Marques A, Mathieux F, Pant R, Sala S, Zampori L (2020) Abiotic and biotic resources impact categories in LCA: development of new approaches. European commission joint research centre. Luxembourg 10:232839. ISBN 9789276172277. https://doi.org/10.2760/232839

Blomsma F, Brennan G (2017) The emergence of circular economy: a new framing around prolonging resource productivity. J Ind Ecol 21(3):603–614. https://onlinelibrary.wiley.com/doi/full/10.1111/jiec.12603

Blomsma F, Bauwens T, Weissbrod I, Kirchherr J (2022) The need for speed: towards circular disruption—what it is, how to make it happen and how to know it's happening. Bus Strategy Environ 32(3):1010–1031. https://onlinelibrary.wiley.com/doi/full/10.1002/bse.3106

BMUV Federal Ministry for the Environment, Nature Conservation, Nuclear Safety and Consumer Protection (2023) Fundamentals of the national circular economy strategy (NKWS): the process of transforming to a circular economy. https://www.bmuv.de/en/download/fundamentals-of-the-national-circular-economy-strategy-nkws#:~:text=The%20German%20government%20is%20drawing

Bocken NM, Short SW, Rana P, Evans S (2014) A literature and practice review to develop sustainable business model archetypes. J Cleaner Prod 65:42–56. https://www.sciencedirect.com/science/article/pii/S0959652613008032

Bossavie L, Cho Y, Heath R (2023) The effects of international scrutiny on manufacturing workers: evidence from the Rana Plaza collapse in Bangladesh. J Dev Econ 163:103107. https://www.sciencedirect.com/science/article/abs/pii/S0304387823000627

Broega AC, Jordão C, Martins SB (2017) Textile sustainability: reuse of clean waste from the textile and apparel industry. IOP Conf Ser: Mater Sci Eng 254(19):192006. https://iopscience.iop.org/article/10.1088/1757-899X/254/19/192006/meta. IOP Publishing

Centre for Policy Dialogue (CPD) (n.d.) Research and advocacy for sustainable development. https://cpd.org.bd/

Chauhan C, Parida V, Dhir A (2022) Linking circular economy and digitalisation technologies: a systematic literature review of past achievements and future promises. Technol Forecast Soc Chang 177:121508

Cicconi P (2020) Eco-design and eco-materials: an interactive and collaborative approach. Sustain Mater Technol 23:e00135. https://www.sciencedirect.com/science/article/abs/pii/S2214993719300442
Circular Economy Network (2021) The third circular economy report. https://circulareconomy.europa.eu/platform/en/knowledge/2020-report-circular-economy-italy-focus-bioeconomy
Circular Economy: Lessons from China. Available at: https://www.nature.com/articles/531440a
Clift R, Doig A, Finnveden G (2000) The application of life cycle assessment to integrated solid waste management: part 1—methodology. Process Saf Environ Prot 78(4):279–287. https://www.sciencedirect.com/science/article/abs/pii/S095758200070886X
Colucci M, Vecchi A (2021) Close the loop: evidence on the implementation of the circular economy from the Italian fashion industry. Bus Strategy the Environ 30(2):856–873. https://onlinelibrary.wiley.com/doi/abs/10.1002/bse.2658
Dahllöf L (2003) Life cycle assessment (LCA) applied in the textile sector: the usefulness, limitations and methodological problems–a literature review. https://core.ac.uk/reader/70570760
Department of Environment (2010) National 3R strategy for waste management. Ministry of environment and forests, government of Bangladesh. https://doe.gov.bd/site/publications/322bf6c9-e23e-401f-864a-73efb078cd29/National-3R-Strategy-for-Waste-Management
Duhoux T, Le Blévennec K, Manshoven S, Grossi F, Arnold M, Mortensen LF (2022) Textiles and the environment—the role of design in Europe's circular economy. European environment agency (EEA)
Eco Intelligent Growth (EIG) (2021) Pioneer denim leads the Bangladesh textile industry towards the circular economy. https://ecointelligentgrowth.net/pioneer-denim-leads-the-bangladesh-textile-industry-towards-the-circular-economy/
E-estonia (2021) Reverse resources: the uber of textile and garment waste. https://e-estonia.com/reverse-resources-the-uber-of-textile-and-garment-waste/
Ellen MacArthur Foundation (n.d.) Fashion and a circular economy. https://www.ellenmacarthurfoundation.org/topics/fashion/overview
Ellen MacArthur Foundation (2017a) A new textiles economy: redesigning fashion's future. https://emf.thirdlight.com/file/24/uiwtaHvud8YIG_uiSTauTlJH74/A%20New%20Textiles%20Economy%3A%20Redesigning%20fashion's%20future.pdf
Ellen MacArthur Foundation (2017b) Infographic, circular economy system diagram. https://www.ellenmacarthurfoundation.org/circular-economy/infographic
Ellen MacArthur Foundation (2019) What is a circular economy? https://www.ellenmacarthurfoundation.org/topics/circular-economy-introduction/overview
Ellen MacArthur Foundation (2021) Toxin-free, recyclable clothing: Napapijri circular series. https://www.ellenmacarthurfoundation.org/circular-examples/napapijri-circular-series
European Commission (2018) Directive (EU) 2018/851 of the European parliament and of the council of 30 May 2018 amending directive 2008/98/EC on waste. Office J Eur Union. http://data.europa.eu/eli/dir/2018/851/oj
European Commission (2022) EU strategy for sustainable and circular textiles. European commission. https://environment.ec.europa.eu/strategy/textiles-strategy_en
European Union (2020) España circular 2030: the new circular economy strategy for a #FuturoSostenible in Spain. European circular economy stakeholder platform. https://circulareconomy.europa.eu/platform/en/strategies/espana-circular-2030-new-circular-economy-strategy-futurosostenible-spain
Finkbeiner M, Inaba A, Tan R, Christiansen K, Klüppel HJ (2006) The new international standards for life cycle assessment: ISO 14040 and ISO 14044. Int J Life Cycle Assess 15(1):86–94. https://link.springer.com/article/10.1065/lca2006.02.002
Finnwatch (2023) Just transition in the garment industry in Bangladesh. https://finnwatch.org/images/pdf/Just_Transition_in_the_Garment_Industry_in_Bangladesh.pdf
Forbes (2021) Leading the way to sustainable fashion. https://www.forbes.com/sites/beximcogroup/2021/09/01/leading-the-way-to-sustainable-fashion/?sh=1fdcceaf7dab

Frayer L (2022) How Bangladesh went from an economic miracle to needing IMF help. NPR. https://www.npr.org/2022/11/09/1134543648/bangladesh-economy-imf-loan

Geissdoerfer M, Savaget P, Bocken NM, Hultink EJ (2017) The circular economy–a new sustainability paradigm?. J Cleaner Prod 143:757–768. https://www.sciencedirect.com/science/article/abs/pii/S0959652616321023

Ghisellini P, Cialani C, Ulgiati S (2016) A review on circular economy: the expected transition to a balanced interplay of environmental and economic systems. J Cleaner Prod 114:11–32. https://www.sciencedirect.com/science/article/abs/pii/S0959652615012287

GIZ (2023) The green button helps with purchasing sustainable textiles. https://www.giz.de/en/worldwide/85043.html

Global Fashion Agenda (GFA) (2023) Esteemed experts convene to map out Cambodia's circular textile future. https://globalfashionagenda.org/news-article/esteemed-experts-convene-to-map-out-cambodias-circular-textile-future/

Government of the Netherlands (n.d.) Circular Dutch economy by 2050—circular economy—Government.nl. Ministry of infrastructure and water management. https://www.government.nl/topics/circular-economy/circular-dutch-economy-by-2050#:~:text=The%20government%2Dwide%20programme%20for

Green Policy Platform (GPP) (2008) China circular economy promotion law. https://www.greenpolicyplatform.org/national-documents/china-circular-economy-promotion-law

Guinée JB, Heijungs R, Huppes G, Zamagni A, Masoni P, Buonamici R, Ekvall T, Rydberg T (2011) Life cycle assessment: past, present, and future. https://doi.org/10.1021/es101316v

Guinée JB (2002) Handbook on life cycle assessment: operational guide to the ISO standards, vol 7. Springer Science and Business Media. https://doi.org/10.1007/0-306-48055-7

Hai HT, Quang ND, Thang NT, Nam NH (2020) Circular economy in Vietnam. Circular economy: global perspective. Springer, Singapore, pp 423–452. https://doi.org/10.1007/978-981-15-1052-6_22

Harsanto B, Primiana I, Sarasi V, Satyakti Y (2023) Sustainability innovation in the textile industry: a systematic review. Sustainability 15(2):1549. https://www.mdpi.com/2071-1050/15/2/1549?uid=a3b3b457bd?uid=ca87b3afb6

Hasan M, Hossain M (2020) Skills and capacity building challenges for sustainability in Bangladesh's textile industry. J Environ Manage 268:110677

Hossain MS, Islam MS, Hossain M (2021) Technical challenges of implementing life cycle assessment in Bangladesh's textile industry. Environ Sci Pollut Res 28(18):22567–22580

Hossain K (2020) Sustainable system transition of Bangladeshi textile industry. https://lutpub.lut.fi/handle/10024/161574

Hossain S (2023) RMG workers' unrest continues, labour ministry urges owners to revise wage proposals. https://www.dhakatribune.com/business/329910/rmg-workers'-unrest-continues-labour-ministry

Huang J (2021) Bangladesh publishes E-waste management rules. Chemlinked. https://chemical.chemlinked.com/news/chemical-news/bangladesh-publishes-e-waste-management-rules

Husain NH (2023) Catalyzing green investments for sustainability: emerging opportunities for circular economy—key note address by Syed Naved Husain. FBCCI (Bangladesh business summit). https://fbcci.org/public/storage/upload/resource/230717085833-2569BEXIMCO_Presentation,%20Keynote%20parallel%208%20by%20Mr.%20Syed%20Naved%20Husain.pdf

IFC (2019) Safety first: Bangladesh garment industry rebounds. https://www.ifc.org/en/stories/2010/bangladesh-garment-industry

International Standard Organization ISO 14040 (2006) Environmental management-life cycle assessment-principles and framework, 2nd edn. Geneva, Switzerland, International Standard Organization

International Standard Organization ISO 14044 (2006) Environmental management-life cycle assessment-principles and framework, 2nd edn. Geneva, Switzerland, International Standard Organization

Islam MR, Ali SM (2018) Technological barriers to sustainable production in Bangladesh's textile industry. Int J Sustain Manufact 2(1):35–47
Islam MT, Jabber PMA (n.d.) Challenges of establishing a circular economy: a case study of readymade garments sector in Bangladesh. DHAKA Univ J Manag. https://managementdu.ac.bd/wp-content/uploads/2022/12/Challenges-of-Establishing-a-Circular-Economy.pdf
IUCN (2021) Kick-off community waste management project in Koh Libong sub-district, Trang province, Thailand. https://www.iucn.org/news/thailand/202109/kick-community-waste-management-project-koh-libong-sub-district-trang-province-thailand
Jia F, Yin S, Chen L, Chen X (2020) The circular economy in the textile and apparel industry: a systematic literature review. J Cleaner Prod 259:120728. https://www.sciencedirect.com/science/article/abs/pii/S0959652620307757
Kazancoglu I, Sagnak M, Kumar Mangla S, Kazancoglu Y (2021) Circular economy and the policy: a framework for improving the corporate environmental management in supply chains. Bus Strategy Environ 30(1):590–608. https://onlinelibrary.wiley.com/doi/full/10.1002/bse.2641
Khan SJ, Badghish S, Kaur P, Sharma R, Dhir A (2023) What motivates the purchasing of green apparel products? A systematic review and future research agenda. Bus Strategy Environ 32(7):4183–4201. https://onlinelibrary.wiley.com/doi/full/10.1002/bse.3360
Kim EJ (2017) Thailand's community-based eco-industrial town development. Global Green Growth Institute. https://www.greenpolicyplatform.org/sites/default/files/downloads/best-practices//GGGI%20Case%20Study_Thailand%27s%20Community%20Based%20Eco-Industrial%20Town%20Devt_June%202017.pdf
Kirchherr J, Reike D, Hekkert M (2017) Conceptualizing the circular economy: an analysis of 114 definitions. Resour, Conserv Recycl 127:221–232. https://www.sciencedirect.com/science/article/pii/S0921344917302835
Konwar M, Boruah RR (2020) Textile industry and its environmental impacts: a review. Indian J Pure Appl Biosci 8:134–139. https://pdfs.semanticscholar.org/234a/e70aefc4b38c6872ed3b1a8c4f47b21b9b77.pdf
Lieder M, Rashid A (2016) Towards circular economy implementation: a comprehensive review in context of manufacturing industry. J Cleaner Prod 115:36–51. https://www.sciencedirect.com/science/article/abs/pii/S0959652615018661
Mahanakorn Partners Group (MPG) (2021) The bio-circular-green economic model. MPG. https://mahanakornpartners.com/the-bio-circular-green-economic-model/
Masum M (2016) The Bangladesh textile-clothing industry: a demand-supply review. Soc Syst Stud 33(5):109–139. https://ritsumei.repo.nii.ac.jp/?action=repository_action_common_download&item_id=3973&item_no=1&attribute_id=22&file_no=1
Mathews JA, Tan H (2016) Circular economy: lessons from China. Nature 531(7595):440–442. https://doi.org/10.1038/531440a
Ministry of Environment, Forest and Climate Change (MoEFCC) (2020) Environmental conservation act and rules. Government of Bangladesh
Ministry of Industry and Trade of the Socialist Republic of Vietnam (MOIT) (n.d.) Vietnam national action plan on sustainable consumption and production (2021–2030). https://www.switch-asia.eu/site/assets/files/2533/national_action_plan_on_scp_vietnam_pdf_pdf.pdf
Mongkolnchaiarunya J (2005) Promoting a community-based solid-waste management initiative in local government: Yala municipality, Thailand. Habitat Int 29(1):27–40. https://doi.org/10.1016/S0197-3975(03)00060-2
Munasinghe P, Druckman A, Dissanayake DGK (2021) A systematic review of the life cycle inventory of clothing. J Cleaner Prod 320:128852. https://www.sciencedirect.com/science/article/pii/S0959652621030481?via=ihub
Muthu SS (2017) Textile and clothing sustainability: recycled and upcycled textiles and fashion. Springer, Berlin/Heidelberg, Germany. ISBN 978-981-10-2146-6. https://doi.org/10.1007/978-981-10-2146-6

National Council on Green Growth (NCGG) (2013) National policy on green growth 2013–2030. Royal Government of Cambodia. https://ncsd.moe.gov.kh/sites/default/files/2019-10/National%20Strategic%20Plan%20on%20Green%20Growth%202013-2030_En.pdf

Nikolina SAJN (2019) Environmental impact of the textile and clothing industry: what consumers need to know. European parliamentary research service (EPRS). https://policycommons.net/artifacts/1335345/environmental-impact-of-the-textile-and-clothing-industry/1941806/

NIST (2022) Circular economy. National institute of standards and technology. Available online at: https://www.nist.gov/circular-economy

Office for Social Responsibility of the China National Textile and Apparel Council (CNTAC) (2020) Make fashion circular: outlook for a new textiles economy in China. https://www.switch-asia.eu/site/assets/files/2628/make_fashion_circular-outlook_for_a_new_textiles_economy-_-20201201.pdf

Oregon State Legislature (2021) HB 3271: establishes carpet stewardship program. https://olis.oregonlegislature.gov/liz/2021R1/Measures/Overview/HB3271

Panda SKBC, Sen K, Mukhopadhyay S (2021) Sustainable pretreatments in textile wet processing. J Cleaner Prod 329:129725. https://www.sciencedirect.com/science/article/abs/pii/S0959652621039020

Panitchart N (2022) Bangkok is ready to register Saleng and receive 12 proposals from the antique buyers association. https://www.thebangkokinsight.com/news/politics-general/politics/929854/

Pheakdey DV, Quan NV, Khanh TD, Xuan TD (2022) Challenges and priorities of municipal solid waste management in Cambodia. Int J Environ Res Public Health 19(14):8458. https://doi.org/10.3390/ijerph19148458

Piribauer B, Bartl A (2019) Textile recycling processes, state of the art and current developments: a mini review. Waste Manage Res 37(2):112–119. https://journals.sagepub.com/doi/full/10.1177/0734242X18819277

Rahman M, Khan MM (2020) Policy and regulatory barriers to sustainable practices in Bangladesh's textile industry. Sustain Sci 15(5):1431–1444

Rainville A (2021) Stimulating a more circular economy through public procurement: roles and dynamics of intermediation. Res Policy 50(4):104193. https://www.sciencedirect.com/science/article/abs/pii/S0048733320302675

Rizos V, Behrens A, Van der Gaast W, Hofman E, Ioannou A, Kafyeke T et al (2016) Implementation of circular economy business models by small and medium-sized enterprises (SMEs): barriers and enablers. Sustainability 8(11):1212. https://doi.org/10.3390/su8111212

Saha K, Dey PK, Papagiannaki E (2022) Implementing circular economy in the textile and clothing industry. In: Supply chain sustainability in small and medium sized enterprises. Routledge. ISBN 9781003018551. https://www.taylorfrancis.com/chapters/edit/10.4324/9781003018551-10/implementing-circular-economy-textile-clothing-industry-krishnendu-saha-prasanta-kumar-dey-eleni-papagiannaki

Salo HH, Suikkanen J, Nissinen A (2020) Eco-innovation motivations and ecodesign tool implementation in companies in the Nordic textile and information technology sectors. Bus Strateg Environ 29(6):2654–2667. https://doi.org/10.1002/bse.2527

Scaling Circularity Report (2021) McKinsey and company and global fashion agenda. https://globalfashionagenda.org/resource/scaling-circularity-report/

Schmidt-Bleek F (1998) The MIPS-concept. Environmental behaviour and quality of life: ecological, sociological, psychological and communicational indicators of sustainability, 27. https://www.academia.edu/download/30933101/Eurokonferenz_1.pdf#page=28

SGS (2023) Profitability and textile durability in the circular economy. https://www.sgs.com/en/news/2023/10/cc-q3-2023-profitability-and-textile-durability-in-the-circular-economy#section

Smith P, Baille J, McHattie LS (2017) Sustainable design futures: an open design vision for the circular economy in fashion and textiles. Des J 20:S1938–S1947. https://doi.org/10.1080/14606925.2017.1352712

Stahel WR (2016) The circular economy. Nature 531(7595):435–438. https://doi.org/10.1038/531435a

STEAM Team (2021) Thailand circular economy ecosystem highlight. STEAM platform. https://www.steamplatform.org/post/thailand-circular-economy-ecosystem-highlight

Stratsea (2023) Thailand's waste crisis and circular economy. https://stratsea.com/thailands-waste-crisis-and-circular-economy/

Subramanian V, Kannan G (2016) Sustainability in the textile industry: a systematic review on metrics, technologies, and future research avenues. J Clean Prod 133:55–72

SwitchAsia (2017) The sustainable consumption and production roadmap 2017–2036. https://www.switch-asia.eu/site/assets/files/2170/th_scp_roadmap.pdf

Textile Today (2018) CYCLO recycled yarn from SIMCO to contribute in closing the loop. https://www.textiletoday.com.bd/cyclo-recycled-yarn-simco-contribute-closing-loop

Tillman AM (2000) Significance of decision-making for LCA methodology. Environ Impact Assess Rev 20(1):113–123. https://doi.org/10.1016/S0195-9255(99)00035-9

Tran CC, Nguyen GTH (2023) The road to circular economy in Vietnam: the policy review of Vietnam green growth strategy. In: Conference on contemporary economic issues in Asian countries. Springer Nature Singapore, Singapore, pp 263–280. https://doi.org/10.1007/978-981-99-0490-7_16

Tukker A (2015) Product services for a resource-efficient and circular economy–a review. J Clean Prod 97:76–91. https://doi.org/10.1016/j.jclepro.2013.11.049

UNCCD (2018) Measuring fashion: environmental impact of the global apparel and footwear industries study. https://library.unccd.int/Details/books/1359

UNCTAD (2014) World investment report 2014: investing in the SDGs: an action plan. https://unctad.org/system/files/official-document/wir2014_en.pdf

UNDP (2021) Circular economy strategy and action plan: building an enabling environment for sustainable development. Kingdom of Cambodia. Available at https://www.undp.org/sites/g/files/zskgke326/files/2022-06/Circular%20Economy%20Strategy%20and%20Action%20Plan%202021%20%282%29.pdf

United Nations (2022) Goal 12: ensure sustainable consumption and production patterns. UN sustainable development goals. https://www.un.org/sustainabledevelopment/sustainable-consumption-production/

Van der Ven C (2022) Circular innovation and ecodesign in the textiles sector. Towards a sustainable and inclusive transition. Sitra Stud. https://www.sitra.fi/app/uploads/2022/09/sitra-circular-innovation-and-ecodesign-in-the-textiles-sector.pdf

VITAS (2023) Vietnam textile and garment industry development strategy to 2030, vision to 2035. http://www.vietnamtextile.org.vn/vietnam-textile-and-garment-industry-development-strategy-to-2030-vis_p1_1-1_2-2_3-686_4-6611.html

Wiedemann SG, Nguyen QV, Clarke SJ (2022) Using LCA and circularity indicators to measure the sustainability of textiles—examples of renewable and non-renewable fibres. Sustainability 14(24):16683. https://doi.org/10.3390/su142416683

Woo S (2023) The circular economy in France: making mendability mainstream. HKTDC. https://research.hktdc.com/en/article/MTU2MTM5MzY5OA

World Bank (2021) Market study for Vietnam: plastics circularity opportunities and barriers. World Bank. https://www.worldbank.org/en/country/vietnam/publication/market-study-for-vietnam-plastics-circularity-opportunities-and-barriers

World Bank (2022) Squaring the circle: policies from Europe's circular economy transition. World Bank. https://www.worldbank.org/en/region/eca/publication/squaring-circle-europe-circular-economy-transition

Yousuf TB (2014) 3R (Reduce, reuse and recycle) in Bangladesh. In: Municipal solid waste management in Asia and the Pacific Islands: environmental science and engineering. Springer Singapore, Singapore. https://doi.org/10.1007/978-981-4451-73-4_4

Life Cycle Assessment (LCA) and Circular Economy (CE) Indicators for Sugarcane Biorefinery Sustainability

Gael Francisco García-Merino and Noé Aguilar-Rivera

Abstract Sustainable development and the goals of the 2030 Agenda for Sustainable Development are robust transdisciplinary frameworks applicable to all activities within the Anthropocene to generate changes at local and global levels increasing population welfare, reducing externalities and dependence on non-renewable raw materials with biomass to improve the performance of socio-ecological systems. For the transition to sustainability, the development of indicators is a fundamental tool for measuring progress towards more sustainable systems. Sugarcane biorefineries have been one of the fastest growing socio-ecological systems, as one of the main agro-industries worldwide, taking as a starting point a highly efficient crop in the conversion of biomass and generating in its processing numerous bioproducts, such as sucrose, bioelectricity, bioethanol, bioplastics and various by-products in large quantities of organic nature with high bioconversion potential. Nevertheless, increasing the sustainability of the conventional sugar industry and cane biorefineries has been a slow process due to the complexity and interaction of socioeconomic, environmental, political, cultural, and technological indicators. Therefore, life cycle analysis in its environmental, cost, and social components and recently the circular economy has generated worldwide indicators that will allow stakeholders and decision-makers to develop technological, social, and public policy actions for more sustainable systems. The objective of this chapter was to use bibliometric mapping to analyze the cases generated globally in sugarcane-producing countries, about the application of life cycle assessment (LCA) and circular economy (CE) to the components of the sugarcane biorefinery (field, transportation, processing, by-product management, society, etc.) for the analysis of indicators and to establish retrospectively and prospectively improvements in the processes to increase the sustainability of this productive system.

Keywords Bioproducts · Cane agribusiness · Research networks · Socio-ecological system

---

G. F. García-Merino · N. Aguilar-Rivera (✉)
Facultad de Ciencias Biológicas y Agropecuarias, Universidad Veracruzana, Córdoba, Veracruz, México
e-mail: naguilar@uv.mx

L. Brandli et al. (eds.), *The Contribution of Life Cycle Analyses and Circular Economy to the Sustainable Development Goals*, World Sustainability Series,
https://doi.org/10.1007/978-3-031-85300-5_16

Abbreviations

CE	Circular economy
GHG	Generates greenhouse gas
LCA	Life cycle assessment
PNPB	National Program of Biodiesel Production and Use
Proálcool	National Alcohol Program
RenovaBio	Brazilian National Biofuel Policy
UK	United Kingdom
US	United States

1 Introduction

The production of sugar cane (*Saccharum officinarum*) and its processing generate a millenary basic food product of high purity, sucrose, which is used worldwide in various industrial applications such as food, pharmaceuticals, chemicals, biotechnology, and biorefineries. However, this product is controversial due to its potential impact on health, the low utilization of its chemical properties as a biomolecule, and the environmental impacts generated from the production of the raw material, the industrial transformation, and the low processing of by-products. These constraints have led to numerous investigations worldwide to identify key points to minimize environmental impacts, and increase sustainability.

Life cycle assessment (LCA) for sugarcane production, and its agroindustry is essential to identify and mitigate environmental impacts along its entire value chain (Khatri and Pandit 2022). By integrating these multidisciplinary analyses with circular economy (CE) principles, it is possible to promote more sustainable, efficient, and resilient production that maximizes the value of resources used and minimizes waste and emissions. Throughout the last decade, the term LCA has enabled a comprehensive analysis of environmental impacts around the world, providing a complete picture of the effects of each stage of the sugarcane life cycle (Hiloidhari et al. 2021). An example in LCA can be seen with the intensive use of fertilizers and pesticides in the vegetative growth stage of the crop, identified as a significant source of soil and water pollution, as well as emissions of nitrogen oxides, a potent greenhouse gas (Huang et al. 2021).

Nonetheless, LCA not only detects problems, but also helps to identify best practices for sustainable production (Ita-Nagy et al. 2020; Katakojwala and Mohan 2020; Yang et al. 2022). An example of this is bagasse, a fibrous by-product of sugar cane, which can be used to generate energy through boiler combustion and cogeneration systems, reducing dependence on fossil fuels and carbon emissions (Zhang et al. 2021). This practice not only optimizes resource use but also improves the energy efficiency of the process. LCA also provides essential quantitative data for informed decision-making by farmers, entrepreneurs, and policymakers (Souza et al. 2021).

These data can lead to policies and strategies that minimize the environmental impact and promote the economic and social sustainability of sugarcane production.

Concerning CE, its goals are developed to maximize the reuse and recycling of materials while minimizing waste (Yang et al. 2023). In this integrative context, it is relevant to have interdisciplinary integration between LCA and CE principles. For example, sugarcane bagasse or trash can be reused not only for energy generation but also for the manufacture of paper, fiberboard, bioplastics, biofuels, and other industrial products with various technologies and levels, closing the cycle of material use and fostering environmental CE (Miranda et al. 2021). LCA can identify valuable by-products that can be recovered and reintroduced into the economic cycle as potential feedstock or energy generators.

Molasses, another by-product of sugarcane processing, can be used in the production of ethanol, a renewable biofuel, contributing to the reduction of greenhouse gas emissions and dependence on fossil fuels (Arshad et al. 2019). Adopting a CE in sugarcane production can also lead to significant reductions in emissions and waste (Gopinath et al. 2018; Karp et al. 2022; de Carvalho et al. 2023). For example, the implementation of composting techniques for organic crop residues (trash) or processing (filter mud, ashes, vinasses) can improve soil quality and reduce the need for chemical fertilizers, thereby reducing soil and water pollution (Sarkar et al. 2020). Finally, CE incentivizes innovation and sustainable development. This interaction between LCA and CE contributes to environmental sustainability but also enhances the long-term economic and social viability of the sugarcane sector.

Nowadays, scientific publication and information on these two major paradigms of the Anthropocene has grown exponentially, however, there is still a low level of application in industrial processes.

To understand what has been published and to organize the growth of scientific production in general, a data-driven approach known as bibliometric mapping was developed. Bibliometric mapping as a quantitative research methodology allows the identification of trends from scientific data in various forms of publications, including scientific articles, book chapters, reviews and conference papers, among other documents, using statistical techniques (Wu et al. 2021). However, this will depend on the databases used, so it is important to select the appropriate database for each analysis individually. This methodology works through the collection of data obtained by keywords provided by authors and allowing the modification of the year, type of publications, and words of concurrence found in bibliographic databases such as Scopus, Dimensions, Lens, PubMed and Web of Science, among others. The use of bibliometric mapping has contributed to various scientific and social activities within the sugarcane sector in the last decade (Figueroa-Rodríguez et al. 2019; Martini et al. 2020; Hasan et al. 2023). The results obtained are generated through different software products such as VOSviewer and CiteSpace that generate visual representations and related information for further analysis.

Such visual representations facilitate the examination of the interactions between keywords, the countries involved in the research, the authors, and even the institutions from which the scientific publications originate. The visual representations of theses interactions are created by clustering algorithms that are mostly branched structures linked with curved or smooth lines, which allow immediate analysis of the interactions by researchers.

Assessing sugarcane LCA and its interaction with CE through bibliometric mapping is crucial to identifying trends, evaluating scientific impact, and detecting collaborations and knowledge gaps. This reveals priority research areas and shows how information has evolved, providing possible policy and strategic decisions.

The analysis is divided into two stages, where the first one details and discusses the analysis of the countries involved in providing scientific publications and information over time, with the use of the keywords sugar cane, LCA, and circular economy. The second stage shows the results and discussion of the concurrence analysis related to the aforementioned keywords.

The data obtained in the two stages of the analysis is extracted from the Scopus database maintained by the scientific publisher Elsevier. This database was selected for its broad coverage and accuracy, as well as its inclusion of peer-reviewed scientific literature and its high level of recorded citations. However, this database may be limited to certain publications such as conference proceedings and peer-reviewed book chapters. Bibliographic network maps are used to visualize the connections between different elements in a research field (Sahni and Kaurav 2023). In these maps, each node represents a country and the size of the node indicates the amount of research that country has produced on the sugarcane agroindustry and its relationship to LCA and CE. The connections between the nodes show how these countries are related to each other in terms of collaboration and citations in research. A smaller distance between two nodes on the map suggests that there is a stronger relationship between them, meaning that countries frequently collaborate or cite each other in their studies.

2 Network of Countries Involved in the Co-authoring of Publications on Life Cycle Assessment of Sugarcane and Its Interaction with the Circular Economy

The first stage of the analysis is divided into two parts and presents a network of a network of countries involved in the co-authoring of publications using VOSviewer where the keywords 'sugarcane and LCA' were used for part (i) and 'sugarcane and circular economy' for part (ii). The search for publications was conducted in May 2024. A total of 441 publications were retrieved for part (i) and 141 for part (ii). At this point, no filtering had been performed. After analysis of the data, the following publications were excluded: editorials, errata, letters, notes, reports, short surveys, conference reviews, and data articles. This resulted in 434 articles for part (i) and 139 articles for part (ii), both of which included different subject areas such as environmental sciences, energy, agriculture and life sciences, social sciences, business, management and accounting, economics, econometrics, and finance. Also, the VOSviewer program was modified to allow a minimum of 1 article per country and 10 bibliographic citations.

2.1 Part 1 of Analysis: Results and Discussion

As a result, in part (i) of the analysis a total of 46 countries were identified, grouped into 8 different categories (Fig. 1a). In this part of the analysis Brazil is the leader with 132 articles and 4059 citations, followed by the United States with 48 articles and 1969 citations, India with 39 articles and 685 citations, the Netherlands with 33 articles and 1397 citations, and Thailand with 33 articles and 987 citations. Four of the world's leading sugarcane producers are among the countries mentioned. This is important because it demonstrates their interest in increasing scientific production on sugarcane Life Cycle Assessment. With this knowledge, environmental impacts can be identified and mitigated throughout the value chain, benefiting everyone involved in agribusiness, from the producer to the manager at the factory or biorefinery (Chauhan et al. 2011).

The networks uncovered show varied behaviors due to the different structures they present. Among the most notable results is Brazil's interaction with the United States (US). Brazil is the main producer of sugar cane in the world and has been able to transform and capitalize a large part of the scientific and social knowledge about the sugar cane agro-industry and biorefineries, thus developing important collaborations with the main economic players in the world (Medina and Costa 2023). In this regard the US considers it important to ensure more efficient, sustainable, and competitive production in all crops where it leads production (Siegner et al. 2018).

Since Brazil is a world leader in the production of ethanol from sugarcane and has shared relevant knowledge with the US, which is interested in diversifying its energy sources and reducing dependence on fossil fuels, while sharing this knowledge opens an important gap to develop its own biofuels industry (Mekonnen et al. 2018; Whistance et al. 2024). Also, this strong interaction may have been most notable because the US is constantly searching for information on successful agricultural practices to improve its production for reasons such as increasing efficiency and productivity, the use of more resistant varieties, and better irrigation and fertilization (Carlisle et al. 2019).

In this regard, Brazil has developed significant technology and innovation, from advanced machinery for sugarcane crop management to biotechnological processes in sugar mills, distilleries, and biorefineries (Carrer et al. 2022; Grandis et al. 2024), which could be adopted and adapted to the demands of US production. Another notable interaction that can be observed in the results of part (i) of the analysis is between India and Thailand, both major sugarcane producers, and who have collaborated on LCA management through the exchange of agricultural technology and sustainable practices (de Matos et al. 2020).

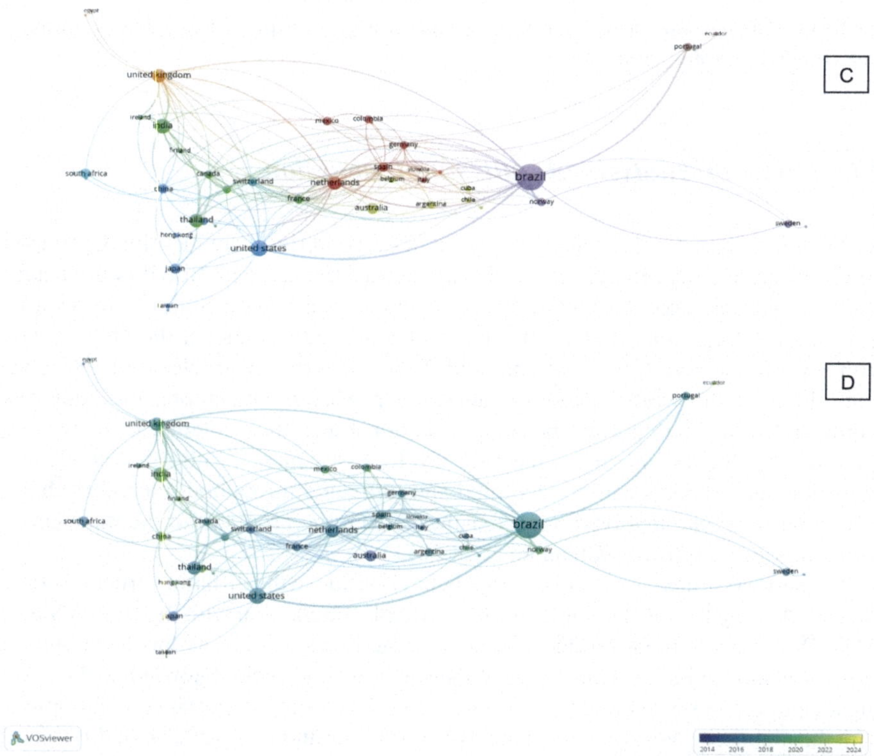

Fig. 1 The international network of countries with publications related to scientific research highlighting the keywords 'sugarcane' and 'life cycle assessment'. **a** The number of articles and **b** by articles published in the last 10 years

Additionally, it is essential to acquire knowledge of resistant varieties, advanced irrigation, and soil management techniques, sustainable practices that reduce emissions, and use by-products to generate energy (Sawaengsak et al. 2021). Sugarcane-producing countries have signed collaboration agreements and participate in international forums to exchange progress and address challenges in sugarcane agribusiness and other crops of interest to all participating nations (Wani and Raju 2021). In this regard they have also collaborated on research and development of biofuels from sugarcane, exchanging technologies and processes for ethanol production and bioenergy generation (Chanthawong and Dhakal 2016; Coelho and Goldemberg 2019). Such interactions have improved their agricultural and technological practices, thus promoting a more efficient and sustainable sugar industry (Sawaengsak et al. 2021; Solomon and Swapna 2022).

In Fig. 1a, b, it can be observed through the period from 2014 to 2024 which countries have been productive in publishing scientific research recently articles related to sugarcane and LCA. It can be observed that China and India, both members of the BRICS economic, political, and social alliance are generating scientific articles on LCA in sugarcane. This is due to its economic importance and the need to improve sustainability and efficiency in agricultural production (Pata 2021). The state-of-the-art, innovations and technologies allow scientists to develop more resistant varieties, optimize the use of resources, and adopt advanced technologies such as biotechnology and precision agriculture. Furthermore, improving sugar cane production is crucial for their global competitiveness and energy diversification, especially in biofuel production (Koizumi 2023). These efforts reflect their commitment to agricultural innovation and sustainability in both countries.

2.2 *Part 2 of Analysis: Results and Discussion*

In part 2 of the analysis of the network of countries involved in the co-authoring of publications on life cycle assessment of sugarcane and its interaction with the circular economy (Fig. 2a). 30 countries were identified, which were grouped into 8 different categories. Brazil tops the list with 31 articles and 477 citations, followed by India with 28 articles and 414 citations, Portugal with 11 articles and 77 citations, Spain with 10 articles and 212 citations, and Thailand with 8 articles and 4 citations complete the Top 5. However, for example Portugal and Spain have published scientific literature related to the keywords 'sugarcane and circular economy', probably since there collaborative research around the world, meaning that scientific research is not limited to the borders of a single country (Pawlak and Kołodziejczak 2020).

Portugal and Spain could collaborate with other sugarcane-producing countries to explore how to integrate sugarcane into the circular economy. Although Portugal and Spain do not produce large quantities of sugarcane, these countries may import products derived from sugarcane (García 2018). They may also be interested in understanding how biofuels can be more efficiently integrated into their circular economies, even if they are not direct producers (Ferreira et al. 2019). Likewise, research on sugarcane and the CE could be related to the search for new trade or investment opportunities in bioeconomy-related sectors.

Another interesting observation is the close interaction between the United Kingdom (UK) and China. This collaborative network could have a background related to the shared interest in developing a more sustainable bioeconomy. Sugar cane is an important feedstock to produce biofuels, bioplastics, and other bio-based products, which are key components of a circular economy (Jesus et al. 2023). Similarly, environmental, and economic policies can be a factor in the relationship because the UK and China have policies and targets related to reducing carbon emissions, mitigating climate change, and promoting sustainable practices, both nations signed the Paris Agreement in 2015 under the United Nations Framework Convention on Climate Change (UNFCCC).

The shared goal is to limit global warming to less than 2 °C, preferably at 1.5 °C, compared to pre-industrial levels (Höhne et al. 2021). In addition, the UK and China have been involved in bilateral and multilateral dialogues on climate change through

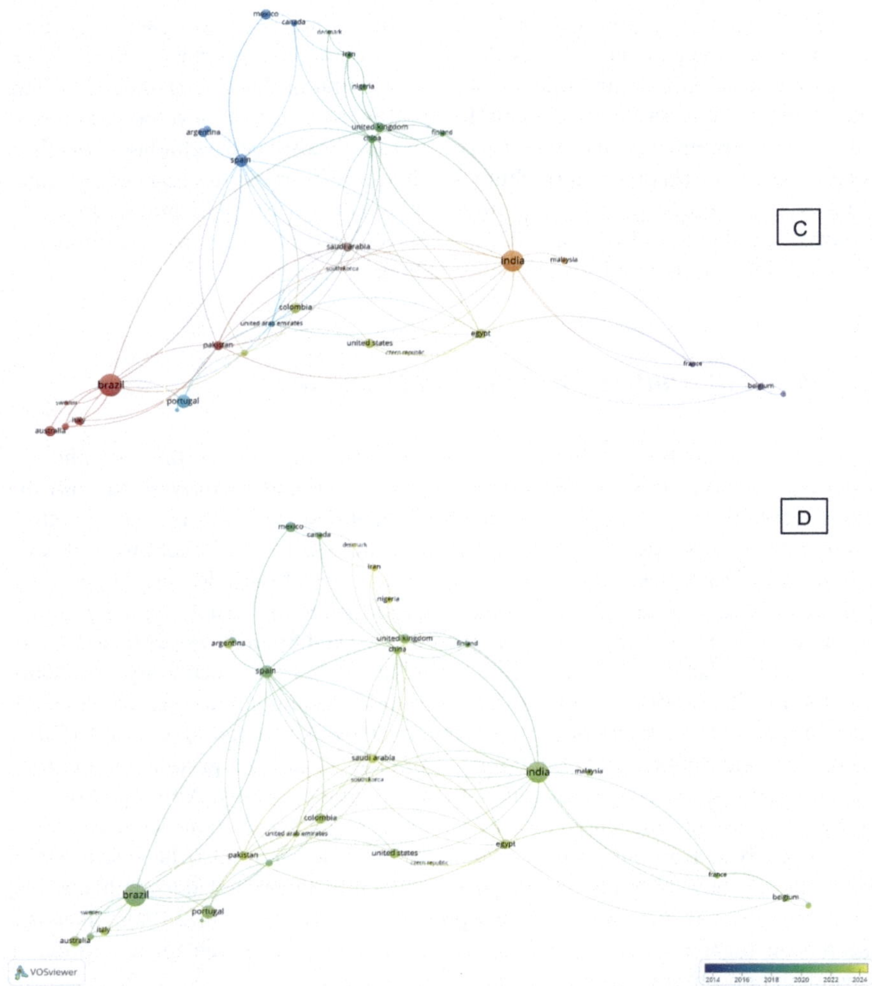

Fig. 2 The international network of countries with publications related to scientific research highlighting the words 'sugarcane' and 'circular economy'. **a** By the number of articles and **b** by articles published in the last 10 years

various platforms, such as the G20 and other international forums. Collaboration in research on sugarcane, LCA, and CE can help both countries achieve these objectives more effectively.

In contrast to the results obtained in part 1 of the analysis, part 2 of the analysis reveals an increase in the number of scientific articles over time. A yellow coloring can be observed that dominates the blue, indicating that in the last 5 years, the different countries involved have been more involved in producing scientific articles on topics that relate to the CE and less in those topics that involve assessing the life cycle of crops (Fig. 2b).

This may be triggered by several reasons. It may be due to the fact that CE provides a more comprehensive and holistic approach to addressing environmental challenges and promoting sustainability in the short to medium term compared to LCA (Yang et al. 2023). This involves looking beyond assessing a product at a specific point in its life cycle, as often done in the LCA of crops, and instead looking at how resources are used, how products are produced, how they are consumed, and how waste is managed in the whole system. Another reason could be that CE implies the need for innovative and technological solutions to close material and resource cycles (Van Fan et al. 2019). Besides the development of new recycling technologies, it is crucial to promote the reuse of products and materials and the implementation of sustainable design practices. These actions are essential to foster a CE, where the value of resources is maximized throughout their life cycle, from production to final disposal. In the context of the LCA, these practices ensure that waste is minimized and valorized, contributing to more efficient and sustainable agricultural production (Formann et al. 2020).

These innovations can not only help reduce the environmental impact of production and consumption but can also boost the economy by creating new business opportunities and jobs in CE-related sectors. It should also be noted that waste management is a key component of CE, and countries are increasingly interested in finding ways to reduce, recycle, and reuse waste to minimize its environmental impact (Meghana and Shastri 2020). developing new recycling technologies, promoting the reuse of products and materials, and implementing sustainable design practices (Formann et al. 2020). These innovations can not only help reduce the environmental impact of production and consumption but can also boost the economy by creating new business opportunities and jobs in CE-related sectors.

The emerging set of solutions requires the implementation of more efficient waste management policies, the promotion of source separation and recycling, and investment in more advanced waste treatment infrastructure (Ungureanu et al. 2022). By focusing on waste management within the broader CE context, countries can more effectively address the challenges associated with waste accumulation and environmental pollution. Finally, another alternative reason could be that the growing market and societal demand for more sustainable products and services is driving interest in the CE. Consumers and businesses are increasingly aware of the environmental impacts of their consumption choices and are looking for more environmentally friendly alternatives (Eggleston et al. 2021). As a result, countries are responding to these demands by focusing their research efforts on areas that promote sustainability, with a resulting increase in the number of scientific articles on related topics, and increased collaboration between countries.

3 Co-authorship Keywords Network Involved in the Evaluation of the Life Cycle in Sugarcane and Its Interaction with the Circular Economy

For the part of the bibliometric analysis focused on co-authorship, the keywords 'sugarcane', 'life cycle assessment', and 'circular economy' were used, applying the criterion that each term had to be repeated at least twice. A set of 32 terms were categorized into five different groups. Each of these groups has a distinctive color for each theme, purple for the direct relationship between CE and LCA, green for ethanol production and derivatives, blue for waste recovery, yellow for greenhouse gas issues, and red for climate change and environmental impact (Fig. 3a).

Three key terms LCA, CE, and recycling emerged within the purple cluster, suggesting a holistic, systemic, and practical approach to sustainability and resource management research. This reflects a growing recognition of the importance of considering multiple aspects and practices related to sustainability to effectively address environmental challenges (Arora et al. 2018). The interaction between these words also suggests a systemic research approach, where synergies between different sustainability-related practices and concepts are considered. This may reflect a deeper understanding of environmental challenges and the need for integrated solutions that address various aspects of the life cycle of products and production systems.

Among the green group are terms that underline the relevance of producing ethanol, derivatives, and by-products. These results reflect that the presence of keywords such as 'ethanol' and 'bioethanol', suggesting a significant interest in the production and use of biofuels. This could reflect a focus on the search for more sustainable and renewable alternatives to fossil fuels in the energy sector, such as sugar cane, which is consistent with the principles of the circular economy. Likewise, the occurrence of the terms 'saccharification' and 'anaerobic digestion' indicates an interest in biomass conversion technologies to produce ethanol and other products from sugar cane.

The referred to technology of saccharification, is a process that converts sugars present in biomass into fermentable glucose, while anaerobic digestion involves the decomposition of organic matter in the absence of oxygen to produce biogas and digested by-products (Guo et al. 2018). These technologies are critical for maximizing yield and efficiency in the production of biofuels from sugarcane, trash or bagasse, and other by-products (Bala and Singh 2019; Kucharska et al. 2020). Also present in this group is the term 'economics', suggesting an interest in the economic considerations associated with the production and utilization of ethanol from sugarcane. This could include aspects such as cost–benefit analysis, financial viability of biofuel production technologies, government incentives, and market policies that affect the demand and price of ethanol (Bastos Lima 2021).

For the blue group, the outstanding terms are related to the valuation of waste produced by the cane agroindustry with a key term that stands out from others which is 'sugarcane bagasse' as fibrous byproduct obtained as a residue from the extraction of sugarcane juice in the sugar or bioethanol manufacturing process (Torgbo et al.

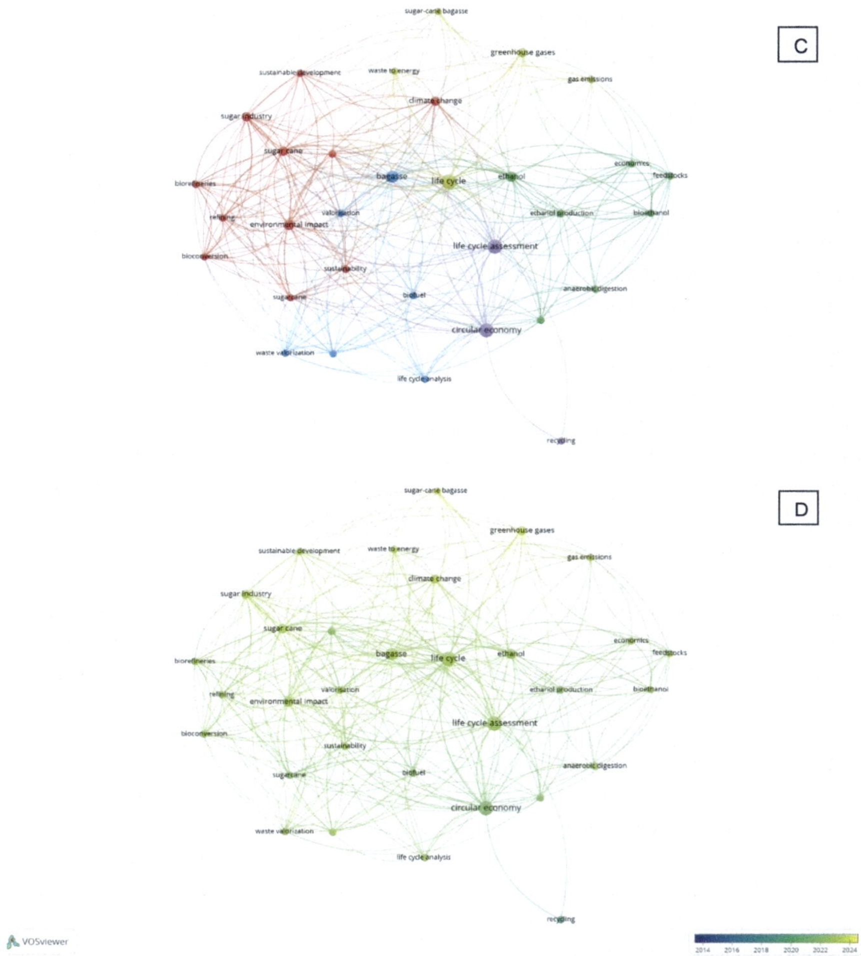

Fig. 3 The bibliometric map shows the concurrence involved in the evaluation of the life cycle in sugar cane and its interaction with the circular economy. **a** Network of separated words by groups with keywords in common, **b** words of concurrence through 2014–2024

2021). This material consists mainly of cellulose, hemicellulose, and lignin fibers, and is presented in the form of a fibrous pulp that remains after squeezing the cane juice (Kumar et al. 2021). The bagasse has multiple applications and is a valuable resource in the context of the CE because one of the most common applications is solid biofuel for the generation of energy in sugar or ethanol processing plants (Tony 2020). Through the combustion of bagasse in boilers by cogeneration technology, steam can be produced and used to generate electricity and heat, which allows industrial facilities to have more self-sufficient energy, reducing their dependence on non-renewable energy sources, and sell energy to the public grid (Kabeyi and Olanrewaju (2023).

Through anaerobic digestion or fermentation processes, the cellulose fibers present in bagasse can be decomposed to produce biogas, which can be used as fuel, or bioethanol as an alternative fuel in transportation (Tony 2020; Fioranelli and Bizzo 2023). In addition to its value as a source of energy and biofuels, bagasse can also be used as a raw material to produce paper, cardboard, fiberboard, construction materials, and bioplastics (Azmin and Nor 2020; Sharma et al. 2023). These biomaterials can help reduce dependence on non-renewable resources and contribute to the transition towards a more circular and sustainable economy.

In the yellow group, issues related to air pollution stand out, terms such as 'greenhouse gas emissions' and 'energy waste' are directly involved with LCA, suggesting that several researchers are analyzing strategies to mitigate these problems partially generated by the production of sugarcane and its by-products, such as sugar and ethanol. Thereby, the sugarcane agroindustry generates greenhouse gas (GHG) emissions, such as carbon dioxide (CO_2), methane (CH_4), and nitrous oxide (N_2O) (Renouf 2016). These emissions can occur at various stages of the sugarcane life cycle, including land preparation, planting, cultivation, harvesting, burning of cane fields, processing, and transportation in addition to environmental factors (Vasconcelos et al. 2018). Sugarcane LCA analyzes the amount and type of GHG emissions associated with each of these stages, as well as their contribution to climate change.

Likewise, during the processing of sugar cane for sugar extraction or ethanol production, waste is generated mainly by filter mud, bagasse, and vinasse (Hoarau et al. 2018). These wastes are composed of organic matter and nutrients; if not managed properly, they can generate greenhouse gas emissions, such as methane, during their decomposition (Dos Reis et al. 2015). Improper management of these wastes can have negative impacts on water and soil quality (Ogura et al. 2022).

For the last group highlighted in red, information about network connections was established with key terms related to climate change and environmental impact. In this group more keywords were obtained compared to the previous ones, these words range from 'biorefinery', 'sugar industry', 'bioconversion' and 'sustainability', leading to the terms 'climate change' and 'environmental impact'. In this context, biorefineries are facilities that apply bioconversion processes to transform sugarcane biomass into a variety of bioproducts, such as biofuels, chemicals, and materials (Freitas et al. 2021). Thereby, bioconversion involves a series of biochemical and thermochemical processes, such as fermentation, hydrolysis, and gasification, to break down biomass into basic components and convert them into value-added products (Perez et al. 2020).

The integration of processes in biorefineries allows for maximizing efficiency and versatility in the use of cane as raw material. The sugar industry is an integral part of the regional economy and plays an important role in the transition to a more sustainable and circular economy (Jesus et al. 2023). Sustainability in this context covers environmental, social, and economic aspects, and implies the adoption of practices and technologies that minimize the use of natural resources, reduce greenhouse gas emissions, and promote social equity in the supply chain (Aguilar-Rivera 2022).

Moreover, the implementation of environmental management systems, such as ISO 14001 certification, and the adoption of sustainable agricultural practices, such as conservation agriculture and agroecological pest management, are some strategies used to improve sustainability in the sugar industry (Chouhan et al. 2022). Finally, climate change and its effects on the environment are central concerns in sugarcane LCA. Greenhouse gas emissions associated with sugarcane production and processing contribute to global warming and climate change. In addition, the expansion of sugarcane agribusiness can lead to deforestation and degradation of natural ecosystems, which has negative impacts on biodiversity and ecosystem services (Postal et al. 2020).

Figure 3b shows the issues related to sugarcane, LCA, and CE that have been present in the last 5 years, the bibliometric map highlights the trend towards the yellow color which indicates closer proximity to 2024, which is due to several key factors. One is sustainability in the agribusinesses of the different nations. Due to the growing concern about climate change, this has prompted agribusinesses to search for ways to reduce their environmental impact. Resource efficiency is also a key factor because LCA allows the environmental impact of a product to be assessed from production to disposal. This allows for the identification of areas where efficiency must be improved and waste can be reduced, promoting a CE where resources are reused and recycled instead of discarded. Another reason why scientific articles are increasingly focusing on issues related to LCA and CE is the policies and regulations of countries. Many countries are implementing more stringent policies and regulations on carbon emissions and waste management.

Through these policies, the use of more sustainable practices and the adoption of clean technologies are encouraged (Ita-Nagy et al. 2020; Yang et al. 2022). As discussed above Brazil is an excellent example of how research and implementation of sustainable practices in sugarcane production have been highly relevant in the last 5 years, as the ethanol industry in Brazil is an example of CE and sustainability. Various investigations in Brazil have shown that sugarcane ethanol has a positive carbon balance, meaning that it reduces greenhouse gas emissions compared to traditional fossil fuels (Bordonal et al. 2018; Lefebvre et al. 2020). The Brazilian government has strongly supported the ethanol industry through policies such as the National Program of Biodiesel Production and Use (PNPB), the National Alcohol Program (Proálcool) and recently the Brazilian National Biofuel Policy (RenovaBio) for bioethanol, biodiesel, biomethane, biokerosene for aviation and others. These initiatives have promoted research and development of new technologies related to sugarcane and have contributed to positioning Brazil as a leader in biofuels from sugarcane, with a resulting increase in scientific articles on this topic.

It should be noted that the analysis undertaken is based solely on the amount of scientific information available and does not fully reflect the research efforts in each country. It is important to note that this quantitative approach may omit crucial qualitative aspects. For example, the quality and impact of the research analysed have not been considered in this ranking. In addition, there may be research of high value and relevance that has not been widely published or is not available in the databases used. Therefore, although the analysis provides an overview of scientific productivity, it should not be interpreted as an absolute measure of the research effort or capacity of each country. It is essential to complement this type of analysis with qualitative assessments to obtain a completer and more accurate picture of the scientific research landscape at a global level.

4 Conclusions

It is possible that in the short term a greater interest in sugarcane crop research will be experienced, due to the restructuring of its value chain and the emergence of bioproducts derived from its by-products. This is not only for sugar production but also focused on the creation of biofuels and new biobased economy. The interaction with the different stages of LCA and the multifactorial approach of CE will make this crop continue to have a higher priority than other crops. Although the analysis presented in this chapter found that there is currently more scientific interest in aspects of CE compared to those of sugarcane LCA, it must be kept in mind that crop LCA will remain relevant and necessary to understand the specific environmental impacts of agricultural production and to inform evidence-based decision-making. CE has become an increasingly important focus on the global agenda, with greater recognition of the need to minimize waste and maximize resource efficiency in all sectors, including agriculture. Furthermore, CE provides a broader and more holistic framework for addressing sustainability challenges, which may be attractive to researchers, policymakers, and companies pursuing integrated and sustainable solutions. Therefore, both approaches are likely to remain important areas of research in the future, with further integration and collaboration between them to address sustainability challenges more effectively in the sugarcane agribusiness.

References

Aguilar-Rivera N (2022) Bioindicators for the sustainability of sugar agro-industry. Sugar Tech 24(3):651–661. https://doi.org/10.1007/s12355-021-01105-z

Arora NK, Fatima T, Mishra I, Verma M, Mishra J, Mishra V (2018) Environmental sustainability: challenges and viable solutions. Environ Sustain 1:309–340. https://doi.org/10.1007/s42398-018-00038-w

Arshad M, Abbas M, Iqbal M (2019) Ethanol production from molasses: environmental and socioeconomic prospects in Pakistan: feasibility and economic analysis. Environ Technol Innov 14:100317. https://doi.org/10.1016/j.eti.2019.100317

Azmin SNHM, Nor MSM (2020) Development and characterization of food packaging bioplastic film from cocoa pod husk cellulose incorporated with sugarcane bagasse fibre. J Bioresour Bioprod 5(4):248–255. https://doi.org/10.1016/j.jobab.2020.10.003

Bala A, Singh B (2019) Development of an environmental-benign process for efficient pretreatment and saccharification of Saccharum biomasses for bioethanol production. Renew Energy 130:12–24. https://doi.org/10.1016/j.renene.2018.06.033

Bastos Lima MG (2021) Bioeconomy lessons from biofuel policies in emerging countries. In: The politics of bioeconomy and sustainability. Springer, Cham. https://doi.org/10.1007/978-3-030-66838-9_8

Bordonal RDO, Carvalho JLN, Lal R, De Figueiredo EB, De Oliveira BG, La Scala N (2018) Sustainability of sugarcane production in Brazil. A review. Agron Sustain Dev 38:1–23. https://doi.org/10.1007/s13593-018-0490-x

Carlisle L, De Wit MM, DeLonge MS, Calo A, Getz C, Ory J, Press D (2019) Securing the future of US agriculture: the case for investing in new entry sustainable farmers. Elem Sci Anth 7:17. https://doi.org/10.1525/elementa.356

Carrer MJ, de Souza Filho HM, Vinholis MDMB, Mozambani CI (2022) Precision agriculture adoption and technical efficiency: an analysis of sugarcane farms in Brazil. Technol Forecast Soc Chang 177:121510. https://doi.org/10.1016/j.techfore.2022.121510

Chanthawong A, Dhakal S (2016) Liquid biofuels development in Southeast Asian countries: an analysis of market, policies and challenges. Waste Biomass Valor 7:157–173. https://doi.org/10.1007/s12649-015-9433-9

Chauhan MK, Chaudhary S, Kumar S (2011) Life cycle assessment of sugar industry: a review. Renew Sustain Energy Rev 15(7):3445–3453. https://doi.org/10.1016/j.rser.2011.04.033

Chouhan VK, Khan SH, Hajiaghaei-Keshteli M (2022) Sustainable planning and decision-making model for sugarcane mills considering environmental issues. J Environ Manage 303:114252. https://doi.org/10.1016/j.jenvman.2021.114252

Coelho ST, Goldemberg J. (2019). Sustainability and environmental impacts of sugarcane biofuels. Sugarcane biofuels: status, potential, and prospects of the sweet crop to fuel the world, pp 409–444. https://doi.org/10.1007/978-3-030-18597-8_18

de Carvalho JC, de Souza Vandenberghe LP, Sydney EB, Karp SG, Magalhães AI Jr, Martinez-Burgos WJ, Soccol CR (2023) Biomethane production from sugarcane vinasse in a circular economy: developments and innovations. Fermentation 9(4):349. https://doi.org/10.3390/fermentation9040349

de Matos M, Santos F, Eichler P (2020) Sugarcane world scenario. In: Sugarcane biorefinery, technology and perspectives, pp 1–19. https://doi.org/10.1016/B978-0-12-814236-3.00001-9

Dos Reis CM, Carosia MF, Sakamoto IK, Varesche MBA, Silva EL (2015) Evaluation of hydrogen and methane production from sugarcane vinasse in an anaerobic fluidized bed reactor. Int J Hydrogen Energy 40(27):8498–8509. https://doi.org/10.1016/j.ijhydene.2015.04.136

Eggleston G, Aita G, Triplett A (2021) Circular sustainability of sugarcane: natural, nutritious, and functional unrefined sweeteners that meet new consumer demands. Sugar Tech 23(5):964–973. https://doi.org/10.1007/s12355-021-00994-4

Ferreira IDA, de Castro Fraga M, Godina R, Souto Barreiros M, Carvalho H (2019) A proposed index of the implementation and maturity of circular economy practices the case of the pulp and paper industries of Portugal and Spain. Sustainability 11(6):1722. https://doi.org/10.3390/su11061722

Figueroa-Rodríguez KA, Hernández-Rosas F, Figueroa-Sandoval B, Velasco-Velasco J, Aguilar RN (2019) What has been the focus of sugarcane research? A bibliometric overview. Int J Environ Res Public Health 16(18):3326. https://doi.org/10.3390/ijerph16183326

Fioranelli A, Bizzo WA (2023) Generation of surplus electricity in sugarcane mills from sugarcane bagasse and straw: challenges, failures and opportunities. Renew Sustain Energy Rev 186:113647. https://doi.org/10.1016/j.rser.2023.113647

Formann S, Hahn A, Janke L, Stinner W, Sträuber H, Logroño W, Nikolausz M (2020) Beyond sugar and ethanol production: value generation opportunities through sugarcane residues. Front Energy Res 8:579577. https://doi.org/10.3389/fenrg.2020.579577

Freitas JV, Bilatto S, Squinca P, Pinto AS, Brondi MG, Bondancia TJ, Farinas CS (2021) Sugarcane biorefineries: potential opportunities towards shifting from wastes to products. Ind Crops Prod 172:114057. https://doi.org/10.1016/j.indcrop.2021.114057

García AF (2018) Commercial crop or plantation system? Sugar cane production from the Mediterranean to the Atlantic. In: From Al-Andalus to the Americas (13th–17th centuries), pp 301–331. https://doi.org/10.1163/9789004365773_011

Gopinath A, Bahurudeen A, Appari S, Nanthagopalan P (2018) A circular framework for the valorisation of sugar industry wastes: review on the industrial symbiosis between sugar, construction and energy industries. J Clean Prod 203:89–108. https://doi.org/10.1016/j.jclepro.2018.08.252

Grandis A, Fortirer JS, Navarro BV, de Oliveira LP, Buckeridge MS (2024) Biotechnologies to improve sugarcane productivity in a climate change scenario. Bioenergy Res 17(1):1–26. https://doi.org/10.1007/s12155-023-10649-9

Guo H, Chang Y, Lee DJ (2018) Enzymatic saccharification of lignocellulosic biorefinery: research focuses. Biores Technol 252:198–215. https://doi.org/10.1016/j.biortech.2017.12.062

Hasan M, Abedin MZ, Amin MB, Nekmahmud M, Oláh J (2023) Sustainable biofuel economy: a mapping through bibliometric research. J Environ Manage 336:117644. https://doi.org/10.1016/j.jenvman.2023.117644

Hiloidhari M, Banerjee R, Rao AB (2021) Life cycle assessment of sugar and electricity production under different sugarcane cultivation and cogeneration scenarios in India. J Clean Prod 290:125170. https://doi.org/10.1016/j.jclepro.2020.125170

Hoarau J, Caro Y, Grondin I, Petit T (2018) Sugarcane vinasse processing: toward a status shift from waste to valuable resource. A review. J Water Process Eng 24:11–25. https://doi.org/10.1016/j.jwpe.2018.05.003

Höhne N, Kuramochi T, Warnecke C, Röser F, Fekete H, Hagemann M, Gonzales S (2021) The Paris agreement: resolving the inconsistency between global goals and national contributions. In: Climate policy after the 2015 Paris climate conference, pp 30–46. eBook ISBN 9781003191582

Huang W, Lu Y, Chen L, Sun D, An Y (2021) Impact of pesticide/fertilizer mixtures on the rhizosphere microbial community of field-grown sugarcane. 3 Biotech 11:1–10. https://doi.org/10.1007/s13205-021-02770-3

Ita-Nagy D, Vázquez-Rowe I, Kahhat R, Quispe I, Chinga-Carrasco G, Clauser NM, Area MC (2020) Life cycle assessment of bagasse fiber reinforced biocomposites. Sci Total Environ 720:137586. https://doi.org/10.1016/j.scitotenv.2020.137586

Jesus GMK, Jugend D, Paes LAB, Siqueira RM, Leandrin MA (2023) Barriers to the adoption of the circular economy in the Brazilian sugarcane ethanol sector. Clean Technol Environ Policy 25(2):381–395. https://doi.org/10.1007/s10098-021-02129-5

Kabeyi MJB, Olanrewaju OA (2023) Bagasse electricity potential of conventional sugarcane factories. J Energy. https://doi.org/10.1155/2023/5749122

Karp SG, Burgos WJ, Vandenberghe LP, Diestra KV, Torres LA, Woiciechowski AL, Soccol CR (2022) Sugarcane: a promising source of green carbon in the circular bioeconomy. Sugar Tech 24(4):1230–1245. https://doi.org/10.1007/s12355-022-01161-z

Katakojwala R, Mohan SV (2020) Microcrystalline cellulose production from sugarcane bagasse: Sustainable process development and life cycle assessment. J Clean Prod 249:119342. https://doi.org/10.1016/j.jclepro.2019.119342

Khatri P, Pandit AB (2022) Systematic review of life cycle assessments applied to sugarcane bagasse utilization alternatives. Biomass Bioenerg 158:106365. https://doi.org/10.1016/j.biombioe.2022.106365

Koizumi T (2023) World biofuel continuum: issues and challenges. Environ Sustain Biofuels 69–85. https://doi.org/10.1016/B978-0-323-91159-7.00009-6

Kucharska K, Słupek E, Cieśliński H, Kamiński M (2020) Advantageous conditions of saccharification of lignocellulosic biomass for biofuels generation via fermentation processes. Chem Pap 74:1199–1209. https://doi.org/10.1007/s11696-019-00960-1

Kumar A, Kumar V, Singh B (2021) Cellulosic and hemicellulosic fractions of sugarcane bagasse: potential, challenges and future perspective. Int J Biol Macromol 169:564–582. https://doi.org/10.1016/j.ijbiomac.2020.12.175

Lefebvre D, Williams A, Meersmans J, Kirk GJ, Sohi S, Goglio P, Smith P (2020) Modelling the potential for soil carbon sequestration using biochar from sugarcane residues in Brazil. Sci Rep 10(1):19479. https://doi.org/10.1038/s41598-020-76470-y

Martini AF, Valani GP, Boschi RS, Bovi RC, da Silva LFS, Cooper M (2020) Is soil quality a concern in sugarcane cultivation? A bibliometric review. Soil Tillage Res 204:104751. https://doi.org/10.1016/j.still.2020.104751

Medina GDS, Costa RBD (2023) Building agro-industrial capabilities in the sugarcane supply chain in Brazil. Logistics 7(4):71. https://doi.org/10.3390/logistics7040071

Meghana M, Shastri Y (2020) Sustainable valorization of sugar industry waste: status, opportunities, and challenges. Biores Technol 303:122929. https://doi.org/10.1016/j.biortech.2020.122929

Mekonnen MM, Romanelli TL, Ray C, Hoekstra AY, Liska AJ, Neale CM (2018) Water, energy, and carbon footprints of bioethanol from the US and Brazil. Environ Sci Technol 52(24):14508–14518. https://doi.org/10.1021/acs.est.8b03359

Miranda NT, Motta IL, Maciel Filho R, Maciel MRW (2021) Sugarcane bagasse pyrolysis: a review of operating conditions and products properties. Renew Sustain Energy Rev 149:111394. https://doi.org/10.1016/j.rser.2021.111394

Ogura AP, Moreira RA, da Silva LCM, Negro GS, Freitas JS, da Silva Pinto TJ, Espíndola ELG (2022) Irrigation with water contaminated by sugarcane pesticides and vinasse can inhibit seed germination and crops initial growth. Arch Environ Contam Toxicol 82(3):330–340. https://doi.org/10.1007/s00244-022-00914-x

Pata UK (2021) Linking renewable energy, globalization, agriculture, CO_2 emissions and ecological footprint in BRIC countries: a sustainability perspective. Renew Energy 173:197–208. https://doi.org/10.1016/j.renene.2021.03.125

Pawlak K, Kołodziejczak M (2020) The role of agriculture in ensuring food security in developing countries: Considerations in the context of the problem of sustainable food production. Sustainability 12(13):5488. https://doi.org/10.3390/su12135488

Perez VH, da Silva NRF, Junior EGS, Rocha DC, Justo OR, David GF, Garcia-Perez M (2020) Integrated process of biomass thermochemical conversion to obtain pyrolytic sugars for biofuels and bioproducts. In: Lignocellulosic biorefining technologies, pp 285–311. https://doi.org/10.1002/9781119568858.ch13

Postal AM, Kamali FP, Asveld L, Osseweijer P, da Silveira JMF (2020) The impact of sugarcane expansion in Brazil: local stakeholders' perceptions. J Rural Stud 73:147–162. https://doi.org/10.1016/j.jrurstud.2019.10.041

Renouf MA (2016) Greenhouse gas abatement from sugarcane bioenergy, biofuels, and biomaterials. In: Sugarcane-based biofuels and bioproducts, pp 333–362. https://doi.org/10.1002/9781118719862.ch13

Sahni S, Kaurav RPS (2023). What? Why? When? How? Where? of technology-based bibliometric review. In: Advancing methodologies of conducting literature review in management domain, vol 2, pp 79–101. https://doi.org/10.1108/S2754-586520230000002005

Sarkar S, Skalicky M, Hossain A, Brestic M, Saha S, Garai S, Brahmachari K (2020) Management of crop residues for improving input use efficiency and agricultural sustainability. Sustainability 12(23):9808. https://doi.org/10.3390/su12239808

Sawaengsak W, Prasara-A J, Gheewala SH (2021) Assessing the socio-economic sustainability of sugarcane harvesting in Thailand. Sugar Tech 23(2):263–277. https://doi.org/10.1007/s12355-020-00888-x

Sharma D, Agrawal S, Nagpal R, Mishra OP, Bhardwaj N, Mahajan R (2023) Production of eco-friendly and better-quality sugarcane bagasse paper using crude xylanase and pectinase biopulping strategy. 3 Biotech 13(2):61. https://doi.org/10.1007/s13205-023-03484-4

Siegner A, Sowerwine J, Acey C (2018) Does urban agriculture improve food security? Examining the nexus of food access and distribution of urban produced foods in the United States: a systematic review. Sustainability 10(9):2988. https://doi.org/10.3390/su10092988

Solomon S, Swapna M (2022) Indian sugar industry: towards self-reliance for sustainability. Sugar Tech 24(3):630–650. https://doi.org/10.1007/s12355-022-01123-5

Souza A, Watanabe MD, Cavalett O, Cunha M, Ugaya CM, Bonomi A (2021) A novel social life cycle assessment method for determining workers' human development: a case study of the sugarcane biorefineries in Brazil. Int J Life Cycle Assess 26:2072–2084. https://doi.org/10.1007/s11367-021-01936-8

Tony MA (2020) From biomass residue to solar thermal energy: the potential of bagasse as a heat storage material. Euro-Mediterr J Environ Integr 5(1):17. https://doi.org/10.1007/s41207-020-00158-y

Torgbo S, Quan VM, Sukyai P (2021) Cellulosic value-added products from sugarcane bagasse. Cellulose 28(9):5219–5240. https://doi.org/10.1007/s10570-021-03918-3

Ungureanu N, Vlăduț V, Biriş SŞ (2022) Sustainable valorization of waste and by-products from sugarcane processing. Sustainability 14(17):11089. https://doi.org/10.3390/su141711089

Van Fan Y, Lee CT, Lim JS, Klemeš JJ, Le PTK (2019) Cross-disciplinary approaches towards smart, resilient and sustainable circular economy. J Clean Prod 232:1482–1491. https://doi.org/10.1016/j.jclepro.2019.05.266

Vasconcelos ALS, Cherubin MR, Feigl BJ, Cerri CE, Gmach MR, Siqueira-Neto M (2018) Greenhouse gas emission responses to sugarcane straw removal. Biomass Bioenerg 113:15–21. https://doi.org/10.1016/j.biombioe.2018.03.002

Wani SP, Raju KV (2021) Success stories from scaling-up initiatives with State Governments and Corporate in India, China and Thailand. In: Scaling-up solutions for farmers: technology, partnerships and convergence, pp 415–479. https://doi.org/10.1007/978-3-030-77935-1_13

Whistance J, Debnath D, Thompson W (2024) A biofuels outlook for the US and Brazil and what it might mean for sustainable development goal 7. Front Chem Eng 5:1290763. https://doi.org/10.3389/fceng.2023.1290763

Wu M, Long R, Bai Y, Chen H (2021) Knowledge mapping analysis of international research on environmental communication using bibliometrics. J Environ Manage 298:113475. https://doi.org/10.1016/j.jenvman.2021.113475

Yang L, Zhou Y, Meng B, Li H, Zhan J, Xiong H, Zhang F (2022) Reconciling productivity, profitability and sustainability of small-holder sugarcane farms: a combined life cycle and data envelopment analysis. Agric Syst 199:103392. https://doi.org/10.1016/j.agsy.2022.103392

Yang M, Chen L, Wang J, Msigwa G, Osman AI, Fawzy S, Yap PS (2023) Circular economy strategies for combating climate change and other environmental issues. Environ Chem Lett 21(1):55–80. https://doi.org/10.1007/s10311-022-01499-6

Zhang J, Li G, Borrion A (2021) Life cycle assessment of electricity generation from sugarcane bagasse hydrochar produced by microwave assisted hydrothermal carbonization. J Clean Prod 291:125980. https://doi.org/10.1016/j.jclepro.2021.125980

Life Cycle Assessment for Environmental Sustainability of the Production of Olive Seedlings at an Industrial Nursery

Giovanni Lanzarini, Adriane Lawisch Rodriguez, Rosana de Cássia de Souza Schneider, and Ênio Leandro Machado

Abstract Olive-growing has received particular interest in several regions of the world, including Brazil. There is a "new wave" of developments in this field, especially to improve at all stages of cultivation and processing the concepts of circular economy. This growth is also reflected in a crucial stage of the olive production chain: the nursery activity of seedlings. In the present research, Life Cycle Assessment (LCA) was used as a tool to identify the environmental impacts and establish prognostic scenarios in the context of clean technology in regard to two important aspects related to the production of seedlings in an industrial nursery in southern Brazil: the use of water for irrigation and the of packages for seedlings. The olive grove that is the focus of this research has a reference of four years of production of the Arbequina, Picual, Koroneiki and Frantoio varieties. It is observed that the orchard used as the basis of this study has characteristics representative of orchards in Rio Grande do Sul, falling within the classification of intensive cultivation, with high density without irrigation. In this study, the main hot spots for LCA were researched considering the recommendations of ISOs 14040 and 14044. The LCA methodology was applied using SimaPro 8.0.4 software, database Ecoinvent 3.1 and Recipe methods with Mid and Endpoints. Through analysis with Characterization, Normalization and Damage Assessment procedures, the main impacts were verified. The Scenario consisted of polyethylene pots with recycling and reuse of wastewater from the nurseries of the olive trees, and a functional unit considered as the production of 1000 units of seedling or 4 ha of planted area with 250 seedlings produced. Marine ecotoxicity was 0.84 kg 1,4-DBeq and freshwater ecotoxicity was 0.83 kg Peq. In regards to Damage Assessment, the impact on Human Health was 4.4×10^{-8} DALY and 1.72×10^{-10} species year^{-1} for Ecosystems. The impact on Resources was 2×10^{-3} \$. With two scenarios proposed as Prognostics, the use of biocontainers reduces the impacts on Human Health and Ecosystems by approximately 80% and Resources by 100%, constituting a focus on innovation for the constant improvement of environmental management of olive trees nurseries. Constituting

G. Lanzarini · A. L. Rodriguez · R. de Cássia de Souza Schneider · Ê. L. Machado (✉)
Graduation Program in Environmental Technology, University of Santa Cruz do Sul (UNISC), Santa Cruz do Sul, RS, Brazil
e-mail: enio@unisc.br

L. Brandli et al. (eds.), *The Contribution of Life Cycle Analyses and Circular Economy to the Sustainable Development Goals*, World Sustainability Series,
https://doi.org/10.1007/978-3-031-85300-5_17

focus on innovation for the constant improvement of environmental management in the production of olive tree nurseries. As with other crops, the material balance for agricultural production includes, on the one hand, the introduction of nutrients necessary for production; and, on the other, the depletion of these by the harvesting of production. In the case of olive growing, around 85% of production turns into pomace from separation with great potential for replacing the nutrients needed for production. This aspect of environmental sustainability can be added to actions that have their carbon footprint and water footprint accounted for, using the LCA as a methodological reference.

Keywords Life cycle assessment · Life cycle inventory · Olive agriculture · Seedlings · Nursery

1 Introduction

Unlike Europe, where olive-growing has occurred for thousands of years, olive cultivation in the Americas is still incipient. Argentina is currently the largest producer of olive oil in the Americas, followed by the United States. In these countries, there was a complete development of the culture with approximately 105,000 ha of cultivated land and 130 factories of olive products (IOC 2016; Sanchez et al. 2015). According to Coutinho et al. (2009), the olive tree, *Olea europaea* L., was introduced in America initially in the countries of Mexico, Peru and the United States, and later Chile and Argentina became part of the cultivated areas. At the end of the sixties, some commercial plantations appeared in Brazil in the states of São Paulo, Rio de Janeiro, Minas Gerais and Rio Grande do Sul, which is currently the main olive producer. Before that, Brazil only had canned olives and canned olive oil that were imported mainly from Portugal. The authors cited above also mention that in Rio Grande do Sul olive cultivation was officially started in 1948, through the creation of a specialized agency of the Department of Agriculture (olive oil service), with the purpose of supervising and guiding the tasks of promotion and research work.

According to Lipp et al. (2017), the state of Rio Grande do Sul has eight olive oil production units, producing 57,873 L of extra virgin olive oil, involving 20 product brands with 145 producers. The development stage of the olive groves is still in the formation phase and the average cultivated area for each producer is 10.8 ha, with the cultivation in high density varying from 238 to 420 olive trees per hectare (Flôres et al. 2016). The edaphic and climatic conditions are marginal to those found in the Mediterranean, especially in relation to the rainfall index, which raises the requirement for the choice of appropriate cultivars, i.e., more adaptable to the condition of the state (Wrege et al. 2015).

Taking into account the identified differences, the propagation of the olive tree plays a critical role in the success of the crop and the nurseries follow the expansion trend. Currently, only three nurseries that operate in the state of Rio Grande do Sul are registered in the state's olive-growing program (Lipp et al. 2017). Nevertheless, only one of them operates in a mechanized industrial processing model and makes use of a vegetative propagation system with mini-cuttings and a clonal garden. Since this is the company that accepted to be a partner of this study, concern for the environment is integrated into the corporate culture of the company. Examples of this concern are identified through several initiatives that were introduced with the objective of aligning productivity with environmental issues. Among them is the partnership to initiate studies with the use of the Life Cycle Assessment (LCA) tool. In this case, the focus was to measure the environmental performance of productive activities and highlight the impacts of the processing, as well as the purpose of detailing the flows of a product system and establishing prognoses based on their interpretation. This tool has been receiving growing attention in agribusiness and is recognized as an effective and reliable tool for the determination of evaluating criteria of the environmental impacts caused by this activity (Ruviaro et al. 2012). However, the greatest possible diversity of LCA studies for olive cultivation, at different stages and places of cultivation around the world, can aggregate quantitative evaluation of the environmental impacts. This may be associated with a product or process using the inventory of inputs and outputs of the mass and energy flows, allowing a basis for the development of prognostics in the concept of clean technologies (Proietti et al. 2016).

Currently, several publications of the European scientific community approach LCA studies for fruit farming, by considering inventory, analysis and management for the identification of solutions and strategies for the promotion of environmental sustainability. In this case, seedling nurseries were analyzed, in their implantation and maintenance of orchards, as well as their dismantling. In one of the works by Tricase et al. (2017) they investigated a 1 ha cherry orchard considering the pre-production stage with the nursery stages and the growing phase (with a six-year growth period), as well as the maximum yield, with a period between seven and 20 years and planting area. Specifically in the Nursery stage, the treatments for the cultivation, chemical products utilized, and the organic and mineral fertilizations were considered, as well as other resources and energies associated with the process: Diesel; water for irrigation; diluted water and electricity. The identification of sustainable agricultural products as well as environmentally-minded transport logistics are topics pointed out by the authors as fundamental to characterize life cycle analyzes more favorable for several other permanent cropping profiles.

In the case of olive and olive oil production, Avraamides and Fatta (2008) presented an LCA study carried out in Cyprus, highlighting the cradle to the gate impacts related to the planting of olive trees, the cultural treatment, the harvesting and the separation of olive oil (including transport and process waste management). Salomone and Loppolo (2012) published an LCA case study in the Province of Messina, Italy, referring to differences in cultivation stages (conventional and organic), harvesting (manual and mechanized), irrigation (drip irrigation or without irrigation) and olive oil production phase (reuse with additional oil extraction from the waste as well as composting and use of effluents for irrigation). The studies performed by De Gennaro et al. (2012) approached the differences of the high density cropping systems (400 trees/ha grafted trees of Coratin—6 m × 4 m) and over adhesion

(1667 trees/ha rooted Cuttings of Arbequina—4 m × 1.5 m). These authors concluded that the high-density system showed better environmental (specially abiotic depletion—ADP, global warming potential—GWP, ozone layer depletion—ODP and human toxicity—HTP) and economic indicators (€ ha^{-1} $year^{-1}$), with lower use of energy and chemical inputs and greater production of olives.

It should be considered that studies of environmental impacts in olive farming allow for expanding research, from the preparation of seedlings (Lazzerini et al. 2022) to the industrial process, where the possible use of pomace from the separation of olive oil, with a more effective circular economy, especially in emerging countries. This research should include the adaptation of technologies developed in large producing centers to different regional conditions, considering aspects of scale, potential crop growth and regulation. In this context, the challenge is, similar to the European Union project—PEF Olive Oil Pilot—to look for the potential of olive farming in controlling greenhouse gases and, consequently, climate change (Tuomisto et al. 2015; Dias et al. 2024).

Thus, the present study considers the existing gap concerning the olive nurseries to present the evaluation of the environmental aspects and impacts of this productive model. Analyses of the propagation of cloned olive tree seedlings provide for a discussion of prognostic models directed for cleaner olive production and make it possible to implement continuity by targeting clean technologies where activities were performed based on Life Cycle Assessment studies.

2 Sustainability Aspects of Seedling Production for Olive Farming

There are several ways to multiply the olive tree, including seed germination, as these are viable for multiplication. However, the most widespread method for propagation in modern olive farming is cuttings. This method consists of three stages: rooting, acclimatization and seedling formation. For rooting to be successful, the cutting must be kept under controlled temperature and humidity conditions from cutting on the mother plant to insertion into the substrate. Treatment of cuttings with fungicides is convenient to ensure the health of the seedling, as well as the use of auxin, a hormone, in an alcoholic vehicle, to induce rooting (Barranco et al. 2008; Fotia et al. 2022).

In the seedling production system, in general, seedling production begins with the manual harvesting of cuttings in the clonal garden and sending them to the planting pavilion. In this location, on a bench, previously sterilized polyethylene tubes or bags are filled with substrate to manually receive the cuttings—called staking. The sterilization of tubes and bags is carried out in a rotating cylinder with water and steam generated in a wood-fired boiler. Immediately after transplanting the cuttings into the tubes or bags, they are irrigated with water mist and possibly with a fertilizer mixture of appropriate chemical composition for the stage (Prosperato 2014).

It must be considered that on average, the total production cycle of olive tree seedlings is 14 months until they are ready to be distributed to customers in polyethylene bags containing a seedling (for greater sustainability it can be made from biodegradable material made from starch). They are packed inside plastic boxes and later loaded onto the truck. Irrigation is also carried out inside the truck in order to maintain ideal transport conditions.

From the point of view of resource use, the intensive use of water is present in all stages of seedling production, however, for better sustainability, an efficient water reuse system can occur. The floor under the production area can function as a gutter to collect rainwater from the effluent generated. These are dammed in a reservoir for subsequent physical–chemical treatment that guarantees appropriate quality levels for use in production. The treated water is then stored in cisterns from where it is pumped for different uses in the steps described previously (Prosperato 2014).

3 The Steps for Developing the Sustainability of Olive Tree Seedling Production with LCA

Sustainability studies can be much more effective if LCA is applied in its initial phase. These studies are developed in four stages, complying with standards NBR ISO 14040 (2009) and NBR 14044 (2009), where the development of objective and scope are defined; data collection; inventory analysis; impact assessment and data interpretation. Interpretation must be carried out for the main causes of environmental impacts. This is the example of the case study presented here to serve as a reference for creating Scenarios to plan sustainability implementation with LCA. This data can be obtained through partnership with the production sector. This example can be related to research with a seedling nursery located in southern Brazil (results in the item 4), where the production stages were studied. The next items describe the steps from items 3.1 to 3.4

The evaluation followed the criteria of temporal, geographic coverage, confidence, precision, completeness and representativeness to generate the uncertainty related to each item presented in the inventory.

3.1 Goal and Scope Definition

Aiming to verify the environmental impacts and identify the main hot spots of seedling production, the scope of the study was defined to cover seedling processing of the company located in the southern region of Brazil having as limits the system from the cradle to gate.

The objective of the study was the application of Life Cycle Assessment (LCA) with the aim of sustainable development. The main consideration of the work was the creation of scenarios to establish prognoses that potentiate the reduction of environmental impacts, based on the generation of data evaluating human health, ecosystems and resources. The scope of this approach defined as a function the production of olive trees suitable for planting, with 18 months of development. In this case, the functional unit (FU) was the production of 1000 units of olive tree seedlings, that were referred to as seedlings suitable for planting 4.0 ha, having 250 seedlings per hectare, according to planting densities found in Coutinho et al. (2009). Therefore, the reference flow was the production of 1000 units of seedlings.

3.2 *Inventory*

The data were obtained through surveys in the technical literature and observations and questionnaires in the field of study. These are summarized in Tables 1 and 2, and refer to the functional unit and Studied Scenarios. The steps depicted in Fig. 1, such as sanitation, irrigation, steam generation as well as water treatment and transportation, energy consumption, raw materials, and resources that were inventoried indistinctly from the type of the olive tree seedlings produced. However, the raw materials consumed in the filling of the 0.8 L plastic pots, which accommodate the olive seedlings, have an individualized inventory.

In the studied scenarios, the flow chart in Fig. 1 was called industrial nursery, with management references of the alternative use of packages for seedlings: Scenario 1 (Disposal of Polyethylene seedling pots); Scenario 2 (Reuse of Polyethylene seedling pots) *and* Scenario 3 (Degradable Bio seedling container—*eco-container*) the latter

Table 1 Differences among the three possible scenarios in nurseries

Characteristics	Scenario 1	Scenario 2	Scenario 3
Seedling pots	PE, recycling at orchard proximities	PE, reuse 99%	Degradable
Logistics	From nursery to orchard	From nursery to orchard and reverse; 250 km	From nursey to orchard
Sterilization	No	Yes	No
Water recycling	No	Yes	Yes
Water source	Well water	Rain water	Rain water
Effluents	Based on waste of 40% of water used Total Nitrogen = $0.12\ g\ kg^{-1*}$ Total Phosphorus = $0.07\ g\ kg^{-1*}$	No	No

Legend PE polyethylene; *N* total load nitrogen in effluent; *P* total load phosphorous in effluent
*Silveira Thebaldi et al. (2016)

Table 2 Collected information for life cycle inventory with functional unity of 1000 olive trees in 4 ha of planting potential

Inputs flows	Amount	Pedigree	Standard deviation
Avoided products			
Polyethylene, low density, granulate {GLO} | market for | Alloc Rec, U (kg)	50	(4, 1, 1, 1, 1, na)	1.21
Resources			
Vermiculite (kg)	21.80	(1, 1, 1, 1, 1, na)	1.05
Peat (kg)	87.20	(1, 1, 1, 1, 1, na)	1.05
Roundwood, eucalyptus ssp. from sustainable forest management, under bark {GLO} | market for | Alloc Rec, U (m^3)	0.022	(1, 1, 1, 1, 1, na)	1.05
Water, well/rain/reuse (kg)	54.75	(1, 1, 1, 1, 1, na)	1.05
Materials/fuels			
Phosphate fertiliser, as P_2O_5 {RER} | monoammonium phosphate production | Alloc Def, U (kg)	2156.25	(1, 1, 1, 1, 1, na)	1.05
Potassium chloride, as K_2O {GLO} | market for | Alloc Def, U (kg)	703.13	(1, 1, 1, 1, 1, na)	1.05
Magnesium sulfate {GLO} | market for | Alloc Def, U (kg)	42.19	(3, na, 1, 1, 1, na)	1.11
Boric acid, anhydrous, powder {RoW} | production | Alloc Def, U (kg)	37.50	(3, na, 1, 1, 1, na)	1.11
Ammonium sulfate, as N {GLO} | market for | Alloc Def, U (kg)	937.50	(3, na, 1, 1, 1, na)	1.11
Zinc monosulfate {GLO} | market for | Alloc Def, U (kg)	28.13	(3, na, 1, 1, 1, na)	1.11
EDTA, ethylenediaminetetraacetic acid {GLO} | market for | Alloc Def, U (kg)	9.38	(3, na, 1, 1, 1, na)	1.11
Copper sulfate {GLO} | production | Alloc Def, U	14.06	(3, na, 1, 1, 1, na)	1.11
Polyethylene, low density, granulate {GLO} | market for | Alloc Rec, U (kg)	50	(3, na, 1, 1, 1, na)	1.11
Manganese sulfate {GLO} | production | Alloc Def, U (kg)	23.44	(3, na, 1, 1, 1, na)	1.11
Molybdenum trioxide {GLO} | market for | Alloc Def, U (g)	4.68	(3, na, 1, 1, 1, na)	1.11
Aluminium sulfate, powder {GLO} | market for | Alloc Def, U (kg)	0.16	(3, na, 1, 1, 1, na)	1.11
Diesel {RoW} | market for | Alloc Def, U (kg)	0.37	(3, na, 1, 1, 1, na)	1.11

(continued)

Table 2 (continued)

Inputs flows	Amount	Pedigree	Standard deviation
Ellepot (p)	1000	(3, na, 1, 1, 1, na)	1.11
Transport, freight, light commercial vehicle {GLO} ǀ market for ǀ Alloc Rec, U (tkm)	8.25	(5, na, 1, 2, 1, 3)	1.51
Electricity/heat			
Electricity, medium voltage {BR} ǀ market for ǀ Alloc Rec, U (kW)	3	(1, 1, 1, 1, 1, na)	1.05
Output flows			
Emissions to air			
Carbon dioxide, biogenic	938	(4, 1, 1, 4, 4, na)	1.57
Carbon dioxide, fóssil	42.22	(4, 1, 1, 4, 4, na)	1.57
Methane, biogenic	0.012	(4, 1, 1, 4, 4, na)	1.57
Nitrogen oxides	0.0016	(4, 1, 1, 4, 4, na)	1.57
Waste to treatment			
Average incineration residue {GLO} ǀ market for ǀ Alloc Rec, U (kg)	0.5	4, 1, 1, 4, 4, na	1.57

being known as *Ellepot*. The eco-container, named Ellepot, was considered in the input data (Table 2).

It must be pointed out that Brazil is very well positioned in the area of recycling plastic packaging, being the leader in reverse logistics with a 94% return rate of these packages (Brasil 2016).

The material used to produce the *Ellepot* (*eco-containers)* is produced by the Danish company Ellegard A/S. These containers consist of punched paper and are glued to keep the holes open. The composition consists of paper reinforced with biodegradable polyester. The raw material of the wood fibers as well as the paper, are natural, sustainable and degradable (Ellepot 2017; Pedersen 2016; Viégas et al. 2018).

In the absence of local data for inventory formation, the Ecoinvent 3.01 database (Wernet et al. 2016) was used to collect data from a global base (GLO). This database was used in conjunction with the Simapro 8.04 Ph.D. program. After collection, data were double checked with international publications and technical bulletins, as well as, congress presentations as references. Afterwards, these data were evaluated using the Pedigree Matrix (Frischknecht et al. 2005; Weidema et al. 2013). The evaluation followed the criteria of temporal coverage, geography, confidence, precision, completeness, and representativeness to generate the uncertainty related to the items presented in the inventory, resulting in the standard deviation of each item of the

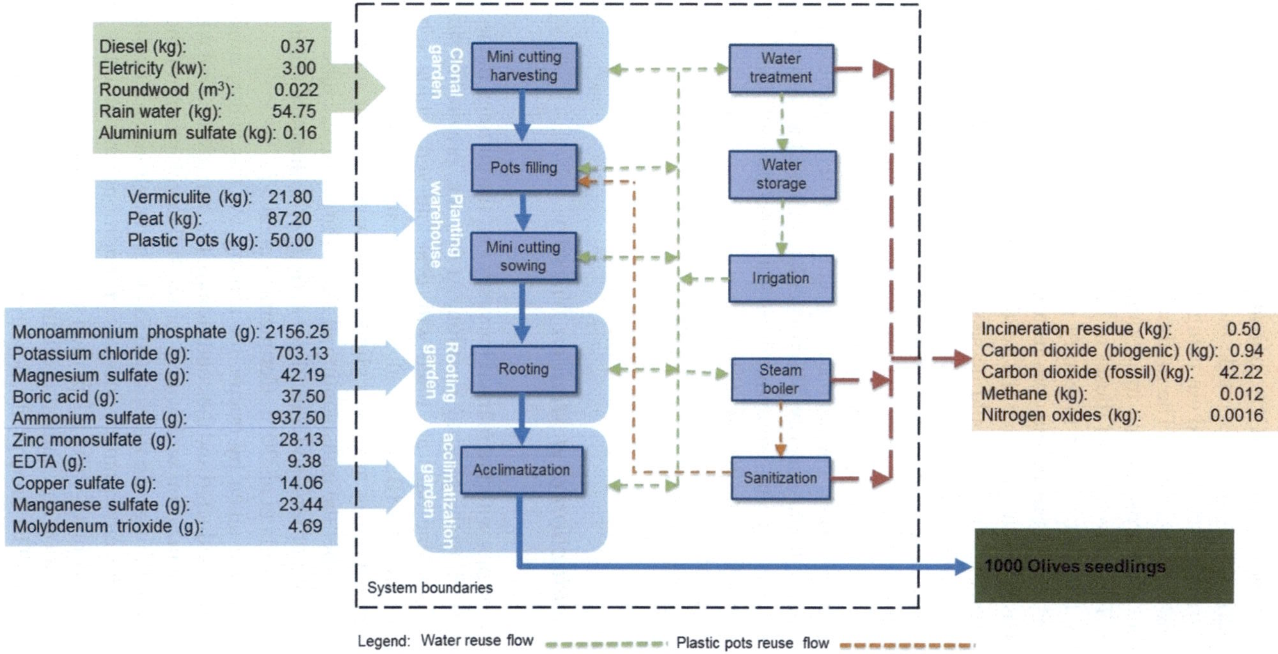

Fig. 1 Inputs and outputs flow of the processing of the olive tree seedlings in the investigated industrial nursery. Functional unit (FU) was referred to suitable for planting 4.0 ha, having 1000 units

inventory. The Pedigree matrix and the standard deviation were considered according to the availability of the spreadsheet associated to SimaPro 8.0.4.

The calculation of the emissions (CO_2 equivalent—biogenic and fossil) related to transport of materials and seedlings inside the nursery (Diesel consumption by a John Deere 5078 tractor) was carried out using the methodology of the Brazilian Program "GHG Protocol—Scope 1". The methodology is the same in scope as the Kyoto Protocol, ISO standards and the methodologies of the IPCC (Brasil 2017; Torres et al. 2017). The conversion of diesel in kg CO_2 equivalent was the choice as quantities of this fuel had its own inventory precisely controlled by the operators at the nursery rather than using the options available in the Ecoinvent database such as transport based on the distance the seedlings were transported. By using the Ecoinvent database, the results would be less precise as they are very different in terms of equipment (tractors) than the ones available at the nursery. The emissions, in terms of greenhouse gases, were calculated using the GHG Protocol tool to supply the amount of wood for direct burning in a stationary source, (with a density of 0.479 g cm^{-3} for eucalyptus grandis, considering CO_2 equivalents in tons and emission factor in kg/l of 1817). The source of the emission of greenhouse gases in the GHG Protocol was stationary energy. The particulate material was characterized as less than 730 mg Nm^{-3}, being below the limit for generators of up to 10 MW using wood (Resolution Conama 382 of 2006).

The consumption of electric energy was inventoried directly in the unit with an employee questionnaire, as well as the consumption of all materials.

3.3 Seedling Production at a Modern Nursery

Seedling production begins with the manual harvesting of the cuttings in the clonal garden and sending them to the planting pavilion. At this site, previously sterilized plastic tubes or polyethylene bags are filled with peat and vermiculite substrates to receive the stakes manually (Badilla et al. 2017). The sterilization of the plastic tubes and bags is carried out in a rotary cylinder with water and steam generated by a wood-fired boiler. After the transplanting of the cuttings into the tubes or bags, they are irrigated with water mist and a fertilizer syringe of a chemical composition suitable for this step (Badilla et al. 2017; Cerqueira et al. 2017).

Then, the seedlings are taken by tractor for rooting. At this stage, subsequent phytosanitary control is conducted and the conditions of the nursery are kept with high relative humidity, by nebulization with water, for the full development of the roots. Thereafter, the rooted seedlings are transported to the growth house for acclimatization and rustification. This environment operates in conditions similar to the natural environment, in order to favor their adaptation to transplanting. Here, under the conditions of water irrigation and natural insolation, the seedlings grow to the ideal height of at least 1.5 m. The total production cycle for olive tree seedlings takes 14 months. At the completion of the cycle, the seedlings are ready to be distributed to customers in polyethylene bags containing one seedling. The seedlings are packed in plastic boxes and then loaded into a truck equipped with an irrigation system in order to maintain the ideal conditions while in transport. After transplantation to the soil, the pots are returned to the factory, with the return and reuse rate of 99%.

From the point of view of resource depletion, the intensive use of water is present in all stages of seedling production; however, the efficient water reuse system developed by the company recovers 90% of the water used. The soil beneath the production area works as a collecting channel for rainwater of the effluents generated in the production process. These are dammed in the lake for further coagulation/decantation treatment (Dortmund type decanter with multiple sequential stages) that assures the appropriate quality levels for the reuse in the production stages. The treated water is then stored in cisterns and then pumped to the different areas for reuse in the steps described above.

Atmospheric emissions were calculated as previously quoted (applying the GHG Protocol 2017), considering equal values for the proposed Studied Scenarios. The same was done for the other outputs: ashes and heat. The differences in wastewater emissions can be seen in Table 2, with Scenario 1 only having the effluents inventoried. These data were obtained with follow-up and interviews with operators from the studied unit.

Based on the explained nursery operation, all data were obtained through surveys in the technical literature and authors observations in the study field. The steps depicted in Fig. 1, such as: Irrigation, reuse water treatment system, inside transport, energy, steam production, residues, effluents and emissions were inventoried indistinctly and are representing the inputs and outputs of the nursery operation. However, the raw materials consumed in the container filling, sowing, rooting and acclimatization have an individualized inventory. The total quantity of chemicals applied presented in Fig. 1 was calculated with based volume of solution applied (V_s), concentration of each chemical (C_f), number of applications per week (A_w) and the growing weeks (G_w) using the Eq. (1):

$$T_a = C_f \times V_s \times A_w \times G_w \tag{1}$$

where: T_a: Total of chemical (g) in considered time; C_f: Concentration of chemical (g/L); V_s: Volume applied per seedling (L) A_w: Number of applications per week and G_w: Growing time in weeks.

Thus, with the data from Eq. (1), the values presented in Table 3 can be considered.

Reverse logistics was included in the inventory using an estimated distance between nursery and a 125 km orchard. In this case, the transportation takes place using a medium-sized truck that transports seedlings to the orchard and, after transplanting the seedlings, the containers are returned to the nursery. The masses of seedlings were not considered since this is common to all scenarios, that is, what differs is the return of the empty containers to the nursery. Mass transported equivalent to 1000 containers is 33 kg and the distance traveled for logistics was 250 km (round trip). In the SimaPro 8.0.4 the transport process is expressed in ton-kilometers (tkm): One ton-kilometer means the transport of one ton over 1 km (Goedkoop et al. 2009). Then the reverse logistics were calculated as:

Table 3 Materials consumed in the container filling, sowing, rooting and acclimatization processing of the olive tree seedlings in the investigated industrial nursery with individualized inventory

Substance	C_f	A_w	G_w	V_s	T_a
Monoammonium phosphate	4.60	1	72	0.0065	2156.25
Potassium chloride	1.50	1	72	0.0065	703.13
Magnesium sulfate	0.090	1	72	0.0065	42.19
Boric acid	0.080	1	72	0.0065	37.50
Ammonium sulfate	2.00	1	72	0.0065	937.50
Zinc monosulfate	0.060	1	72	0.0065	28.13
EDTA	0.020	1	72	0.0065	9.38
Copper sulfate	0.030	1	72	0.0065	14.06
Manganese sulfate	0.050	1	72	0.0065	23.44
Molybdenum trioxide	0.01	1	72	0.0065	4.68

$$T = Wt \times Dt \quad (2)$$

where: T: Transport expressed in tkm; Wt: Mass transported in tons; Dt: Traveled distance in kilometers and $T = 0.033 \times 250.00 = 8.25$ tkm.

3.4 Methods and Procedures for LCA

Life-cycle assessments were developed with a 100-year baseline model and applied for the extension of damages in the hierarchical scale. Special consideration was given to the contribution of the chemical inputs and the Brazilian energy matrix. From the SimaPro 8.0.4 software, the ReCiPe (Midpoint and Endpoint—H) method was applied, mainly exploiting the database with reference to the (S) systems for considering the study process illustrated in Fig. 1. The exceptions were electricity and heat, where unit (U) was used. The broader coverage of the data profile also considered the RoW, RER and GLO choices. The Hierarchical version of the ReCiPe was chosen to give a more balanced coverage of the environmental impact determination response. It is the classic method of greater comprehensiveness and balance to determine causes and damages. This choice is referenced in the works of Yi et al. (2014) where the organization of the data in the hierarchical scale was recommended to show balance between midpoint and endpoint parameters. The global representativity (GLO) parameters, including logistic and losses (market for), with allocation by classification cut (Alloc Rec) and unit processes with uncertainty (U) were also considered when possible.

4 Results

4.1 LCA Characteristics for Scenario 1

The first impact assessments for the LCA are presented in Figs. 2 and 3. The most significant negative impacts are associated with PE seedling pots and Full Disposal PE seedling pots. The most significant negative impact for Nursery processing is eutrophication, with 6.82×10^{-5} kg Peq. It is only 3.11% of the total of the negative impact when compared to the impacts caused by the management of plastic materials. However, the positive impact of water depletion at the studied seedling production unit (93.4% when compared to Full Disposal PE seedling pots Nursery processing).

The Normalization of Fig. 3 shows that ecotoxicity and eutrophication, in both marine and freshwater, are the most significant impacts. This is due to the generation of the effluents recorded in Table 1.

When considering the contribution of materials and the processing of Scenario 1, the main impact elements can be observed in Figs. 4 and 5. Here the pots are highlighted in the petrochemical matrix of the packages, along with Diesel.

The positive impact effect of water depletion can best be understood in the network illustration of Fig. 5.

The environmental gain with injection molding processing is sustainable due to the energy matrix for the supply of the operation, which counteracts the negative environmental impacts of the use of polyethylene, low density, granulate (GLO). Thus, the environmental gain of PE seedling pots influences water depletion favorably in

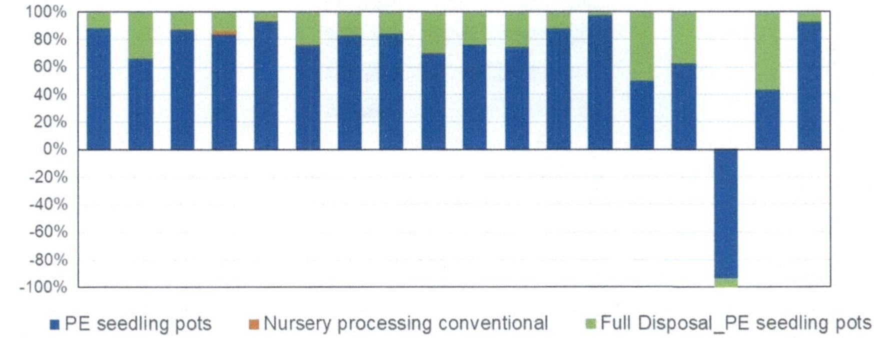

Fig. 2 Characterization of Scenario 1 for the impact assessments

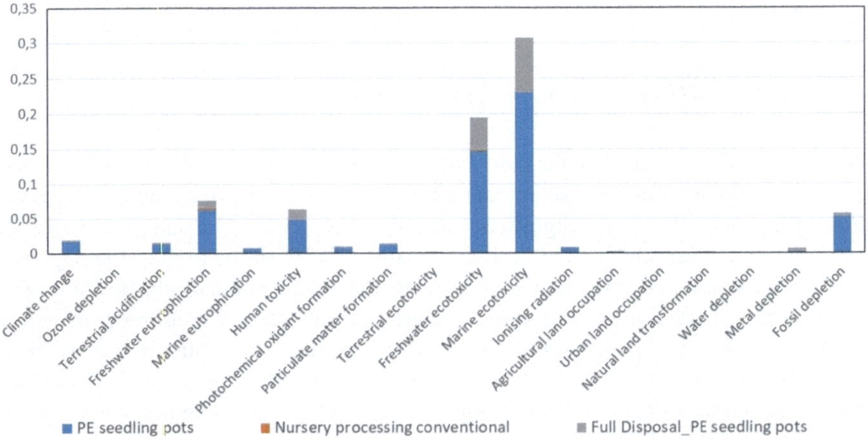

Fig. 3 Normalization of Scenario 1 for the impact assessments

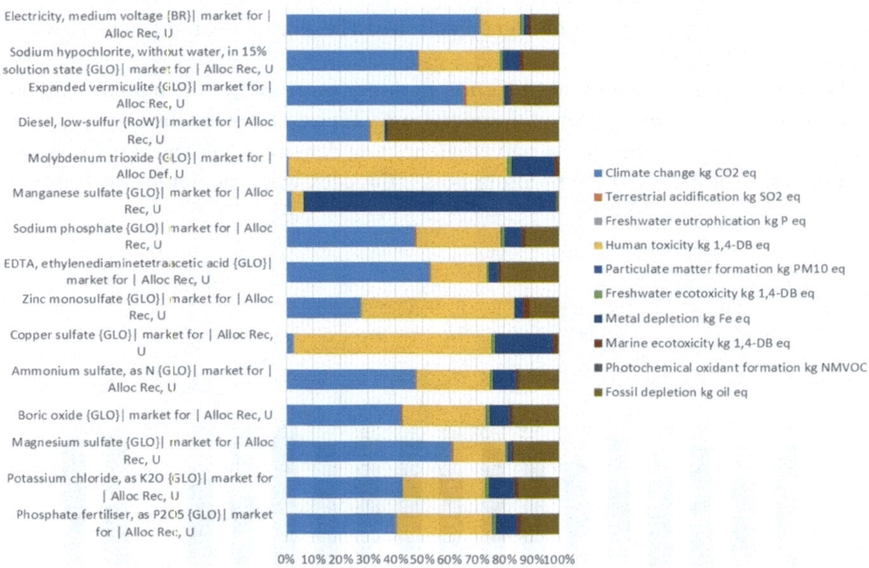

Fig. 4 Characterization for input contribution for olive nursery processing in Scenario 1

Scenario 1. In the case of the other significant impacts presented in the Standardization, the Marine and Freshwater ecotoxicity has only preponderance of negative impacts associated especially with polyethylene, low density, granulate (GLO). Similar behavior may be associated with human toxicity, freshwater eutrophication, climate change and fossil depletion impacts.

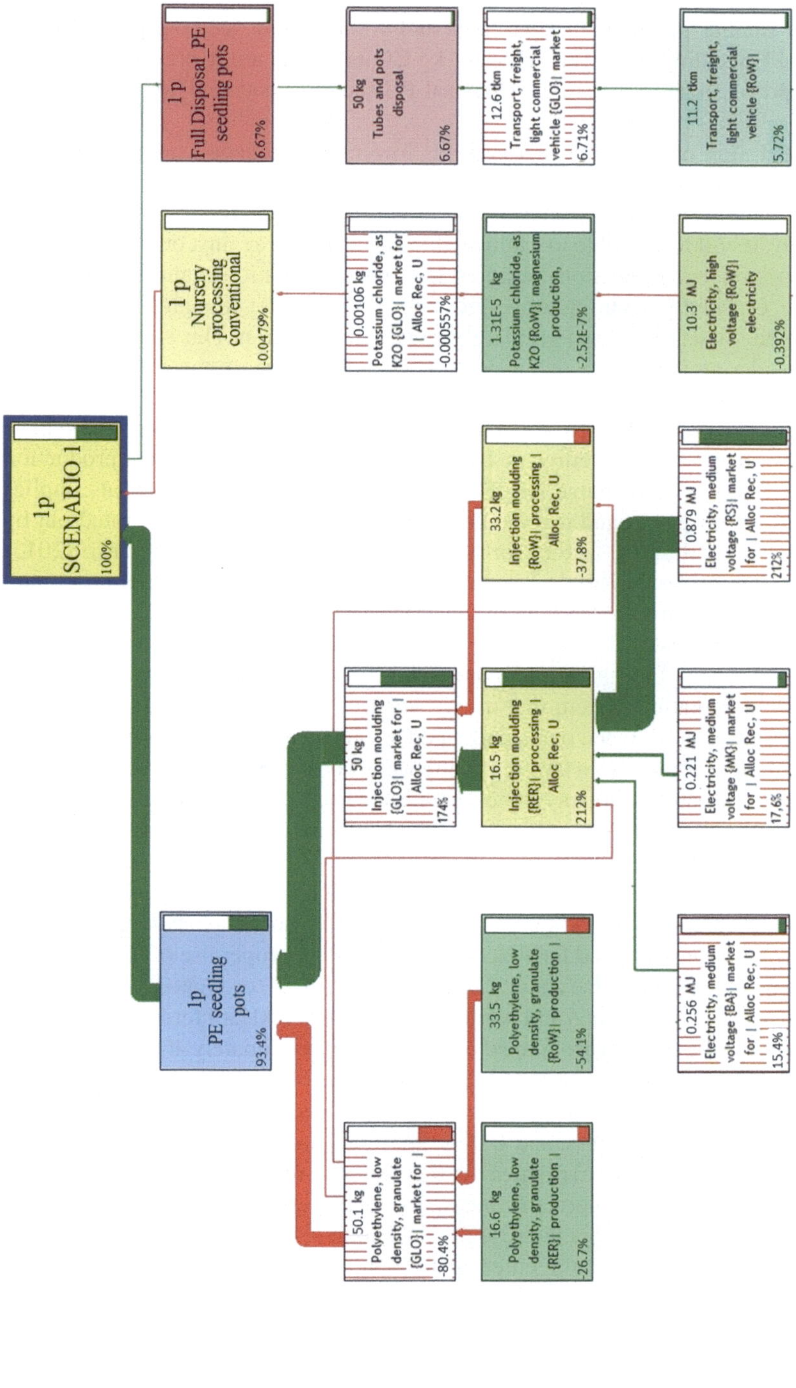

Fig. 5 Network for water depletion contribution in olive nursery processing in Scenario 1 with Cut Point of 4.1%

Other important aspects of the results of Fig. 5 can be extended to all other Characterization parameters: all impacts associated with Nursery processing materials are much smaller in values of kgeq, area and volume, than seedling pots and Full Disposal PE seedling pots. The values of kg CO_2eq for example present 0.0704 for Nursery processing, 24.3 for Full Disposal PE seedling pots and 181 for seedling pots PE.

The operation of a nursery that presents lower environmental impacts does not represent what happens in most nurseries operating in Brazil. Thus, it is necessary to compare the results with scenarios that might indicate the reality of the industry. In this way, it will be possible to evaluate this performance against others that do not adopt environmental protection measures. Data on the operation of nurseries in Brazil are scarce, especially regarding olive-growing. There are not enough publications to define a model and the dimensions of a characteristic operation. This is especially true when considering the operational aspects, such as the reuse of pots and wastewater (with the associated nutrient load).

The problems related to reuse can also be observed with plastic pots, which have been replaced by black polyethylene bags as definitive containers for the production of the seedlings in most nurseries. The reuse of bags or plastic materials implies the need for sanitization and reverse logistics. These conditions are pointed out by the producers as limiting factors for their reuse. In another study, Alonso (2013) surveyed 70 nurseries in Rio de Janeiro and concluded that most of them are small and lack infrastructure for seedling production.

In general, there is predominance for the use of plastic bags as seedling containers and a lack/absence of technical and managerial control over production. These nurseries have an irrigation system that uses water from several sources: spring water, well water, streams or rivers, dams and piped water, with the latter coming from the local distribution company. In fact, regarding water consumption, Embrapa (Brazilian Agricultural Research Company) recommends the use of well water for irrigation. In the case of treated water, it must have a residual chlorine level of 5 ppm (De Oliveira and Scivittaro 2003).

The profile of nurseries for the seedling production of olive trees is comparable to the clonal nurseries for the genus *Eucalyptus*. This sector is characterized as a major consumer of water and fertilizers. This makes the proper use of these inputs a crucial issue of the productive chain. Cerqueira et al. (2017) evaluated the fertigation process of seedling production and verified that for every 200 L of fertigation solution applied daily in the semi-hydroponic clonal garden, approximately 40% was drained. Therefore, the solution became an effluent, with losses of nutrients through leaching. According to the authors, measured values of the generated effluent showed the presence of 0.12 g kg^{-1} of N and 0.07 g kg^{-1} of phosphorus. In a similar study, Silveira Thebaldi et al. (2016) pointed out that water losses can range from 49 to 72% depending on the irrigation process and may contain significant amounts of unused fertilizers. This wastewater is the result of irrigation lost by leaching, transforming the water into effluents with the potential for contamination of surface water in the vicinity of the nurseries.

Regarding the water consumption necessary for the growth of seedlings, Macedo (1993) highlighted that a nursery that produces 100,000 seedlings per year will need approximately 10,000 L of water per day (54.75 L to complete the seedling production cycle). A major producer of seedlings, which has a water reuse system, reported its water consumption information for a period of 60 days on its website. According to the data provided by this producer, approximately 5.7 L of water are required to complete the production cycle of a seedling (Tecnoplanta 2017).

It becomes evident that enterprises that do not reuse wastewater present a technological or operational gap, which has relevance to the environmental context of seedling production. These results are in line with the hypothesis of Silveira Thebaldi et al. (2016) that report through LCA the contamination of surface water with the aspects of eutrophication and water ecotoxicity.

The technological gap of the nurseries pointed out in the study carried out by Cerqueira et al. (2017) places them in an environmental disadvantage when compared to the nursery referenced in the present study, which has a system for the capture and reuse of wastewater produced by irrigation. The impact on water eutrophication due to wastewater disposal is very significant. The obtained results show that a modern nursery contributes only 20% in terms of eutrophication when compared to the operation of a conventional nursery, since the latter uses steam generation for heating and sterilization. When considering the question of bags, pots and polyethylene tubes, the publication by Schafer and Dornelles (2000) presented the conditions of delivery for citrus seedlings in the state of Rio Grande do Sul. At that time, these were predominantly commercialized in bare root.

According to data surveyed in the presented study, the use of plastic bags for seedling development has improved, since it is currently a standard procedure in most nurseries. Considering that the waste generated is rural solid waste, these nurseries follow the national solid waste plan (Brasil 2010). Therefore, the operating licenses are obtained locally, issued by the state environmental agency (FEPAM RS 2014). This being the case, the final destination for the solid waste should be in landfills or recycling centers. A parallel with plastic bags can be established, due to their similarity in final destination and their use as transport and a protection agent. The alternatives pointed out, from several life cycle assessment (LCA) studies and/or their environmental impact, show that the more sustainable option, in the case of bags, is reuse. Saving energy and raw materials are mentioned in studies by Santos et al. (2012).

4.2 Scenarios 2 and 3 for Cleaner Production of Olive Seedlings at an Industrial Nursery

The prognostic stage compared Scenarios 1, 2 and 3 to see the improvements in the olive seedling unit at an industrial nursery. The results are presented in Figs. 6 and 7.

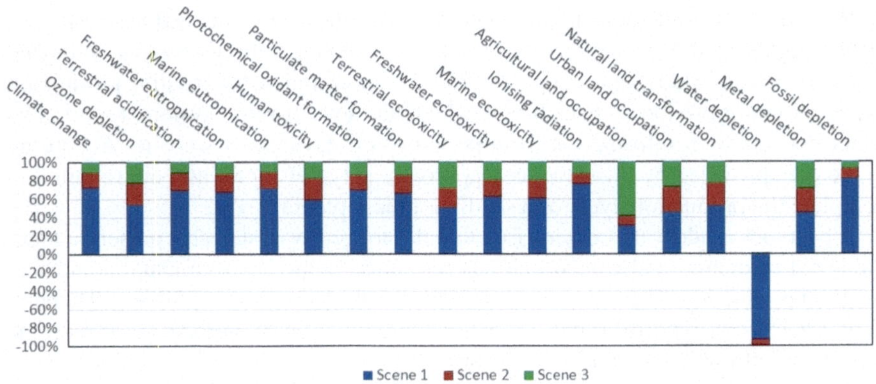

Fig. 6 Characterization of the Scenarios 1, 2 and 3 for clean technologies propositions in the seedling nursery of olive trees

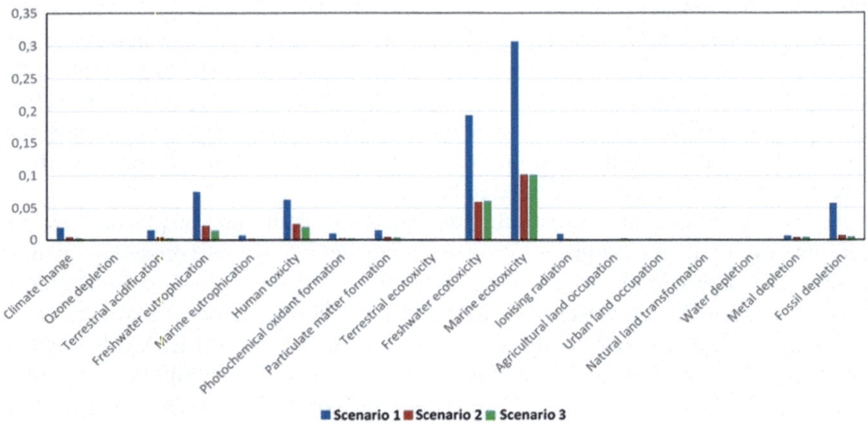

Fig. 7 Normalization of the Scenarios 1, 2 and 3 for clean technologies propositions in the seedling nursery of olive trees

It is possible to observe the environmental gains of Scenarios 2 and 3, in the analysis of Characterization and Normalization. This means that the only parameters that showed fewer environmental gains than Scenario 1 were water depletion and agricultural land use.

The environmental gains in Scenario 3 demonstrate why the use of biodegradable packaging is indeed a cleaner technology option for nurseries. The production and application of eco-containers has been constantly increasing. These have been rapidly replacing the polyethylene tubes (Teodoro et al. 2017). Although there is still no production of olive trees with the use of these more sustainable materials, our study added a scenario where these seedlings develop in an *"ecological"* container that

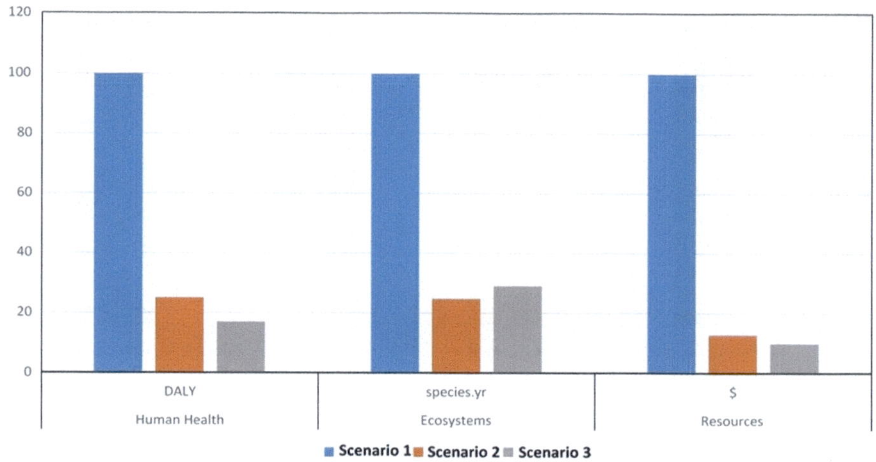

Fig. 8 Damages evaluations of the Scenarios 1, 2 and 3 for clean technologies propositions in the seedling nursery of olive trees

is later transplanted to the soil in conjunction with the seedling. Besides biodegradability aspects, one of the advantages of using this support for the seedlings, when compared to the plastic tubes, is the absence of reverse logistics and sterilization.

Many nursery workers point out these processes as barriers to the reuse of plastic pots. In this scenario, the dimensions of a plastic pots—6 cm height × 3.5 cm diameter ~ 14.4 cm^2—and—15 cm height × 8 cm width × 6.5 cm width lower ~ 477 cm^2—were transformed into data for use in the SimaPro 8.04 software. These results can be seen in Fig. 8.

The data shows that the only area to have no gains with the use of the eco-container was Damage to Ecosystems. The explanation for this behavior can be seen in Fig. 9.

Transport logistics is the area that weighs the most for negative damages to Ecosystems (Fig. 8). However, the local or regional production strategy can reduce these negative impacts. For packages intended for the production of seedlings, there is no consensus among the scientific community and the production sector that biodegradability is the solution. Santos et al. (2012) published a collection of LCA articles suggesting that paper packages cause higher energy and environmental impacts than polyolefin due to the effluents and gaseous emissions generated in pulping and bleaching processes of paper production. On the other hand, reforestation areas are carbon fixers and consist of an important tool to earn carbon credits.

However, subsequent cutting of paper-making trees removes the carbon-fixing agent and contributes to land degradation, ecosystem damage, fuel use, and carbon emissions. In addition, trees from reforested areas present negative impacts arising from monoculture and land use, fertilizers and water. Bags have a higher degradation rate than plastic and are responsible for higher methane emissions (Santos et al. 2012). The results of this LCA study indicate that *biocontainer* and pot reuse presented

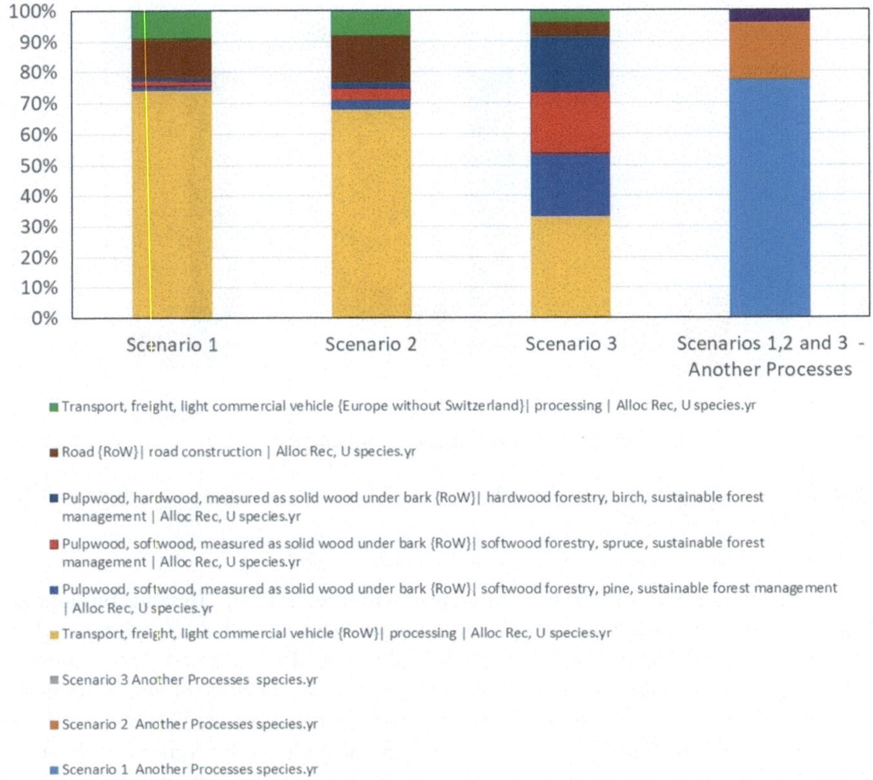

Fig. 9 Characterization of the processing contribution of the Scenarios 1, 2 and 3 for clean technologies propositions in the seedling nursery of olive trees

advantages in the environmental context (except for aspects associated with soil occupation related to reforestation).

Polyethylene (PE) pots use more material (50 kg PE/1000 seedlings) when compared to *eco-containers* (3.68 kg paper/1000 seedlings). This mass ratio is directly responsible for the difference in LCA. Through sensitivity analysis, which includes an increase of paper equivalent to that of PE mass, it was possible to observe that the reuse becomes environmentally better aligned with the studies on bags.

Considering that seedling production is only a part of the production chain which involves different stages such as orchard implantation, handling, harvesting and processing, and that the data related to the impacts of the nursery generated the amount of equivalent carbon dioxide, it was possible to compare these steps. In this case, the stage of seedling production was compared in terms of impact in the global warming category, where more data are available, with the other stages of the olive production.

Table 4 presents the CO_2eq emissions normalized to one ton of olives for the other steps and considers recent studies on orchards in other countries. The results, however, indicate certain variations between the studies, making them difficult to compare. On the other hand, through this study, it was possible to verify that the impact of the nursery on the production chain is minimal. The variations shown in Table 4 can be mainly attributed to the LCA approach, differences between agricultural practices, materials and energy consumption from different sources, and logistics.

In fact, these variations, which occur with the results presented in Table 4, are important discussion topics that several authors address. Eicker et al. (2010) compared the Life Cycle Inventory (LCI) of the triple superphosphate, an agrochemical widely used in agriculture, made in Brazil, with the LCI calculated in Europe and found that the results of the evaluation using the European LCI tend to be higher. Maciel et al. (2015), in their publication on soybean cultivation in Rio Grande do Sul, goes beyond the question of the database and raises the regional differences existing in the country and the lack of regional data.

Table 4 Comparison of the global warming impact of seedling production with the other stages of the production chain

Context	This study (Scenarios 1, 2 and 3)	(Salomone and Loppolo 2012)	(Tsarouhas et al. 2015)	(De Gennaro et al. 2012)
Step	Seedling production	Most relevant agricultural stage	Most relevant agricultural stage	Only the agricultural stage that was analyzed
Functional unit	4 ha (1000 seedlings)	1 ton olives	1 L of olive oil	1 ton of olives
kg CO_2 eq (characterization model)	18.44 (Scenario 3) 30.88 (Scenario 2) 148.8 (Scenario 1) (Recipe midpoint)	481–496 (CML 2000)	0.438 (eco indicator 199)	542 (CML 200)
Value normalized for 1 ton of olives kg CO_2 eq	0.276* (Scenario 3) 0.464* (Scenario 2) 2.232* (Scenario 1)	–	281.8*	–

*Considers the production of 60 kg of olives (15% olive) per plant in an orchard of 1000 plants in 4 ha

5 Analysis

As for the indirect Life Cycle Inventory (LCI), from the Ecoinvent database, it is important to note that the origin of the data is predominantly from European processes. Although it was built with a database preferably global in scope, in the absence of data from Brazil, there are important discussions about the impacts of using data from European processes on LCA results.

The results presented must consider limitations and exclusions in the inventory stage, highlighting the issue of agrochemical packaging. Even though this study included dealing with the issue of agrochemical packaging, regulated by Law No. 9974/2000 and Decree 4074/2002, including the definitions of the National Solid Waste Policy (PNRS 12305/2014) (Brasil 2010). It was decided to This is not an approach, understanding that the country is positioned very well in this regard, being a leader in reverse logistics with a 94% return rate for these packages (Brasil 2016). Among the difficulties of LCA is one of the main residues in the chain, bagasse, which does not exist in the database. Also, importantly, there was a lack of information about the inert substances that make up pesticides. These substances have significant quantities from the mass point of view of the composition of agrochemicals.

Data collection on seedling processing in the studied unit must consider what happens in many seedling production nurseries. Both eucalyptus and olive tree seedlings are produced in these units (typical occurrence for the southern region of Brazil). Thus, to overcome the distribution of the use of resources related to the operation of the nursery, which are shared, it was decided to make the allocation based on the number of seedlings produced.

6 Conclusions

The use of *eco-containers* (Scenario 3) could be an alternative with better environmental performance than the reuse of tubes and their disposal in landfills. This type of container eliminates the need for the return of pots and sterilization, which involves the use of boilers, presenting energy consumption and atmospheric emissions. However, concerning the reuse of plastic tubes, the distance traveled in the reverse logistics should be considered, because the transportation involved appears as a relevant factor in relation to environmental impacts.

Although there is a lack of studies showing the impacts for all the stages of olive cultivation, a global warming category was used as a basis for comparing the stages involved in the crop. The production of seedlings represents only a small part of the impacts of olive production. Even in the worst-case scenario, this stage produces 0.558 kg eq. CO_2 per ton of olives, while the other stages vary between 218 and 542 kg CO_2 in the base table. With regard to the environmental impacts of olive growing caused by the production of seedlings in other stages of olive production, it can be concluded that this has a much lower impact when compared to the stages that occur in the field.

However, the gain of the research developed organized the inventory and prognosis of improvements regarding the sustainable environmental management, which can be applied to other nurseries of seedlings, being then, initial stage to propose cleaner production or clean technologies of the activities.

Considering the relationship of this study with environmental aspects and impacts, it is recommended to expand research on the possible use of pomace from olive oil separation. This research must include the adaptation of technologies developed in large producer centers to local conditions, considering aspects of scale, potential crop growth and regulation. In this context, it is suggested to expand the study following the example of the European Union project—PEF Olive Oil Pilot—which seeks to understand the role of olive farming in controlling greenhouse gases and, consequently, climate change.

Acknowledgements The authors gratefully thanks Conselho Nacional de Pesquisa—CNPq, Prosperato and Fapergs

References

Alonso JM (2013) Analysis of nurseries and Brazilian legislation on native forest seeds and seedlings in the state of Rio de Janeiro. Dissertation (Postgraduate Program in Environmental and Forest Sciences). Original in portuguese: análise dos viveiros e da legislação brasileira sobre sementes e mudas florestais nativas no estado do Rio de Janeiro. Dissertação (Programa de Pós-Graduação em Ciências Ambientais e Florestais)—Universidade Federal Rural do Rio de Janeiro, [lSeropédica-RJ], p 78

Avraamides M, Fatta D (2008) Resource consumption and emissions from olive oil production: a life cycle inventory case study in Cyprus. J Clean Prod 16:809–821

Badilla Y, Xavier A, Gamboa OM (2017) Storage time effect on mini-cuttings rooting in tectona grandis Tectona grandis linn f. clones. Revista Árvore 41

Barranco D, Fernández-Escobar R, Rallo L (2008) El cultivo del olivo. Coed. Junta de Andalucía/ Mundi-Prensa, Madrid

Brasil (2006) Resolution Conama number 382, of December 26, Published in the DOU nº 1, of January 2, 2007, Section 1, pp 131–137. Original in portuguese: Resolução Conama número 382, de 26 de dezembro de 2006 Publicada no DOU nº 1, de 2 de janeiro de 2007, Seção 1, pp 131–137

Brasil (2010) Lei Nº 12.305—National policy on solid waste. Original in portuguese: Política Nacional de Resíduos Sólidos, 3 de agosto de 2010 ed. Diário Oficial da União, Brasília, DF

Brasil (2016) 5ª official opening of the olive harvest shows growth in the sector. Original in portuguese: Abertura Oficial da Colheita de Oliva mostra crescimento do setor. EMATER/ RS—ASCAR, Porto Alegre

Brasil (2017) GHG protocol—calculation tool. Original in portuguese: GHG protocol—Ferramenta de cálculo. http://www.ghgprotocolbrasil.com.br/ferramenta-de-calculo. Accessed 20 Apr 2018

Coutinho EF, Ribeiro FC, Cappellaro TH (2009) Cultivation of Olive (*Olea europaea* L.)—(production system). Original in portuguese: Cultivo de Oliveira (*Olea europaea* L.)—(Sistema de Produção). Pelotas: Embrapa Clima Temperado. ISSN 1806-9207, p 125

De Cerqueira PHA, de Azevedo GB, de Souza AM, Azevedo GTOS (2017) Residual fertilization in the production of clonal eucalyptus seedlings. Original in portuguese: Adubação residual na produção de mudas clonais de eucalipto. Pesquisa Florestal Brasileira 37:119–129

De Eicker MO, Hischier R, Kulay LA, Lehmann M, Zah R, Hurni H (2010) The applicability of non-local LCI data for LCA. Environ Impact Assess Rev 30:192–199

De Gennaro B, Notarnicola B, Roselli L, Tassielli G (2012) Innovative olive-growing models: an environmental and economic assessment. J Clean Prod 28:70–80

De Oliveira RP, Scivittaro WB (2003) Norms and standards for production of certified citrus seedlings in partnership with Embrapa. Original in portuguese: Normas e padrões para produção de mudas certificadas de citros em parceria com a Embrapa. Embrapa Clima Temperado-Documentos (INFOTECA-E), p 18

Dias S, Pino-Hernández E, Gonçalves D, Rego D, Redondo L, Alves M (2024) Challenges and opportunities for pilot scaling-up extraction of olive oil assisted by pulsed electric fields: process, product, and economic evaluation. Appl Sci 14(9):3638. https://doi.org/10.3390/app14093638

Ellepot (2017) Ellepot fruit trees paper. Retrieved 01 Sept 2018, 2018, from https://ellepot.com/ellepot-products/ellepot-paper/paper-programme/fruit-trees-paper/

Flôres CA, Coutinho EF, Haerter JA, Alba JMF, Wrege MS, Sousa ROD, Costa VB (2016) Olive trees: technical aspects and cultivation in southern Brazil. Original in portuguese: Oliveira: Aspectos técnicos e cultivo no sul do Brasil. EMBRAPA, Brasília

Fotia K, Nanos G, Barouchas P, Giannelos M, Linardi A, Vallianatou A, Mpeza P, Tsirogiannis I (2022) Growth development, physiological status and water footprint assessment of nursery young olive trees (*Olea europaea* L. 'Konservolea') irrigated with urban treated wastewater. Resources 11(5):40. https://doi.org/10.3390/resources11050040

Frischknecht R, Jungbluth N, Althaus H-J, Doka G, Dones R, Heck T, Hellweg S, Hischier R, Nemecek T, Rebitzer G, Spielmann M (2005) The ecoinvent database: overview and methodological framework. Int J Life Cycle Assess 10:3–9

Goedkoop M, Heijungs R, Huijbregts M, De Schryver A, Struijs J, van Zelm R (2009) ReCiPe 2008 a life cycle impact assessment method which comprises harmonised category indicators at the midpoint and the endpoint level. PRé Consultants, University of Leiden, Radboud University, RIVM. https://www.rivm.nl/dsresource?objectid=76975579-adab-4ce5-9b22-3cbc073ea5b. Accsessed in 2018

IOC (2016) World olive oil figures. International Olive Oil Council, Madrid

Lazzerini G, Manzini J, Lucchetti S, Nin S, Nicese FP (2022) Greenhouse gas emissions and carbon sequestration from conventional and organic olive tree nurseries in Tuscany, Italy. Sustainability 14(24):16526. https://doi.org/10.3390/su142416526La

Lipp JP, de Almeida GTF, Ambrosini LB (2017) Technical note: olive register. Original in portuguese: nota Técnica: Cadastro Olivícola, Secretaria da Agricultura, Pecuária e Irrigação Câmara Setorial das Oliveiras, Brasil, p 5

Macedo AC (1993) Seedling production in forest nurseries: native species. Original in Portuguese: Produção de Mudas em viveiros florestais: espécies nativas/A.C. Macedo; revisado e ampliado por Paulo Y. Kageyama, Luiz G. S. da Costa. Fundação Florestal, São Paulo, p 18

Maciel VG, Zortea RB, Menezes da Silva W, Cybis LFA, Einloft S, Seferin M (2015) Life cycle inventory for the agricultural stages of soybean production in the state of Rio Grande do Sul, Brazil. J Clean Prod 93:65–74

NBR ISO 14040 (2009) Environmental management—life cycle assessment—principles and structure. Brazil. Brazilian Association of Technical Standards (ABNT)

NBR ISO 14044 (2009) Environmental management—life cycle assessment—requirements and guidelines. Brazil. Brazilian Association of Technical Standards (ABNT)

Pedersen BB (2016) Use of an organic permeable composite sheet material for propagation plugs for organic production. Denmark, Ellepot A/S 12

Proietti P, Sdringola P, Brunori A, Ilarioni L, Nasini L, Regni L, Pelleri F, Desideri U, Proietti S (2016) Assessment of carbon balance in intensive and extensive tree cultivation systems for oak, olive, poplar and walnut plantation. J Clean Prod 112:2613–2624

Prosperato (2014) In Englsh: Tecnolivas harvests its second harvest. Barra do Ribeiro. Original in Portuguese: Tecnolivas colhe sua segunda safra. Barra do Ribeiro, RS. Disponível em: http://www.prosperato.com.br/2014/03/03/tecnolivas-colhe-sua-segunda-safra-de-azeitonas/

Rio Grande do Sul State (2014) Plan of solid waste of Rio Grande do Sul State 2015–2034. In: FEPAM (ed) Original in portuguese: Plano Estadual de Resíduos Sólidos do Rio Grande do Sul 2015–2034, Porto Alegre, p 559

Ruviaro CF, Gianezini M, Brandão FS, Winck CA, Dewes H (2012) Life cycle assessment in Brazilian agriculture facing worldwide trends. J Clean Prod 28:9–24

Salomone R, Ioppolo G (2012) Environmental impacts of olive oil production: a life cycle assessment case study in the province of Messina (Sicily). J Clean Prod 28:88–100

Sanchez EL, Lettry RN, Godoy JP, de Paz MC (2015) Mesoeconomic costs in the olive chain of Mendoza. Original in spanish: Costos Mesoeconómicos en la cadena olivícola de Mendoza, XXXVIII Congreso Argentino de profesores universitários de costos, San Juan, p 23

Santos AS, Freire FHO, da Costa BL, Manrich S (2012) Plastic bags: sustainable destinations and substitution alternatives. Original in portuguese: Sacolas plásticas: destinações sustentáveis e alternativas de substituição. Ciência e Tecnologia, Polímeros, p 22

Schafer G, Dornelles ALC (2000) Citrus seedlings production in Rio Grande do Sul: diagnosis of the producing region. Original in portuguese: produção de mudas cítricas no Rio Grande do Sul: diagnóstico da região produtora. Ciencia Rural. Santa Maria 30(4):587–592

Silveira Thebaldi M, Lima LA, da Silva AC, de Figueiredo Braga Colares M, Terra Lima PL (2016) Efficiency of irrigation systems in seedlings of native forest species produced in tubes. Original in portuguese: eficiência de sistemas de irrigação em mudas de espécies florestais nativas produzidas em tubetes. Ciênc Flores 26(2):401–410

Tecnoplanta (2017) The best infrastructure in the gaucho market. Original in portuguese: a melhor infra-estrutura do mercado gaúcho. Tecnoplanta, Barra do Ribeiro, RS

Teodoro LL, da Silva AJ, Silva JG, de Sá Fernandes M (2017) Productivity of *Corymbia citriodora* seedlings in different types of tubers. Original in portuguese: produtividade de mudas de *Corymbia citriodora* em diferentes tipos de tubetes. Anais Colóquio Estadual de Pesquisa Multidisciplinar, p 1. ISSN-2527-2500

Torres RDCS, Andrade C, Gomes SMDS (2017) Construction of greenhouse gas inventory for a Brazilian distribution electricity company. Latin Am J Manag Sustain Dev 3:261–288

Tricase C, Rana R, Andriano AM, Ingrao C (2017) An input flow analysis for improved environmental sustainability and management of cherry orchards: a case study in the Apulia region. J Clean Prod 156:766–774

Tsarouhas P, Achillas C, Aidonis D, Folinas D, Maslis V (2015) Life cycle assessment of olive oil production in Greece. J Clean Prod 93:75–83

Tuomisto H, Michalopoulos G, Pattara C, Polo Palomino J, Russo C, Cappelletti G, Nicoletti G, Saouter E (2015) Developing product environmental footprint category rules for olive oil. In: International conference on Life Cycle Assessment as reference methodology for assessing supply chains and supporting global sustainability challenges ENEA, JRC96585, pp 41–44

Weidema BP, Bauer C, Hischier R, Mutel C, Nemecek T, Reinhard J, Vadenbo C, Wernet G (2013) Overview and methodology: data quality guideline for the ecoinvent database version 3. Swiss Centre for Life Cycle Inventories

Viégas LB, Simões D, de Lima KB, da Silva MR (2018) Water management for *Schinus terebinthifolius* Raddi seedlings in degradable containers. Floresta Ambient 25(1). https://doi.org/10.1590/2179-8087.008416

Wernet G, Bauer C, Steubing B, Reinhard J, Moreno-Ruiz E, Weidema B (2016) The ecoinvent database version 3 (part I): overview and methodology. Int J Life Cycle Assess 21:1218–1230

Wrege MS, Coutinho EF, Pantano AP, Jorge RO (2015) Potential distribution of olive trees in Brazil and worldwide. Original in portuguese: Distribuição potencial de oliveiras no Brasil e no mundo. Rev Bras Frutic 37:656–666

Yi S, Kurisu KH, Hanaki K (2014) Application of LCA by using midpoint and endpoint interpretations for urban solid waste management Sora. J Environ Prot 5:1091–1103

Life Cycle Assessment of Bioethanol Production from Microalgae in a Biorefinery Approach

João Felipe Freitag, Victória Dutra Fagundes, Viviane Simon, Rosana de Cássia de Souza Schneider, and Luciane Maria Colla

Abstract This study aimed to analyze the life cycle of bioprocesses involved in the evaluation of microalgal biomass until the generation of bioethanol. SimaPro® 8.5.0 software was used to analyze the life cycle impacts, applying the Impact 2002+ methodology for the global warming impact category (Kg CO_2 eq.). The impacts were analyzed without considering the energy consumption of equipment, in order to understand the inputs and outputs of the processes from the perspective of materials and inputs. Data were collected for the inventory of microalgae cultivation, biomass harvesting, drying, pre-treatment, enzymatic hydrolysis, fermentation, and distillation. The results showed that the fermentation step had the highest global warming impact, with 43.52 kg CO_2 eq. It is also worth noting that the microalgae cultivation process using effluent as a medium had a negative carbon footprint (−20.69 kg CO_2 eq.), as 30 L of cattle effluent was utilized during microalgae cultivation. Therefore, it was observed that the conversion of microalgae to bioethanol has sustainable potential through the utilization of waste and by-products from other processes.

Keywords LCA · *S. platensis* · Impact 2002+ · Global warming

1 Introduction

Microalgae emerge as strategic microorganisms to face climate change (Olabi et al. 2023), as they value waste and effluents in addition to incorporating CO_2 during the cultivation phase through photosynthesis. This breadth of use of microalgae in different sectors is related to the cell's ability to grow in different environmental and

J. F. Freitag (✉) · V. D. Fagundes · V. Simon · L. M. Colla
Civil and Environmental Engineering, Universidade de Passo Fundo, Bairro São José, Passo Fundo, Brazil
e-mail: 158662@upf.br

R. de Cássia de Souza Schneider
Postgraduate Program in Environmental Technology, Universidade de Santa Cruz Do Sul, Santa Cruz Do Sul, Brazil

L. Brandli et al. (eds.), *The Contribution of Life Cycle Analyses and Circular Economy to the Sustainable Development Goals*, World Sustainability Series,
https://doi.org/10.1007/978-3-031-85300-5_18

nutritional scenarios, being able to direct its metabolism to obtain bioproducts of interest (Ma et al. 2022). The concept of microalgae biorefineries is associated with the context of circular economy (Mahari et al. 2022), as it aims to maximize the use of biomass to generate bioproducts of interest and energy.

To analyze microalgae biorefinery processes with the Sustainable Development Goals (SDGs), some studies focused on this topic stand out. Olabi et al. (2023) discuss the aspects of microalgae bioprocesses that contribute to the 17 SDGs and the circular economy. Highlights include the relationship with SDG 6 "Clean water and sanitation" (related to the potential for bioremediation of wastewater by microalgae), SDG 7 "Affordable and clean energy" (related to the uses of microalgal biomass to generate bioenergy), SDG 9 "Industry, innovation and infrastructure" (due to research and development of technologies for bioethanol production) and SDG 13 "Climate action" (related to the process of capturing CO_2 during the cultivation of microalgae cells and generating numerous bioproducts).

Life cycle analysis (LCA) is a methodology that addresses all stages of production of a given product, standardized through ISO 14.040/2009. These steps include objective and scope, inventory analysis, life cycle impact analysis and interpretation of results. As positive points of applying the LCA methodology, the integration and connection of different processes and analysis of different boundaries stand out, such as: cradle-to-gate, gate-to-gate, gate-to-grave and cradle-to-serious.

Over the years, Life Cycle Assessment (LCA), which was initially used solely to map the inputs and outputs of industrial products, has been applied to a range of other processes. The LCA of microalgae biorefineries has been reported through integrated schemes, which consider stages such as cultivation, harvesting, biocompound extraction, digestion, and waste management (Soh et al. 2014). LCA research on microalgae addresses the production process and its impact on various areas, including eutrophication, scarcity of fossil resources, and global warming (Ubando et al. 2022). In terms of the LCA of biofuel processes using microalgae cells, bioethanol has been found to have a reduced carbon footprint due to the high accumulation of intracellular carbohydrates (Hossain et al. 2019). This study aims to analyze the life cycle of bioprocesses involved in the assessment of microalgal biomass, leading to the production of bioethanol in a biorefinery approach.

2 Material and Methods

2.1 *Goal and Scope Definition*

The scope of the LCA carried out in this study includes the activities and processes of obtaining bioethanol from the cultivation of Spirulina platensis microalgae in a medium containing cattle effluent. The study deepened the "cradle-to-gate" approach, in which the stages of production and generation of raw materials are considered until the gate of bioethanol generation, without considering the impacts of its uses.

The scope of the study as well as its purpose is justified by the question of pointing out possible process improvements, based on the environmental impacts generated in each of the stages of its production. The limits of LCA comprise the upstream and downstream stages of processes involving microalgae biotechnological routes. In this way, inputs and outputs of microalgal cultivation, biomass harvesting, biotechnological conversion processes up to the generation of bioethanol will be understood.

For this LCA, the logistical processes involved in the transport of materials, as well as the environmental impacts generated by the use of bioethanol, will not be assessed. Furthermore, it should be noted that entries relating to the energy consumption of equipment during laboratory experiments will be disregarded. The strategy becomes interesting when considering the conception of the life cycle in terms of introducing only inputs, excluding the noise caused by energy consumption in the process.

The ACV's functional unit will generate bioethanol from a total cultivation volume of 130 L in the raceway tank. At the stage of defining the scope and objectives, the quality of the data used will adhere to temporal, geographic, and technological coverage. This will consider aspects such as precision, integrity, representativeness, consistency, and reproducibility according to LCA methods (ABNT ISO, 2009). Priority was given to information from the database with an attributional approach (APOS) and the geographic composition in the context of the rest of the world (RoW), as it provides more reliable data that represents flows in Brazil.

2.2 *Life Cycle Inventory*

Inventory data collection was carried out using data from studies carried out by the research group of the Biochemistry and Bioprocesses Laboratory of the University of Passo Fundo, highlighting the following works:

(a) Magro et al. (2022): Cultivation of *S. platensis* in a medium with cattle effluent;
b) Rempel et al. (2019): Enzymatic hydrolysis and bioethanol productivity from *S. platensis* microalgae.

Among the data linked to upstream processes, the use and consumption of wastewater with the consumption of nutrients and additional source of nutrients stands out. The stages observed in the data collection for the inventory correspond to input data (material resources), avoided/valued products (anaerobic biodigestion effluent) and outputs (microalgal biomass, cultivation effluent).

For downstream processes, biomass is harvested using centrifuge methodologies. Next, there is drying of the biomass and the specific route of bioethanol, in this case bioethanol from microalgal biomass harvested via centrifugation.

In the same way as for preparing the inventory for upstream processes, each stage considered the input stages (inputs), valued products, and outputs of each process (waste and process effluent, final bioproduct). Figure 1 provides a general overview of the processes involved in the conversion of microalgal biomass to bioethanol in

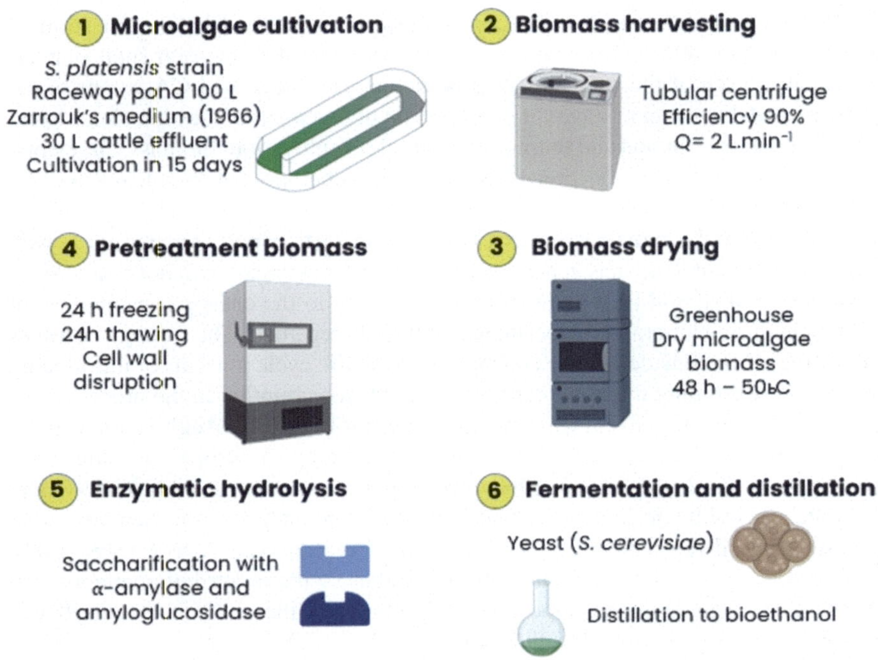

Fig. 1 Overview of bioethanol generation from microalgae cells

six steps. The efficiencies of 90% for biomass harvesting via centrifugation and 82% for saccharification efficiency of hydrolyzed sugars were taken into consideration (Rempel et al. 2019).

2.3 *Life Cycle Assessment*

The LCIA stage in accordance with ISO 14.042 (ABNT 2009), comprises the assessment of the significance of potential environmental impacts, highlighting the following phases: selection of impact categories, category indicators and characterization models, inventory results (classification), calculation of category indicator results (characterization), normalization, weighting, and analysis of data quality. For this LCIA, the Impact 2002+ method was used focused on the global warming (GW) impact category, in the unit of Kg CO_2 eq with damage in the climate change category. Impact 2002+ methodology proposes a feasible implementation of a combined midpoint/damage approach, linking all types of life cycle inventory results via 14 midpoint categories to four damage categories (Fig. 2) (SimaPro 2020).

For the classification and characterization of impacts, the SimaPro® software was used. The application of the methodology combined with the database present in the

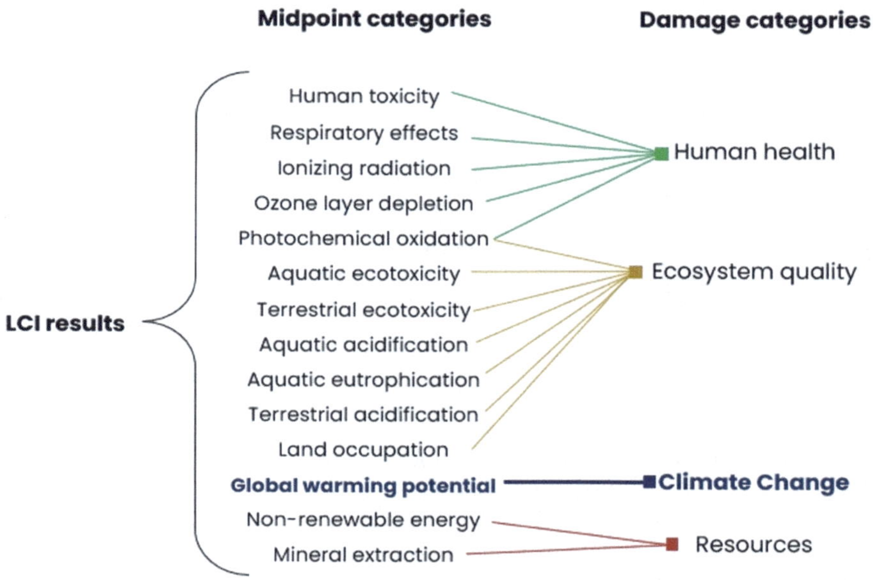

Fig. 2 Overall scheme of the IMPACT 2002+ framework highlighting GW

software is justified to include LCA studies in the treatment of effluents and biofuels, thus covering the approach to microalgae biorefineries. Furthermore, through the use of software, the results of the impact analysis are presented in tables and graphs.

3 Results and Discussions

The results indicate that the process evaluated in this life cycle analysis generates GW (global warming) impacts of 29.73 kg CO_2 eq., with emphasis on the fermentation and distillation stage, which had a contribution of 43.52 kg CO_2 equivalent (Fig. 3).

3.1 Cultivation, Harvesting and Drying

For the biomass cultivation, harvesting and drying stage, GW results of -20.69 kg CO_2 eq. (Fig. 4). Therefore, ungenerated impacts (with negative values) are related to the entry of cattle effluent into the cultivation, with microalgal cultivation being used as a technology for the final treatment of the effluent (Fig. 4). Furthermore, one of the factors that may result in the non-generation of impacts associated with the use of cattle effluent includes the non-generation of greenhouse gases (GHG), such as methane (CH_4) and nitrous oxide (N_2O). The GW impact category is related to

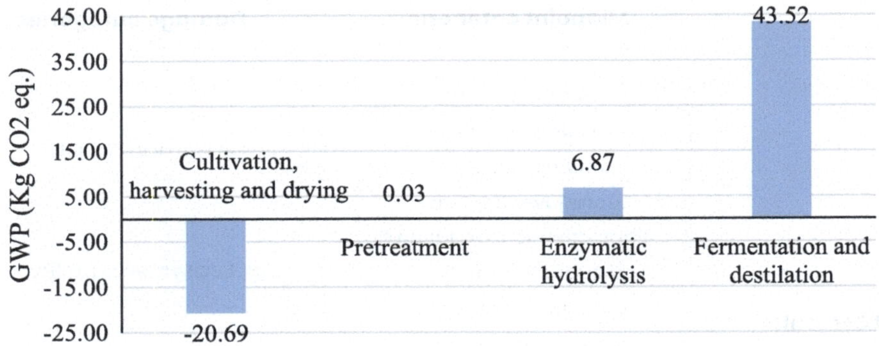

Fig. 3 Global warming (GW) results for life cycle analysis processes

CO_2 emissions from the burning of fossil fuels. In this sense, the negative results of GW are related to the capture of CO_2 during the photosynthesis phase in microalgal cultivation, having a negative carbon footprint (Müller et al. 2020).

Arashiro et al. (2022) researched the impacts associated with microalgae systems in the treatment of effluents, as well as the simultaneous recovery of bioproducts. Three different systems were compared: (a) domestic effluent; (b) food industry effluent; (c) conventional microalgae cultivation system. The results indicated that both processes with the treatment of effluent from microalgal cultivation presented better environmental performance, mainly related to the reduced consumption of chemical nutrients during cultivation.

The research is in line with the study by Magro et al. (2022), who also used industrial effluent during microalgal cultivation. For Arashiro et al. (2022) an also preponderant factor for the low carbon footprint was the production of biogas in the process, seeing an organic load in the industrial effluent of 2250 mg/L, highlighting the viability of biogas production.

In the scenario studied by Moradiya and Marathe (2023), researchers evaluated the cultivation and harvesting of marine microalgae biomass from the perspective of life cycle analysis. The approach used was cradle to gate, with the functional unit

Fig. 4 Results of GWP for the cultivation, harvesting and drying processes

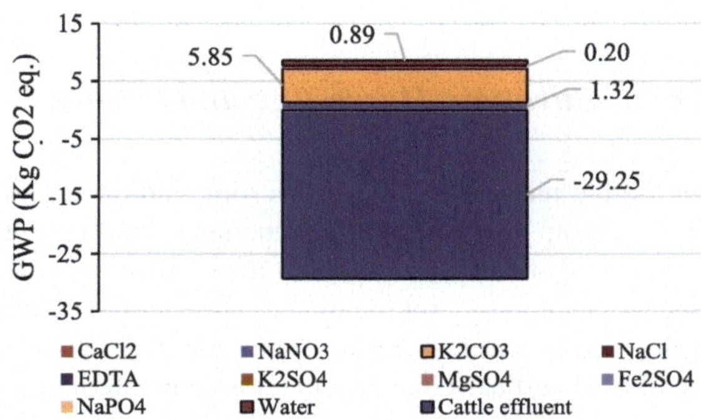

being 1 kg of algae on a dry basis. As a result, it is highlighted that for the global warming potential for daily biomass productivity of 35 g/m^2, the impact was -1.34 kg CO_2 eq/kg of algae. The study approach also aimed to change the basis of the energy matrix, for a scenario with 25, 50 and 100% solar energy, the mitigation of CO_2 generation reached up to 24%.

According to Maga (2017), greenhouse gas emissions per cubic meter of water treated by the microalgae biorefinery are 0.28 kg CO_2-eq. and those from wastewater treatment using conventional treatment models represent 0.47 kg CO_2-eq. Therefore, the sustainability of microalgal cultivation involves the introduction and simultaneous treatment of wastewater to increase cell concentration in the cultivation.

3.2 Pretreatment Stage

The pre-treatment stage presented impact results of 0.03 kg CO_2 eq., characterizing the stage as being close to carbon neutral. Optimizing the processes involving the pre-treatment of microalgae biomass involves reducing the introduction of inputs and reducing energy consumption, while simultaneously maintaining the biomass pre-treatment efficiency (Sankaran et al. 2020).

In the study by Wang et al. (2017), researchers analyzed the life cycle of microalgae biomass pretreatment by sonication, acid treatment and microwaves targeting pyrolysis. The research hypothesis proposed that pre-treated microalgae biomass requires less activation energy than biomass without pre-treatment. Overall, pre-treated samples showed higher greenhouse gas generations, highlighting that the process during biofuel production and extraction of valuable chemicals can be further researched to improve energy efficiency and cost-effectiveness.

3.3 Enzymatic Hydrolysis Stage

In relation to the enzymatic hydrolysis stage, the process resulted in 6.87 kg CO_2 eq. The only entry in this step is related to the commercial enzymes α-amylase and amyloglucosidase, used in the enzymatic hydrolysis process. Aiming for greater sustainability of the process, technological advances are necessary aimed at greater productivity of enzymes by microorganisms (Kamaraj and Subramaniam 2020) while combining strategies for valuing agro-industrial solid waste during the cultivation of these microorganisms (Almanaa et al. 2020; Leite et al. 2021).

In the study by Huang et al. (2020), the authors examined the productivity of bioethanol generation from microalgae through four different routes. They compared the life cycles of each process. Among the routes considered, the alternative that had the most significant impacts was the route involving enzymatic hydrolysis, liquefaction, and saccharification. This was especially true when considering the impact category of human toxicity. On the other hand, the route with the lowest impacts was

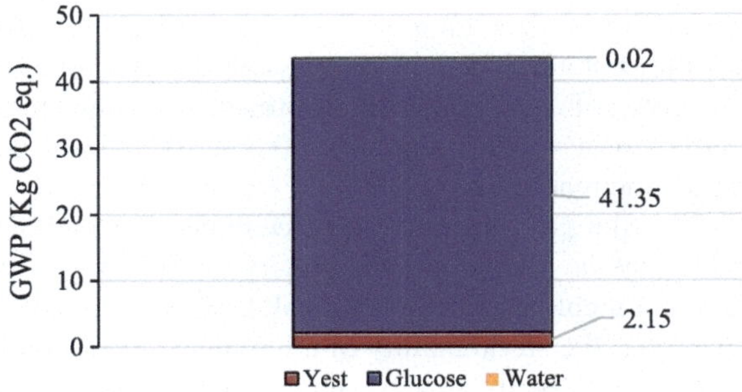

Fig. 5 Results of GWP for the pretreatment process

the extraction of pigments through sonication and acetone. According to the authors, the impact category of human toxicity is demonstrated by the emission of Se, Mg, Ba, As, and Mo into the environment. For instance, Se found in microalgal cells may be absorbed by aquatic organisms at a lower trophic level, potentially causing human toxicity through food consumption.

3.4 Fermentation and Distillation

The fermentation and distillation stage includes the preparation of the yeast inoculum, with inputs such as S. cerevisiae yeast, peptone and glucose (Fig. 5). The greatest contribution of impacts is related to glucose with 41.35 kg CO_2 eq. An, Evelien and Katrien (2015) report that the agricultural phase for preparing liquid glucose is the most significant contribution to CO_2 emissions. Furthermore, depending on the glucose production process, energy consumption, enzyme input, and glucose yield significantly affect the environmental load (Abu-Bakar et al. 2023).

In the study by Azevedo et al. (2016) researchers analyzed the LCA of the ethanol production process from lignocellulosic material contained in cattle manure. As a result, it is highlighted that the fermentation and distillation stage presented greater impacts associated with the carcinogenic impact category. Overall, when evaluating all stages, distillation and fermentation ranked second and fourth, respectively, among the stages with the greatest impact on the generation of bioethanol.

4 Conclusions

Life cycle analysis applied to biotechnological processes of microalgae allows a detailed interpretation of each input and output focused on the stages of the processes. Considering third-generation biofuels, this approach makes it possible to analyze the potential impacts in different categories and, thus, define improvement and consolidation strategies with lower costs and greater productivity.

When considering these aspects, it was concluded that the strategy of bioethanol routes from microalgae biorefineries stands out as promising. The main aspect that contributes to the sustainability of the process is related to the use of effluents during the microalgal cultivation phase. However, the fermentation and distillation stages still require more in-depth studies to reduce the impacts associated with global warming potential. Therefore, the stages of strain selection and microalgal cultivation have greater consolidation of technological aspects, while downstream processes still require further studies aimed at environmental and financial sustainability.

References

Abu-Bakar NA, Roslan AM, Hassan MA, Rahman MHA, Ibrahim KN, Abd Rahman MD, Mohamad R (2023) Environmental impact assessment of rice mill waste valorisation to glucose through biorefinery platform. Sci Rep 13(1):14767. https://doi.org/10.1038/s41598-023-28487-2

Almanaa TN, Vijayaraghavan P, Alharbi NS, Kadaikunnan S, Khaled JM, Alyahya SA (2020) Solid state fermentation of amylase production from Bacillus subtilis D19 using agro-residues. J King Saud Univ-Sci 32(2):1555–1561. https://doi.org/10.1016/j.jksus.2019.12.011

ASSOCIAÇÃO BRASILEIRA DE NORMAS TÉCNICAS (ABNT) (2009) ISO 14.040: Gestão ambiental. Avaliação do ciclo de vida (ACV): princípios e estrutura. Rio de Janeiro, RJ

de Azevedo A, de Cassia de Souza Schneider R, de Souza D, Hoeltz M (2016) Análise do ciclo de vida aplicada à produção de bioetanol a partir de material Lignocelulósico remanescente em dejeto bovino. Tecno-Lógica 20(2):118–128. https://doi.org/10.17058/tecnolog.v20i2.7681

Hossain N, Zaini J, Mahlia TMI (2019) Life cycle assessment, energy balance and sensitivity analysis of bioethanol production from microalgae in a tropical country. Renew Sustain Energy Rev 115:109371. https://doi.org/10.1016/j.rser.2019.109371

Huang X, Bai S, Liu Z, Hasunuma T, Kondo A, Ho SH (2020) Fermentation of pigment-extracted microalgal residue using yeast cell-surface display: direct high-density ethanol production with competitive life cycle impacts. Green Chem 22(1):153–162. https://doi.org/10.1039/C9GC02634G

Kamaraj M, Subramaniam D (2020) Amylase production by Aspergillus niger in submerged cultivation using cassava. J Appl Biol Biotechnol 8(6):82–87. https://doi.org/10.7324/JABB.2020.80613

Leite P, Sousa D, Fernandes H, Ferreira M, Costa AR, Filipe D, Salgado JM (2021) Recent advances in production of lignocellulolytic enzymes by solid-state fermentation of agro-industrial wastes. Curr Opin Green Sustain Chem 27:100407. https://doi.org/10.1016/j.cogsc.2020.100407

Ma X, Mi Y, Zhao C, Wei Q (2022) A comprehensive review on carbon source effect of microalgae lipid accumulation for biofuel production. Sci Total Environ 806:151387. https://doi.org/10.1016/j.scitotenv.2021.151387

Maga D (2017) Life cycle assessment of biomethane produced from microalgae grown in municipal wastewater. Biomass Convers Biorefin 7:1–10. https://doi.org/10.1007/s13399-016-0208-8

Magro FG, Freitag JF, Bergoli A, Costa JAV, Colla LM (2022) Microalgae consortia for post-treating effluent of anaerobic digestion of cattle waste and evaluation of biochemical composition of biomass. Bioenerg Res 15:371–384. https://doi.org/10.1007/s12155-021-10270-8

Mahari WAW, Razali WAW, Manan H, Hersi MA, Ishak SD, Cheah W, Lam SS (2022) Recent advances on microalgae cultivation for simultaneous biomass production and removal of wastewater pollutants to achieve circular economy. Bioresour Technol 128085. https://doi.org/10.1016/j.biortech.2022.128085

Moradiya KK, Marathe KV (2023) Life cycle assessment (LCA) of marine microalgae cultivation and harvesting process for the Indian context. Sustain Energy Technol Assess 56:103063. https://doi.org/10.1016/j.seta.2023.103063

Müller LJ, Kätelhön A, Bachmann M, Zimmermann A, Sternberg A, Bardow A (2020) A guideline for life cycle assessment of carbon capture and utilization. Front Energy Res 8. https://doi.org/10.3389/fenrg.2020.00015

Olabi AG, Shehata N, Sayed ET, Rodriguez C, Anyanwu RC, Russell C, Abdelkareem MA (2023) Role of microalgae in achieving sustainable development goals and circular economy. Sci Total Environ 854:158689. https://doi.org/10.1016/j.scitotenv.2022.158689

Rempel A, de Souza Sossella F, Margarites AC, Astolfi AL, Steinmetz RL, Kunz A, Treichel H, Colla LM (2019) Bioethanol from Spirulina platensis biomass and the use of residuals to produce biomethane: An energy efficient approach. Bioresour Technol 288:121588. https://doi.org/10.1016/j.biortech.2019.121588

Sankaran R, Cruz RAP, Pakalapati H, Show PL, Ling TC, Chen WH, Tao Y (2020) Recent advances in the pretreatment of microalgal and lignocellulosic biomass: a comprehensive review. Biores Technol 298:122476. https://doi.org/10.1016/j.biortech.2019.122476

SimaPro (2020) Database manual: methods library. PRé, version 4.15

Soh L, Montazeri M, Haznedaroglu BZ, Kelly C, Peccia J, Eckelman MJ, Zimmerman JB (2014) Evaluating microalgal integrated biorefinery schemes: empirical controlled growth studies and life cycle assessment. Bioresour Technol 151:19–27. https://doi.org/10.1016/j.biortech.2013.10.012

Ubando AT, Ng EAS, Chen WH, Culaba AB, Kwon EE (2022) Life cycle assessment of microalgal biorefinery: a state-of-the-art review. Bioresour Technol 127615. https://doi.org/10.1016/j.biortech.2022.127615

Wang X, Guo F, Li Y, Yang X (2017) Effect of pretreatment on microalgae pyrolysis: kinetics, biocrude yield and quality, and life cycle assessment. Energy Convers Manage 132:161–171. https://doi.org/10.1016/j.enconman.2016.11.006

Life Cycle Assessment of Microalgal Biomass Production in Photobioelectrochemical System

Vanessa Rosana Ribeiro, Anna Giullia Quintana Soder, Dariane Severgnini, Enio Leandro Machado, Lisianne Brittes Benitez, and Rosana de Cassia de Souza Schneider

Abstract The Microalgal Microbial Fuel Cell (MMFC), a type of photobioelectrochemical system, possesses considerable potential for treating effluents, generating biomass, and producing bioenergy. This study assessed the environmental performance of microalgal biomass production within a photobioelectrochemical system, contrasting it with a conventional bench-scale production system. Our findings emphasize the noteworthy role of electricity consumption in providing light, facilitating cathode agitation, and enabling anode medium recirculation, all of which significantly influence the environmental impact. Therefore, the LCA study revealed essential insights about the MMFC system, demonstrating its potential to mitigate environmental impacts and to provide biomass recovery and bioelectricity generation. The MMFC system exhibited superior biomass production and reduced environmental repercussions compared to the conventional system, which reduced the impact in all assessed categories. The results present the importance of MMFCs in sustainable microalgal biomass production, providing valuable insights into their potential for efficient wastewater treatment and renewable energy generation. In this context, the LCA can guide the better way to conduct the MMFC systems, allowing for the identification of high efficiency in terms of bioproducts and less environmental impacts.

Keywords Microalgae · Microalgal microbial fuel cell · Photobioelectrochemical system · Effluent · Biomass · Environmental impacts

V. R. Ribeiro · A. G. Q. Soder · D. Severgnini · E. L. Machado · L. B. Benitez · R. de Cassia de Souza Schneider (✉)
Centre of Excellence in Oleochemical and Biotechnological Products and Processes, University of Santa Cruz do Sul, Santa Cruz do Sul, Rio Grande do Sul, Brazil
e-mail: rosana@unisc.br

Environmental Technology Post Graduation Program, University of Santa Cruz Do Sul, Santa Cruz do Sul, Rio Grande do Sul, Brazil

V. R. Ribeiro
e-mail: vanessa.rrib@hotmail.com

L. Brandli et al. (eds.), *The Contribution of Life Cycle Analyses and Circular Economy to the Sustainable Development Goals*, World Sustainability Series,
https://doi.org/10.1007/978-3-031-85300-5_19

1 Introduction

There is a global challenge in reducing the use of fossil fuel-based resources to decrease the carbon footprint. Researchers from several fields concerned about the future have been striving to develop technologies capable of replacing conventional polluting methods with alternative sources of clean and renewable energy. The shift from a fossil fuel-dependent linear economy to a circular bioeconomy is imperative amidst worldwide apprehensions regarding the limited availability of environmental resources (Igbokwe et al. 2022).

Photobioelectrochemical Systems (PBES) have recently gained significant attention, as they primarily harness natural environmental resources. Effluents serve as a nutrient source for microorganisms, which can effectively bioremediate and generate high-value bioproducts.

The microorganisms present in PBES metabolize organic material in the anodic chamber under aerobic and anaerobic conditions. These microorganisms, known as electricigens, transfer the generated electrons directly to the cathodic chamber, which is essential in generating electricity in the system (Fan and Xue 2016). This innovative approach holds great promise in advancing sustainable energy solutions and addressing environmental concerns on a global scale (Cetinkaya et al. 2017, 2022). Therefore, these systems can convert the organic matter in wastewater into bioenergy, such as biohydrogen and bioelectricity, through the catalytic activity of bacteria and fungi. Additionally, they play a crucial role in the production of microalgae biomass. This multifaceted approach not only addresses the urgent need for cleaner energy production but also contributes to sustainable wastewater treatment and the cultivation of valuable bioresources (Cetinkaya et al. 2017; Ribeiro et al. 2021; Hosseinzadeh et al. 2022).

Microalgae are autotrophic microorganisms that exhibit great metabolic flexibility and rapid growth under different cultivation conditions. They inhabit marine ecosystems, freshwater, and soil. During photosynthesis, they transform nutrients, carbon dioxide, and sunlight into organic compounds (Dolganyuk et al. 2020).

The role of microalgae in these systems has been extensively studied, mainly through Microalgal Microbial Fuel Cells (MMFCs), due to the remarkable capacity of these photosynthetic organisms to metabolize various compounds. Consequently, they exhibit high efficiency in pollutant remediation. The integration of microalgae into such systems not only enhances their environmental remediation capabilities but also offers a promising avenue for sustainable bioenergy production and resource recovery (Hendroko Setyobudi et al. 2021; Ribeiro et al. 2022).

Microalgae are broadly used to enhance MMFC performance due to (1) their potential to catalyze the oxygen reduction reaction (ORR), (2) low space requirements, (3) rapid growth, and (4) the production of high-value-added products, including bioelectricity. (Ahirwar et al. 2023; Sharma et al. 2024a, b). Many studies aim to improve the microalgae efficiency in these systems, using wild and genetically modified types to determine the best approaches for enhancing these innovative technologies (Sharma et al. 2024a, b).

Microalgae are highly effective at cleaning water due to their high capacity to degrade various pollutants, including chemicals (Deng et al. 2022; Hou et al. 2024) and pharmaceuticals (Kumar et al. 2024), commonly found in the environment. Their bioremediation efficiency is even higher when integrated into MMFCs, resulting in increased biomass production.

Furthermore, the residual microalgae biomass generated in these systems can be utilized for various purposes, given its known efficiency in biofuel production. This multifunctional approach highlights the potential of MMFCs with several microalgae for efficient environmental remediation, sustainable biofuel production, and resource utilization (Lee et al. 2015; Kusmayadi et al. 2020).

Despite the great diversity of photosynthetic organisms with use potential, algae have emerged as one of the most techno-economically viable options in photosynthetic microbial fuel cells for wastewater treatment and bioelectricity generation (Ahirwar et al. 2023).

However, there is a significant need for research and technological advancement aimed at scaling up to substantially increase the production of bioproducts and address the challenges associated with the overall economic viability of the process (Savla et al. 2021).

Life Cycle Assessment (LCA) is a widely used tool in various fields of study. LCA can aid in decision-making by enabling the adaptation and improvement of the stages of a process to achieve better economic and environmental performance (Mälkki and Alanne 2017; Çetinkaya et al. 2018). LCA is a quantitative methodology that examines the environmental consequences of a product or process's life cycle. It encompasses standardized stages, clear objectives and scope definition, life cycle inventory, life cycle impact assessment, and the interpretation of the obtained results (International Organisation for Standardisation 2006). Thus, the system boundaries and the data quality comprising the input and output flow of materials and energy are crucial factors (Savla et al. 2021; Chin et al. 2022).

According to Razman et al. (2022), the number of LCA studies on wastewater treatment technologies has been increasing, showing that in the last decade (2011–2020), there was a significant leap compared to the previous decade (2000–2010). In 2020, the number of studies doubled compared to previous years, demonstrating a growing interest in LCA as an assessment tool in wastewater treatment systems.

Some research efforts have already been directed toward developing LCA methodologies for various bioelectrochemical systems, as evidenced by studies conducted by Zhang et al. (2019a, b), Singh Mathuriya et al. (2020), Cetinkaya et al. (2022), Chin et al. (2022). These studies have analyzed these systems to seek compelling insights into their performance. However, there is still a lack of studies specifically focusing on developing LCA methodology based on MMFC system operation, which adds an innovative dimension to this study.

As the demand for sustainable and environmentally friendly technologies continues growing, understanding the life cycle impacts of these emerging bioelectrochemical systems becomes paramount. The MMFC became an alternative technology with high performance regarding pollutant bioremediation and bioproduct generation. Therefore, focusing on research in this field is increasing, and understanding the best way to conduct these technologies is extremely important to reducing environmental impacts. The gaps in this system can be recognized by presenting a comprehensive LCA approach to MMFC, shedding light on their environmental and economic implications, and contributing to the ongoing efforts towards sustainable bioenergy production and wastewater treatment.

2 Objective and Scope

This study's main goal was to estimate the environmental impacts of the operational parameters of the MMFC system compared to conventional microalgae cultivation in the cradle-to-gate process. The data collection related to the installation, operation, and final disposal product of an MMFC, applied to the LCA tool, aims to comprehend its environmental performance in terms of carbon footprint.

LCA has become a widely adopted tool in environmental discussions, allowing comprehensive insights into processes and technologies and the understanding of the potential impacts primarily associated with carbon sequestration. The MMFC technology was developed to bring environmental benefits through the products it generated, such as bioethanol, bioelectricity, biogas, high-value-added biomass, and bioremediation (Ribeiro et al. 2021). However, there is still a gap in research in this area to fully understand the entire process involved in the functioning of these systems.

The MMFC works to bioremediate wastewater and, consequently, produce bioenergy. Bacteria promote the degradation of pollutants while simultaneously producing electrons and hydrogen ions. The electrons pass from the anodic side to the cathodic side via the copper wire bridge that connects the electrodes, and the hydrogen ions pass through the Nafion membrane (Ribeiro et al. 2022). The microalgae use these ions and electrons to release water. This chemical species improves the microalgae bioremediation performance, which act as a biological electron acceptor. The main reactions were in Eq. 1 (anode) and 2 and 3 (cathode) (Christwardana et al. 2020).

$$\text{organic carbon} \xrightarrow{\text{Bacteria}} CO_2 + H^+ + e^- \tag{1}$$

$$6CO_2 + 12H^+ + \text{light} + 12e^- \xrightarrow{\text{Microalgae}} \text{biomass} + 3O_2 \tag{2}$$

$$O_2 + 4H^+ + 4e^- \rightarrow 2H_2O \tag{3}$$

These systems can be applied to various contaminants, whether urban or industrial wastewaters, easily biodegradable molecules, or persistent ones. This means that the potential applications of these systems for wastewater treatment and bioenergy generation are extensive. The studied MMFC uses various microalgae and bacteria to provide bioremediation, bioelectricity, and biomass production.

The acquisition of data will provide empirical knowledge about the MMFC and the most relevant impacts for future decision-making, with the aim of scaling up. In the context of this research, the LCA system boundary compares two systems (Fig. 1). The MMFC system initiates the continuous production of microalgae inoculum, which is consistently introduced into the cathodic chamber. The initial bacteria inoculum utilized in the anodic chamber does not necessitate regeneration, as the bacterial sludge from the chamber can be recycled for subsequent treatments. The conventional system was similar to the cathodic chamber but lacked electrodes or connections to other chambers.

The MMFC utilized in this research comprises two glass chambers (one anode and one cathode), each with a combined volume of 300 mL, separated by a thin proton-exchange membrane (specifically, Nafion® NR212), which permits the selective transport of hydrogen ions from the anode chamber to the cathode chamber. These chambers are securely fastened together by a polypropylene clip designed for this system.

A hermetic acrylic reservoir with a maximum capacity of 1,300 mL was employed, connected to the anodic side of the system. The purpose was to induce liquid recirculation within the chamber using a submerged aquarium pump with a maximum flow

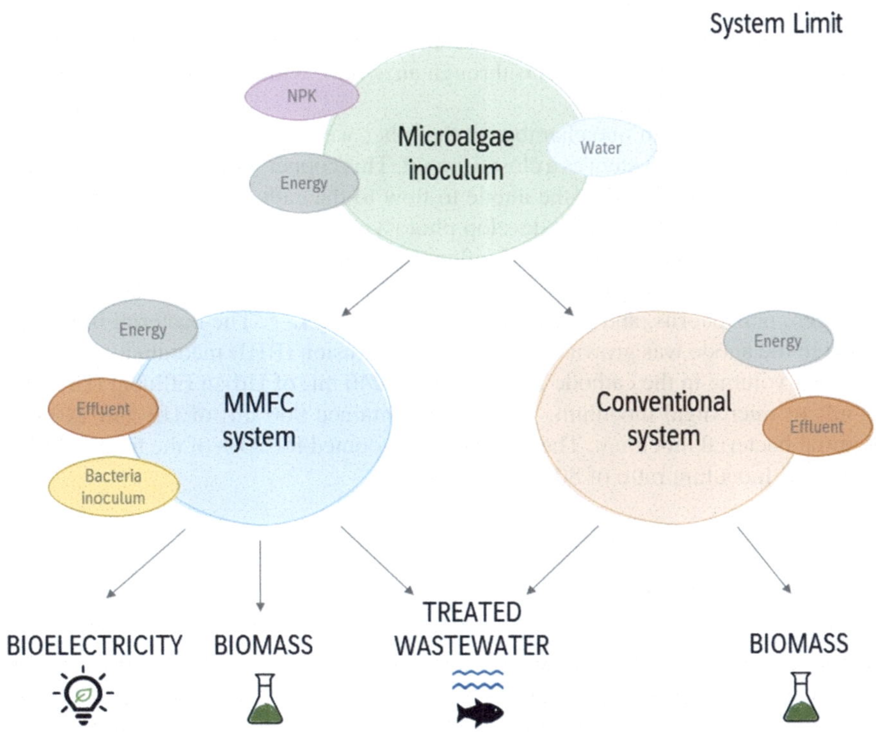

Fig. 1 System limit studied by LCIA

Fig. 2 Operational MMFC system

rate of 180 Lh^{-1} and a power rating of 2.5 W. The connection between the chamber and the reservoir is established through Teflon hoses. An airlock gas release filter was attached to the reservoir lid, allowing for the release of potential gases produced by bacteria while preventing the entry of oxygen.

Graphite bar electrodes (10.0 × 1.0 × 0.2 cm) were adapted onto the chamber lids. On the anodic side, the electrode was affixed with a thermosetting adhesive that hardens when mixed with epoxy resin (Durepóxi, Loctite). Additionally, a small opening was made in the cathode lid to allow for the passage of a silicone hose, facilitating the introduction of CO_2 through an aerator with a capacity of up to 40 L and a power rating of 2.5 W.

The electrodes from one chamber to the other were interconnected with copper wire, establishing the system as a closed circuit. This copper wire enables the conduction of electrons produced at the anode to flow to the cathode. Direct illumination was provided to the cathode to develop photosynthetic organisms. The anode was covered with aluminum foil to avoid influencing bacterial growth (Fig. 2).

The photosynthetic organisms introduced to the cathode were cultivated in an NPK (nitrogen, phosphorus, and potassium) medium at 3 g L^{-1}. The bacterial inoculum added to the anode was grown in a Brain Heart Infusion (BHI) medium at 37 g L^{-1}. The total volume in the cathode corresponds to 240 mL of Urban Effluent (UE) and 60 mL of microalgal inoculum. The anode contained 960 mL of UE and 240 mL of initial bacterial inoculum. The inoculums accounted for 20% of the total volume, with a UE:Inoculum ratio of 80:20 (v/v).

3 Life Cycle Inventory

In order to understand the environmental performance of the tested MMFC system, all input and output values of a bench-scale MMFC system operated for up to 28 days under direct artificial illumination (LED lamps) were listed and quantified. The SimaPro software was used to generate LCA data. The graphite electrodes had a mass of 1.8 g each, and the mass of the Nafion membrane was not considered, as it

Table 1 Volumes used in the MMFC and Conventional System

MMFC	Anode	Cathode
Urban effluent	960 mL	240 mL
Bacterial inoculum	240 mL (BHI* + bacteria)	–
Microalgal inoculum	–	60 mL (NPK (12:11:18) 0.5 g L^{-1} + microalga)
Total volume	1.200 mL	300 mL
Total energy	1.064 Wh	
Biomass	–	1.799 g L^{-1}
Avoided water	1200 mL	240 mL
Conventional	Effluent	Inoculum
Composition	240 mL	60 mL (NPK (12:11:18) 0.5 g L^{-1} + microalga)
Avoided water	240 mL	–
Biomass	0.481 g L^{-1}	
Total energy	1.288 kW	

* Not considered as it is a medium used only in the initial first inoculum

is reused and approximately 2 cm in diameter. The voltage used to power the LED lamps, aquarium pump, and aerators was 220 V, operating 24 h daily.

For comparative purposes, an analysis of a separate conventional bench-scale system was conducted using the same volumes involved in the cathode of the MMFC system (Table 1).

The energy consumption for production was calculated based on the duration the system remained operational, considering a photoperiod of 16:8 (light/dark) for both the MMFC system and the conventional bench-scale system. Two lamps (9 W) and four aeration pumps (2.5 W) were used for light supply. Energy and input consumption calculations were performed to produce 1 L of medium, resulting in 1064 W and 1288 W consumption values for the MMFC and conventional bench-scale systems, respectively.

The biomass obtained in each system was determined after the completion of the process, resulting in 1.799 g L^{-1} for the MMFC system and 0.481 g L^{-1} for the conventional bench-scale system.

4 Life Cycle Impact Assessment (LCIA)

MMFCs are technologies capable of producing various bioenergy-related products. They are renewable resources that can be used for clean energy production (Pant et al. 2011). LCIA was an essential tool for recognizing the impacts associated with each input or output of the process, as can be observed in Table 3, based on the inventory conducted and secondary data obtained from the Ecoinvent 3.0 database using SimaPro 9.5.0.1 software. The selected method for impact assessment was Environmental Footprint 3.1 (adapted for Simapro substances). The LCIA was applied to compare the systems (MMFC and conventional cultivation) considering normalized damages and uncertainties (excluding avoided energy and water) obtained from Monte Carlo analysis using $n = 1000$ iterations with a 95% confidence level.

The most considerable impact of electrical energy usage is evident, standing out among the other inputs, as expected, since the system uses electrical energy throughout the process for photoperiod, aeration, and recirculation. In these data, energy consumption is the most significant contributor to water use, eutrophication, climate change, ionizing radiation, ecotoxicity, land use, and fossil resources. These categories impact human health, ecosystems, and resources, making it an input that affects all evaluated dimensions.

MMFC was developed to produce bioelectricity, remediate effluents, and produce biomass. However, despite mitigating potential adverse environmental effects, inputs are associated with inoculum production to ensure that the produced biomass primarily comprises inoculated microalgae. This includes the consumption of inorganic fertilizers, which also contribute to impacts associated with their life cycle, with a concerning relationship between the nutrients extracted and what is utilized for cultivation (Panagos et al. 2022).

In our system, we did not account for the bioelectricity produced, as its value is significantly low compared to the energy inventory required for microalgae production in the bench-scale MMFC system, as observed by Sahaq and Hadiyanto (2019) and Ribeiro et al. (2022).

In the evaluation using the Environmental Footprint 3.0 method, the 27 categories presented in Table 2 were assessed. The results highlighted the impact of electricity on water use, ecotoxicity, and climate change categories. Additionally, a greater impact on ecotoxicity was observed due to nutrient use in the microalgae chamber. The results indicate that eutrophication has a greater impact on terrestrial eutrophication than marine and freshwater eutrophication.

Regarding nutrients, using NPK results in impacts associated with K, even though it is not a limiting nutrient for microalgae production. To reduce this impact, using other NPK compositions can be considered. Reducing K and increasing P makes higher productivity possible (Rocha et al. 2024). Besides reducing the impactful input, higher productivity can result in a lower overall environmental impact. From these categories, those with the highest values were selected after normalization based on the damages shown in Fig. 3.

Figure 3a illustrates the impact categories with a high contribution from energy use, considering the electricity used in Southern Brazil, as per the life cycle data from the Ecoinvent 3.0 database. For the MMFC, the consumption of equipment used for lighting 16 h per day and pumps 24 h per day was considered. The electricity consumption in the system was associated with the days required for treatment and microalgae biomass production. Energy consumption can be an inventory input that

Table 2 Impacts identified by categories using the environmental footprint method 3.1

Impact category	Unit	Total	N	P	K	Water	Graphite	Electricity	Water avoided
Water use	m^3 depriv	3.55E + 02	4.09E−02	1.21E−02	6.37E−03	−3.50E−01	5.34E−03	4.73E + 02	−8.96E + 00
Human toxicity, cancer—organics	CTUh	9.69E−08	7.29E−12	2.90E−12	2.89E−12	9.76E−11	6.11E−12	9.93E−08	−2.50E−09
Human toxicity, non-cancer—organics	CTUh	1.64E−07	2.66E−11	2.18E−11	1.19E−11	1.02E−10	1.03E−11	1.66E−07	−2.60E−09
Human toxicity, cancer—inorganics	CTUh	3.62E−08	1.78E−11	9.33E−12	5.27E−12	2.57E−10	8.25E−12	4.25E−08	−6.58E−09
Human toxicity, cancer	CTUh	1.33E−07	2.50E−11	1.22E−11	8.16E−12	3.54E−10	1.44E−11	1.42E−07	−9.08E−09
Ozone depletion	kg CFC11 eq	3.03E−06	6.31E−10	2.13E−10	1.62E−10	2.74E−09	2.06E−10	3.09E−06	−7.03E−08
Human toxicity, non-cancer—inorganics	CTUh	2.93E−06	6.63E−10	5.54E−10	2.13E−10	5.71E−09	3.33E−10	3.07E−06	−1.46E−07
Human toxicity, non-cancer	CTUh	3.10E−06	6.89E−10	5.76E−10	2.25E−10	5.82E−09	3.43E−10	3.24E−06	−1.49E−07
Particulate matter	disease inc	2.54E−05	2.27E−09	1.59E−09	9.05E−10	7.17E−09	3.53E−09	2.56E−05	−1.84E−07
Resource use, minerals, and metals	kg Sb eq	2.43E−03	1.04E−06	4.96E−07	2.28E−07	1.96E−06	6.00E−08	2.48E−03	−5.02E−05
Eutrophication, freshwater	kg P eq	2.08E−02	1.27E−05	9.81E−06	3.15E−06	2.71E−05	1.38E−05	2.14E−02	−6.96E−04
Eutrophication, marine	kg N eq	1.49E−01	4.55E−05	2.10E−05	1.43E−05	8.13E−05	5.43E−05	1.50E−01	−2.08E−03
Photochemical ozone formation	kg NMVOC eq	4.40E−01	1.33E−04	6.92E−05	5.00E−05	3.07E−04	1.67E−04	4.47E−01	−7.87E−03
Acidification	mol H^+ eq	6.19E−01	2.56E−04	1.50E−04	6.48E−05	5.66E−04	2.62E−04	6.32E−01	−1.45E−02
Eutrophication, terrestrial	mol N eq	1.51E + 00	6.15E−04	2.72E−04	1.72E−04	8.54E−04	6.28E−04	1.53E + 00	−2.19E−02

(continued)

Table 2 (continued)

Impact category	Unit	Total	N	P	K	Water	Graphite	Electricity	Water avoided
Climate change—Land use and LU change	kg CO_2 eq	1.24E + 01	8.83E−05	8.41E−05	1.67E−05	1.52E−03	5.50E−05	1.25E + 01	−3.88E−02
Climate change—Biogenic	kg CO_2 eq	1.83E + 01	3.94E−05	2.16E−05	2.04E−05	2.14E−03	4.78E−05	1.83E + 01	−5.47E−02
Ionizing radiation	kBq U-235 eq	2.19E + 01	1.24E−03	8.69E−04	4.10E−04	3.95E−03	2.56E−03	2.20E + 01	−1.01E−01
Ecotoxicity, freshwater—organics—p.1	CTUe	1.19E + 02	6.82E−03	5.05E−03	5.41E−03	2.09E−02	3.43E−03	1.20E + 02	−5.35E−01
Ecotoxicity, freshwater—organics—p.2	CTUe	8.05E + 01	3.21E−03	1.88E−03	1.78E−03	3.00E−02	4.55E−03	8.12E + 01	−7.68E−01
Climate change—Fossil	kg CO_2 eq	8.10E + 01	3.99E−02	1.62E−02	1.00E−02	6.41E−02	3.78E−02	8.25E + 01	−1.64E + 00
Climate change	kg CO_2 eq	1.12E + 02	4.01E−02	1.63E−02	1.01E−02	6.78E−02	3.79E−02	1.13E + 02	−1.74E + 00
Ecotoxicity, freshwater—part 1	CTUe	1.84E + 02	2.10E−01	5.77E−02	4.12E + 00	2.25E−01	1.01E−01	1.85E + 02	−5.75E + 00
Ecotoxicity, freshwater—part 2	CTUe	3.98E + 02	4.89E−01	5.81E−02	3.00E−02	2.56E−01	2.99E−02	4.04E + 02	−6.56E + 00
Ecotoxicity, freshwater—inorganics	CTUe	3.82E + 02	6.88E−01	1.09E−01	4.14E + 00	4.30E−01	1.23E−01	3.88E + 02	−1.10E + 01
Land use	Pt	1.06E + 03	1.73E−01	2.00E−01	1.33E−01	6.49E−01	1.32E−01	1.08E + 03	−1.66E + 01
Resource use, fossils	MJ	1.41E + 03	5.10E−01	1.86E−01	1.33E−01	7.44E−01	4.17E−01	1.43E + 03	−1.91E + 01

Table 3 Environmental impacts associated with the main damage categories and their respective uncertainties (Excluding avoided energy and water) obtained from Monte Carlo analysis using n = 1000 iterations with a 95% confidence level

Damage categories	Impact (%)	MMFC		Conventional	
	Conventional > MMFC	Average	Uncertainty	Average	Uncertainty
Acidification (mol H^+ eq)	100	1.0E−03	9.0E−04	3.0E−03	9.9E−02
Climate change (kg CO_2 eq)	100	1.3E−01	9.0E−04	3.6E−01	1.1E−01
Ecotoxicity, freshwater (CTUe)	47.7	3.3E + 05	2.3E−01	1.0E + 06	2.7E + 05
Eutrophication, freshwater (kg P eq)	100	0.0E + 00	4.0E−03	1.0E−04	3.4E−01
Eutrophication, marine (kg N eq)	100	2.0E−04	1.0E−03	4.0E−04	1.1E−01
Eutrophication, terrestrial (mol N eq)	100	2.0E−03	1.0E−03	5.2E−03	1.1E−01
Land use (Pt)	100	8.2E−01	1.6E−03	2.7E + 04	1.8E−01
Photochem. O_3 formation (kg NMVOC eq)	100	5.0E−04	1.0E−03	1.3E−03	1.2E−01
Resource use, fossils (MJ)	100	1.5E + 00	1.3E−03	4.2E + 04	1.4E−01
Water use (m^3 depriv.)	99.7	−1.3E + 02	−4.0E−04	−9.0E + 05	−1.8E−01

can be reduced, considering that the type of microalgae and the biomass profile may be different at the end of the process.

The energy was higher because the experiment was conducted on a lab scale. On an industrial scale, the ratio of energy consumed to biomass produced is expected to be lower. Obtaining more biomass per unit of energy and nutrient consumption is crucial to mitigate these impacts. Water impacts can be reduced when production is conducted in wastewater (avoiding using tap water). Additionally, other inputs become less significant due to the high impact value of electricity across all damage categories. Some categories are impacted by the life cycle of energy production in Brazil. These categories are no longer affected when energy is removed as an input. Figure 3b shows the impacts without energy. These results highlight the importance of using wastewater as a culture medium, resulting in a negative impact, which can be described as a suppressed impact. This balances the impact contributions associated with the life cycle of the other inventory inputs.

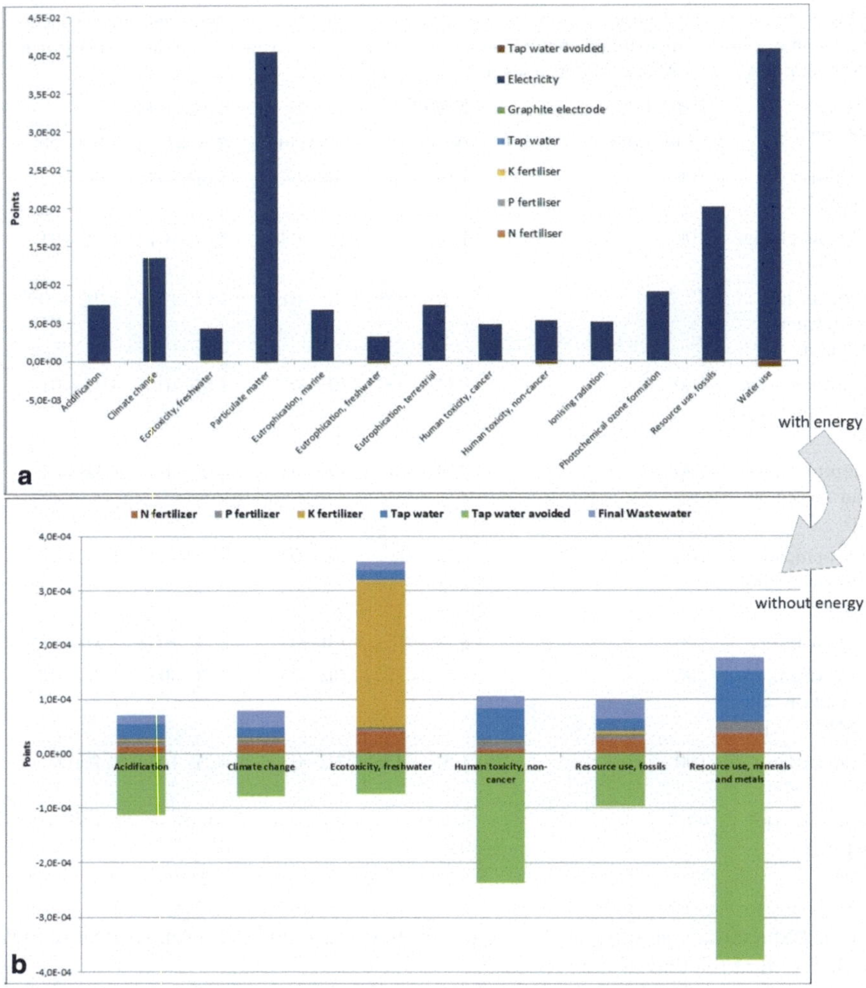

Fig. 3 Main damage categories of the MMFC. **a** Normalized impacts of the system with energy and water treated avoided due to effluent use, and **b** impacts without energy and water treated avoided

Furthermore, an environmental advantage becomes evident when considering the significant reduction in water consumption for microalgal biomass production when effluent is employed as the cultivation medium. This approach of avoiding treated water use in the system primarily contributes to reductions in water, land, and resource utilization and mitigating climate change and ecotoxicity impacts. Moreover, MMFCs can also integrate atmospheric carbon reduction and enhance bioenergy production in water treatment systems (Kusmayadi et al. 2020).

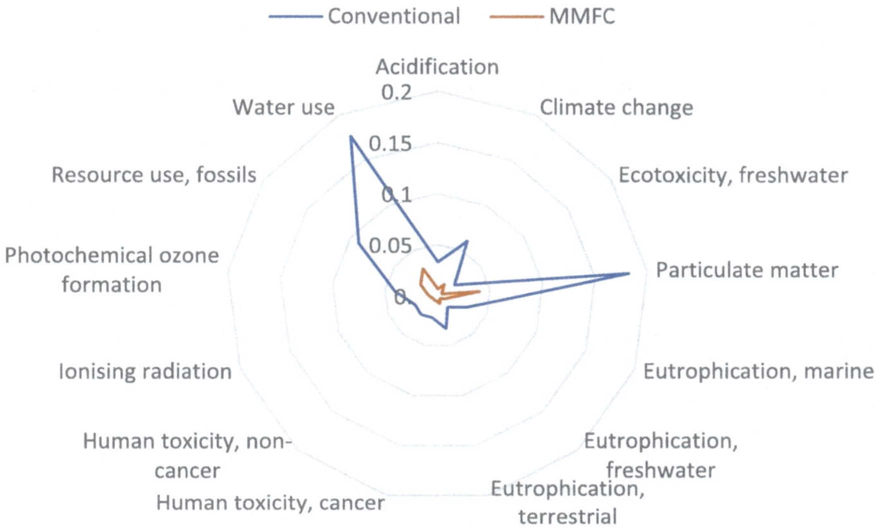

Fig. 4 Normalized Impacts for the Conventional System and MMFC

It can be observed that the impacts were lower in the MMFC since the system yielded 3.7 times more biomass than the conventional bench-scale cultivation. Figure 4 illustrates the impact reduction through a PBES for all the analyzed damage categories due to a better relation between the obtained biomass mass and the energy consumed. According to Zhang et al. (2019a, b), superior environmental performance was also noted when wastewater treatment was exclusively executed using microbial fuel cells (MFCs).

Reducing environmental impacts in the MMFC system presents benefits associated with microbial electrochemical reactions. These reactions convert residual carbon into biomass without extern energy (Sánchez et al. 2020), nitrogen, nor phosphorus inputs (Cerrillo et al. 2021; Lee et al. 2022). They can produce bioelectricity, reducing the energy impact when scaling up the production to industrial levels. Furthermore, producing bioenergy with the recovery of inorganic or organic carbon is attractive for mitigating environmental problems, reducing carbon footprint by storing carbon in biomass, and supporting negative carbon pathways (Wang et al. 2024).

The observed energy impacts are associated with the life cycle of inputs, biomass separation, aeration, and artificial light. These inputs also have economic impacts, as using NPK and energy adds costs to the system. Since carbon recovery is associated with production, there is a complex potential to contribute to a circular economy in addition to the direct environmental benefits (Chen et al. 2022). When scaled up to a pilot or industrial level, the MMFC carries the same environmental and economic impacts as the conventional system; however, using the MMFC mitigates these impacts. With the life cycle analysis, we could not recognize a reduction of impacts with the MMFC system that assists in economic impacts.

Although we achieved more biomass and less environmental impact, the bioelectricity, which is a determining factor for the system's feasibility, is still low. Conversely, anaerobic digestion at the anode is a highly efficient and cost-effective biological treatment method (Ribeiro et al. 2022), as it not only treats the water but also avoids using potable water for production. It plays a vital role in the circular economy because it can digest various organic wastes and produce bioenergy (Chen et al. 2022).

An uncertainty estimation was performed considering the damages associated with inputs in microalgae production in the compared systems. Low values were obtained using the Monte Carlo approach in the Simapro software (Scrucca et al. 2020). Table 3 shows that only in the Ecotoxicity in the freshwater category were the values higher, but still below 0.3, considered adequate for result uncertainty.

The uncertainty values found may be associated with aspects discussed by Beloin-Saint-Pierre et al. (2020), who argued that temporal considerations are often overlooked in life cycle inventory calculations, resulting in potential environmental impacts with greater uncertainty.

5 Conclusion

This LCIA study provides valuable insights into the environmental impacts of microalgal biomass production within an MMFC system, mainly impacts associated with energy (soil use, ecotoxicity, and climate change) and nutrient use (ecotoxicity). Using wastewater avoided the need for tap water and reduced impacts. However, the energy consumption on a lab scale is higher and outweighs other impacts. Additionally, all environmental impacts were lower than the conventional system since 3.7 times more biomass was produced. It underscores the system's pivotal role in environmental preservation. By simultaneously remediating urban effluents and generating valuable microalgal biomass, the MMFC system demonstrates its potential to mitigate environmental impacts, with the main impacts primarily associated with energy consumption. This highlights the importance of optimizing energy use in MMFC operations to enhance its environmental sustainability and efficiency further. Additionally, the MMFC system exhibits superior environmental impact reduction and biomass production performance compared to conventional microalgae production systems. Continuous technological advancements in the system have the potential to bring even greater benefits, including bioelectricity production.

Acknowledgements The authors are grateful to the CNPq (protocol 306216/2022-1, 311531/2021-0), Capes (code 001 and 88881710390/2022-1) and also to FAPERGS (protocol 23/2551-0000/61-6) for the financial support and fellowships.

References

Ahirwar A, Das S, Das S, Yang Y-H, Bhatia SK, Vinayak V, Ghangrekar MM (2023) Photosynthetic microbial fuel cell for bioenergy and valuable production: a review of circular bio-economy approach. Algal Res 70:102973. https://doi.org/10.1016/j.algal.2023.102973

Beloin-Saint-Pierre D, Albers A, Helias A, Tiruta-Barna L, Fantke P, Levasseur A, Benetto E, Benoist A, Collet P (2020) Addressing temporal considerations in life cycle assessment. Sci Total Environ 743:140700. https://doi.org/10.1016/j.scitotenv.2020.140700

Cerrillo M, Burgos L, Noguerol J, Riau V, Bonmatí A (2021) Ammonium and phosphate recovery in a three chambered microbial electrolysis cell: towards obtaining struvite from livestock manure. Processes 9. https://doi.org/10.3390/pr9111916

Çetinkaya AY, Bilgili L, Kuzu SL (2018) Life cycle assessment and greenhouse gas emission evaluation from Aksaray solid waste disposal facility. Air Qual Atmos Health 11(5):549–558. https://doi.org/10.1007/s11869-018-0559-3

Cetinkaya AY, Kuzu SL, Bilgili L (2022) Development of an MFC-biosensor for determination of Pb(+2): an assessment from computational fluid dynamics and life cycle assessment perspectives. Environ Monit Assess 194(4):245. https://doi.org/10.1007/s10661-022-09894-w

Cetinkaya AY, Ozdemir OK, Demir A, Ozkaya B (2017) Electricity production and characterization of high-strength industrial wastewaters in microbial fuel cell. Appl Biochem Biotechnol 182(2):468–481. https://doi.org/10.1007/s12010-016-2338-7

Chen J, Dai L, Mataya D, Cobb K, Chen P, Ruan R (2022) Enhanced sustainable integration of CO_2 utilization and wastewater treatment using microalgae in circular economy concept. Biores Technol 366:128188. https://doi.org/10.1016/j.biortech.2022.128188

Chin MY, Phuang ZX, Woon KS, Hanafiah MM, Zhang Z, Liu X (2022) Life cycle assessment of bioelectrochemical and integrated microbial fuel cell systems for sustainable wastewater treatment and resource recovery. J Environ Manage 320:115778. https://doi.org/10.1016/j.jenvman.2022.115778

Christwardana M, Hadiyanto H, Motto SA, Sudarno S, Haryani K (2020) Performance evaluation of yeast-assisted microalgal microbial fuel cells on bioremediation of cafeteria wastewater for electricity generation and microalgae biomass production. Biomass Bioenerg 139:105617. https://doi.org/10.1016/j.biombioe.2020.105617

Deng Z, Zhu J, Yang L, Zhang Z, Li B, Xia L, Wu L (2022) Microalgae fuel cells enhanced biodegradation of imidacloprid by Chlorella sp. Biochem Eng J 179:108327. https://doi.org/10.1016/j.bej.2021.108327

Dolganyuk V, Belova D, Babich O, Prosekov A, Ivanova S, Katserov D, Patyukov N, Sukhikh S (2020) Microalgae: a promising source of valuable bioproducts. Biomolecules 10(8):1153

Fan L, Xue S (2016) Overview on electricigens for microbial fuel cell. Open Biotechnol J 10:398–406. https://doi.org/10.2174/1874070701610010398

Hendroko Setyobudi R, Faizi Polontalo N, Alifsmara Joelyna F, Malik Islam Filardli A, Hadiyanto H, Akmar Zakaria Z, Winaya A, Burlakovs J, Mel M, Anne O (2021) Microalgae microbial fuel cell (MMFC) using chlorella vulgaris and "Batik" wastewater as bioelectricity. E3S Web Conf 226:00032. https://doi.org/10.1051/e3sconf/202122600032

Hosseinzadeh A, Zhou JL, Li X, Afsari M, Altaee A (2022) Techno-economic and environmental impact assessment of hydrogen production processes using bio-waste as renewable energy resource. Renew Sustain Energy Rev 156:111991. https://doi.org/10.1016/j.rser.2021.111991

Hou C, Zhao J, Huang B, Zhou X, Zhang Y (2024) Microalgae-based technologies for carbon neutralization and pollutant remediation: a comprehensive and systematic review. Resour Conserv Recycl 202:107323. https://doi.org/10.1016/j.resconrec.2023.107323

Igbokwe VC, Ezugworie FN, Onwosi CO, Aliyu GO, Obi CJ (2022) Biochemical biorefinery: a low-cost and non-waste concept for promoting sustainable circular bioeconomy. J Environ Manage 305:114333. https://doi.org/10.1016/j.jenvman.2021.114333

International Organisation for Standardisation (2006) Environmental management—life cycle impact assessment—requirements and guidelines

Kumar N, Saini K, Shyam K, Sharma G, Chandel H, Murugesan S, Saxena G (2024) Chapter 9: Microalgae in removal of emerging contaminants from wastewater environment: a case of potentially toxic pharmaceuticals and personal care products. In: Shah MP, Ghosh P (eds) Development in wastewater treatment research and processes. Elsevier, pp 167–197

Kusmayadi A, Leong YK, Yen HW, Huang CY, Dong CD, Chang JS (2020) Microalgae-microbial fuel cell (mMFC): an integrated process for electricity generation, wastewater treatment, CO_2 sequestration and biomass production. Int J Energy Res 44(12):9254–9265. https://doi.org/10.1002/er.5531

Lee DJ, Chang JS, Lai JY (2015) Microalgae-microbial fuel cell: a mini review. Bioresour Technol 198:891–895. https://doi.org/10.1016/j.biortech.2015.09.061

Lee Y-J, Lin B-L, Xue M, Tsunemi K (2022) Ammonia/ammonium removal/recovery from wastewaters using bioelectrochemical systems (BES): a review. Biores Technol 363:127927. https://doi.org/10.1016/j.biortech.2022.127927

Mälkki H, Alanne K (2017) An overview of life cycle assessment (LCA) and research-based teaching in renewable and sustainable energy education. Renew Sustain Energy Rev 69:218–231. https://doi.org/10.1016/j.rser.2016.11.176

Panagos P, Koningner J, Ballabio C, Liakos L, Muntwyler A, Borrelli P, Lugato E (2022) Improving the phosphorus budget of European agricultural soils. Sci Total Environ 853:158706. https://doi.org/10.1016/j.scitotenv.2022.158706

Pant D, Singh A, Van Bogaert G, Gallego YA, Diels L, Vanbroekhoven K (2011) An introduction to the life cycle assessment (LCA) of bioelectrochemical systems (BES) for sustainable energy and product generation: relevance and key aspects. Renew Sustain Energy Rev 15(2):1305–1313. https://doi.org/10.1016/j.rser.2010.10.005

Razman KK, Hanafiah MM, Mohammad AW (2022) An overview of LCA applied to various membrane technologies: progress, challenges, and harmonization. Environ Technol Innov 27:102803. https://doi.org/10.1016/j.eti.2022.102803

Ribeiro V, Pacheco MM, Machado ÊL, Benitez L, Leal VL, Machado GdRM, Kliemann M, Fuentefria A (2021) Photobioelectrochemical systems as an alternative for bioenergies production and bioremediation—clean technologies approach. Ciências. https://doi.org/10.22533/at.ed.491211207

Ribeiro VR, Osório HDD, Ulrich AC, Rizzetti TM, Barrios AS, de Cassia de Souza Schneider R, Benitez LB (2022) The use of microalgae-microbial fuel cells in wastewater bioremediation and bioelectricity generation. J Water Process Eng 48. https://doi.org/10.1016/j.jwpe.2022.102882

Rocha GS, Lopes LFP, Melão MGG (2024) Phosphorus limitation combined with aluminum triggers synergistic responses on the freshwater microalgae Raphidocelis subcapitata (Chlorophyceae). Chemosphere 352:141320. https://doi.org/10.1016/j.chemosphere.2024.141320

Sahaq AB, Hadiyanto (2019) Bioelectricity production from Tofu wastewater using microbial-microalgae fuel cell technology (MMFC). AIP Conf Proc 2202(1). https://doi.org/10.1063/1.5141684

Sánchez C, Dessì P, Duffy M, Lens PNL (2020) Microbial electrochemical technologies: electronic circuitry and characterization tools. Biosens Bioelectron 150:111884. https://doi.org/10.1016/j.bios.2019.111884

Savla, N, Suman, Pandit S, Verma JP, Awasthi AK, Sana SS, Prasad R (2021) Techno-economical evaluation and life cycle assessment of microbial electrochemical systems: a review. Curr Res Green Sustain Chem 4. https://doi.org/10.1016/j.crgsc.2021.100111

Scrucca F, Baldassarri C, Baldinelli G, Bonamente E, Rinaldi S, Rotili A, Barbanera M (2020) Uncertainty in LCA: an estimation of practitioner-related effects. J Clean Prod 268:122304. https://doi.org/10.1016/j.jclepro.2020.122304

Sharma A, Chhabra M, Kumar S (2024a) Performance evaluation of genetically modified microalgae in photosynthetic microbial fuel cells for carotenoids and power generation. J Environ Chem Eng 12(3):112751. https://doi.org/10.1016/j.jece.2024.112751

Sharma M, Jalalah M, Alsareii SA, Harraz FA, Xue W, Thakur N, Salama E-S, Li X (2024b) Microalgal cycling in the cathode of microbial fuel cells (MFCs) induced oxygen reduction

reaction (ORR) and electricity: a biocatalytic process for clean energy. Chem Eng J 479:147431. https://doi.org/10.1016/j.cej.2023.147431
Singh Mathuriya A, Hiloidhari M, Gware P, Singh A, Pant D (2020) Development and life cycle assessment of an auto circulating bio-electrochemical reactor for energy positive continuous wastewater treatment. Bioresour Technol 304:122959. https://doi.org/10.1016/j.biortech.2020.122959
Wang J, Zheng Y, He S, Yan J, Zeng X, Li S, Tian Z, Lei L, Chen Y, Deng S (2024) Can bioenergy with carbon capture and storage deliver negative emissions? A critical review of life cycle assessment. J Clean Prod 434:139839. https://doi.org/10.1016/j.jclepro.2023.139839
Zhang J, Yuan H, Abu-Reesh IM, He Z, Yuan C (2019a) Life cycle environmental impact comparison of bioelectrochemical systems for wastewater treatment. Procedia CIRP 80:382–388. https://doi.org/10.1016/j.procir.2019.01.075
Zhang J, Yuan H, Deng Y, Abu-Reesh IM, He Z, Yuan C (2019b) Life cycle assessment of osmotic microbial fuel cells for simultaneous wastewater treatment and resource recovery. Int J Life Cycle Assess 24(11):1962–1975. https://doi.org/10.1007/s11367-019-01626-6

Life Cycle Assessment, Circular Economy, and Sustainable Development in the Context of Africa's Major Regions: A Reflective Review of Research Contributions—2015–2023

Rudi Pretorius, Melanie Nicolau, Lindokuhle Sibiya, Zongho Kom, and Leslie Adriaanse

Abstract Recently published reviews on the topics of circular economy (CE) and of life cycle assessment (LCA) provide limited evidence of implementation and progress in the Global South and specifically Africa when compared globally. This chapter reports on a review of the literature based on a systematic search conducted on Google Scholar, to reflect on the research undertaken since 2015 for specifically Africa on CE and LCA, with focus on their connections, and relevance for sustainable development (SD). Aspects of the bibliometric analysis of the review are presented, but the focus of the chapter is to provide a synthesis of the themes covered by research on CE and LCA in Africa since 2015. Comparative findings are set out with reference to the five major regions of Africa, thus providing a more in-depth view compared to consideration of the continent as a whole. The regional focus is unique and adds insight to the body of knowledge on CE, LCA, their connections, and SD—not only for Africa, but also globally. Most importantly, the results presented in this paper indicate a definite contribution for Africa in terms of research in this field and of a more significant nature than generally reported up to now. This finding highlights that results of literature reviews are dependent on selection of data bases and search strategies, which might unintentionally lead to under- or over-representation of research contributions from certain parts of the world and thus not providing an accurate representation of reality.

Keywords African regions · Circular economy · Life cycle assessment · Sustainable development · Thematic synthesis · Bibliometric overview

R. Pretorius (✉) · M. Nicolau · L. Sibiya · Z. Kom
Department of Geography, University of South Africa, Florida, South Africa
e-mail: pretorw@unisa.ac.za

L. Adriaanse
Unisa Library, University of South Africa, Florida, South Africa

L. Brandli et al. (eds.), *The Contribution of Life Cycle Analyses and Circular Economy to the Sustainable Development Goals*, World Sustainability Series,
https://doi.org/10.1007/978-3-031-85300-5_20

1 Introduction

In light of the continued increased rate of extraction, processing, and use of natural resources, with confirmed links to negative impacts on the human–environment system (UNEP IRP 2019), circular economy (CE) approaches are receiving growing attention (Peña et al. 2021). Classically, CE is regarded as a system where materials never become waste, nature is regenerated, and products and materials kept in circulation through maintenance, reuse, refurbishment, remanufacture, recycle and composting (Godfrey et al. 2021). This view of CE has been adapted by adding that the utility and value of products and materials should be maintained for as long as possible and at the highest possible level, while minimising and designing out waste and working towards regenerating natural systems (Peña et al. 2021; UNEP 2020).

While CE is increasingly gaining ground to support sustainable development (Haupt and Zschokke 2017), the development of tools for evaluation, monitoring and management of environmental impacts of CE lags behind (Spreafico 2022). Of these, LCA stands out, aiming to assess potential benefits and trade-offs across the entire value chain. In this way the most appropriate CE practices for products or services can be identified. The practice to pass on negative impacts through the life cycle stages of products or services should be avoided as far as possible (Godfrey et al. 2021). However, such planning is not straightforward since several studies show that the best CE option is not necessarily the most preferable from an environmental point of view (Haupt and Zschokke 2017).

CE and LCA, their connections (including LCA as a tool for evaluating CE), and their relevance for sustainable development can be regarded as a developing study field. Reviews of the associated research production are therefore required to summarize the current state of knowledge about this study field and to provide insight into research directions (Visentin et al. 2020). Such reviews can be bibliometric, with a quantitative focus on aspects such as authors, publications, institutions, countries and frequency of keywords. It can also take the form of a systematic review, with a qualitative analysis of the selected literature, comprising, for example, a thematic analysis (Visentin et al. 2020). Reviews frequently cover bibliometric and systematic aspects, which is the approach in this chapter.

This chapter reflects on the results of an Africa focused review of published research from 2015 on CE, LCA, their connections, and their relevance for sustainable development. The results are presented for the five regions of Africa, which allows for cross-regional comparisons of the themes covered by research on CE and LCA in Africa, and provides more insights than a continental scale review. The value of this chapter lies in the insight provided into the African contribution to CE, LCA, their connections, and sustainable development. After the introduction, the literature on reviews of CE and of LCA is reflected upon, and then the methodology for the review is explained, followed by the results of the bibliometric and systematic components of the review. A regional comparison and conclusion close the chapter.

2 A Brief Overview of Previous Reviews of CE and LCA

An abundance of studies is available that provide reviews of research production on CE and on LCA. Some have a broad scope and cover all publications in these fields (e.g. Merli et al. 2018), while others focus on themes and specific applications (e.g. Munaro et al. 2020). The majority of these studies provide reviews of publications focused on either CE or LCA, while only a few provide reviews on the use of LCA in CE (e.g. Teixeira 2020). Furthermore, these studies can be broadly categorized as either bibliometric or systematic reviews, although some of them indeed combine these two approaches (e.g. Visentin et al. 2019). Although most of these studies include publications from all over the world, a few of them provide reviews for specific geographic areas, for instance Latin America, Africa and the European Union (e.g. Karkour et al. 2021), or in a few cases, even a specific country (e.g. Onungwe et al. 2023). Databases that are most often used for reviews appear to be Scopus and Web of Science (WoS), which are generally regarded as the most comprehensive, although other databases are also used, including Google Scholar, Science Direct, ProQuest and EBSCO Host (Merli et al. 2018).

Concerning CE, the study by Merli et al. (2018) is an example of a review covering the overall research production on CE. This study found that until 2017, publications on this theme mainly originated in China and Europe, with waste management as the most relevant subsector, while work connected with sustainability also featured prominently. The most explored practices could be linked to cleaner production. Reviews of research on CE applications or themes are common, with examples including Singh et al. (2021)—CE and e-waste research, Tsai et al. (2020)—municipal solid waste management in CE, Esposito et al. (2020)—transformation of the agri-food sector to CE, and Munaro et al. (2020)—CE in the built environment. For this chapter, reviews focused on Africa are of particular relevance. An example is the review of Mhlanga et al. (2022) on CE and the built environment in Africa. According to their review, dominant concepts explored included sustainability, alternative materials, waste management, lifecycle, demolition and climate change. They conclude that CE research output in Africa is low. They relate this to the novelty of the concept, or that its implementation is facing resistance.

Studies with reviews of the research production on LCA follow the same pattern as for CE. The study by He and Yu (2020) covers the overall research production on LCA from 1999 to 2018. They observed an increase in the number of publications per year, with the United States leading, followed by China, but a relatively low contribution by the rest of the developing world. Environmental impact and sustainable development formed the main content, with increased attention to carbon footprints, climate change, and industrial ecology. Reviews of research on LCA applications or themes are common, with examples including Nwodo et al. (2019)—LCA of buildings, Blanco et al. (2022)—LCA of the olive oil supply chain, and Esteves et al. (2019)—LCA of manure biogas production. For this chapter, the review on LCA for Africa (2000–2020) by Karkour et al. (2021) is particularly important. Although the number of publications is still relatively low, an increased number of publications can be observed over recent years. Agriculture is the most prominent topic, followed by electricity and energy. Most of the publications originated from South Africa, followed by Egypt and then Tunisia.

3 Study Area and Methodology for the Review

The Africa focused review of published research from 2015 to 2023 on CE and on LCA reported in this chapter was conducted with Google Scholar. The reason for this is related to the observation by the authors that most of the recently published reviews on CE and on LCA (for examples, see the reviews referred to in the previous section) provide limited evidence for Africa when compared globally. These reviews have mostly been conducted with WoS and Scopus, which are known for bias and a selective approach in creating a curated collection of publications, but with limited coverage in some study fields, publications not in English, and publications other than journal articles. Google Scholar has a more inclusive approach and indexes any scholarly document on the academic web (Martín-Martín et al. 2018). In this regard Martín-Martín et al. (2021) found that Google Scholar could locate most citations in Social Sciences articles (94%), while WoS and Scopus could respectively locate only 35% and 43%.

Although Google Scholar follows an inclusive approach and has greater coverage, its search functionality has shortcomings such as limited support of Boolean operators, which impose constraints to formulate search equations (Gusenbauer and Haddaway 2020). Several trial runs were conducted to design a search strategy that would produce an optimal selection of publications with Africa-based case studies of LCA and of CE. Initial searches based on full texts, comprehensive use of key words and Boolean operators, delivered lengthy lists of publications, but with limited suitability due to inclusion based on incidental reference to Africa or a country in Africa as an example, and not being based on a case study. To rule out this problem, it was reasoned that if a publication is a case study on LCA or CE for an African country, the title would include the country's name and "Life Cycle Assessment" (LCA) or "Circular Economy" (CE). The search equation was structured accordingly, and the search repeated 55 times for each of the countries in Africa, limited to the period 2015–2023. The search was conducted from 23 to 29 September 2023.

To compile the bibliographic portfolio for this research, the references of the publications, which were retrieved per country from Google Scholar, were exported to corresponding country folders on the online reference management tool, Mendeley. Duplicates were then removed, and the full text of the remaining 285 publications was added to the country folders for perusal and analysis. To obtain a proper grasp of each of the case studies, a manual check was conducted on each of these publications to determine their suitability for the research, leading to 75 exclusions in terms of publications behind pay walls, regional/country comparisons, edited collections by multiple authors and further duplicates.

Since the aim of the research is to analyse the research production on LCA and on CE for the different regions in Africa, the 210 remaining publications in the bibliographic portfolio were grouped according to the five major regions in Africa (see Fig. 1). They were then analysed in detail to extract information on authors, affiliations, publication types (journal, book, chapter, thesis, dissertation, report, etc.), theme covered and classification (specific project, general overview, literature review, model development, viewpoint, other). In addition, for LCA publications a note was made if and how CE is included, and vice versa for CE publications. The same was done for reference to sustainable development/sustainable development goals. The results of this analysis are presented and discussed in the following two sections.

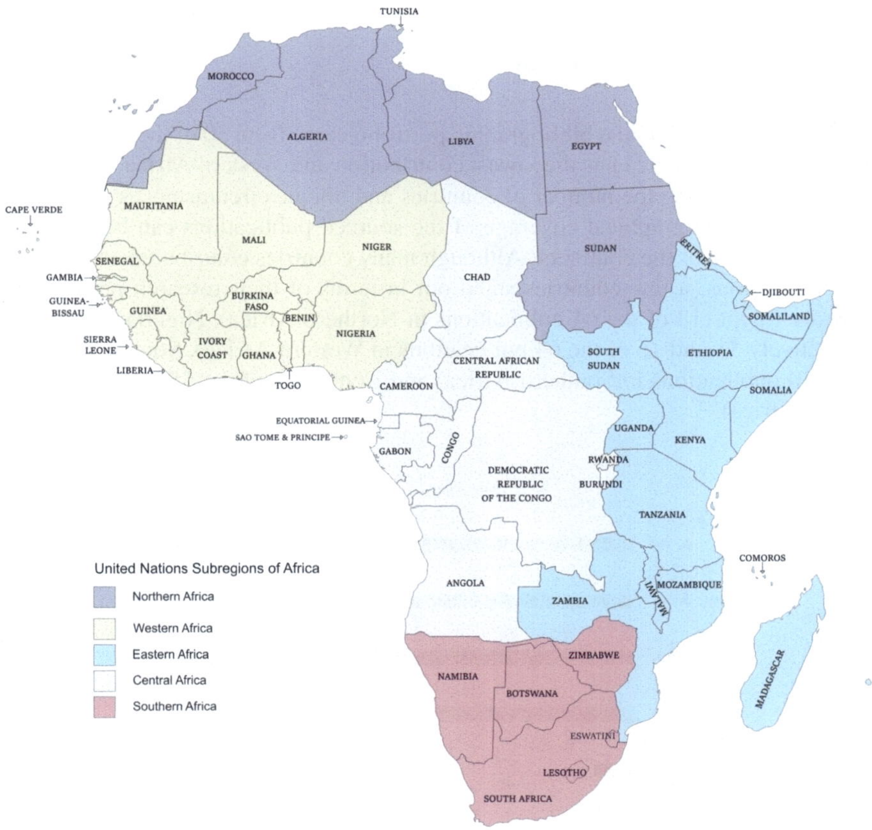

Fig. 1 Regional division of Africa. *Source* Adapted from FREEWORLDMAPS.NET and Wikimedia Commons)

4 Bibliometric Overview

This section provides an overview of the results of the bibliometric analysis of the publications in the bibliographic portfolio extracted from Google Scholar. The purpose of this overview is not to provide a detailed analysis of the data on matters such as authors, affiliations and publications, but rather to set the scene for the next section and therefore to provide information to inform the regional synthesis of themes covered by research on CE and on LCA in Africa. Due to their corresponding trends, the results for CE and LCA are reported jointly, but with unique aspects pointed out if necessary.

4.1 Geographical Coverage

The 210 publications in the bibliographic portfolio came from 30 of the 55 countries in Africa. Central Africa has the lowest contribution and Western Africa the highest (Fig. 2). Considering the number of countries and unique circumstances associated with each, the geographical coverage of the sourced publications can be regarded as representative of the continent. Although many countries contributed only one or two publications, a few countries stand out in terms of their research production. They are Egypt (18 of the 64 publications in Northern Africa), Nigeria and Ghana (respectively 31 and 16 of the 64 publications in Western Africa), South Africa (27 of the 30 publications in Southern Africa) and Kenya (21 of the 55 publications in Eastern Africa).

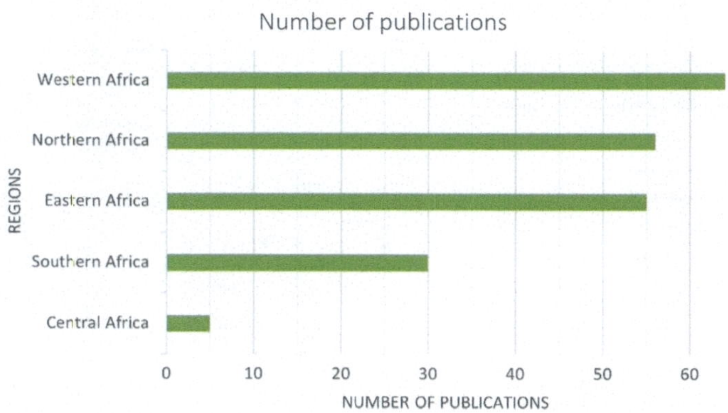

Fig. 2 Regional distribution of publications

4.2 Coverage of CE Versus LCA

For all the regions together, there are slightly fewer CE publications (96) than LCA publications (114), with only one publication (from Southern Africa) with a fully integrated approach to CE and LCA. However, this general pattern is not reflected in the same way in each region. While CE and LCA publications are almost the same in number for Western Africa, LCA papers are almost double the number for CE for Northern Africa and almost three times as many for Southern Africa. Eastern Africa differs with almost double the number of CE papers compared to LCE. All five papers for Central Africa are for LCA. The following countries stand out in terms of their research production: For CE—Nigeria (20), Kenya (16) and Ghana (9); for LCA—South Africa (20), Nigeria (17), Egypt (11) and Tunisia (9).

4.3 Author Information

A total of 664 authors were involved with the 210 publications in the bibliographical portfolio, reflecting the occurrence of multiple co-authorships. This implies an average of three authors per publication, but with apparent variations. While 59% of the authors are affiliated with only an African institution/s, 41% have some form of international affiliation (mostly combined with an African affiliation). This varies per region, with Eastern and Western Africa having the highest number of publications with international connections, respectively 51% and 43%, and Southern Africa the lowest (27%). The affiliations referred to are primarily with higher education institutions (e.g. 77% for Southern Africa), while research institutes, government bodies, businesses, and Non-Government Organisations (NGOs) make up the rest (e.g. 23% for Southern Africa).

4.4 Publication Information

Publications in scientific journals formed most of the entries in the bibliographic portfolio (about 71%), but varied from as low as 52% (Southern Africa) to as high as 80% (Western Africa). Second in line at about 13% is theses/dissertations, with the variation from a low of 8% (Western Africa) to a high of 20% (for both Eastern and Central Africa). In third place at about 10% is conference proceedings with the lowest contribution of 4% (Eastern Africa) and the highest of 19% (Southern Africa). In fourth place are reports (4%) and then book chapters (2%). The low number of book chapters in the bibliographic portfolio relates to excluding a number of these due to access only after payment. The relative high contribution by publications other than scientific journals is noteworthy, and partly explains the underreporting of the research contribution by Africa by reviews based on Scopus and WoS.

4.5 Themes Addressed

A manual analysis of each of the publications in the bibliographical portfolio revealed the themes addressed by research on CE and on LCA in the different regions of Africa. The results of this analysis can be viewed in the bar graph shown in Fig. 3. The results for Central Africa, based on only five publications, are not shown in this figure, but are considered in the following discussion. The overall trend reveals that publications focused on the follow applications of CE and of LCA dominates the scene (arranged in a decreasing sequence: 1st—agriculture, 2nd—waste management, 3rd—construction, 4th—overviews of CE or LCA, 5th—biofuel and biomass, 6th—energy use and production, and 7th—manufacturing).

Some exceptions to the above trend can be observed, for example Southern Africa, with relatively more attention given to publications on plastics production and recycling, as well as to mining. In the case of Northern Africa, such an exception is provided by environmental management. If distinguishing between CE and LCA, the same general trends can be observed as in the case where CE and LCA are considered jointly, although slight changes in emphasis can be observed. The following examples serve as an illustration: publications covering plastics production and those on overviews are more dominant for CE than for LCA, while publications on mining, and especially energy use and production, are more dominant for LCA than for CE.

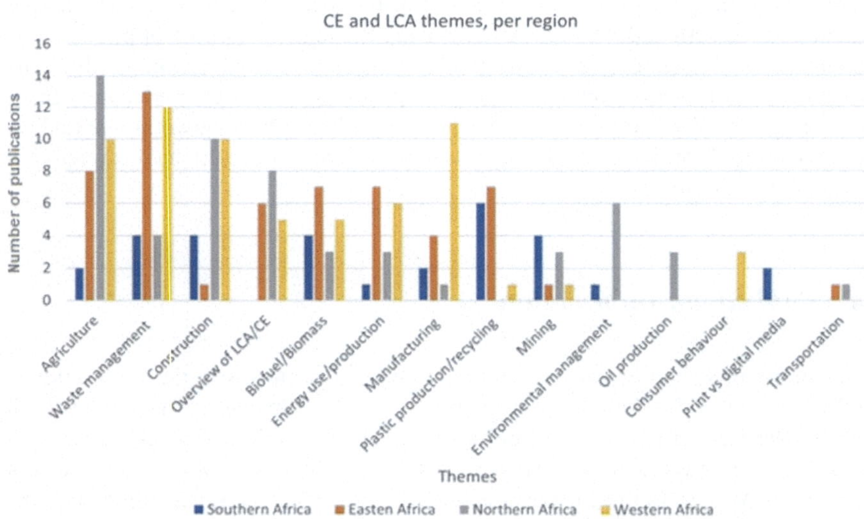

Fig. 3 Bar graph indicating the number of publications per theme (jointly for CE and LCA), and per region. The data for Central Africa, based on only five publications, is not shown

4.6 Connections Between CE and LCA

Limited evidence of integration between CE and LCA could be found in the publications comprising the bibliographic portfolio. No such CE publications could be found, while 9 CE publications referred significantly to LCA, although not an integral part of the contribution, and another 16 CE publications referred to LCA, but in an ad hoc, insignificant way. For LCA one such publication could be found, while 5 LCA publications referred significantly to CE (although not an integral part of the contribution), and incidental reference to CE could be found in another 11 LCA publications. The probability of locating a reference to LCA in a CE publication is relatively higher than for locating a reference to CE in an LCA publication.

4.7 Connections with Sustainable Development (SD)

The analysis of the bibliographic portfolio also focused on identifying publications in which CE/LCA were applied to promote or achieve SD. In total, 19 publications were found showcasing an integrated approach between SD and CE/LCA, with 16 focused on CE and only three on LCA. The one publication with an integrated approach between CE and LCA did not address SD at all. Almost all the CE publications referred in some way to SD, although not forming an integral part of the contribution or making just an incidental reference. Although very few of the LCA publications showcased integration with SD, most of these publications also referred in some way to SD. The aspect of integration between CE/LCA and SD will be further highlighted in the next section.

5 Regional Thematic Synthesis and Analysis

The publications selected for this review included research applications across a vast continent. However, to avoid generalisation of the CE and LCA publications, a sub-regional approach is followed to illustrate the different themes (Fig. 3) that dominated the research documented in the publications included in this bibliographic portfolio. The examples provided are to inform the reader and do not represent the full bibliometric portfolio. They should thus be seen as illustrative samples of the sub-regions, and where possible, the reason for selection was related to the degree of SD integration. This section concludes with some cross-regional comparisons with comparable research published for Africa.

5.1 Northern Africa

Approximately 249 million people live in Northern Africa (World Atlas 2023). Despite progress in human development, the region faces challenges of inequality, exclusion and gender discrimination (ILO 2015). In Morocco and Egypt, agriculture dominates the economy, while petroleum and natural gas exports provide most of Libya's revenues, but the rest of the region also contains deposits of these resources.

The dominant themes of research in Northern Africa were agriculture; an overview of the application of CE; and construction. However, there are skewed disparities on a finer scale with 11 of 14 contributions on the theme of agriculture and 9 of 10 contributions on construction based only on LCA. The significant contributions to achieving SD were dominantly CE related.

Pradeleix et al. (2023) assessed the LCA of farming systems and farming-activity contributions to regional impacts and identify environmental hotspots in Tunisia. Azeb et al. (2020) used LCA to analyse the impact of water reuse irrigation and farmers' practices in greenhouse cucumber production in Algeria. Morsi et al. (2020) assessed the environmental impact of a residential building in Egypt from cradle to cradle.

Six publications that focused their research on SD were all related to CE. Benamraoui and Berbar (2021) analysed the contribution that CE could make to SD in Algeria, while (Fersi et al. 2021) discussed the relevance of the transition to a CE in a more sustainable Tunisia. Campitelli et al. (2023) integrated SD into their research on waste management systems in Morocco. Research undertaken in Egypt should also be noted, where authors (Hossam and Hussein 2023;; Abdelzaher and Awad 2022) analysed the impact of CE on SD and economic development respectively, while (Mostafa and Youssef 2022) examined the various CE strategies for sustainable management of medical waste.

5.2 Western Africa

Sixteen countries located in Western Africa are in either a semi-arid or a savanna environment. The World Atlas (2023) estimates that 444 million people live in the region. In 2019, the Economic Community of West African States (ECOWA) contributed 25 per cent to the GDP of Africa and had an economic growth of 3.6 per cent. UNECA (2023) indicated that only 15 per cent of the SDGs are on track in the region and point towards challenges related to the COVID-19 pandemic, climate change and geopolitical conflicts.

Dominant research themes on LCA and CE in Western Africa include waste management, agriculture, construction, and manufacturing, and originated mainly from two countries, namely Ghana and Nigeria. Waste management dominated the CE research that was undertaken predominantly in Nigeria and ranged from the conversion of municipal solid waste to energy (Dickson et al. 2023), to the handling of hazardous waste in terms of CE practices (Mokuolu and Timothy 2021). The LCA contributions were mostly within the manufacturing theme and include applications of LCA. The industries researched included the manufacturing of food, cosmetics, and sports industries in Ghana, while in Nigeria research focused on the palm oil industry (Anyaoha and Zhang 2021, 2023), conversion of municipal waste to energy (Nubi et al. 2021) and environmental impacts of cowpea production, storage and disposal (Ogunjirin et al. 2020).

The research that linked CE specifically with SD can be noted. Kwarteng et al. (2022) documented the accounting practices within a CE, while Boon and Anuga (2020) used the 6R framework (reduce, reuse, refuse, rethink, repair and recycle) to analyse the CE of agricultural crops in Ghana. The research contributions of Salami (2022) explored the link between CE and the fashion industry; Omeiza (2021) identified major factors needed for the transition to a sustainable CE economy; and Ogunmakinde (2019) focused on the development of a CE framework for the minimisation of construction material waste for Nigeria and presented integrated links with the achievement of the SDGs.

5.3 Central Africa

The nine countries located in Central Africa have an estimated population of 184 million (World Atlas 2023). The region has experienced recurring coups d'état, civil wars and rebellions since independence resulting in challenges that negatively impact the achievement of the SDGs, including high levels of poverty, inequality, corruption, environmental degradation, food insecurity, and low human development. The countries depend on natural resources—such as oil, minerals, timber, and cocoa—and are vulnerable to price fluctuations and resource depletion. This is aggravated by inefficient and often corrupt public sectors and relatively weak private sectors. Five papers were identified from only two countries, namely Cameroon and the Democratic Republic of Congo (DRC) and are related to applications of LCA and none on CE.

The publications span five themes including: an assessment of the environmental impact of Cameroon's thermal power production technology (Mfetoum et al. 2023); identification of critical hotspots that lower the environmental impact of tea production (Tagne et al. 2023); the establishment of the environmental impact of second-generation biofuels production from agricultural residue (Tagne et al. 2022); an assessment of cobalt mining (Orola 2021); and a determination of the environmental impact of municipal solid waste management (Kang et al. 2023). Only one paper (Tagne et al. 2023) integrated the concepts of sustainability of the environmental impact of tea production; while the aim of the paper explicitly stated that the LCA could make a significant contribution to the achievement of the SDGs, no mention is made with reference to specific SDGs.

Compared to the other sub-regions in Africa, the contribution of this region to the body of knowledge of CE and LCA in relation to SD is negligible. This could be associated with the ongoing conflict in this sub-region related to socio-political and cultural challenges rooted in colonialism, instability, or violence.

5.4 Eastern Africa

Eastern Africa extends from the Horn of Africa to the Great Lakes along the Indian Ocean, with a population count of 488 million (World Atlas 2023). The region is vulnerable to external challenges including changes in the world economy; fluctuating commodity prices; and the ongoing Russian invasion of Ukraine. Domestic challenges include the impact of climate change; financial imbalances; political unrest; and infrastructure gaps such as energy shortages and limited access to water. There is an economic dependency on natural resources and agriculture in the region.

Most publications included in the bibliographic portfolio from Eastern Africa focused on the CE in waste management, agriculture, biofuel/biomass, plastic production, and recycling. Muriithi and Nagre (2023) examined the attempts of the CE policy landscape in Kenya by scrutinising the CE policy environment to document how Kenya has taken to adopting CE concepts. Gatari and Mutiso (2022) examined the impact of CE techniques on the long-term viability of Kenyan companies that produce rubber and plastic. The CE research of Yhdego and Kingu (2016) found that the urban solid waste management in Tanzania challenged local municipal authorities. While the majority of LCA papers included in the bibliographic portfolio did not present a significant connection to SD, the work of Shabani et al. (2023) found that in Zimbabwe, LCA applications played a crucial role in solid waste management in informing decision-makers to adopt practices.

Two publications included in the bibliographic portfolio for this region, both from Kenya, integrated the drive towards SD into their research. Turing's (2021) doctoral research highlighted that CE routinely applied in the CE in the Global North is a relatively a new application in the Global South and makes a case for continued application of the concepts of CE in the drive for Africa to achieve meaningful SD. Muriithi and Nagre (2023) advocate for the increased application of CE in Kenya as an avenue to SD through improved resource efficiency.

5.5 Southern Africa

The World Atlas (2023) describes Southern Africa as the least populous sub-region of Africa with a population count of 70 million, with most living in South Africa. The sub-region experiences high levels of unemployment; inequality; poverty; HIV/AIDS; and climate change. The economy in the region relies on agriculture, mining, tourism and, to a lesser degree, manufacturing. The dominant themes of the contributions from CE or LCA in Southern Africa were plastic production and recycling; and a relatively even spilt between waste management, construction, biofuel/biomass and mining. Research related to mining and the plastics industry was the dominant theme in the CE category, while plastic production and recycling; biofuel/biomass; and construction dominated the LCA research.

Nahman et al. (2021, 2022) applied CE concepts in the South African plastics industry. Goga et al. (2022) and Stafford et al. (2022) investigated the LCA of plastic polymer production and recycling; this included the use of alternative shopping bags. Mdhluli and Harding (2021) applied an LCA on extracting electricity from maize and wheat. In Botswana, Ishimoto et al. (2018) evaluated the LCA of using drought resistant indigenous plants for biofuel production. Saint et al. (2023) undertook an LCA of sandbag housing construction in South Africa. Harvey (2019) presented an overview of greener technology and innovation to reduce the carbon footprint of mining in South Africa. One contribution addressed both LCA and CE (Charles et al. 2019) by investigating more sustainable energy storage for solar home systems in South Africa.

Three publications (Harvey 2019; Nahman et al. 2021, 2022) presented their research within the mining and plastics industry in South Africa, and while they did refer to SD generally, their findings do not integrate SD significantly. The three publications of (Godfrey et al. 2021; Owojori et al. 2022; Von Blottnitz et al. 2022); ; undertook research that was related to general manufacturing; solid waste management; and the development of CE indicators for South Africa, and their findings were integrated and contextualised within SD. Naicker (2015) undertook an LCA comparison of the impact of print media compared to digital media by infusing SD throughout the documented research.

5.6 Cross-Regional Perspectives

While 114 publications included in this bibliographic portfolio were LCA applications, over 96 of the publications included an application of CE. Karkour et al. (2021) reported on the LCA research undertaken in Africa over two decades, and while their research reported a relatively low output in LCA research from the continent, the sample for this chapter shows that while still relatively low, the research in LCA application is gaining momentum on the continent and exceeds the research on CE. As was the case with Karkour et al. (2021), most publications (27) for LCA were applied within the context of agriculture, unlike the findings of Karkour et al. (2021) where the themes ranked second for applications related to construction (18). The production of electricity and energy (16 publications) closely followed by biofuel/ biomass production (12 publications) were ranked third and fourth respectively. The publications included in this bibliographic portfolio are represented by 16 countries across all sub-regions of the continent, and not just from three as reported by Karkour et al. (2021).

While the research of Mhlanga et al. (2022) was limited to the CE of the built environment in Africa, and this chapter has cast a wider net to document more research themes undertaken on CE, the findings of Mhlanga et al. (2022) documented that

CE research output is low; this is affirmed in the review undertaken for this chapter. The assumptions of the authors of this chapter are thus aligned to Mhlanga et al. (2022) that the CE concept remains a novelty and its implementation by researchers in Africa could face resistance. It should, however, be noted that research from 17 countries from the sub-regions (excluding Central Africa) contributed to the publications included in this bibliometric portfolio. Only one paper from South Africa (Charles et al. 2019) combined a CE and an LCA into their research objectives in their investigation on sustainable energy storage in solar powered home systems, however, they did not integrate SD into their findings.

While LCA and CE are inherently sufficiently developed to have a definite application in any research that strives for SD, only 19 of 210 papers identified for this review presented their research with a significant link to sustainable development. From a sub-regional perspective, only six countries in the sub-Saharan regions presented research that had integrated SD into their findings. In contrast, research reported in four countries in Northern Africa integrated their research within the context of SD. The majority of the six publications included in the bibliographic portfolio for Northern Africa that had an SD focus were undertaken with multiple authors, many of which were affiliated to institutions in the Global North, while very few of the 13 publications with an SD focus from the other sub-regions were a result of collaboration with researchers in the Global North. The focus on SD in Northern Africa could be associated with a trend documented by UNECA (2023), which indicated that the achievement of the SDGs in the Northern African countries was comparable to average achievements globally, and higher achievements in comparison to other African countries; this is opposed to the below average achievement of the SDGs in the remaining sub-regions (UNECA 2023).

6 Conclusion

This chapter reflected on the results of an Africa focused review of published research from 2015 to 2023 on CE, LCA, their connections and their relevance for SD. The results were presented and analyzed in the context of the five regions of Africa and allowed for a more in-depth view than with a continental scale review. A more substantial contribution from Africa-based research is revealed compared to the results of many recently published global reviews on the topic. This could be due to a significant portion of the research production from Africa on CE/LCA constituted from conference proceedings, theses/dissertations, reports and book chapters, while these are not necessarily indexed by the most often used search engines, that is, Scopus and WoS. The use of Google Scholar, despite limitations, allowed for a more comprehensive view on published research from Africa on this topic. This review, using a finer geographical demarcation, has the further advantage of highlighting a more nuanced research contribution from the continent, which is often lost in the reporting of results from global reviews, where references to the top ten to twenty performers in terms of matters such as countries, authors and institutions are made.

Apart from foregrounding the contribution from Africa, the results reported in this chapter furthermore not only confirmed, but also expanded on results of previous Africa focused reviews, while specifically highlighting research in which CE/LCA is applied to support SD, which is an encouraging development.

Some limitations that could have influenced the results reported in this chapter can be highlighted. First and foremost, the fact that only Google Scholar has been used as a search engine to source the publications included in the bibliographical portfolio, could have excluded some publications. In addition, the nature of the search strategy and the assumption that the presence of certain keywords in the title of a publication would be indicative of a case study on LCA or CE, could also have excluded some publications. Despite these potential limitations, it was possible to compile a substantial bibliographic portfolio for this research, which presented good coverage of the five regions of Africa in terms of CE and LCA research. Lastly, it must be noted that although the regional demarcation of the results holds distinct advantages, it can also have led to over- or under-emphasis of trends, and inclusion of, or emphasis on certain cases at the expense of others. However, the level of detail obtained with this type of approach overrides this potential limitation by far. The need for future work on this topic in the African context cannot be emphasized enough; to mind come a specialized and detailed full bibliometric review, and a similar systematic review, with the use of more databases than only Google Scholar and the use of software for reviews such as VOSviewer, which will add new dimensions and expand insights.

References

Abdelzaher MA, Awad MM (2022) Sustainable development goals for the circular economy and the water-food nexus: full implementation of new drip irrigation technologies in upper Egypt. Sustainability 14(21):13883

Anyaoha KE, Zhang L (2021) Renewable energy for environmental protection: Life cycle inventory of Nigeria's palm oil production. Resour Conserv Recycl 174:105797

Anyaoha KE, Zhang L (2023) Technology-based comparative life cycle assessment for palm oil industry: the case of Nigeria. Environ Dev Sustain 25(5):4575–4595

Azeb L, Hartani T, Mouheb N, Pradeleix L, Hajjaji N, Aribi S (2020) Life cycle assessment of cucumber irrigation: unplanned water reuse versus groundwater resources in Tipaza (Algeria). J Water Reuse Desalination 10(3)

Benamraoui F, Berrached Berbar W (2021) The reality of the circular economy in Algeria and its contribution to sustainable development. Revue d'Economie et de Management 20(2):92–109

Blanco I, de Bellis L, Luvisi A (2022) Bibliometric mapping of research on life cycle assessment of olive oil supply chain. Sustainability 14(7):3747

Boon EK, Anuga SW (2020) Circular economy and its relevance for improving food and nutrition security in Sub-Saharan Africa: the case of Ghana. Mater Circ Econ 2:1–14

Campitelli A, Aryoug O, Ouazzani N, Bockreis A, Schebek L (2023) Assessing the performance of a waste management system towards a circular economy in the Global South: the case of Marrakech (Morocco). Waste Manage 166:259–269

Charles RG, Davies ML, Douglas P, Hallin IL, Mabbett I (2019) Sustainable energy storage for solar home systems in rural Sub-Saharan Africa—a comparative examination of lifecycle aspects of battery technologies for circular economy, with emphasis on the South African context. Energy 166:1207–1215

Dickson EM, Hastings A, Smith J (2023) Energy production from municipal solid waste in low to middle income countries: a case study of how to build a circular economy in Abuja, Nigeria. Front Sustain 4:1173474

Esposito B, Sessa MR, Sica D, Malandrino O (2020) Towards circular economy in the agri-food sector. A systematic literature review. Sustainability 12(18):7401

Esteves EMM, Herrera AMN, Esteves VPP, Morgado CDRV (2019) Life cycle assessment of manure biogas production: a review. J Clean Prod 219:411–423

Fersi C, Salah IB, Medimagh R (2021) Circular economy in Tunisia. In: Circular economy: recent trends in global perspective, pp 113–144. https://doi.org/10.1007/978-981-16-0913

FREEWORLDMAPS.NET (2023) Free printable maps of Africa. https://www.freeworldmaps.net/printable/africa/

Gatari C, Mutiso S (2022) Effect of circular economy practices on sustainable performance of plastic and rubber manufacturing firms in Kenya. IJARIIE 8(3):2640–2656

Godfrey L, Roman H, Smout S, Maserumule R, Mpofu A, Ryan G, Mokoena K (2021) Unlocking the opportunities of a circular economy in South Africa. In: Ghosh SK, Ghosh SK (eds) Circular economy: recent trends in global perspective. Springer, Singapore, pp 145–180

Goga T, Harding K, Russo V, von Blottnitz H (2022) What material flow analysis and life cycle assessment reveal about plastic polymer production and recycling in South Africa. S Afr J Sci 118(SPE):1–5

Gusenbauer M, Haddaway NR (2020) Which academic search systems are suitable for systematic reviews or meta-analyses? Evaluating retrieval qualities of Google Scholar, PubMed, and 26 other resources. Res Synth Methods 11(2):181–217

Harvey R (2019) Mining for a circular economy in the age of the fourth industrial revolution: The case of South Africa. https://africaportal.org/publication/mining-circular-economy-age-fourth-industrial-revolution-case-south-africa/

Haupt M, Zschokke M (2017) How can LCA support the circular economy? 63rd discussion forum on life cycle assessment, Zurich, Switzerland, November 30, 2016. Tint J Life Cycle Assess 22:832–837

He X, Yu D (2020) Research trends in life cycle assessment research: a 20-year bibliometric analysis (1999–2018). Environ Impact Assess Rev 85:106461

Hossam S, Hussein A (2023) The impact of circular economy on sustainable development (applied on Egypt). J Bus Res 45(1):137–161

ILO (2015) Employment for stability and socio-economic progress in North Africa. Strategy for North Africa 2011–2015. https://www.ilo.org/wcmsp5/groups/public/---africa/documents/meetingdocument/wcms_178194.pdf

Ishimoto Y, Yabuta S, Kgokong S, Motsepe M, Tominaga J, Teramoto S, Konaka T, Mmopelwa G, Kawamitsu Y, Akashi K, Ueno M (2018) Environmental evaluation with greenhouse gas emissions and absorption based on life cycle assessment for a Jatropha cultivation system in frost-and drought-prone regions of Botswana. Biomass Bioenerg 110:33–40

Kang YO, Yabar H, Mizunoya T, Higano Y (2023) Environmental and economic performances of municipal solid waste management strategies based on LCA method: a case study of Kinshasa. Heliyon 9(3):e14372

Karkour S, Rachid S, Maaoui M, Lin CC, Itsubo N (2021) Status of life cycle assessment (LCA) in Africa. Environments 8(2):10

Kwarteng A, Agyenim-Boateng C, Simpson SNY (2022) The barriers to adapting accounting practices to circular economy implementation: evidence from Ghana. J Global 14(1):1–26

Martín-Martín A, Orduna-Malea E, Delgado López-Cózar E (2018) Coverage of highly-cited documents in Google Scholar, Web of Science, and Scopus: a multidisciplinary comparison. Scientometrics 116(3):2175–2188

Martín-Martín A, Thelwall M, Orduna-Malea E, Delgado López-Cózar E (2021) Google Scholar, Microsoft Academic, Scopus, Dimensions, Web of Science, and OpenCitations' COCI: a multidisciplinary comparison of coverage via citations. Scientometrics 126(1):871–906

Mdhluli FT, Harding KG (2021) Comparative life-cycle assessment of maize cobs, maize stover and wheat stalks for the production of electricity through gasification vs traditional coal power electricity in South Africa. Clean Environ Syst 3:100046

Merli R, Preziosi M, Acampora A (2018) How do scholars approach the circular economy? A systematic literature review. J Clean Prod 178:703–722

Mfetoum IM, Ngangue MKN, Ngoh SK, Koffi FLD, Tamba JG, Monkam L (2023) Assessment of green house gas emissions from thermal technologies for electricity generation in Cameroon using life cycle analysis method. Open Access Library Journal 10(8):1–17

Mhlanga J, Haupt TC, Loggia C (2022) Shaping circular economy in the built environment in Africa. A bibliometric analysis. J Eng Des Technol 1726–0531

Mokuolu OA, Timothy RS (2021) Circular economy and waste management actions during the COVID-19 pandemic in Nigeria. J Hum Environ Health Promot 7(1):1–5

Morsi MA, Ismaeel SE, El-Hamed AE (2020) Life cycle assessment of a residential building in Egypt: a case study. In: 13th international conference on civil and architecture engineering ICCAE-13-2020

Mostafa A, Youssef G (2022) COVID-19 and medical waste management in the context of circular economy: a case study of Egypt. J Leg Econ Res 12(82):123–138

Munaro MR, Tavares SF, Bragança L (2020) Towards circular and more sustainable buildings: a systematic literature review on the circular economy in the built environment. J Clean Prod 260:121134

Muriithi JK, Ngare IO (2023) Transitioning circular economy from policy to practice in Kenya. Front Sustain 4:1190470

Nahman A, Godfrey L, Oelofse S, Trotter D (2021) Driving economic growth in South Africa through a low carbon, sustainable and inclusive circular economy. Circular Economy Briefing Note 2021/001, CSIR, Pretoria. https://www.csir.co.za/sites/default/files/Documents/Driving%20economic%20growth%20in%20South%20Africa%20through%20a%20low%20carbon%2C%20sustainable.pdf

Nahman A, Oelofse SH, Palfreman J (2022) Recommendations for advancing the circular plastics economy in South Africa. In: Conference paper presented at WasteCon 2022, Emperors Palace, Johannesburg

Naicker V (2015) An LCA study comparing e-books read on an Apple Air iPad to printed books in South Africa. Master's thesis, University of Cape Town

Nubi O, Morse S, Murphy RJ (2021) A prospective social life cycle assessment (SLCA) of electricity generation from municipal solid waste in Nigeria. Sustainability 13(18):10177

Ogunjirin OC, Jekayinfa SO, Olaniran JA, Ogunjirin OA (2020) Environmental life cycle assessment of cowpea production, storage and disposal in Ilorin, Kwara State, Nigeria. IOP Conf Ser Earth Environ Science 445(1):12–40

Ogunmakinde OE (2019) Developing a circular-economy-based construction waste minimisation framework for Nigeria. Thesis for Ph.D. in Building, University of Newcastle

Omeiza-Michael S (2021) The transition to a circular economy in Nigeria: identifying the key factors. Intergovernmental Research and Policy Journal, 1:1–14

Onungwe I, Hunt DV, Jefferson I (2023) Transition and implementation of circular economy in municipal solid waste management system in Nigeria: A systematic review of the literature. Sustainability 15(16):12602

Orola A (2021) Social life cycle assessment of cobalt mining in Democratic Republic of Congo. Master's dissertation Lappeenranta-Lathi University of Technology, Finland

Owojori OM, Mulaudzi R, Edokpayi JN (2022) Student's knowledge, attitude, and perception (KAP) to solid waste management: a survey towards a more circular economy from a rural-based tertiary institution in South Africa. Sustainability 14(3):1310

Peña C, Civit B, Gallego-Schmid A, Druckman A, Pires AC, Weidema B, Mieras E, Wang F, Fava J, Canals LMI, Cordella M (2021) Using life cycle assessment to achieve a circular economy. Int J Life Cycle Assess 26:215–220

Pradeleix L, Roux P, Bouarfa S, Bellon-Maurel V (2023) Multilevel life cycle assessment to evaluate prospective agricultural development scenarios in a semi-arid irrigated region of Tunisia. Agric Syst 212:103766

Saint R, Eltaweel A, Adetooto J, Pomponi F, Windapo A (2023) Sandbag housing construction in South Africa: life cycle assessment and operational energy modelling. Int J Life Cycle Assess 1–16

Salami AO (2022) The potential contribution of circular economy to strategic sustainable development framework within the fashion industry—a case study of fashion SMEs in Nigeria. Thesis for MSc, University of Graz

Shabani T, Jerie S, Shabani T (2023) Applicability of the life cycle assessment model in solid waste management in Zimbabwe. Circ Econ Sustain. https://doi.org/10.1007/s43615-023-00268-z

Singh S, Trivedi B, Dasgupta MS, Routroy S (2021) A bibliometric analysis of circular economy concept in E-waste research during the period 2008–2020. Mater Today Proc 46:8519–8524

Spreafico C (2022) An analysis of design strategies for circular economy through life cycle assessment. Environ Monit Assess 194(3):180

Stafford W, Russo V, Nahman A (2022) A comparative cradle-to-grave life cycle assessment of single-use plastic shopping bags and various alternatives available in South Africa. Int J Life Cycle Assess 27(9–11):1213–1227

Tagne RFT, Santagata R, Tchuifon DRT, Atangana JAK, Ateba FR, Vintila T, Ndifor-Angwagor NG, Anagho SG, Ulgiati S (2022) Environmental impact of second-generation biofuels production from agricultural residues in Cameroon: a life-cycle assessment study. J Clean Prod 378:134630

Tagne RFT, Ncube A, Atangana JAK, Tchuifon DRT, Ateba FR, Azambou IC (2023) Analysis of environmental sustainability of Cameroon tea production: an LCA study. Sustain Food Technol 1(1):116–125

Teixeira WP (2020) Life cycle assessment (LCA) in circular economy systems: a bibliometric literature review. In: International Joint conference on industrial engineering and operations management—ABEPRO-ADINGOR-IISE-AIM-ASEM, IJCIEOM 2020, Rio de Janeiro

Tsai FM, Bui TD, Tseng ML, Lim MK, Hu J (2020) Municipal solid waste management in a circular economy: a data-driven bibliometric analysis. J Clean Prod 275:124132

Turing J (2021) Understanding the circular economy in Kenya. Ph.D. In: International Development University of Edinburg UK

UNECA (2023) West Africa sustainable development report, 2023. UNECA, Sub Regional Office for West Africa

UNEP (United Nations Environmental Programme) (2020) Understanding Circularity. https://buildingcircularity.org/

UNEP IRP (UNEP International Resource Panel) (2019) Global resources outlook—2019: natural resources for the future we want. https://www.resourcepanel.org/reports/globalresources-outlook

Visentin C, da Silva Trentin AW, Braun AB, Thomé A (2019) Application of life cycle assessment as a tool for evaluating the sustainability of contaminated sites remediation: a systematic and bibliographic analysis. Sci Total Environ 672:893–905

Visentin C, da Silva Trentin AW, Braun AB, Thomé A (2020) Life cycle sustainability assessment: a systematic literature review through the application perspective, indicators, and methodologies. J Clean Prod 270:122509

Von Blottnitz H, Virag D, Wiedenhofer D, Haas W (2022) An economy-wide material flow analysis to develop circular economy indicators for South Africa. Chemical Engineering Department, University of Cape Town and Institute of Social Ecology, University of Natural Resources and Life Sciences (BOKU), Vienna

Wikimedia Commons (2023) Africa map regions. https://en.wikipedia.org/wiki/File:Africa_map_regions.svg

World Atlas (2023) World Atlas. https://www.worldatlas.com/geography/regions-of-africa.html
Yhdego M, Kingu A (2016) Solid waste management in urban canters of Tanzania. Leapfrogging towards a circular economy. Report 1, Environmental Resources Consultancy

Rudi Pretorius is an associate professor in the Department of Geography, Unisa. He holds a Ph.D. in Geography, and his present research focus is on education for sustainability, sustainability in higher education, Geography education and open distance e-learning.

Melanie Nicolau is a full professor in the Department of Geography, Unisa. She holds a Ph.D. in Geography, and her present research focus is on post-development Geography, poverty and inequality studies, sustainability studies and indigenous science.

Lindokuhle Sibiya is a temporary lecturer in the Department of Geography, Unisa. He holds a MSc in Geography and is presently registered for a Ph.D. in Geography.

Zongho Kom is a post-doctoral research fellow in the Department of Geography, Unisa. He holds a Ph.D. in Geography, and his present research focus is on climate change and adaptation, small-scale farming, rural development, indigenous knowledge systems, sustainable livelihoods, geospatial analysis, and demography.

Leslie Adriaanse is a personal librarian for the College of Agriculture and Environmental Sciences, Unisa. She holds a Ph.D. in Information Management, and her present research focus is on knowledge management, research metrics, e-visibility, citation literacy, and citation analysis.

Linking Life Cycle Sustainability Assessment (LCSA) to Sustainable Development Goals (SDGs): Methodology and Applications

Arnaud Diemer, Tipawan Durand, and Henri Sourgou

Abstract In the early 90s, the international and scientific community set out to assess the environmental, economic and social impacts associated with the various stages of Life Cycle Assessment (LCA). This period also coincided with the rise of the concept of sustainable development, popularized by the Brundtland Report (1987). Today, the two approaches seem to be converging, with the aim of leading both governments and companies along the path to sustainability. On the one hand, life cycle assessment now comprises four stages (definition of the objective and scope of the assessment, inventory analysis, impact assessment, interpretation) and several forms of sustainability (LCSA: ELCA, LCC, SLCA, GLCA). On the other hand, the United Nations (2015) has launched a roadmap around the Sustainable Development Goals (SDGs). Linking sustainable life cycle assessment and the SDGs is now a new step in the 2030–2050 agenda. This means identifying targets and indicators for a sustainable approach, and incorporating this new approach into the roadmaps (circular economy) of the different states.

Keywords G-LCA · E-LCA · LCSA · LCC · SDGs · SLCA

1 Introduction

The family of LCA ISO Standards is used to evaluate environmental impacts associated with all the different stages of the life cycle analysis of a product, service or process. A LCA study introduces four basic steps (defining goal and scope of

A. Diemer (✉) · T. Durand · H. Sourgou
CERDI, ERASME, University Clermont Auvergne, Clermont-Ferrand, France
e-mail: diemera@aol.com; arnaud.diemer@uca.fr

French Chapter of System Dynamics Society, San Diego, USA

T. Durand
e-mail: tipawan.durand@uca.fr

H. Sourgou
e-mail: henri.sourgou@uca.fr

L. Brandli et al. (eds.), *The Contribution of Life Cycle Analyses and Circular Economy to the Sustainable Development Goals*, World Sustainability Series,
https://doi.org/10.1007/978-3-031-85300-5_21

the assessment, inventory analysis, impact assessment, interpretation) to identify the environmental consequences of a decision (Hauschild et al. 2018). Although ecological impacts have traditionally been presented as the main objective of an LCA, recent studies have also emphasized the social and sustainable dimensions of these impacts. Social Life Cycle Assessment (SLCA) identifies social and socio-economic impact along the lifecycle of a product or a service on different stakeholders. This tool has been built on the ISO 26000 (Guidelines for Social Responsibility 2010) and Global Reporting Initiative (GRI 1999). More recently, the Life Cycle Sustainability Assessment (LCSA) has been introduced to challenge environmental, social and economic impacts. As many companies published sustainability report, corporate social responsibility report or governance report, LCSA enables practitioners to understand the dynamics of complex systems (with economic, social and ecological dimensions), to identity the trade-offs between the three dimensions for providing a sustainable roadmap, to support guiding principles for designing the framework of the supply chain (innovation, product, services, stakeholders …). These evaluations of all economic, social and environmental impacts are more and more connected with SDGs reporting, initiated by the United Nations (2015). The aim of this chapter is to demonstrate the convergence of the two approaches, while underlining the need (i) to identify targets and indicators for a sustainable approach, and (ii) to use companies SDGs Reporting to provide an interesting reading grid for transition pathways.

2 Life Cycle Sustainability Assessment (LCSA): Framework and Methodology

The development of Life Cycle Assessment (LCA) methodology is related to the works of the Society of Environmental Toxicology and Chemistry (SETAC) and the ISO standardization process (Diemer 2023; Marsmann 2000; Ryding 1999; Hunt and Franklin 1997). From 1990 to 1993, SETAC North America and SETAC Europe shaped the basic structure of LCA in a series of workshops. The workshop, 'A Technical Framework for Life Cycle Assessments', held August 18–23, 1990, at Smugglers Notch, Vermont, was organized by SETAC to develop a framework and consensus on the current state of LCA and research needs for conducting life cycle assessment (Fava et al. 2014). Although life cycle assessments have been used before the name was coined, this workshop report is the first document which presented the name of the method. The following workshops, Leiden (1991), Sandestin (1992) and Wintergreen (1992) formed a step by step process (Gabathuler 1997) culminating in the *Code of Practice* workshop of Sesimbra (1993). The results of these workshops are illustrated by the SETAC Triangle (Fava et al. 1992). LCA is structured around four main areas: (i) Goal definition and scoping, (ii) Inventory Analysis, (iii) Impact Assessment, and (iv) Improvement Assessment. Shortly after the workshop, in November 1993 (Paris), the ISO standardization process (ISO Technical Committee, TC 207, Subcommittee SC5) was initiated (Marsmann 1997). The structure proposed by ISO (ISO 1996) differs from the SETAC structure (Lecouls

1999) only in the last element which is called "*interpretation*" in the international standard 14040 (Klöpffer 1997a). For Klöpffer and Grahl (2014, p. 9), two reasons could explain this replacement: (1) the fear by industry that an improvement assessment may become obligatory for all LCAs; (2) the fact that an LCA may also serve different purposes, not only product improvements. The double arrows symbolize that the different phases of the LCA may be modified if new aspects emerge during the performance of an LCA. ISO 14040 quickly became the official model of environmental life cycle (Finkbeiner et al. 2006). A success that Pryshlakivsky and Searcy (2013) also attribute to the position of 14040 ISO series vis-à-vis the overall ISO 14000 family of standards. The ISO 14040 was conceived as a supplementary tool of an overall Environmental Management System (EMS) platform centered on 14001: "*It may be stated that a useful goal, of ISO 14040 series is to inform ISO 14001 regarding the latter standard's requirement of continuous improvement to maintain compliance with the statut*" (2013, p. 116) (Fig. 1).

Soon after ISO 14040 standard publications, many criticisms emerged and have been discussed at the International Life Cycle Assessment and Management Conference in Seattle (2003). A draft list of improvements for a new achievement of the ISO 14040 has been proposed. New topics (new standard 14025 'Environmental Labelling and Declaration Type III') and revisions of the ISO Standards (14040, 14041, 14042, 14043) were formulated in Mexico City (January 2005) by the ISO Technical Committee 207 (Klüppel 2005). A new version of ISO 14040 was published in 2006. The family of ISO 14040 standards frames the requirements for conducting Life Cycle Assessments (LCA) while leaving the actual mechanics of analysis—data collection, normalization, calculation, interpretation, etc.—to the practitioner (many sections are transcriptions from 1997 to 2000 requirements). ISO 14040 (2006) specifies the principles and framework for carrying out life cycle assessments, including: the definition of LCA objectives and scope, the life cycle inventory phase, the life cycle impact assessment phase, the life cycle interpretation phase, the communication and critical review of the life cycle assessment, the limitations of life cycle assessment, the relationship between life cycle assessment phases, and the conditions for use of value choices and optional elements (ISO 2006). But the most significant element coming from the new revision of ISO 14040 series was the publications of

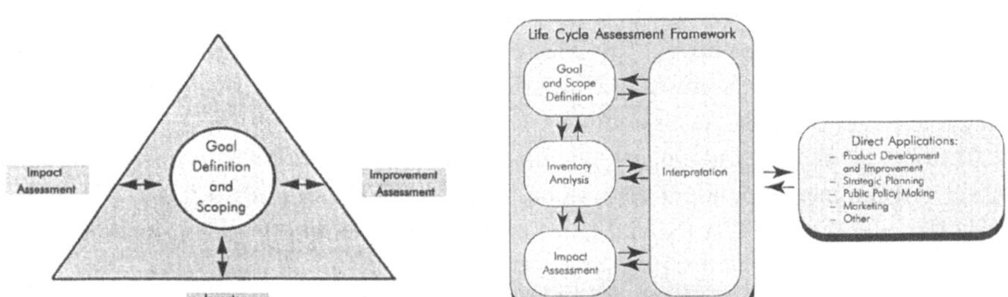

Fig. 1 SETAC and ISO standard definition of LCA. *Sources* Fava et al. (1991, p. 1) and Klöpffer (1997b, p. 223)

LCA case studies, proving that LCA was a relevant tool for sustainability assessment (Hellweg and Mila 2014).

Parker et al. (2007, p. 1120) noted that "*LCAs have been applied to a wide variety of energy technologies including nuclear, wind, coal as well as electrical networks*". Each stage of the product life cycle is evaluated in detail. Data on the energy and emissions from each stage are then gathered and, where not available, justifiable assumptions are made. For example, Pehnt (2006, p. 62) used dynamic LCA to challenge renewable energy technologies (this concerns the development of products, their production processes, as well as their technical performance and the development of so-called background systems). Schau and Fet (2008) explored the suitable functional units, system boundaries and allocation procedures for LCA in food production in general, and the product category rules (PCR) and environmental product declaration (EPD) for food products specifically. The term EDP is used on environmental product declarations that belong to a type III-programme which requires an LCA according to the ISO 14040—standards, an approval of the LCA and a third-party verification (ISO 14025 2006). Fava et al. (2009) presented the increasing demands for LCA in North America, especially with a focus on the integration of life cycle approaches into greener buildings, the development of life cycle-based carbon footprint protocols and the rapid development of requirements from retails companies demanding environmental performance of consumer goods (Pinto et al. 2019).

In the 2010s, a lot of international organizations and institutions recognised the importance of LCA. In its report *The Future We Want* (2012), the United Nations considered "*the importance of adopting a life cycle approach and of further development and implementation of policies for resource efficiency and environmentally sound waste management*". UNEP (2009, 2012) published the guidelines on social LCA (SLCA) and the framework on Life Cycle Sustainability Assessment (LCSA). LCSA refers to the evaluation of all environmental, social and economic negative impacts and benefits in decision making processes towards more sustainable products throughout their life cycle. Heijungs et al. (2010) have explored the scientific basis of life cycle based tools for sustainability, and then the concept of LCA. Using the framework of Huppes and Ishikawa (2009), they identify different models (technical, physical, environmental, micro-economic, meso and macro-economic, cultural, ethical and societal values, integrated) to propose a deeper and broader LCA, separating the empirical knowledge (the facts), the normative position (the values), and the transdisciplinary (integration). Figure 2 contains a framework for Lifecycle Sustainability Assessment.

To grasp all the dimensions of sustainability, Kloepffer (2008) proposed to split LCSA in three interdependent approaches: ELCA (Environmental LCA), LCC (Life Cycle Costing) and SLCA (Social Life Cycle Assessment).

$$\mathrm{LCSA} = \mathrm{ELCA} + \mathrm{LCC} + \mathrm{SLCA}$$

For UNEP (2022, 2012), it could be possible to integrate techniques from ELCA, LCC and SLCA (all based on the ISO 14040 (2006) into an overarching life cycle

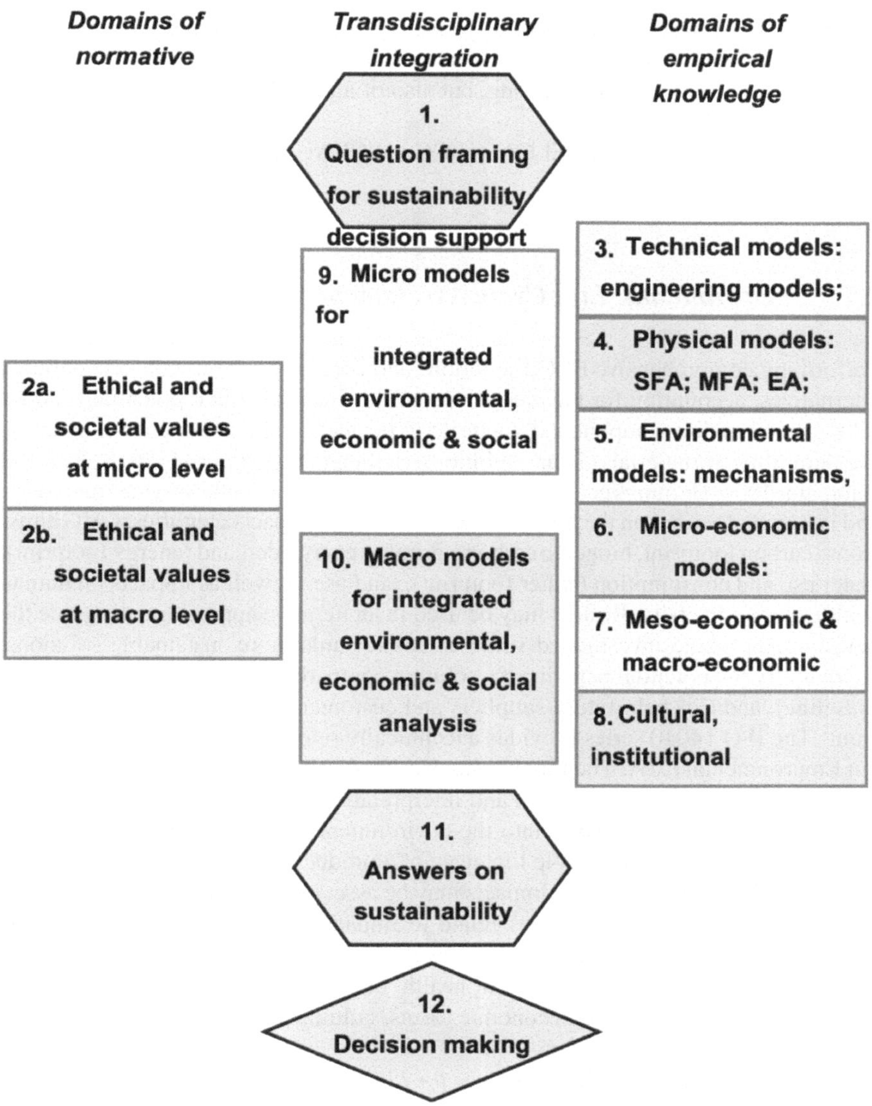

Fig. 2 Life cycle sustainability assessment. *Source* Heijungs et al. (2010, p. 426)

sustainability assessment (LCSA). So LCSA refers "*to the evaluation of all environmental, social and economic negative impacts and benefits in decision-making processes towards more sustainable products throughout their life cycle*" (UNEP 2012, p. 3). To this definition of life-cycle assessment, we will add a further component, geopolitical life-cycle assessment (GLCA). Many products (batteries, electric cars, solar panels, wind turbines, cell phones, etc.) require many rare metals or natural resources whose reserves are held by just a few countries. Lithium and cobalt, for example, are used in the design of electric car batteries. Thus, life-cycle assessment not only raises questions of recycling, but also of national security.

$$\text{LCSA} = \text{ELCA} + \text{LCC} + \text{SCLA} + \text{GLCA}$$

2.1 Environmental Life Cycle Assessment (ELCA)

Performing comprehensive ELCA solutions and comparing them with conventional alternatives, accounting for the complexity of the materials (incl. potential recyclability of end-of-life components) and criteria for technical performance (impact on the durability of materials) is an ambitious challenge (Leung et al. 2015; Rex and Baumann 2004; Heijungs et al. 1992). The ELCA follows a cradle-to-gate framework and is usually focused on the following environmental impact categories: GHG emissions (carbon footprint, biogenic carbon), primary energy demand (energy footprint), water use and consumption (water footprint), land use, as well as impacts on human health and biodiversity. Results may be used in an iterative approach to (i) guide the development of the investigated value chains towards more sustainable solutions, (ii) quantify the potential benefits of a solution compared to alternative technologies (baseline), and (iii) help future suppliers and customers make more informed decisions. The ISO 14040 series provides a technically rigorous framework for carrying out Environmental LCA. The four phases (goal and scope, inventory of resources use and emissions, impact assessment and interpretation) are interdependent. Figure 3 presents all emissions released into the environment and resources extracted from the environment along the whole life cycle of a product are grouped in an inventory (it's a list of elementary flows). Impacts may be assessed at the midpoint or endpoint level. Environmental interventions relate to impact damages (midpoints) such as ozone depletion, climate change, GHG emissions,[1] human toxicity… and damage categories (endpoints) such as human health, ecosystem quality, natural resources and ecosystem services, socioeconomic assets, cultural heritage or natural heritage (UNEP 2012).

In its report (2016), *Global Guidance for Life Impact Assessment Indicators* (vol 1), UNEP introduced few recommendations: (i) the overall framework should distinguish between intrinsic, instrumental and cultural values and the damage categories; (ii) the damage categories indicators are disability adjusted life years (DALY, human health) and biodiversity loss including measures of vulnerability (ecosystem quality), (iii) two climate change impact categories are introduced (short term and long term),

[1] In 2018, ISO 14064 specified principles and requirements at the organization level for the quantification and reporting of greenhouse gas (GHG) emissions and removals. It includes requirements for the design, development, management, reporting and verification of an organization's GHG inventory.

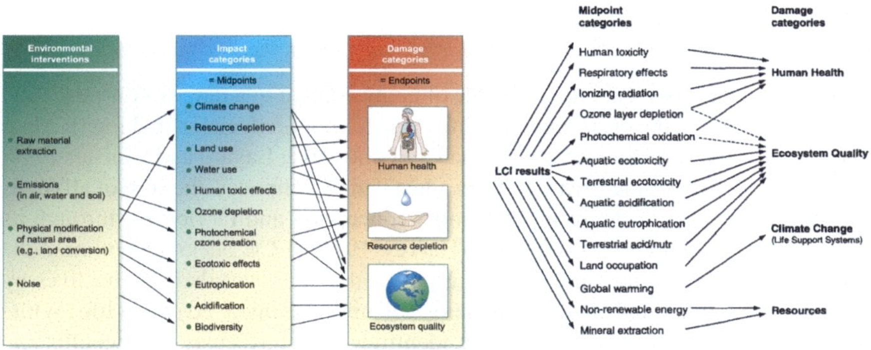

Fig. 3 ECLA with midpoints and endpoints. *Sources* Jolliet et al. (2003) and UNEP (2012)

(iv) fine particulate matter health impacts (PM 2.5 and CFs secondary) in rural and urban areas, (v) Water use impacts (malnutrition caused by lack of water for irrigated food production), (vi) Land use impacts (global potential species loss from land use). In 2019, UNEP produced a second report on the indicators for guidance, including few innovations. For Human toxicity, three indicators have been recommended: severity for cancer, reproductive/developmental, and other non-cancer effects. For ecotoxicity, it suggested considering the effects of chemicals on organisms living in coastal waters, soil freshwater and freshwater sediments. For Acidification and eutrophication, recommendations are to (i) use spatially explicit models with global coverage, (ii) aggregate CFs using agricultural, non-agricultural or overall emissions weighting, (iii) use existing environmental concentrations for effect modelling. One of the big challenges of the indicators is the capacity to capture uncertainties across impact categories at midpoint level (Fig. 4).

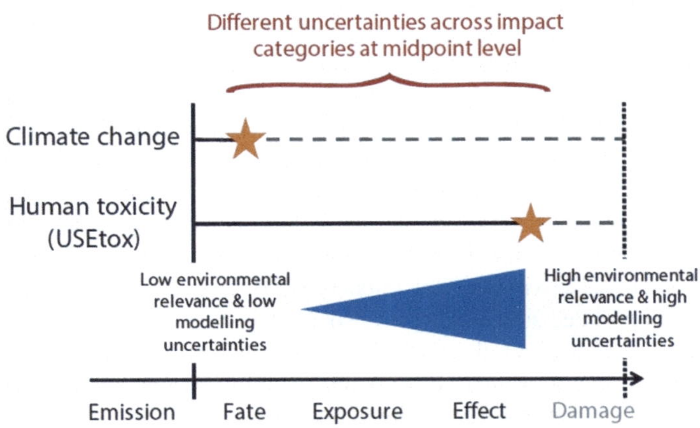

Fig. 4 Different uncertainties across indicators for midpoint impact categories. *Source* UNEP (2019, p. 46)

2.2 Life Cycle Costing

Life Cycle Costing is a technique capable of being applied to a range of purposes and at different stages in the project or asset life cycle to support decision making (Rödger et al. 2018; Manzo and Salling 2016; Quah and Tho 2012; Layard and Glaister 1994; Stevens 1976). Life Cycle Costing of a product can be defined as the sum of all funds expended in support of the product from its conception and fabrication through its operation to the end of its useful life (Harvey 1976). LCC seeks to optimize the cost of acquiring, owning, and operating physical assets over their useful lives by attempting to identify and quantify all the significant costs involved in that life (White and Ostwald 1976). The LCC approach identifies all future costs and benefits and reduces them to their present value using the discounting technique (Woodward 1997). It is possible to present a non-exhaustive list of elements of the LCC such as the initial capital costs, the life of the asset, the discount rate, the operating and maintenance costs, the disposal cost, information and feedback, and uncertainty and sensitivity analysis. To facilitate the evaluation, Kaufman (1970) proposed a LCC procedure based on an eight step approach: (1) establish the operating profile OP (description of the periodic cycle through which the equipment will go), (2) establish the utilization factors (indicates in what way equipment will be functioning within each mode of the OP), (3) identify all the cost elements (every cost must be identified), (4) determine the critical cost parameters (time period between failures, time period for repairs, energy use rate, etc.), (5) calculate all costs at current prices, (6) escalate current costs at assumed inflation rates (example of interest rates for physical assets), (7) discount all costs to the base period (money has a time value and the cash flows occurring in different time periods should be discounted back to the base period), and (8) sum discounted costs to establish the net present value. More recently, three variants of the LCC method have been developed. *Conventional LCC* (financial LCC) is the original method. LCC is applied as a decision-making tool to support acquisition of capital equipment and long lasting products with high investment costs (Martinez-Sanchez et al. 2015). The Convention LCC is useful for a single actor or from the manufacturer's point of view (Hunkeler et al. 2008). In this last case, the life cycle cost breaks down with specific focus on the production stages. *Environmental Life Cycle Costing* (eLCC) is connected with the ISO Standard 14040 and 14044 on LCA. This method considers the perspective of a functional unit and the whole life cycle, including all actors of the supply chain or life cycle (Steen 2005; Hoogmartens et al. 2014). eLCC has been developed to support LCA in the sense that it covers the economic dimension and helps to identify hot-spots in terms of cost and environmental impacts (Rödger et al. 2018). Societal LCC (sLLC) is to support decision making on a societal level, including governments and public authorities (Clroth et al. 2015). Societal assessment is advocated as one of the three pillars in the movement toward sustainability (Hunkeler 2006). It permits relative product comparisons rather than absolute analysis. Societal LLC includes selected external costs by assigning a monetary value to them. This process is a monetization of costs and impacts. A sLLC goes very far because it monetarizes

Tool/Method	LCA	LCC
Purpose	Compare relative environmental performance of alternative product systems for meeting the same end-use function, from a broad, societal perspective	Determine cost-effectiveness of alternative investments and business decisions, from the perspective of an economic decision maker such as a manufacturing firm or a consumer
Activities which are considered part of the 'Life Cycle'	All processes causally connected to the physical life cycle of the product; including the entire pre-usage supply chain; use and the processes supplying use; end-of-life and the processes supplying end-of-life steps	Activities causing direct costs or benefits to the decision maker during the economic life of the investment, as a result of the investment
Flows considered	Pollutants, resources, and inter-process flows of materials and energy	Cost and benefit monetary flows directly impacting decision maker
Units for tracking flows	Primarily mass and energy; occasionally volume, other physical units	Monetary units (e.g., dollars, euro, etc.)
Time treatment and scope	The timing of processes and their release or consumption flows is traditionally ignored; impact assessment may address a fixed time window of impacts (e.g., 100-year time horizon for assessing global warming potentials) but future impacts are generally not discounted	Timing is critical. Present valuing (discounting) of costs and benefits. Specific time horizon scope is adopted, and any costs or benefits occurring outside that scope are ignored

Fig. 5 Differences between LCA and LCC. *Source* Norris (2001, p. 118)

all environmental and social impacts (job creation, job quality, working conditions, etc.). Societal LCC has some connection with Corporate Social Responsibility (CSR). Societal LCC offers the advantage of presenting results in one single monetary unit, integrating all the three pillars of sustainability, and supporting policy decisions in a combined Life Cycle Sustainability Assessment (Timonen et al. 2017). Integrating societal Life Cycle Cost (sLLC) and LCSA removes all pre-existing methodological differences between LLC and LCA. Initially, LCC and LCA were used to answer different questions (Klöpffer and Ciroth 2011). LCC compares the cost effectiveness of alternative investments or business decisions from the perspective of an economic decision maker (producer or consumer). LCA evaluates the relative environmental performance of alternative product systems for providing the same function (Norris 2001). This environmental performance considers all important process casualties, resources, and consumption flows regardless of their impact (Brent 2009). Figure 5 presents the différences between LCA and LCC.

2.3 Social Life Cycle Assessment

Assessment of social dimension introduces two components: the method (Social Life Cycle Assessment, SLCA) to catch social impacts of innovations and leverage points (potential issues for social acceptance) to implement in a better way the innovation. SLCA is a social impact assessment method (Wur et al. 2014; Macombe et al. 2011; Norris 2006) that aims to assess the social aspects of products and their positive and negative impacts along their life cycle encompassing extraction and processing of raw materials, manufacturing, distribution, use, reuse, maintenance, recycling, and final disposal (Petti et al. 2016). We define social aspects as all parameters, indicators, or issues related to the social relations between individuals and between individuals and society, and we apply social aspect classifications established in the Social Life Cycle Assessment (SLCA) methodology (Weidema 2006;

Jorgensen et al. 2008; Ellen MacArthur Foundation 2013). The 2000s produced a lot of studies about SLCA (Dreyer et al. 2006), including a multitude of impacts (ranging from direct impacts on workers to broad societal issues), several types of decisions makers and the question of data access. So, a lot of methodologies have been developed without a specified target group of users. The Global Reporting Initiative has developed a flexible framework for integrating SLCA. The GRI Standards are structured as a set of interrelated standards. There are three universal standards that apply to every organization preparing a sustainability report. An organization selects from the set of topic-specific GRI standards for reporting on its material topics. The topic-specific GRI standards are organized into three series: 200 (economic topics), 300 (environmental topics), and 400 (social topics). The economic dimension of sustainability concerns an organization's impacts on the economic conditions of its stakeholders and on economic systems at local, national, and global levels. It does not focus on the financial condition of an organization. The environmental dimension of sustainability concerns an organization's impacts on living and non-living natural systems, including land, air, water, and ecosystems. The social dimension of sustainability (Fig. 6) concerns an organization's impacts on the social systems within which it operates. Social dimension includes employment, labor/management relations, occupational health and safety, training and education, child labor, security practices, human rights assessment, local communities, marketing and labeling, customer privacy, etc.

JRC technical reports (2015) proposed to present (1) the state of the art in social Life Cycle Assessment, illustrating the main theoretical and methodological elements under scientific literature; (2) the overlaps and the synergies with traditional LCA, towards a common and integrated assessment framework; and (3) examples of applications of SLCA at macro scale (EU-28) and at sector scale. The JRC report

Fig. 6 Social dimension in the global reporting initiative. *Source* GRI (2007, p. 11)

insisted on databases and indicators for SLCA. Data sources for implementing social impact assessment are crucial and in many cases, very difficult to be built and maintained. Two relevant databases have been described: the social hotspots Database (SHDB), including a global input–output model, a worker hours' model and data on social risks and opportunities, and the Product Social Impact Life Cycle Assessment Database (PSILCA). PSILCA is a database showing how social data can be organized, assessed and finally used for SLCA or Life Cycle Sustainability Assessment (Table 1). PSILCA is a one of reference for SLCA indicators in research (JRC 2015; Diemer 2023; Christina et al. 2018).

2.4 Geopolitical Life Cycle Assessment

Impact analysis is not confined to the environmental, social or economic spheres, but is increasingly incorporating geopolitical issues. The latter are directly linked to national security. Life-cycle analysis (from extraction to recycling or reuse of the natural resource) must also take geopolitical risk into account in impact assessments. Many products (batteries, electric cars, solar panels, wind turbines, cell phones, etc.) require many rare metals or natural resources whose reserves are held by just a few countries. Lithium and cobalt, for example, are used in the design of electric car batteries. Political, commercial or simply belligerent conflicts (such as the war between Russia and Ukraine) can cause value chains to break down. States and companies alike need to build up reserves, but also to integrate geopolitical risks into their life-cycle assessment (a change of supplier can result in a drop in raw material quality, a civil war can block imports of natural resources, recycling sites—particularly when they are different from processing sites—can be the subject of political tensions, and control of supplies of an imported resource can jeopardize political choices).

In 2015, The European Commission set up a monitoring framework to keep track of progress towards a circular economy. This framework provided a holistic view as measuring direct and indirect benefits of becoming circular, valuing the contribution of a circular economy in living well within the limits of the planet, addressing energy and material supply risks (Pena et al. 2021). Five thematic sections (production and consumption, waste management, secondary raw materials, competitiveness and Innovation, global sustainability and resilience) with a total of 11 statistical indicators were selected to capture the main elements of a circular economy. A self-sufficiency table for raw materials and rare metals has been proposed to help individual EU members to secure their supply chain. All these materials have been identified as strategic (Fig. 7) because of geopolitical issues in the regions where they are sourced, geologically availability is limited to few locations or their production in only a few countries (Gemechu et al. 2015a).

In the case of rare metals, China accounts for nearly 88% of mining production. Neodymium and dysprosium are used in wind turbines. Lanthane, Cerium, praseodymium and Neodymium are present in automobile catalytic converters.

Table 1 List of social indicators for PSILCA

Stakeholder	Subcategory	Indicator	Unit of measurement
Workers	Child labour	Children in employment, male (% of male children ages 7–14)	%
		Children in employment, female (% of female children ages 7–14)	%
		Children in employment, total (% of all children ages 7–14)	%
	Forced labour	Evidence of forced labour	Text
		Frequency of forced labour	%
	Fair salary	Living wage, per month	Local currency
		Minimum wage, per month	Local currency
		Sector average wage, per month	Local currency
	Working time	Hours of work per employee, per day	h
		Hours of work per employee, per week	h
		Standard weekly hours	h
		Standard daily hours	h
	Discrimination	Occurrence of discrimination	Text
		Women in the labour force (% of economically active female population)	%
		Men in the labour force (% of economically active male population)	%
		Ratio of salary of women wages to men	%
	Health and safety	Accident rate at workplace	#/100,000 workers
		Fatal accidents at workplace	#/100,000 workers
		Occupational risks	Text
		DALY due to indoor and outdoor air and water pollution	DALY/1000 persons
		Presence of sufficient safety measures	# of security incidents
	Social benefits, legal issues	Social security expenditures out of the total GDP	%
		Evidence of violations of laws and employment regulations	#/year h
		% of workers with a contract	%
	Freedom of association, collective bargaining, right to strike	Trade union density (% of employees organised in trade unions)	%
		Right of association	Index value
		Right of collective bargaining	Index value

(continued)

Table 1 (continued)

Stakeholder	Subcategory	Indicator	Unit of measurement
		Right to strike	Index value
		Existence of standard rates	Y/N
Local community	Access to material resources	Level of industrial water use (% of total withdrawal)	%
		Level of industrial water use (% of total actual renewable)	%
		Extraction of material resources (fossil fuels, biomass, ores, minerals)	t/capita
		Presence of certified environmental management systems	#
		Description of (potential) material resource conflicts	Text
	Respect of indigenous rights	Presence of indigenous population	Y/N
		Human rights issues faced by indigenous people	Text
		Respect of indigenous rights	Text
	Safe and healthy living conditions	Pollution level of the country	Index value
		Contribution of the sector to environmental load	Text
		Drinking water coverage (% of the population)	%
		Sanitation coverage (% of the population)	%
	Local employment	Unemployment rate in the country	%
		Work force hired locally	%
		Percentage of spending on locally based suppliers	%
	Migration	Migrant workers in the sector	%
Society	Contribution to economic development	Economic situation of the country	Index value
		Contribution of the sector to economic development (in % of total GDP)	%
		Public expenditure on education (% of GDP)	%
	Education	Illiteracy rate, male (% of male population)	%
		Illiteracy rate, female (% of female population)	%
		Illiteracy rate, total (% of total population)	%

(continued)

Table 1 (continued)

Stakeholder	Subcategory	Indicator	Unit of measurement
	Health and safety	Health expenditure out of the total GDP of the country	%
		People affected by natural disasters (as % of population)	%
		Life expectancy at birth	Years
	Prevention and mitigation of conflicts	Risk of conflicts with regard to the sector	Text
Value chain actor	Fair competition	Presence of anti-competitive behaviour or violation of anti-trust and monopoly legislation	Text
		Presence of policies to prevent anti-competitive behaviour	Y/N
	Corruption	Corruption index of country	Index value
		Evidence of an active involvement of the enterprises in corruption and bribery	%
	Promoting social responsibility	Presence of codes of conduct that protect human rights of workers among suppliers	Index value
		Membership in an initiative that promotes social responsibility along the supply chain (number of enterprises)	#
	Supplier relationships	Interaction of the companies with suppliers (payment on time, sufficient lead time, reasonable volume fluctuations, appropriate communication…)	Text
Consumers	Health and safety	Presence of management measures to assess consumer health and safety	Y/N
	Transparency	Presence of certifications or labels for the product/sites sector	Y/N
	End of life responsibility	Strength of national legislation covering product disposal and recycling	Text

Source JRC (2015, p. 44)

Geopolitical impacts are not so easy to identify and quantify. Supply risks and vulnerability to supply disruptions are the main impacts of geopolitical risks. Gemechu et al. (2015b) identified what they called the geopolitical-related supply risk "GeoPolRisk" which is calculated by the following equation:

$$\mathrm{SR}_{c,i} = \left[\left(\sum_{k=1}^{n} s_k^2 \right) \times \left(\sum_{k=1}^{n} g_k \times f_{i,k} \right) \right]$$

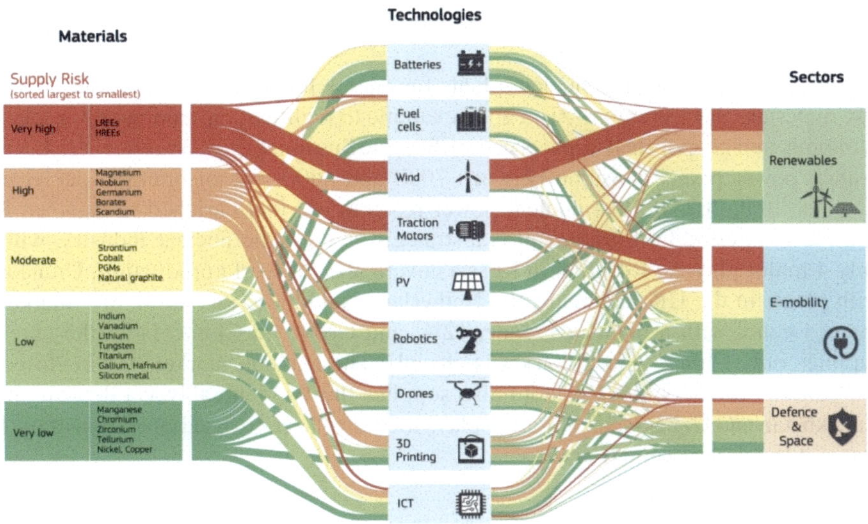

Fig. 7 Flows of raw materials and their current risks. *Source* JRC (2020)

SR_{cj} is the supply risk of country i for commodity c; s_k is the global production share of country k of the commodity c, g_k is the political instability indicator of country k, $f_{i,k}$ is an import share of country k in the supply chain of country i. The GeoPolRisk is expressed as a socioeconomic risk oriented midpoint indicator with values between 0 and 1.

Caldara and Iacoviello (2022) presented a measure of geopolitical risk based on the timeliness of adverse geopolitical events and associated risks. The Geopolitical Risk Index (GPR) is an indicator designed to measure the impact of geopolitical tensions on global economic stability. It uses newspaper articles to quantify the intensity of geopolitical risks at any given time. It has four features: (i) it is based on the number and tone of articles related to geopolitical tensions in major international newspapers. It uses algorithms to analyse the text of articles and assess the level of risk. (ii) it is generally updated monthly and offers a historical overview of geopolitical risks over several decades, enabling current periods of tension to be compared with historical periods; (iii) investors and decision-makers can use this index to assess how geopolitical tensions could affect financial markets and economic stability; (iv) the index considers various events such as wars, diplomatic tensions, terrorist acts and other crises that can have a significant impact on global geopolitics. An elevated geopolitical risk index (GPR) predicts lower investment and employment, and is associated with a higher probability of catastrophe and greater downside risks. The negative consequences for the GPR index are due to both the threat and realization of adverse geopolitical events. The GPR index is also applied at industry and company level. Investments decrease more in sectors exposed to global geopolitical risk. Higher geopolitical risk at company level is associated with lower investment at company level.

For few sectors, such as nuclear plants for energy transition, importations of the raw material—here uranium—is a good example to highlight environmental, social, economic and geopolitical impacts in the life cycle Sustainability Assessment (LCA). French power plants consume between 7000 and 8000 tons of uranium per year (Diemer et al. 2024). Most nuclear power reactors use fuel enriched to between 3 and 5% uranium-235. Enrichment is a strategic step for the nuclear industry, with only 4 major players: Rosatom in Russia covers 46% of world production; Urenco, with plants in the UK, Germany, the Netherlands and the USA, has a 30% market share; Orano, with a plant in France, 12%; and CNNC in China, 11% (the latter supplies only China, whose market is closed). The Russian market is also closed to Western enrichers, but Rosatom also supplies the West, notably via deposits in Kazakhstan and Uzbekistan. Rosatom plays the role of a control variable (constraint) that is difficult to escape (Fig. 8).

3 Sustainable Development Goals (SDGs)

The United Nations Sustainable Development Goals (SDGs) have become a reference point for policy-making processes worldwide. The goal of SDGs was first published in 1987, where the main goal was to meet the needs of present and future generations in full compliance with the natural environment (WCED 1987). The concept of sustainable development has evolved over time to the Millennium Sustainable Development Goals (MDGs) in 1992 and to the current version named 2030 Agenda goals which was drawn up in 2015 with a 15-year period of implementation planned. It outlined 17 Sustainable Development goals (SDGs), 169 associated targets and 231 unique indicators (Table 2).

Traditionally developed with national governing organisations for applications on a policy level, SDGs became a great challenge for companies, especially for those who wanted to contribute to these goals. SDGs provide a shared vision for sustainable development which essentially indicate the need to protect the social and environmental equilibrium in the process of economic development. Their implementation is possible due to the global goals set, which in a precise manner indicate the directions in which organisations and people should follow to care for both the environment and each other (Aneta 2023). According to the UN Global Compact Strategy (2024–2025), the strategy for business practice on the SDGs can be divided into three groups. (1) 'Lead and shape', (2) 'Cooperate with others', and (3) 'Follow and amplify' (Fig. 9).

There are several studies researching SDGs in business practice. KPMG (2022) investigated the world's 250 largest companies across all sectors. The result reveals that Three SDGs remain the most popular for companies: 8: Decent Work and Economic Growth; 12: Responsible Consumption and Production; and 13: Climate Action. Annex presents the three goals addressed (8, 12 and 13) with their targets

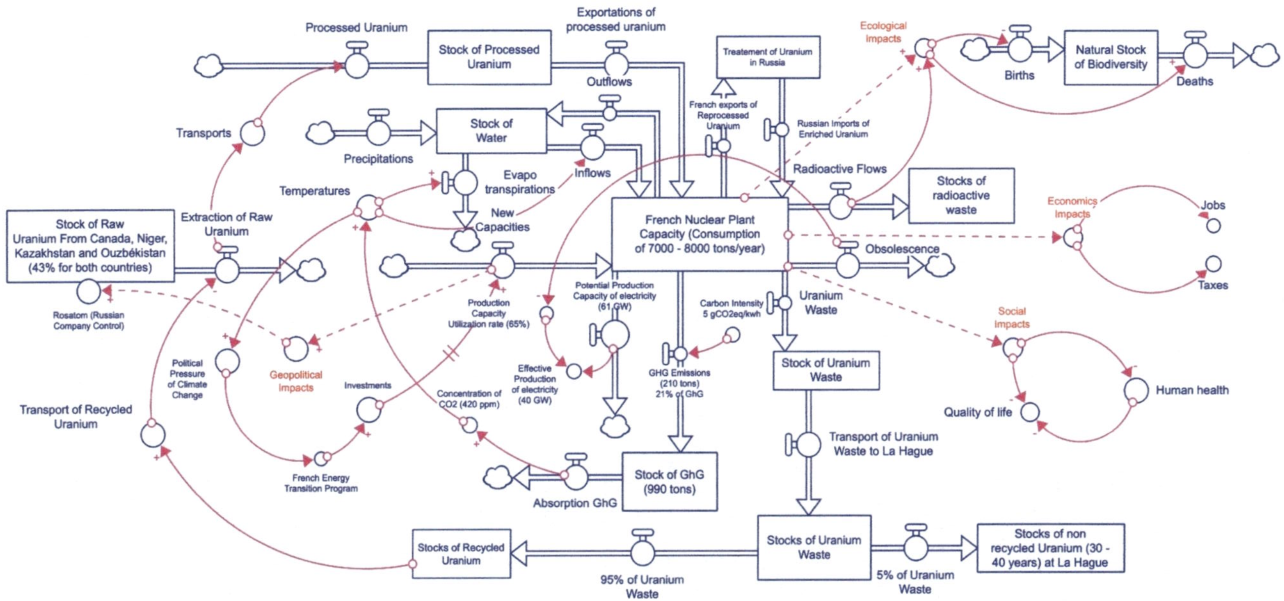

Fig. 8 Environmental, social, economic and geopolitical impacts in LCSA. *Source* Diemer et al. (2024)

Table 2 SDGs, targets and indicators

Goals and targets (from the 2030 agenda for sustainable development)	Number of targets	Number of indicators
Goal 1. End poverty in all its forms everywhere	7	13
Goal 2. End hunger, achieve food security and improved nutrition and promote sustainable agriculture	8	14
Goal 3. Ensure healthy lives and promote well-being for all at all ages	13	28
Goal 4. Ensure inclusive and equitable quality education and promote lifelong learning opportunities for all	10	12
Goal 5. Achieve gender equality and empower all women and girls	9	14
Goal 6. Ensure availability and sustainable management of water and sanitation for all	8	11
Goal 7. Ensure access to affordable, reliable, sustainable and modem energy for all	5	6
Goal 8. Promote sustained, inclusive and sustainable economic growth, full and productive employment and decent work for all	12	16
Goal 9. Build resilient infrastructure, promote inclusive and sustainable industrialization and foster innovation	8	12
Goal 10. Reduce inequality within and among countries	10	14
Goal 11. Make cities and human settlements inclusive, safe, resilient and sustainable	10	15
Goal 12. Ensure sustainable consumption and production patterns	11	13
Goal 13. Take urgent action to combat climate change and its impacts 3	5	8
Goal 14. Conserve and sustainably use the oceans, seas and marine resources for sustainable development	10	10
Goal 15. Protect, restore and promote sustainable use of terrestrial ecosystems, sustainably manage forests, combat desertification, and halt and reverse land degradation and halt biodiversity loss	12	14
Goal 16. Promote peaceful and inclusive societies for sustainable development, provide access to justice for all and build effective, accountable and inclusive institutions at all levels	12	24
Goal 17. Strengthen the means of implementation and revitalize the Global Partnership for Sustainable Development	19	24
Total	169	248[a]

Source https://unstats.un.org/sdgs/indicators/indicators-list/

[a]The global indicator framework includes **231** unique indicators. Please note that the total number of indicators listed in the global indicator framework of SDG indicators is 248. However, thirteen indicators repeat under two or three different targets

Fig. 9 Summarises SDGs priorities in three general categories. *Source* UN Global Compact Strategy (2024–2025)

and respective indicators to monitor their achievement. Several guidelines of steps to define SDGs in company strategies and sustainable reporting can be observed from the Business reporting on the SDGs as shown in Fig. 10 (https://www.unglobalcompact.org) and Fig. 11 (Morao and Visser 2022) proposed steps to implement SDGs from impact areas analysis.

Examples of how companies contribute to the SDGs by addressing them in their sustainability or performance reports are illustrated in Figs. 12, 13 and 14.

To achieve its 2050 Net Zero Ambition, Total Energies defined its sustainability ambitions and targets around four axes: (i) Climate and Sustainable Energy, (ii) Caring for the Environment, (iii) Acting for the well-being of employees, (iv) having a positive impact on stakeholders. In addition to its strategy of contributing to the 17 SDGs, LVMH has designed its business model on SDGs 8, 9 and 10 (targets 8.2, 8.3, 8.9, 9.2 and 10.2), especially by improving its economic footprint and its commitment as a responsible employer. Its Ethics program has been driven by a Code of Conduct for Suppliers and Commercial Partners (SDG 8, target 8.7). For example, LVMH made main contributions to gender equality initiatives, diversity and inclusion policy or discrimination.

Contributions of Michelin and Veolia to Sustainable Development Goals (SDGs) have been included in the United Nations Global Compact, called *Forward Faster Initiative*, which is focused on five major areas for action: gender equality, climate

Fig. 10 The steps outlined to embed the SDGs in existing business and reporting processes. *Source* Pedercena (2018)

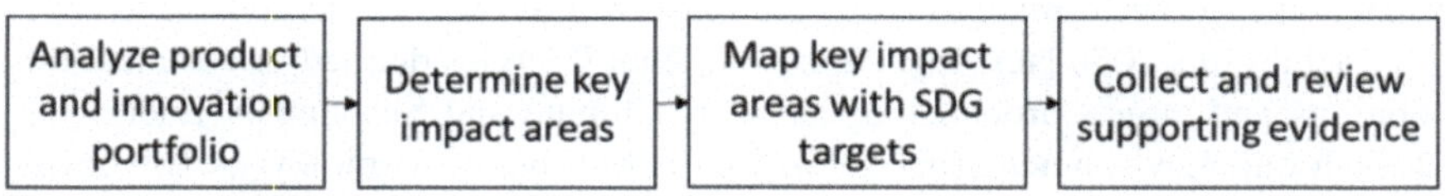

Fig. 11 The steps outlined to embed the SDGs from impact areas analysis. *Source* Morao and Visser (2022)

action, living wage, water resilience and finance and investment. Since 2020, Michelin has participated in a working group in the UN SDGs (Tire Industry Project) and has designed a roadmap identifying the tire's industry main impacts and the leverage points. Beside its key sustainability indicators (waste, energy, water and Environment—ISO 14001), the company expanded the scope of reporting to include a diversity of social responsibility indicators (percentage of women in

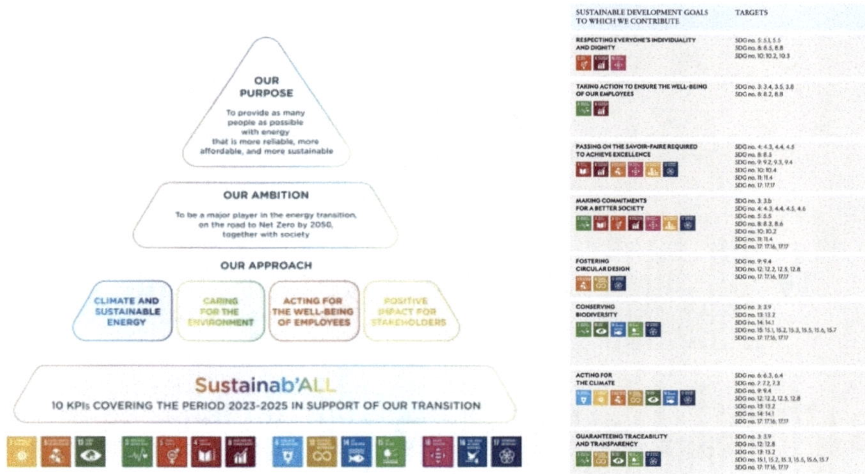

Fig. 12 Total energies and LVMH sustainability reporting *Sources* Total Energies (2023) and LVMH (2023)

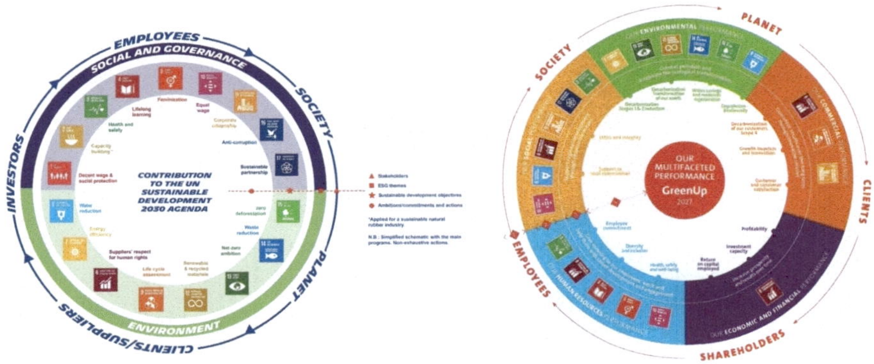

Fig. 13 Michelin and Veolia sustainability reporting *Sources* Michelin (2023) and Veolia (2023)

positions of responsibility, percentage of members with public commitments to diversity and inclusion, percentage of members with responsible sourcing policies). Veolia engaged the Multifaceted Performance (Commercial, Economic and Financial, Human Resources, Social and Environmental Performance) with a direct impact on 13 SDGs.

Carrefour supported the 17 SDGs set by the United Nations, especially challenges associated with its Corporate Social Responsibility (CSR) and Food Transition Index. Targets include social audits, agroecology practices, reducing CO_2 emissions from downstream transport, carbon neutrality ….

Fig. 14 Carrefour sustainability reporting. *Source* Carrefour (2023)

4 Linking UN SDGs to Life Cycle Sustainability Assessment

In the 2020s, *the Life Cycle Initiative* (LCI 2020, 2022, 2023), some public–private, multi-stakeholders' partnership hosted by the United Nations Environment Program (UNEP), provided a global forum to ensure a science-based, consensus building process to support decisions and policies towards the shared vision of sustainability. The initiative facilitated the application of life cycle knowledge in the global sustainable development agenda to achieve global goals faster and more efficiently. This project, which spanned over the course of 3 years, has created a first attempt at making an operational link between the SDGs to company activities on a product level.

Two approaches have been developed in parallel—the life cycle SDG screening (LCSS) and the life cycle SDG assessment (LCSA). The methodology that resulted from this project allows companies to assess and report on their contribution to the SDGs, and therefore substantially helps to streamline and coordinate action and increase efficiency in implementing the 2030 Agenda. For this study, we investigate the adoption and implementation of SDGs and explore the qualitative relationships between LCSA categories and SDGs. The research framework examines five companies (Carrefour, LVMH, Michelin, Total Energies, Veolia) actively engaged in SDGs reporting and coming from different sectors or industries (Energy, Environment, Food distribution, Luxury and Tires). To understand these qualitative relations, a multi-step approach will be used to explore the connections between SDGs and the potential economic, environmental, and social impacts identified by LCSA. The research design follows a three-step process.

Step (1) Company Document Selection and Analysis

This step focuses on collecting information from official company reports. The principal source will be the recent SDG communication reports of the five chosen companies retrieved from the UN Global Compact database (UNGlobal Compact).

Step (2) Mapping LCSA Indicators to SDGs

Based on the documents collected in step 1, this step will involve extracting LCSA indicators relevant to each SDG. A connection table will be created to map these indicators to the 17 SDGs. Finally, the total number of linkages across the five companies will be consolidated. LCSA indicators selection are from the paper recommended (Marwa et al. 2023; Michael and Rudiger 2017; Sabrina et al. 2015).

Step (3) Assessing Degree of Connection and Indicator Measurement

Building on step 2, this step will identify the top LCSA indicator with the strongest relevance to SDGs. The analysis will then focus on how these companies report on the chosen indicator to measure progress towards their SDG targets.

The three-step process enables us to present two tables showing the connections (points of contact) between the environmental, social and economic impacts of the Life Cycle Sustainability Assessment (LCSA) and the 17 SDGs (Tables 3 and 4).

A summary of connection between LCSA indicators and SDG link in Table 3, reveals that LCSA topics that are highly relevant for the SDG is CO_2 emission. For LCSA indicators that can break down into several SDG's (6SDGs concerned) are Climate change and CO_2 emission.

From Table 4, it shows that the top 3 SDGs that are well covered by LCSA are Decent work (SDG8), Consumption and prod (SDG12) and Climate action (SDG13). To understand how companies apply SDGs in their targets, this study focuses on climate change and CO_2 emission measurement.

From five companies report, climate change and CO_2 emission has a connection with seven SDGs that are: SDG 6—Clean water and sanitation, SDG 7—Affordable and clean energy, SDG 9—Industry, innovation and infrastructure, SDG 11—Sustainable cities and communities, SDG 12—Responsible consumption and production, SDG 13—Climate action and SDG 15—Life on land. CO_2 emissions are a key contributor to climate change, making them a direct link to SDG 13. By providing a specific measurement of climate impact, CO_2 emissions data helps companies understand their contribution to climate change and align their actions with SDG 13. All five companies have integrated climate change and CO_2 emission into their policies with the scope and methodology as shown in Table 5. The unit of measurement as mass of accumulated CO_2 emissions along the supply chain, LCA indicator, can be a representative of the amount of greenhouse gases (GHG) produced during the life cycle of a product, a process, or a service. For most companies, getting to know GHG emissions is one of the first steps in taking corporate climate change. This is a way to understand where the opportunities for reducing are, and how to make business decisions and investments for the future. GHG emissions are considered either "direct" (scope 1, emissions are direct from sources that are owned or controlled by the company) or "indirect" (scope 2 when emissions are generated at another facility such as a power station for energy and scope 3 when emissions occurred across the supply chain). Figure 15 gives an overview of the scope's classification and the different GHG. The GHG inventory contains six different greenhouse gases: carbon dioxide (CO_2), methane (CH_4), nitrous oxide (N_2O), hydrofluorocarbons (HFCs), perfluorocarbons (PFCs), and sulfur hexafluoride (SF_6). GHG emissions and the different scopes are well documented in the indicators of Life Sustainability Cycle Assessment, but also play a key role in the integration of SDG13 and other SDGs (Fig. 16).

The analysis of five companies within this case study reveals that the translation of Sustainable Development Goals (SDGs) can be effectively mapped onto various Life Cycle Sustainability Assessment (LCSA) indicators. Given that SDGs are primarily designed to enhance sustainability at a national or global scale, integrating LCSA indicators—which are defined at the micro-level with specific units of measurement—provides a pragmatic framework for companies to align their operations and policies with SDG targets. By adopting LCSA indicators (Weidema 2021), companies can quantify and track their progress towards sustainability goals with precision.

Table 3 The connection between LCSA economy, environmental indicators and the 17 SDGs

SDG/LCSA		Economy			
		Share of spending with local stakeholders	Investment in research for circular economy	Green material productivity	Revenue
SDG1	No poverty	0	0	1	0
SDG2	Zero hunger	0	0	1	1
SDG3	Health	0	0	0	0
SDG4	Education	0	0	0	0
SDG5	Gender	0	0	0	0
SDG6	Water	0	0	0	0
SDG7	Clean energy	0	1	0	0
SDG8	Decent work	2	3	1	5
SDG9	Industry	1	0	0	0
SDG10	Inequality	1	0	0	0
SDG11	Sust. cities	0	0	0	0
SDG12	Consumption and prod	0	1	1	0
SDG13	Climate	0	0	0	0
SDG14	Life below water	0	0	0	0
SDG15	Life on land	0	0	0	0
SDG16	Peace justice strong institutions	0	0	0	0
SDG17	Partnerships	0	0	0	0
	Sum of number of links	4	5	4	6
	SDG link to this topic	3	3	4	2

(continued)

Table 3 (continued)

SDG/ LCSA	Environment										
	Climate change	Energy consumption and efficient	Water consumption and quality	Total wastewater generated	Total solid waste generated	CO_2 emission	Products energy efficiency	Renewable and recycled materials rate	Increase/ decrease in total renewable energy consumption	Products and services with any kind of eco-label	Total use of Renewable primary energy
SDG1	0	0	0	0	0	0	0	0	0	0	0
SDG2	0	0	0	0	1	0	0	0	0	1	0
SDG3	0	0	1	0	0	0	0	0	0	0	0
SDG4	0	0	0	0	0	0	0	0	0	0	0
SDG5	0	0	0	0	0	0	0	0	0	0	0
SDG6	2	0	4	1	0	0	0	1	0	0	0
SDG7	2	1	0	0	0	3	1	1	2	0	1
SDG8	0	0	0	0	0	0	0	1	0	0	0
SDG9	0	0	0	0	0	2	2	0	1	1	1
SDG10	0	0	0	0	0	0	0	0	0	0	0
SDG11	2	1	0	0	0	1	0	0	0	0	0
SDG12	0	2	1	0	1	5	0	4	0	0	0
SDG13	5	1	0	0	0	5	1	1	0	0	0
SDG14	1	0	0	2	0	0	0	0	0	1	0
SDG15	1	0	0	0	1	1	0	0	0	1	0
SDG16	0	0	0	0	0	0	0	0	0	0	0
SDG17	0	0	0	0	0	0	0	0	0	0	0
	13	5	6	3	3	17	4	8	3	4	2
	6	4	3	2	3	6	3	5	2	4	2

Source The Authors

Table 4 The connection between LCSA Social indicators by categories and the 17 SDGs

SDG. LCSA	Social									
	Society			Workers						
	Corruption	Training for young people	Implementation of a biodiversity management	International empowerment	Freedom of association and bargaining	Employee engagement rate	Number of trainings provided	Woman empowerment	Discrimination	Health and safety
SDG1	0	0	0	0	1	0	0	0	0	0
SDG2	0	0	0	0	0	0	0	0	0	0
SDG3	0	0	1	0	0	1	0	0	0	3
SDG4	0	2	0	0	0	0	4	0	1	0
SDG5	0	0	0	0	0	0	0	5	1	0
SDG6	0	0	0	0	0	0	0	0	0	0
SDG7	0	0	0	0	0	0	0	0	0	0
SDG8	0	1	0	2	0	4	2	0	4	5
SDG9	0	0	0	0	0	0	0	0	0	0
SDG10	1	0	0	1	1	0	0	1	2	0
SDG11	0	0	0	0	0	0	0	0	0	0
SDG12	0	0	0	0	0	0	0	0	0	0
SDG13	0	0	1	0	0	0	0	0	0	0
SDG14	0	0	3	0	0	0	0	0	0	0
SDG15	0	0	4	0	0	0	0	0	0	0
SDG16	3	0	0	0	1	0	0	0	0	0

(continued)

Table 4 (continued)

SDG. LCSA	Social									
	Society			Workers						
	Corruption	Training for young people	Implementation of a biodiversity management	International empowerment	Freedom of association and bargaining	Employee engagement rate	Number of trainings provided	Woman empowerment	Discrimination	Health and safety
SDG17	0	0	1	0	0	0	0	0	0	0
Sum of number of links	4	3	10	3	3	5	6	6	8	8
SDG link to topic	2	2	5	2	3	2	2	2	4	2

SDG. LCSA	Social							
	Consumers			Value chain actors		Local community		Number of link with LCSA
	Safe product	Feedback mechanism	Transparency	Promoting social responsible	Fair trading relationship	Number of local coordinators recruited	Secure living condition	
SDG1	0	0	0	0	0	2	0	4
SDG2	1	0	0	0	0	0	0	5
SDG3	1	0	0	1	0	1	1	10
SDG4	0	0	0	0	0	0	0	7
SDG5	0	0	0	0	0	0	0	6
SDG6	0	0	0	0	0	0	0	8

(continued)

Table 4 (continued)

SDG. LCSA	Social							Number of link with LCSA
	Consumers			Value chain actors		Local community		
	Safe product	Feedback mechanism	Transparency	Promoting social responsible	Fair trading relationship	Number of local coordinators recruited	Secure living condition	
SDG7	0	0	0	0	0	0	0	12
SDG8	0	5	0	0	0	0	0	35
SDG9	0	0	0	0	0	0	0	8
SDG10	0	0	1	0	1	0	0	9
SDG11	1	0	0	0	0	1	1	7
SDG12	0	0	4	0	1	0	0	20
SDG13	0	0	0	0	0	0	0	14
SDG14	0	0	0	0	1	0	0	8
SDG15	0	0	0	0	1	0	0	9
SDG16	0	0	0	0	0	0	1	5
SDG17	1	0	3	4	2	0	0	11
Sum of number of links	4	5	8	5	6	4	3	
SDG link to topic	4	1	3	2	5	3	3	

Source The Authors

Table 5 Scope and methodology for CO_2 emissions

	Total energies	Carrefour	Michelin	LVMH	Veolia
Unit of measurement	Million tons of CO_2 equivalent (Mt CO_2e)	Tonnes (metric tons) of CO_2 equivalent (tCO_2e)	Million tons of CO_2 equivalent (Mt CO_2e)	Tonnes (metric tons) of CO_2 equivalent (tCO_2e)	Million tons of CO_2 equivalent (Mt CO_2e)
Description	1 million metric tons of CO_2 equivalent, accounting for the impact of different GHGs	1 Tonnes or metric ton of CO_2 equivalent, accounting for the impact of different GHGs	1 million metric tons of CO_2 equivalent, accounting for the impact of different GHGs	1 Tonnes or metric ton of CO_2 equivalent, accounting for the impact of different GHGs	1 Million metric tons of CO_2 equivalent, accounting for the impact of different GHGs
Scope 1	Direct emissions of greenhouse gases from sites or activities that are included in the scope of reporting for climate change-related indicators. Direct biogenic CO_2 emissions are excluded from Scope 1 and reported separately	CO_2 emission from gas, fuel consumption and use of refrigerants from its integrated stores	CO_2 emissions from the heating plants at production and R&D sites	CO_2 emissions by business group mainly through the combustion of fuel oil and natural gas	Direct emissions from fixed or mobile facilities within the organizational scope, i.e. emissions from sources held or controlled by the organization, such as: fixed and mobile combustion, industrial processes excluding combustion, biogas from landfills, refrigerant leaks, etc.

(continued)

Table 5 (continued)

	Total energies	Carrefour	Michelin	LVMH	Veolia
Scope 2	Indirect emissions attributable to brought-in energy (electricity, heat, steam), net from potential energy sales, excluding purchased industrial gases (H_2). If not slated otherwise, Total energies reports Scope 2 GHG emissions according to the market based method defined by the GHG protocol	CO_2 emission from electricity	CO_2 emissions from generating the electricity and steam used by the production and R&D sites	CO_2 emissions by business indirectly from energy use, mainly electricity used in stores and at the Group's production sites	Indirect emissions linked to the imported production of electricity, heat or steam for the organization's activities. Veolia also includes losses from electricity and heat distribution networks operated by the Group
Scope 3	Other indirect emissions. If not staled otherwise, it correspond to indirect GHG emissions related to the use of energy products by customers, i.e. from their combustion to obtain energy	CO_2 emission from upstream and downstream activities	CO_2 emissions from the relevant activity categories corresponding to the Group's value chain	CO_2 emissions generated by upstream and downstream transportation	Other emissions indirectly produced by the organization's activities which are not recognized in Scope 2 but which are linked to the complete value chain, such as: the purchase of raw materials (electricity, heat or gas for a retail business, etc.), services or other products (reagents, etc.), business travel, upstream and downstream transport of goods, managing waste generated by the organization's activities, use and end of life of sold products and services, capitalization of production goods and equipment, etc.

(continued)

Table 5 (continued)

	Total energies	Carrefour	Michelin	LVMH	Veolia
Scope 4					Emissions brought about by an organization's activities, products or services when these reductions occur outside the scope of its own business
Guidance/ methodology/ techniques	CDP guidance methodology for estimation category 11 emissions for oil and gas companies	DEFRA	GRI 305-1: direct (Scope 1) GHG emissions GRI 305-2: direct (Scope 2) GHG emissions GRI 305-3: other indirect (Scope 3) GHG emissions	Bilan Carbone® method to measurement	"BREF" (Best available techniques REFerence documents)

Source The Authors

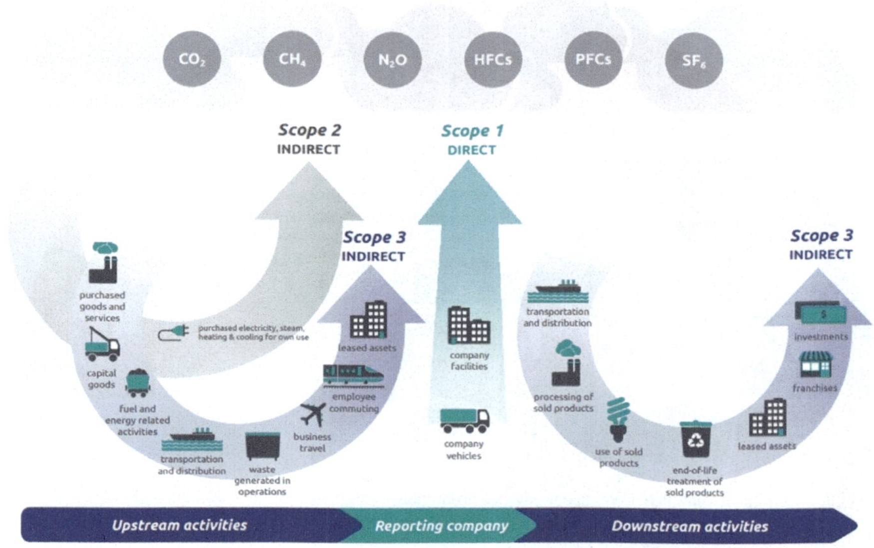

Fig. 15 Overview of GHG protocol scopes and emissions across the value chain. *Source* Corporate-value-chain-accounting-reporing-Standard_041613_2.pdf (ghgprotocol.org)

These indicators facilitate the identification of specific areas where improvements are needed, thus enabling targeted actions that contribute to broader SDG objectives. Furthermore, the application of LCSA indicators ensures that sustainability initiatives are grounded in measurable outcomes, promoting accountability and continuous improvement.

Fig. 16 The link between SDG 13 and other SDGs. *Source* Leal Filho et al. (2023)

5 Conclusion

Since 1990s, the international and scientific community has been working to assess the environmental, economic and social impacts associated with the various stages of Life Cycle Assessment (LCA). LCA practices and methodology have been successively improved: (i) four steps (definition of the objective and scope of the assessment, inventory analysis, impact assessment, interpretation) and (ii) several approaches (Environmental Life Cycle Assessment, Life Cycle Cost, Social Life Cycle Assessment) to map the sustainability (Life Cycle Sustainability Assessment). This period also coincides with the rise of the concept of sustainable development, popularized by the Brundtland Report (1987) and the roadmap of the United Nations (2015) defining the Sustainable Development Goals (SDGs). Today, the two approaches seem to be converging, with the aim of guiding both governments and businesses along the path to sustainability. Linking Life Cycle Sustainability Assessment (LCSA) and the SDGs is now a new step in the 2030–2050 agenda. The aim is to identify targets and indicators for a sustainable approach, and to integrate this new approach into the roadmaps of the different states and companies. Some SDGs, such as SDG 8, SDG 12 and SDG 13, interact more with the LCSA, but sustainable reporting seems to demonstrate a real willingness on the part of companies to adopt a systemic approach, integrating challenges linked to efficiency, resilience, circularity, eco-responsibility, gender…. All these challenges are at the heart of the 14,000 standards (UNEP 2022), which have structured life cycle assessment for the past 30 years.

Annex

Goal 8. Promote sustained, inclusive and sustainable economic growth, full and productive employment and decent work for all	8.1 Sustain per capita economic growth in accordance with national circumstances and, in particular, at least 7% gross domestic product growth per annum in the least developed countries	8.1.1 Annual growth rate of real GDP per capita
	8.2 Achieve higher levels of economic productivity through diversification, technological upgrading and innovation, including through a focus on high-value added and labour-intensive sectors	8.2.1 Annual growth rate of real GDP per employed person
	8.3 Promote development-oriented policies that support productive activities, decent job creation, entrepreneurship, creativity and innovation, and encourage the formalization and growth of micro-, small- and medium-sized enterprises, including through access to financial services	8.3.1 Proportion of informal employment in total employment, by sector and sex

(continued)

(continued)

	8.4 Improve progressively, through 2030, global resource efficiency in consumption and production and endeavour to decouple economic growth from environmental degradation, in accordance with the 10-Year Framework of Programmes on Sustainable Consumption and Production, with developed countries taking the lead	8.4.1 Material footprint, material footprint per capita, and material footprint per GDP
		8.4.2 Domestic material consumption, domestic material consumption per capita, and domestic material consumption per GDP
	8.5 By 2030, achieve full and productive employment and decent work for all women and men, including for young people and persons with disabilities, and equal pay for work of equal value	8.5.1 Average hourly earnings of employees, by sex, age, occupation and persons with disabilities
		8.5.2 Unemployment rate, by sex, age and persons with disabilities
	8.6 By 2020, substantially reduce the proportion of youth not in employment, education or training	8.6.1 Proportion of youth (aged 15–24 years) not in education, employment or training
	8.7 Take immediate and effective measures to eradicate forced labour, end modern slavery and human trafficking and secure the prohibition and elimination of the worst forms of child labour, including recruitment and use of child soldiers, and by 2025 end child labour in all its forms	8.7.1 Proportion and number of children aged 5–17 years engaged in child labour, by sex and age
	8.8 Protect labour rights and promote safe and secure working environments for all workers, including migrant workers, in particular women migrants, and those in precarious employment	8.8.1 Fatal and non-fatal occupational injuries per 100,000 workers, by sex and migrant status
		8.8.2 Level of national compliance with labour rights (freedom of association and collective bargaining) based on International Labour Organization (ILO) textual sources and national legislation, by sex and migrant status

(continued)

(continued)

	8.9 By 2030, devise and implement policies to promote sustainable tourism that creates jobs and promotes local culture and products	8.9.1 Tourism direct GDP as a proportion of total GDP and in growth rate
	8.10 Strengthen the capacity of domestic financial institutions to encourage and expand access to banking, insurance and financial services for all	8.10.1 (a) Number of commercial bank branches per 100,000 adults and (b) number of automated teller machines (ATMs) per 100,000 adults
		8.10.2 Proportion of adults (15 years and older) with an account at a bank or other financial institution or with a mobile-money-service provider
	8.a Increase Aid for Trade support for developing countries, in particular least developed countries, including through the Enhanced Integrated Framework for Trade-related Technical Assistance to Least Developed Countries	8.a.1 Aid for Trade commitments and disbursements
	8.b By 2020, develop and operationalize a global strategy for youth employment and implement the Global Jobs Pact of the International Labour Organization	8.b.1 Existence of a developed and operationalized national strategy for youth employment, as a distinct strategy or as part of a national employment strategy
Goal 12. Ensure sustainable consumption and production patterns	12.1 Implement the 10-Year Framework of Programmes on Sustainable Consumption and Production Patterns, all countries taking action, with developed countries taking the lead, taking into account the development and capabilities of developing countries	12.1.1 Number of countries developing, adopting or implementing policy instruments aimed at supporting the shift to sustainable consumption and production
	12.2 By 2030, achieve the sustainable management and efficient use of natural resources	12.2.1 Material footprint, material footprint per capita, and material footprint per GDP
		12.2.2 Domestic material consumption, domestic material consumption per capita, and domestic material consumption per GDP

(continued)

(continued)

	12.3 By 2030, halve per capita global food waste at the retail and consumer levels and reduce food losses along production and supply chains, including post-harvest losses	12.3.1 (a) Food loss index and (b) food waste index
	12.4 By 2020, achieve the environmentally sound management of chemicals and all wastes throughout their life cycle, in accordance with agreed international frameworks, and significantly reduce their release to air, water and soil in order to minimize their adverse impacts on human health and the environment	12.4.1 Number of parties to international multilateral environmental agreements on hazardous waste, and other chemicals that meet their commitments and obligations in transmitting information as required by each relevant agreement
		12.4.2 (a) Hazardous waste generated per capita; and (b) proportion of hazardous waste treated, by type of treatment
	12.5 By 2030, substantially reduce waste generation through prevention, reduction, recycling and reuse	12.5.1 National recycling rate, tons of material recycled
	12.6 Encourage companies, especially large and transnational companies, to adopt sustainable practices and to integrate sustainability information into their reporting cycle	12.6.1 Number of companies publishing sustainability reports
	12.7 Promote public procurement practices that are sustainable, in accordance with national policies and priorities	12.7.1 Number of countries implementing sustainable public procurement policies and action plans
	12.8 By 2030, ensure that people everywhere have the relevant information and awareness for sustainable development and lifestyles in harmony with nature	12.8.1 Extent to which (i) global citizenship education and (ii) education for sustainable development are mainstreamed in (a) national education policies; (b) curricula; (c) teacher education; and (d) student assessment
	12.a Support developing countries to strengthen their scientific and technological capacity to move towards more sustainable patterns of consumption and production	12.a.1 Installed renewable energy-generating capacity in developing and developed countries (in watts per capita)

(continued)

(continued)

	12.b Develop and implement tools to monitor sustainable development impacts for sustainable tourism that creates jobs and promotes local culture and products	12.b.1 Implementation of standard accounting tools to monitor the economic and environmental aspects of tourism sustainability
	12.c Rationalize inefficient fossil-fuel subsidies that encourage wasteful consumption by removing market distortions, in accordance with national circumstances, including by restructuring taxation and phasing out those harmful subsidies, where they exist, to reflect their environmental impacts, taking fully into account the specific needs and conditions of developing countries and minimizing the possible adverse impacts on their development in a manner that protects the poor and the affected communities	12.c.1 Amount of fossil-fuel subsidies (production and consumption) per unit of GDP
Goal 13. Take urgent action to combat climate change and its impacts	13.1 Strengthen resilience and adaptive capacity to climate-related hazards and natural disasters in all countries	13.1.1 Number of deaths, missing persons and directly affected persons attributed to disasters per 100,000 population
		13.1.2 Number of countries that adopt and implement national disaster risk reduction strategies in line with the Sendai Framework for Disaster Risk Reduction 2015–2030
		13.1.3 Proportion of local governments that adopt and implement local disaster risk reduction strategies in line with national disaster risk reduction strategies

(continued)

(continued)

	13.2 Integrate climate change measures into national policies, strategies and planning	13.2.1 Number of countries with nationally determined contributions, long-term strategies, national adaptation plans and adaptation communications, as reported to the secretariat of the United Nations Framework Convention on Climate Change
		13.2.2 Total greenhouse gas emissions per year
	13.3 Improve education, awareness-raising and human and institutional capacity on climate change mitigation, adaptation, impact reduction and early warning	13.3.1 Extent to which (i) global citizenship education and (ii) education for sustainable development are mainstreamed in (a) national education policies; (b) curricula; (c) teacher education; and (d) student assessment
	13.a Implement the commitment undertaken by developed-country parties to the United Nations Framework Convention on Climate Change to a goal of mobilizing jointly $100 billion annually by 2020 from all sources to address the needs of developing countries in the context of meaningful mitigation actions and transparency on implementation and fully operationalize the Green Climate Fund through its capitalization as soon as possible	13.a.1 Amounts provided and mobilized in United States dollars per year in relation to the continued existing collective mobilization goal of the $100 billion commitment through to 2025
	13.b Promote mechanisms for raising capacity for effective climate change-related planning and management in least developed countries and small island developing States, including focusing on women, youth and local and marginalized communities	13.b.1 Number of least developed countries and small island developing States with nationally determined contributions, long-term strategies, national adaptation plans and adaptation communications, as reported to the secretariat of the United Nations Framework Convention on Climate Change

References

Aneta K (2023) Sustainable development goals—assessment and relationships. In: Organizing sustainable development, pp 18–32

Brent RJ (2009) Handbook of research on cost benefit analysis. EE Elgar Editions

Caldara D, Iacoviello M (2022) Measuring geopolitical risk. Am Econ Rev 112(4):1194–1225

Carrefour (2023) Universal registration document. https://www.carrefour.com/sites/default/files/2024-03/CFR2023_URD_FR_MEL_24.03.29_1.pdf

Christina W, Werker J, Zapp P, Schreiber A, Holger S, Kuckshinrichs W (2018) Sustainable development goals as a guideline for indicator selection in life cycle sustainability assessment. Proc CIRP 69:59–65

Ciroth A, Hildenbrand J, Steen B (2015) Life cycle costing. In: Dewulf J, De Meester S, Alvarenga RAF (eds) Sustainability assessment of renewables based products: methods and case studies. Wiley Editions

Diemer A (2023) From life cycle assessment (LCA) to life cycle sustainability assessment (LCSA), methodological issues and prospects for implementing circular business models. Int J Sci Eng Appl Sci 9(12):1–50

Diemer A, Denave W, Dieterlen N, Gardaz A (2024) Some digressions on the life cycle analysis of a nuclear power plant in France. Curr Opin 4:558–597

Dreyer I, Hauschild M, Schierbeck J (2006) A framework for social life cycle impact assessment. Int J LCA 11(2):88–97

EMF (2013) Towards the circular economy, economic and business rationale for an accelerated transition. Ellen MacArthur Foundation

Fava JA Denison R, Jones B, Curran MA, Vigon B, Selke S, Barnum J (1991) SETAC Workshop report: A technical framework for life cycle assessments. Smugglers Notch, Vermont. SETAC, Washington, DC, 18–23 August 1990

Fava J, Denison R, Jones B, Curran MS, Vigo B, Selke S, Barnum J (1992) A technical framework for life-cycle assessments. Workshop report. Society of Environmental Toxicology and Chemistry, Washington, DC

Fava J, Baer S, Cooper J (2009) Increasing demands for life cycle assessment in North America. J Ind Ecol 13(4):491–494

Fava JA, Smerek A, Heinrich AB, Morisson L (2014) The role of the society of environmental toxicology and chemistry (SETAC) in life cycle assessment (LCA), development and application. In: Kloepffer W (eds) Background and future prospects in life cycle assessment. LCA Compendium. The Complete World of Life Cycle Assessment, Springer

Finkbeiner M, Inaba A, Tan R, Christiansen K, Klüppel J (2006) The new international standards for life assessment: ISO 14040 and ISO 14044. Int J Life Cycle Assess 11:80–85

Gabathuler H (1997) The CML story, how environmental sciences entered the debate on LCA. J Life Cycle Assess 2(4):187–194

Gemechu ED, Sonnemann G, Young SB (2015a) Geopolitical related supply risks assessment as a complement to environmental impact assessment: the case of electric vehicles. Int J Life Cycle Assess. https://doi.org/10.1007/s11367-015-0917-4

Gemechu ED, Helbig C, Sonnemann G (2015b) Import-based indicator for the geopolitical supply risk of raw materials in life cycle sustainability assessments. J Ind Ecol. https://doi.org/10.1111/jiec.12279

Global Reporting Initiative (1999) Sustainability reporting guidelines, exposure draft for public comment and pilot testing. https://www.sustainability-reports.com/wordpress/wp-content/uploads/1999/03/gri_draft1999.pdf

Global Reporting Initiative (2007) Groen EA, Heijungs R, Bokkers EAM, Boer IJM (2017) Methods for global sensitivity analysis in life cycle assessment. Int J Life Cycle Assess 22:1125–1137

Harvey G (1976) Life cycle costing: a review of the technique. Manag Account, pp 343–347

Hauschild MZ, Rosenbaum RK, Olsen SI (2018) Life cycle assessment, theory and practice. Springer

Heijungs R et al (1992) Environmental life cycle assessment of products. Guide, Leiden
Heijungs R, Huppes G, Guinee JB (2010) Life cycle assessment and sustainability analysis of product, materials and technologies. Toward a scientific framework for sustainability life cycle analysis. Polym Degradation and Stability, vol 95, p. 422–428
Hellweg S, Mila L (2014) Emerging approaches, challenges and opportunities in life cycle assessment. Science 344(6188):1109–1113
Hoogmartens R, van Passel S, van Acker K, Dubois M (2014) Bridging the gap between LCA, LCC and CBA as sustainable assessment tools. Environ Impact Assess Rev 48:27–33
Hunkeler DD (2006) Societal LCA methodology and case study. Int J Life Cycle Assess 11:371–382
Hunkeler DD, Lichtenvort K, Rebizer G, Ciroth A (2008) Environmental life cycle costing. CRC Press
Hunt RG, Franklin WE (1997) LCA—how it came about. Int J Life Cycle Assess 1:4–7
Huppes G, Ishikawa M (2009) Eco-efficiency in industry and science. Quantified eco-efficiency, an introduction with applications. Springer
ISO (1996) Technical Committee TC 207/Subcommittee SC 5: environmental management—life cycle assessment—principles and framework. International Standard 14040
ISO (2006) Environmental management, life cycle analysis. International Standard of Organisation
ISO (2010) ISO 26000, guidance on social responsibility. International Standard of Organisation
Jolliet O, Margni M, Charles R, Humbert S, Payet J, Rebitzer G, Rosenbaum R (2003) IMPACT 2002+: a new life cycle impact assessment methodology. Int J Life Cycle Assess 8(6):324–330
Jorgensen A, le Bocq A, Nazarkina L, Hauschild M (2008) Methodologies for social life cycle assessment. Int J Life Cycle Assess 13:96–103
JRC (2015) Social life cycle assessment: state of the art and challenges for supporting product policies. JRC Publications Repository, European Commission. https://publications.jrc.ec.europa.eu/repository/handle/JRC99101
JRC (2020) Critical raw materials for strategic technologies and sectors in the EU, a foresight study. Luxembourg Publications Office of the European Union, p 92
Kaufman RJ (1970) Life cycle costing: a decision-making tool for capital equipment acquisition. Cost Manag 21–38
Kloepffer W (2008) Life cycle sustainability assessment of products. Int J Life Cycle Assess 13:89–95
Klöpffer W (1997a) Peer (expert) review in LCA according to SETAC and ISO 14040: theory, and practice. Int J Life Cycle Assess 2(4):183–184
Klöpffer W (1997b) Life cycle assessment: from the beginning to the current state. Environ Sci Pollut Res 4:223–228
Klöpffer W, Ciroth A (2011) Is LCC relevant in a sustainability assessment? Int J Life Cycle Assess 16:99–101
Klöpffer W, Grahl B (2014) Life cycle assessment (LCA), a guide to best practice. Wiley-VCH. https://toc.library.ethz.ch/objects/pdf03/e05_978-3-527-32986-1_01.pdf
Klüppel HJ (2005) The revision of ISO Standards 14040. Int J Life Cycle Assess 10:165
KPMG (2022) Survey of sustainability reporting, p 60
Layard R, Glaister S (1994) Cost-benefit analysis. Cambridge Press
Leal Filho W, Wall T, Lange Salvia A, Pimenta Dinis MA, Mifsud M (2023) The central role of climate action in achieving the United Nations' Sustainable Development Goals. Sci Rep 13:20582
Lecouls H (1999) ISO 14043: environmental management—life cycle assessment—life cycle, interpretation. Int J Life Cycle Assess 4(5)
Leung W, Noble B, Gunn J, Jaeger JAG (2015) A review of uncertainty research in impact assessment. Environ Impact Assess Rev 50:116–123
Life Cycle Initiative (2020) LCA based assessment of the sustainable development goals. In: Development update and preliminary findings of the project "Linking the UN Sustainable Development Goals to Life Cycle Impact frameworks", LCI—UNEP, p 53
Life Cycle Initiative (2022) Life cycle initiative progress report 2022, p 14

Life Cycle Initiative (2023) Life cycle initiative progress report 2023, p 15
LVMH (2023) Universal registration document. https://r.lvmh-static.com/uploads/2024/03/urd-2023-va_vdef.pdf
Macombe C, Feschet P, Garrabe M, Loeillet D (2011) 2nd international seminar in social life cycle assessment—recent developments in assessing the social impacts of product life cycles. Int J Life Cycle Assess 16:940–943. https://doi.org/10.1007/s11367-011-0331-5
Manzo S, Salling KB (2016) Integrating life-cycle assessment into transport cost benefit analysis. Transp Res Proc 14:273–282
Marsmann M (1997) ISO 14040 e the first project. Int J Life Cycle Assess 2(3):122–123
Marsmann M (2000) The ISO 14040 family. Int J Life Cycle Assess 5(6):317–318
Martinez-Sanchez V, Kromann MA, Astrup TF (2015) Life cycle costing of waste management systems: overview, calculation principles and case studies. Waste Manag 36:343–355
Marwa B, Alejandro P, Getachew A, Ian G (2023) Methodological framework to find links between life cycle sustainability assessment categories and the UN sustainable development goals based on literature. J Ind Ecol 27:707–725
Michael K, Rudiger H (2017) Indicators in social life cycle assessment: a review of frameworks, theories, and empirical experience. J Ind Ecol 21:1547–1565
Michelin (2023) Universal registration document. https://www.michelin.com/investisseur/rapport-annuel-documents-legaux-reglementes
Morao A, Visser D (2022) Creating impact by integrating the SDGs in a company strategy. In: E3S web conference, 10th international conference on life cycle management (LCM 2021), vol 349
Norris GA (2001) Integrating life cycle cost analysis and LCA. Int J LCA 6(2):118–120
Norris GR (2006) Social impacts in product life cycles—towards life cycle attributes assessment. Int J LCA 11(1):97–104
Parker RPM, Harrison GP, Chick JP (2007) Energy and carbon audit of an offshore wave energy converter. Proc Inst Mech Eng Part A J Power Energy 221:1119–1130. https://doi.org/10.1243/09576509JPE483
Pedercena CS (2018) The UN sustainable development goals (SDGs) are a great gift to business! 25th CIRP Life Cycle Engineering (LCE) conference, 30 April–2 May 2018, Procedia CIRP 69:21–24. Copenhagen, Denmark
Pehnt M (2006) Dynamic life cycle assessment (LCA) of renewable energy technologies. Renew Energy 31:55–71
Pena C et al (2021) Using life cycle assessment to achieve a circular economy. Int J Life Cycle Assess. https://doi.org/10.1007/s11367-020-01856-z
Petti L, Serreli M, Di Cesare S (2016) Systematic literature review in social life cycle assessment. Int J Life Cycle Assess. https://doi.org/10.1007/s11367-016-1135-4
Pinto JTM, Sverdrup HU, Diemer A (2019) Integrating the life cycle analysis into system dynamics: the case of steel in Europe. Environ Syst Res 8:1–21. https://doi.org/10.1186/s40068-019-0144-2
Pryshlakivsky J, Seracy C (2013) Fifteen years of ISO 14040: a review. J Clean Prod 57:115–123
Quah E, Toh R (2012) Cost-benefit analysis, cases and materials. Routledge
Rex E, Baumann H (2004) Expanding the green practice of LCA. Centre for Environmental Assessment of Product and Material Systems, Chalmers University of Technology. Sweden
Rödger JM, Kjaer LL, Pagoropoulos A (2018) Life cycle costing: an introduction. In: Hauschild MZ, Rosenbaum RK, Olsen SI (eds) Life cycle assessment, theory and practice. Springer
Ryding S-O (1999) ISO 14042: environmental management—life cycle assessment—life cycle, impact assessment. Int J Life Cycle Assess 4(6):3
Sabrina N, Julia M, Rene S, Matthias F (2015) Enhancing the practical implementation of life cycle sustainability assessment—proposal of a tiered approach. J Clean Prod 102:165–176
Schau EM, Fet AM (2008) LCA studies of food products as background for environmental product declarations. Int J Life Cycle Assess 13(3):255–264
Steen B (2005) Environmental costs and benefits in life cycle costing. Manag Environ Qual 16(2):107–118
Stevens J (1976) The life cycle costing technique. Industrial Purchasing News, pp 55–56

Timonen K, Harrison E, Katajajuuri JM, Kurppa S (2017) Environmental cost accounting methodologies. In: Natural resources and bioeconomy studies. Luke Natural Resources Institute Finland

Total Energies (2023) Universal registration document. https://totalenergies.com/investors/reports

UN (2012) The future we want—outcome document. United Nation Organization. https://sustainabledevelopment.un.org/futurewewant.html

UN (2017) SDG indicators: global indicator framework for the sustainable development goals and targets of the 2030 agenda for sustainable development. https://unstats.un.org/sdgs/indicators/indicators-list/online.

UN Global Compact Strategy (2024–2025) https://ungc-communications-assets.s3.amazonaws.com/docs/publications/UNGC-Strategy-Extension_2024-25.pdf

UNEP (2009) Guidelines for social LCA of products. https://www.lifecycleinitiative.org/wp-content/uploads/2012/12/2009%20-%20Guidelines%20for%20sLCA%20-%20EN.pdf

UNEP (2012) Towards a life cycle sustainability assessment. Making informed choices on products. https://www.lifecycleinitiative.org/wp-content/uploads/2012/12/2011%20-%20Towards%20LCSA.pdf

UNEP (2016) Global guidance for life cycle impact assessment indicators, vol 1. United Nations Environment Programs

UNEP (2019) Global guidance for life cycle impact assessment indicators, vol 2. United Nations Environment Programs

UNEP (2022) Single-use face masks and their alternatives: recommendations from the life cycle assessments. United Nations Environment Programme

Veolia (2023) Universal registration document. https://www.veolia.com/en/newsroom/publications

WCED (1987) Our common future. World Commission on Environment and Development

Weidema BP (2006) The integration of economic and social aspects in the life cycle impact assessment. Int J Life Cycle Assess 1:89–96

Weidema BP (2021) Data Collection guideline for pressure indicators for life cycle based sustainability assessment. LCA Consultants, Aalborg, p 53

White GE, Ostwald PH (1976) Life cycle costing. Manag Account, 39–42

Woodward DG (1997) Life cycle costing—theory, information acquisition and application. Int J Proj Manag 15(6):335–344

Wur R, Yang D, Chen J (2014) Social life cycle assessment revisited. Sustainability 6(7):4200–4226. https://doi.org/10.3390/su6074200

Practical Application of LCA to Optimize Environmental Impacts of Steel–concrete Footbridges

Fernando Luiz Tres Junior, Víctor Yepes, Guilherme Fleith de Medeiros, and Moacir Kripka

Abstract Life cycle assessment (LCA) is an outstanding methodology used to estimate the environmental impacts resulting from the manufacture of a given product or service, examining each stage of its life cycle, from the extraction of natural resources to its disposal. The purpose of this research is to present a practical application of LCA in optimizing the sustainability of steel–concrete composite footbridges, considering the cost of the construction and its environmental impact as objectives. The established design variables for this include the slab's concrete strength and thickness, the dimensions of the welded steel I-girder, and the interaction degree of the composite beam. The Multi-Objective Harmony Search (MOHS) algorithm is applied to address the optimization problem. LCA studies are applied to estimate the environmental impacts associated with the consumption of concrete and steel by the structure, using CO_2 emissions and embodied energy as environmental impact criteria. The results of the multiobjective optimization allow to build associations among different objectives and evaluate viable trade-offs for the project. By applying the estimates of CO_2 emissions and embodied energy provided by LCA studies, it was possible to successfully quantify and evaluate the environmental impact of the structure. Initially, a cost-versus-CO_2 emission comparison is presented, and

F. L. Tres Junior (✉) · V. Yepes
Institute of Concrete Science and Technology (ICITECH), Universitat Politècnica de València, Camí de Vera, Valencia, Spain
e-mail: fltrejun@upv.edu.es

V. Yepes
e-mail: vyepesp@cst.upv.es

G. F. de Medeiros · M. Kripka
Graduate Program in Civil and Environmental Engineering (PPGEng), University of Passo Fundo, Passo Fundo, Brazil
e-mail: guifleith@upf.br

M. Kripka
e-mail: mkripka@utfpr.edu.br

M. Kripka
Civil Engineering Graduate Program (PPGEC), Federal University of Technology-Paraná, Pato Branco, Brazil

L. Brandli et al. (eds.), *The Contribution of Life Cycle Analyses and Circular Economy to the Sustainable Development Goals*, World Sustainability Series,
https://doi.org/10.1007/978-3-031-85300-5_22

among the findings, it was that solutions that perform well in terms of costs are also viable in terms of CO_2 emissions. Following that, the carbon dioxide emissions are compared to the structure's embodied energy, also revealing a linear association. The scarcity of infrastructure, specially in developing countries, in addition to the low economic resources, underline the significance of optimization and LCA as techniques to contribute to the sustainable development of structures with lower expenses while resulting in reduced environmental impacts.

Keywords Carbon dioxide emission · Embodied energy · Multi-objective optimization · Steel–concrete composite footbridge · Sustainability · Life cycle assessment (LCA)

1 Introduction

During the design of structures, the engineer must solve problems while taking into consideration numerous criteria which are frequently in conflict with one another. The traditional structural sizing procedure starts with a pre-design, that is verified and then modified until it meets safety and service requirements, which is quite comparable to a trial-and-error method. However, because it is highly dependent on the designer's previous experiences to define the starting structure, there is no assurance that the final solution is the best performing alternative compared to the enormous number of possible options. More recently, the sustainability of infrastructure projects has gained prominence, as in publications by Orcesi, Cremona and Ta (2018), Milani, Yepes and Kripka (2020), Martínez-Muñoz, Martí and Yepes (2021), Lu et al. (2022) and Tres Junior et al. (2023), particularly with the establishment of the sustainable development goals (SDG) by the United Nations (UN 2015). The civil construction industry is critical to a nation's sustainable development since, in addition to the massive economic and social impacts it generates, it is also responsible for 34% of energy consumption and 37% of CO_2 emissions worldwide (UNEP 2022).

Optimization has shown to be a useful tool for engineers in getting optimal and feasible solutions to a wide range of problems (Mei and Wang 2021). Optimization methods are capable of conducting a systematic and efficient search for solutions independent of the engineer's previous expertise by modeling the project as a mathematical problem with objective functions and constraints to be satisfied. According to Mei and Wang (2021), although various studies employ optimization in structural engineering, many of them are limited to benchmark problems. Beyond that, the great majority of engineers who apply optimization to practical issues are mainly concerned with decreasing the cost or weight of the structure (Mei and Wang 2021), using steel as a structural material.

Nonetheless, in recent years, more complex and practical problems have been investigated as optimization subjects and it is possible to argue that its popularization could have positive effects on the economic and environmental impacts generated by the construction sector (Largaros 2018). It is worth mentioning multi-objective optimization, which considers multiple objective functions to be minimized or maximized at the same time (Collette and Siarry 2004). For example, using this technique, the joint optimization of the three pillars of sustainability can be achieved. Unlike single-objective optimization, multi-objective optimization yields a set of non-dominated, or Pareto-optimal, solutions. These are solutions for which there is no other admissible solution that simultaneously improves all the objective functions, where the improvement of one objective is only possible to the detriment of at least one of the others. As a result, without establishing weights or preferences for each objective, it is impossible to assert that one solution is preferable to another. The set of non-dominated solutions can be utilized to produce a graph that connects the goal functions under consideration, known as the Pareto front, from which reasonable trade-offs between the objectives can be analyzed, assisting in decision-making.

Life cycle assessment (LCA) is an approach for estimating the environmental, social, and economic impacts of a material or product. There are several sorts of applications of LCA discussed in the literature in structural engineering. Concerning exclusively bridge-type structures, Gervásio and Silva (2008) propose a comparative LCA on bridges with a steel–concrete composite structure. Niu and Fink (2018) provide an overview of the literature on the LCA of bridges with modern timber structures, revealing a lack of consensus on the LCA criteria for this type of structure. Milani, Yepes, and Kripka (2020) provide sustainability indicators for bridges based on material manufacturing characteristics in Brazil. According to the investigation, bridges with a steel–concrete composite structure had the lowest environmental impact among the usual alternatives. Penadés-Plà et al. (2020) examine the environmental and social impacts of post-tensioned concrete bridges, concluding that the pre-stressed concrete precast bridge is the most sustainable structural alternative. Martínez-Muñoz, Martí and Yepes (2021) applied the ReCiPE approach to compare the LCA of bridges with concrete structures and steel–concrete composites. When the steel recycling rate is greater than 90%, the composite box-girder construction has the lowest environmental impact among the alternatives. On the other hand, when the steel recycling rate grows, the social impact of the structure is negatively affected (Martínez-Muñoz, Martí and Yepes 2022a, b).

A particularly practical and extremely significant use of LCA is in combination with structural optimization, with the goal of obtaining optimal solutions that take into account, in addition to costs, the project's environmental and social impacts. In this regard, Medeiros and Kripka (2014) can be cited, who apply optimization to the design of reinforced concrete pillars while taking into account cost and CO_2 emissions, comparing the optimum solutions for each objective. The sustainability of steel–concrete composite bridges is thoroughly discussed by Orcesi, Cremona, and Ta (2018), who take social, economic, and environmental factors into consideration. Yepes et al. (2019) optimize a steel–concrete composite box-girder footbridge in terms of cost and CO_2 emissions. The prices were lowered by more than 20% in the study, and it was found that cost-effective solutions had an excellent performance in terms of environmental impact. Tormen et al. (2020) describe a similar approach, optimizing steel–concrete composite beams while accounting for the structure's expenses and CO_2 emissions. Martínez-Muñoz et al. (2022a, b) provide a cost and CO_2 emission optimization of steel–concrete composite bridges, with a correlation of 2,51 kg CO_2 decreased for every euro saved by the optimization. In a subsequent study, the authors optimize the same structure while accounting for embodied energy as a metric of environmental effect (Martínez-Muñoz et al. 2023).

Within the context outlined, the purpose of this study is to apply multi-objective optimization to a steel–concrete composite footbridge, with the cost, CO_2 emissions, and embodied energy of the structure as objective functions of the optimization problem. This research aims to carry out a practical application of LCA to minimize the environmental impact of the footbridge and contribute to the sustainable design of such structures. The optimization results are used to construct Pareto fronts, which are then used to investigate the correlations between the researched objectives. At the conclusion of the study, the findings obtained for alternative scenarios, simulating changes in concrete strength and material costs, are provided in order to analyze modifications in the Pareto front.

2 Methodology

2.1 Formulation of the Optimization Problem

The optimization study object in this research is a steel–concrete composite footbridge located in the northern region of Rio Grande do Sul, the southernmost state in Brazil, with a total span of 17.5 m and a width of 3 m, as shown in Figs. 1 and 2. The structural material of the footbridge was chosen based on research that show composite structures to have lesser environmental impacts (Milani, Yepes, and Kripka 2020), particularly in countries with significant rates of steel recycling (Martínez-Muñoz, Martí and Yepes 2021), such as Brazil. The structure features two supports at both ends and lateral bracing every 2.5 m, as depicted in Fig. 2, which also shows that the steel girders were positioned to get an ideal balance of positive and negative moments in the concrete slab, so that the steel girders are 0.6 m apart from the edges of the cantilever and 1.8 m apart from each other. In both illustrations, d indicates the height of the steel profile and h_t the thickness of the concrete slab, both of which are variables of the optimization problem and are later defined.

The steel profile utilized is welded, allowing for greater versatility in the dimensions of its components, given the possibility of ordering the steel structure with customized size and millimetric precision. ASTM A572 steel is used for the beams considering it has a yield tensile strength of 350 MPa and a modulus of elasticity of 200 GPa. The compressive strength used for the concrete slab is 40 MPa, and for reinforcement, bars with a diameter of 10 mm and CA 50 steel with a tensile strength of 500 MPa are considered. Headed welding stud shear connections with a height of 135 mm and a nominal diameter of 19 mm are utilized to ensure a connection between the two structural components. The shear connectors' steel has a tensile

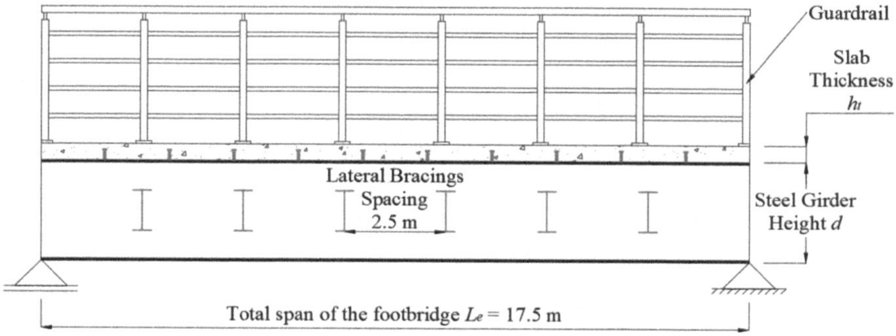

Fig. 1 Longitudinal view of the footbridge

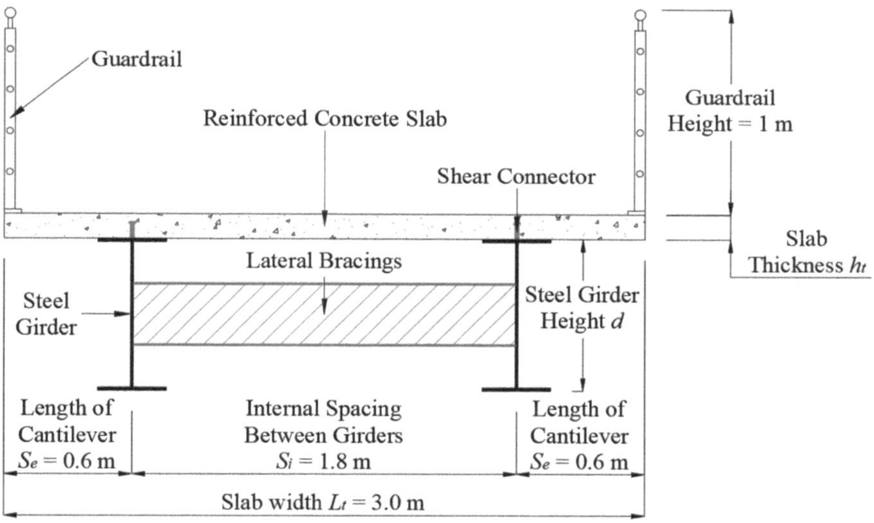

Fig. 2 Cross section of the footbridge

strength of 415 MPa and a modulus of elasticity of 200 GPa. The steel beams are not supported during the building phase of the composite structure.

As design variables, the following were considered: the slab thickness (h_t), the web height (h_w), the web thickness (t_w), the superior flange width (b_{fs}), the superior flange thickness (t_{fs}), the inferior flange width (b_{fi}), the inferior flange thickness (t_{fi}), and the interaction degree (α) of the composite beam, as illustrated in Fig. 3. The discrete values that each variable can assume are within the ranges represented in Eqs. (1)–(5). To demonstrate the complexity of the problem, it is possible to calculate the number of possible solutions through a combinatorial analysis considering the number of possible values that each variable can take on, arriving at a value in the order of 7E + 11 possible combinations (or solutions).

$$x_1 \in [100; 110; 120; \cdots; 180; 190; 200], \text{ in mm} \quad (1)$$

$$x_2 \in [150; 151; 152; \cdots; 1998; 1999; 2000], \text{ in mm} \quad (2)$$

$$x_3, x_5, x_7, \in [4, 75; 6, 35; 8, 0; 9, 5; 12, 5; 16, 0; 19, 0; 22, 4; 25, 0; 28, 5; 31, 5; 37, 5; 44, 5; 50, 0], \text{ in mm} \quad (3)$$

$$x_4, x_6 \in [100; 101; 102; \cdots; 1998; 1999; 2000], \text{ in mm} \quad (4)$$

$$x_8 \in [0.4; 0.5; 0.6; 0.7; 0.8; 0.9; 1.0] \quad (5)$$

The values of the objective functions are obtained using the values of the variables described above. The cost, CO_2 emissions, and embodied energy of the structure are all assessed simultaneously as objectives in this optimization problem. As indicated

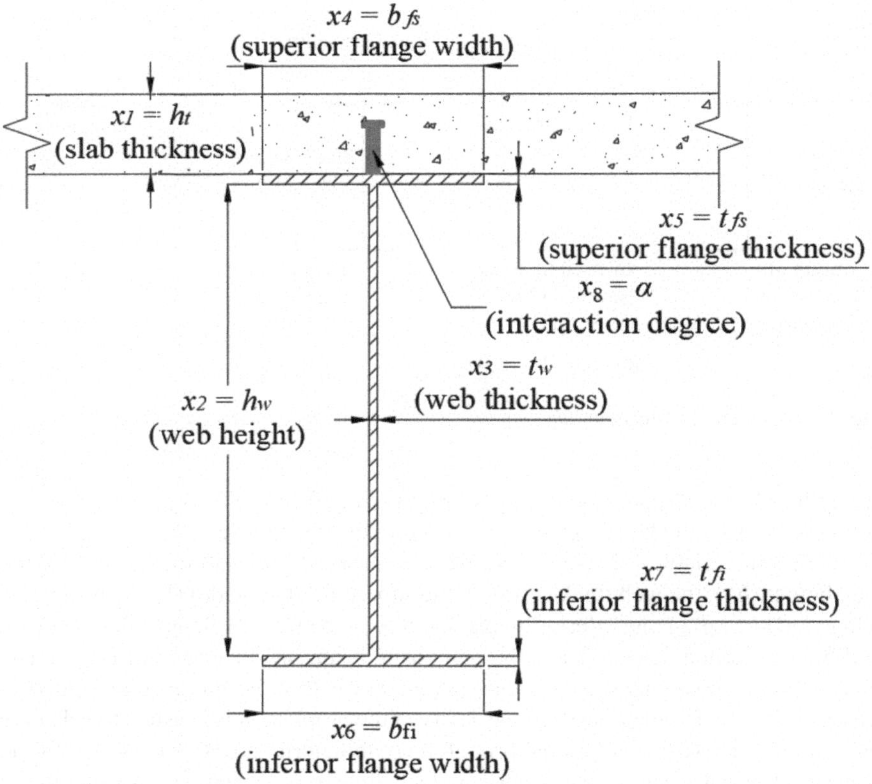

Fig. 3 Design variables

in Eq. (6), the objectives are a function of the consumption measurement of each material (m_i) used in the construction and a unit factor (u_i) that measures the material's cost, unit emission, and embodied energy. The number of materials considered (nm) is three: the steel of the beam, the concrete, and the slab reinforcement.

$$minimizef(x) = \sum_{i=1}^{nm} u_i \cdot m_i(x) \tag{6}$$

2.2 Steel and Concrete LCA Databases

To measure the environmental impacts caused by the consumption of steel and concrete by the structure, two databases applying the LCA methodology were used. The first is an international database available on the Sustainability Indicators 2023 Report by Worldsteel (2023), which provides unit CO_2 emissions and the unit embodied energy of structural steel. The environmental impact indicators are estimated based on worldwide data collection from 94 companies and associations, which account for 53% of the world's raw steel production. These values are used to estimate the environmental impacts generated by welded steel beams of the bridge.

The environmental impact indicators for concrete and its reinforcement used are those provided by the study published by Santoro and Kripka (2020), which consists in an LCA of reinforced concrete materials in a cradle to gate analysis. The publication considers the productive characteristics of a city in southern Brazil, quantifying the environmental impacts to produce reinforced concrete with different compressive strengths.

2.3 Unit Factors and Alternative Scenarios

a) In order to evaluate the influence of the parameters of the optimization problem, alternative scenarios were investigated, to assess the impact of modifying parameters on optimization outcomes. This research evaluated two alternative scenarios, varying the following parameters:Scenario A: involves the use of reinforced concrete with a reduced compressive strength of 20 MPa, at a unit cost that is also lower than 40 MPa concrete.
b) Scenario B: aims to evaluate the influence of possible variations in the unit costs of structural materials. To illustrate this, an example is taken of the price of steel in 2022, when its cost peaked in Brazil as a result of price rises due to the COVID pandemic.

Table 1 Unit factor values employed in this study

Material	Unit	Cost (R$)	Emission ($kgCO_2$)	Embodied Energy (MJ)
Original Scenario				
Concrete 40 MPa	m^3	581.57	182.14	1.34E-04
Reinforcement Steel	kg	8.42	1.05	17.35
I-beam Steel	kg	13.87	1.89	21.31
Alternative Scenario A (20 MPa concrete)				
Concrete 20 MPa	m^3	516.04	157.65	9.68E-05
Reinforcement Steel	kg	8.42	1.05	17.35
I-beam Steel	kg	13.87	1.89	21.31
Alternative Scenario B (2022 costs)				
Concrete 40 MPa	m^3	565.75	182.14	1.34E-04
Reinforcement Steel	kg	9.68	1.05	17.35
I-beam Steel	kg	14.56	1.89	21.31

All material unit costs were acquired from the National Research System for Construction Costs and Indices (SINAPI) report (CAIXA 2023). The unit CO_2 emissions and unit embodied energy of each material were obtained from LCA studies, as described in Sect. 2.2. Table 1 summarizes the unit factor values employed in this study, as well as unit values related to two alternative study scenarios.

2.4 *Structural Verifications*

To make sure that every solution produced by the optimization algorithm satisfies the requirements for security, performance, and usability, the structure is verified using appropriate standards stablished by Brazilian codes. The footbridge structure is verified according to NBR 8800 (2008) for the composite beam, NBR 6118 (2014) for the concrete slab and NBR 5884 (2013) for the welded profile dimensions. Given that the footbridge is exposed to human-induced vibrations, a dynamic analysis of the structure based on NBR 7187 (2021) is also required. The structure analysis was carried out analytically, considering the linear behavior of the materials. The footbridge must meet the demands of different combinations of permanent, variable and exceptional loads. Permanent loads consist of the self-weight of the slab, steel beams and shear connectors, a concrete regularization (estimated as 0.3 kN/m^2) and guardrails (estimated as 1.0 kN/m). The variable loads are the pedestrian live load of 5 kN/m^2 (NBR 7188 2013), a constructive live load of 1 kN/m^2 (NBR 6120 2019), and the horizontal action of the wind (NBR 6123 1988). Finally, the exceptional load is a punctual and horizontal action of 100 kN in the most unfavorable position of the structure (NBR 7188 2013).

The ultimate limit state (ULS) verifications relate to the shear stress and bending moment acting on the composite beam before and after concrete curing, in addition to the bending moment acting on the concrete slab. The lateral buckling before concrete hardening is assessed as the steel beams are unpropped during the construction phase. The geometry of the welded steel beam is also considered as constructability criteria, such as the steel girder slenderness and depth, width, and web thickness ratios. Both top and bottom flanges depth must be bigger than the web thickness, and the depth of the beam must be 50% bigger than the bottom flange width. The composite beam interaction degree must be bigger than the value of 0.4 (40%) and bigger than a minimum value determined by the span and the ratio between the superior and inferior flanges areas. The serviceability limit state (SLS) is also validated, with displacements limited to 1/350 of the total span of the footbridge (NBR 8800 2008). Furthermore, regarding the dynamic performance of the footbridge, its natural frequency must be outside of the range of 1.0 to 2.6 Hz, and vertical accelerations generated by human-induced vibrations should be lesser than 2.5 m/s^2, to avoid the phenomenon of dynamic amplification (NBR 7187 2021).

For the purpose of dealing with the constraints (g_x) described above, they are expressed as normalized inequations, and all of them must be met for the structure to be certified as feasible. If a solution fails to meet any of the sizing requirements, the objective function suffers a penalization, making the solution less attractive to the optimization algorithm. This penalization is calculated by multiplying each restriction value by the penalty factor r. Using prior research and a series of experiments, the value of r in this paper was defined as 10^6 (Tres Junior et al. 2023).

2.5 *Multiobjective Harmony Search (MOHS)*

The Harmony Search algorithm was first published by Geem, Kim, and Loganathan (2001). The algorithm makes an analogy to jazz musical improvisation, in which musicians improvise unique combinations of notes from each instrument in search of perfect harmony. Since then, a number of modifications and enhancements have been suggested to the algorithm in an effort to increase its efficiency and broaden its range of applications. It is worth noting the publication of Ricart et al. (2011), who present two Multiobjective Harmony Search algorithms (MOHS1 and MOHS2) that preserve as much of the method's original structure as possible.

Based on MOHS2, Sivasubramani and Swarup (2011) use an algorithm that additionally incorporates a dynamic modification of the method's internal parameters, an improvement provided for the single-objective HS algorithm by Mahdavi, Fesanghary, and Damangir (2007). Furthermore, the algorithm takes into account the ranking technique proposed by Deb et al. (2002), with the goal of improving the distribution of solutions on the Pareto front. Molina-Pérez et al. (2021) demonstrate that the approach outperforms the previous versions by implementing it in benchmark multi-objective optimization problems. Thus, the MOHS algorithm used in this study was based on these criteria.

The flowchart in Fig. 4 gives a visual description of the adopted method. As can be observed, the MOHS technique is divided into five steps, commencing with the startup of the optimization problem and the method's internal parameters. Table 2 shows the values utilized in this study for each of the parameters. The harmony memory (HM) is then initialized with a set of random solutions within the search space. The third step initiates the main loop of the algorithm, in which new harmonies are improvised, the harmony memory is updated depending on the ranking of solutions, and the stopping criterion is verified. The method returns the obtained solutions and creates the Pareto front once the maximum number of improvisations is attained.

3 Results

As a result of applying optimization to the formulated problem, numerous feasible solutions were obtained. The Pareto front is three-dimensional since three objectives were studied simultaneously during optimization. To make the results clearer to comprehend, they are presented as two-dimensional projections of the Pareto front, contrasting two distinct objectives at a time. The relationship between the cost and CO_2 emission of the footbridge structure is depicted in Fig. 5, where the non-dominated solution is highlighted in a stronger color and the dominated solutions depicted in a more subdued tone.

Within each of the alternatives, only one solution is non-dominated (Pareto optimal), based on the graph provided above. This suggests that, for the problem under consideration, the same solution provides the optimal performance in terms of cost and CO_2 emissions. This finding gives strong evidence that both objectives under examination are not conflicting, i.e., enhancing a single objective also improves the structure's performance regarding the other objectives. This happens as both the cost and CO_2 emissions simultaneously benefit from the footbridge structure's reduced material usage as a result of optimization.

Another aspect which reinforces this conclusion is the existence of a significant linear correlation among the objectives, observed by the tendency formed between the values of the objectives, which rise in tandem. Considering the equation that represents the straight line, it is possible to demonstrate that a reduction in the structure's CO_2 emission of 1 kg is obtained for every R$ 6.56 saved, in each meter of the footbridge. This correlation between the cost of the structure and its environmental impact has also been achieved by Medeiros and Kripka (2014), García-Segura and Yepes (2016), Yepes et al. (2019), Tormen et al. (2020), and Martínez-Muñoz et al. (2022a, b).

When the relationship between CO_2 emissions and the embodied energy of the structure is examined, the objectives are once again linearly related, as depicted in Fig. 6. As an outcome, both criteria for evaluating the environmental impact of the structure are non-conflicting. The equation generated for those objectives demonstrates that 1 kg of CO_2 is saved for every 9.3 MJ minimized per meter of the footbridge. Finally, as implied, there is a linear relationship between the cost

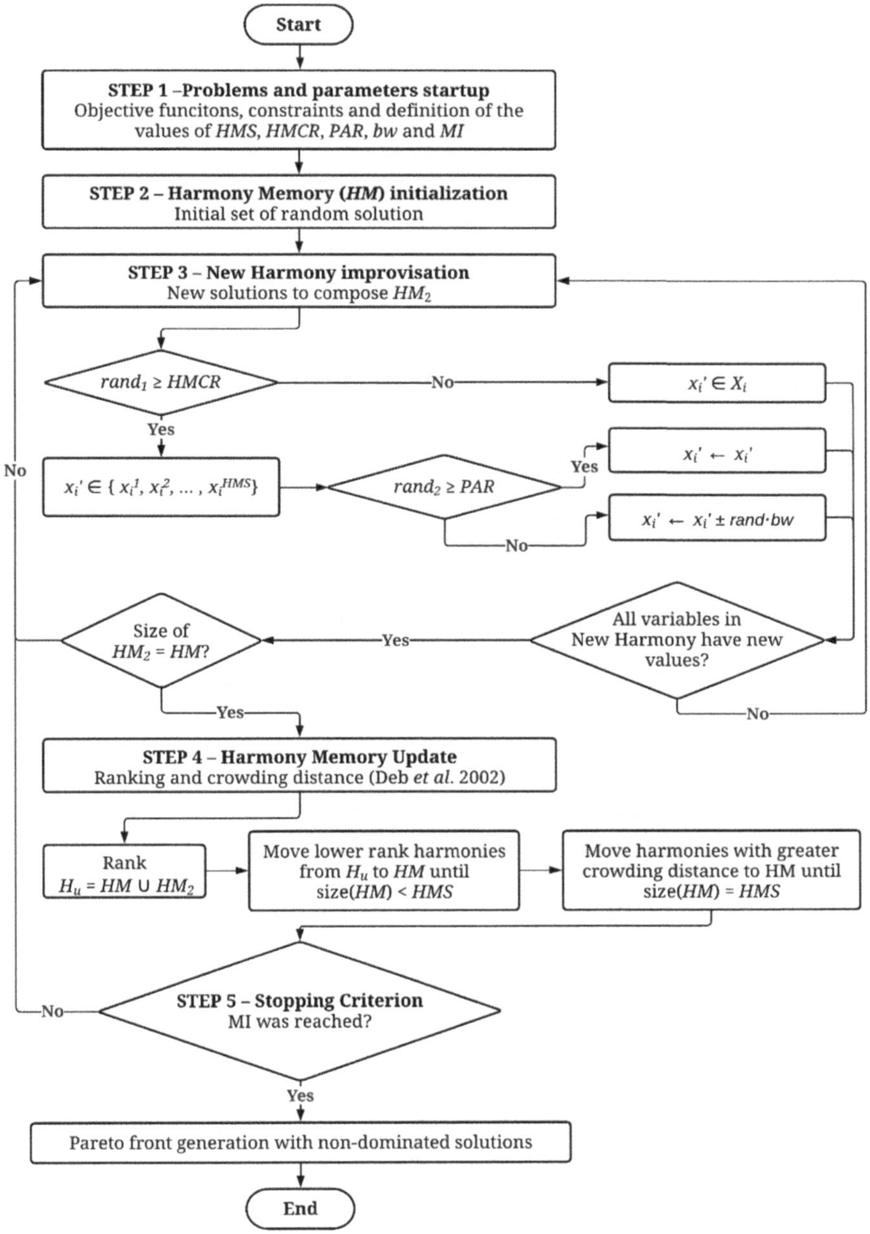

Fig. 4 Multiobjective Harmony Search algorithm

Table 2 Multiobjective Harmony Search parameters

Parameter	Value
HMS	20
HMCR	0.5
PAR_{min}	0.1
PAR_{max}	0.9
bw_{min}	0.1
bw_{max}	0.5
MI	500,000

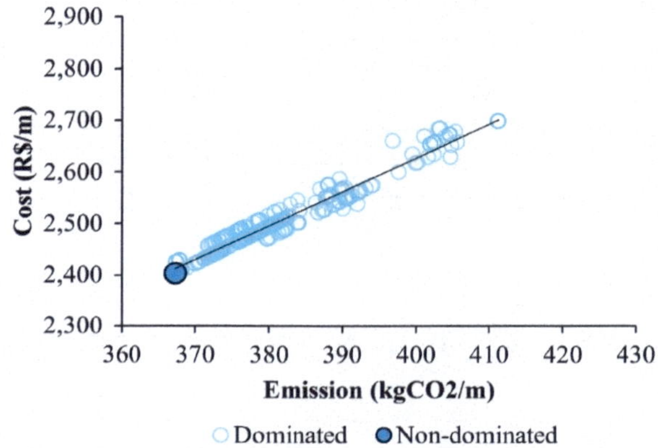

Fig. 5 Cost versus CO_2 emissions

objective and the structure's embodied energy, as illustrated in Fig. 7. Based on the linear proportion, one can assume that 1 MJ is reduced for every R$ 0.70 saved per meter.

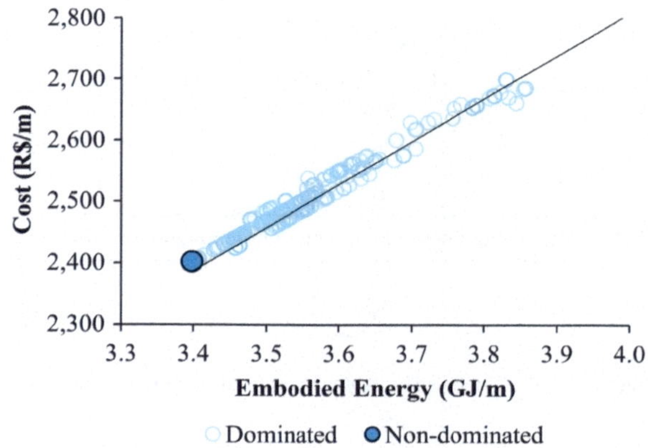

Fig. 6 Cost versus embodied energy

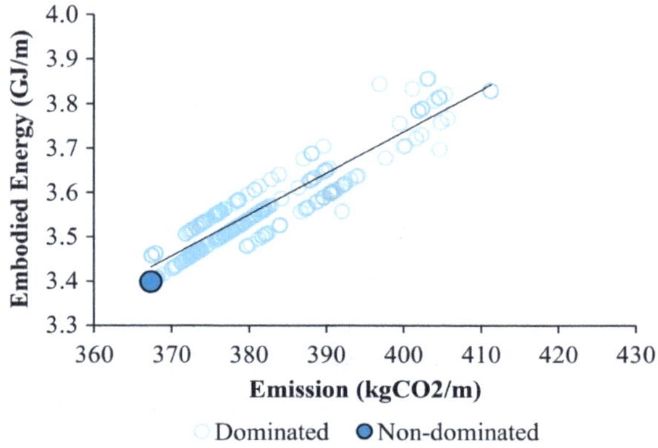

Fig. 7 CO_2 emission versus embodied energy

Two distinct examples were explored in order to investigate the sensitivity of Pareto fronts to changes in optimization problem parameters. The first evaluates the reduction in the slab's concrete strength, which is reduced from 40 to 20 MPa. Figures 8 and 9 and 10 show the findings for this scenario, which relate the cost and CO_2 emission, the CO_2 emission and embodied energy, and the cost and embodied energy, respectively.

Figures 8, 9, and 10, show that the linear relationship between the objectives persists even when the concrete strength class changes. For this scenario, once again, a single non-dominated solution was found, indicating that this is the best performing alternative in terms of costs and environmental impacts. However, while lowering the f_{ck} to 20 MPa implies in an 11% cheaper unit cost of the slab concrete, the total costs of the solutions achieved are around 3% greater. This is due to the requirement to

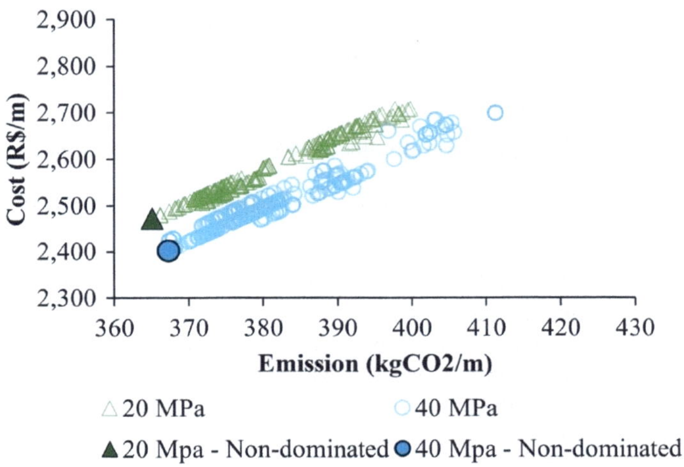

Fig. 8 Cost versus CO_2 emissions for alternative scenario A

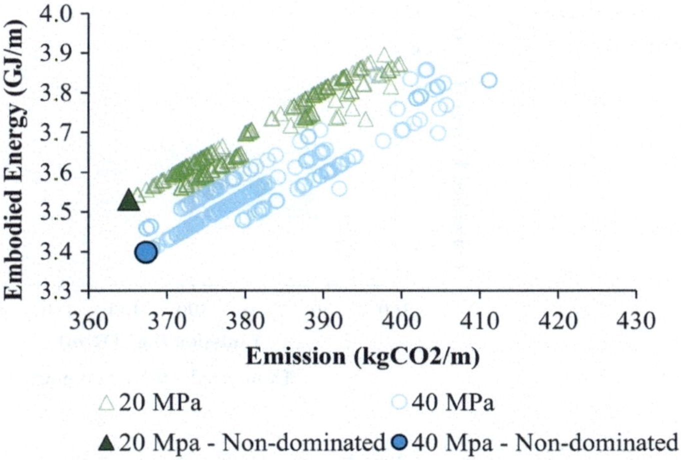

Fig. 9 CO_2 emissions versus embodied energy for alternative scenario A

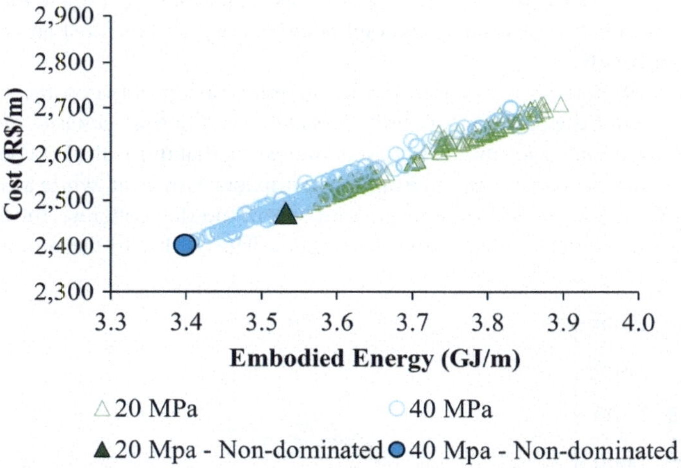

Fig. 10 Cost versus embodied energy for alternative scenario A

compensate for the decrease in composite beam strength produced by the reduction in slab f_{ck}, which demands a rise in steel consumption, a material with a proportional higher cost. It is worth mentioning that the structure's increased steel consumption implies changes in its environmental impact. In terms of CO_2 emissions, there is a minor decrease of less than 1%, however the embodied energy of the structure increases by around 4%.

The second scenario, on the other hand, evaluates a change in material unit costs, comparing the original costs of 2023 with those of the same period in 2022. It is crucial to note that in 2022, expenses in Brazil were still heavily influenced by the pandemic, which caused a considerable spike in the cost of steel. Steel prices in 2023 are up to 13% lower when compared to 2022. Figures 11, 12 and 13 provide the optimization results for this scenario.

In the case of Figs. 11, 12 and 13, multiobjective optimization yields two non-dominated solutions. One has a cheaper cost and embodied energy but produces more CO_2. In turn, the other trades off lower CO_2 emissions for increased cost and embodied energy. This behavior is consistent with prior research, as shown by Tres Junior et al. (2023), that demonstrates that changes in the unit factor of costs and environmental impacts may influence the Pareto front for this type of structure.

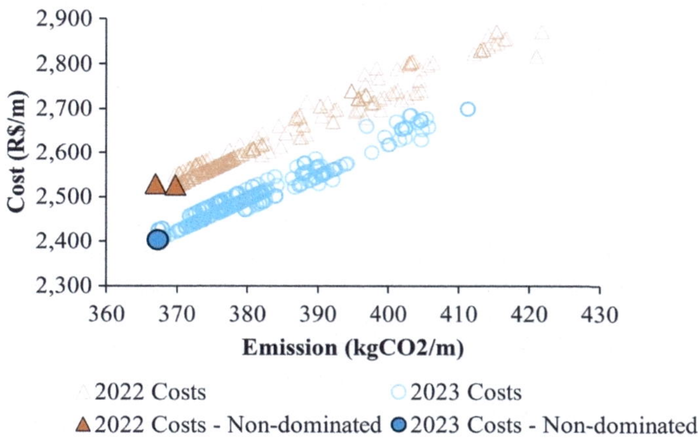

Fig. 11 Cost versus CO_2 emissions for alternative scenario B

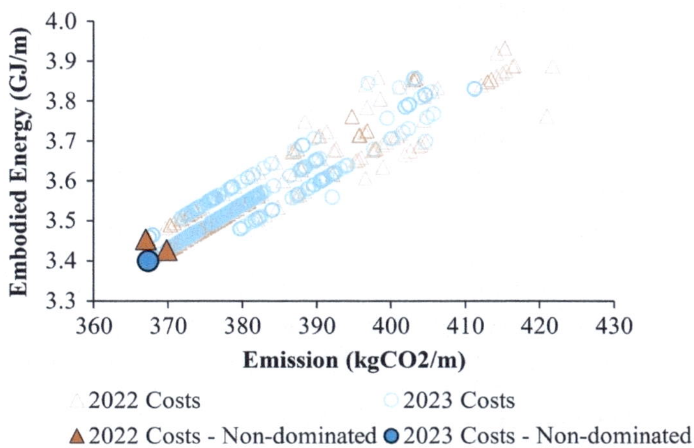

Fig. 12 CO_2 emissions versus embodied energy for alternative scenario B

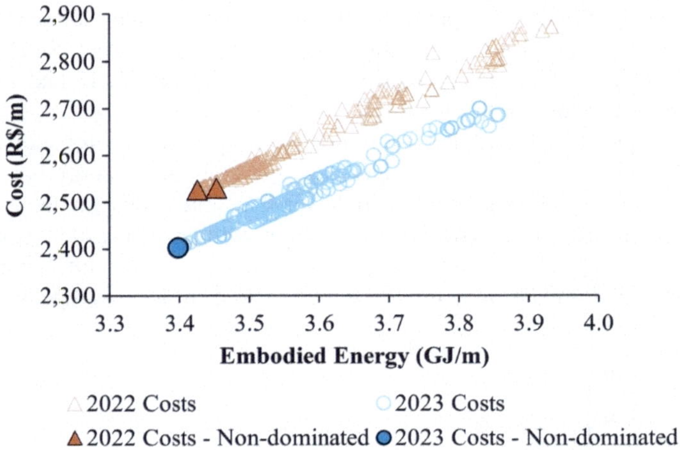

Fig. 13 Cost versus embodied energy for alternative scenario B

However, as shown by the graphs, the discrepancies between the objective values are fairly small, and the linear relationship between the objectives remains in place. In this manner, it is still possible to state that a solution with good cost performance also has excellent environmental performance when evaluated in terms of CO_2 emissions and embodied energy. Using 2023 costs, the final cost of the structure increases by around 5%, with no changes in CO_2 emissions or embodied energy.

The Pareto-optimal cross sections of the composite beams are shown in Fig. 14 for the original scenario and for the alternative scenarios A and B. The optimal values of the design variables, as well as the values of the objective functions, are also represented. As previously demonstrated, while there are changes between the non-dominated solutions for each scenario, these differences are so limited that the cross sections are ultimately very similar to each other. Thus, it is acceptable to support the assertion that, for this sort of construction, minimizing its environmental impacts is likewise advantageous to reducing its cost, or vice versa.

When all Pareto-optimal options are considered, it appears that the slab thickness is 12 cm in virtually all circumstances. The steel I-profile web is 6.35 mm thick and approximately 860 mm tall. This suggests that the optimal ratio between the span and the profile web height is about $L_e/h_w = 20$, a number frequently employed in steel structure pre-dimensioning. While the width of the bottom flange is between 185 and 200 mm, the width of the upper flange is closer to 120 mm. The thickness of the bottom flange is always bigger than the thickness of the top flange. In this sort of structure, a bigger steel area on the bottom flange in comparison to the top flange is remarkably prevalent. The degree of interaction of the composite beam adopted by all solutions has a value of 1.0, which represents a total interaction between the steel profile and the concrete slab.

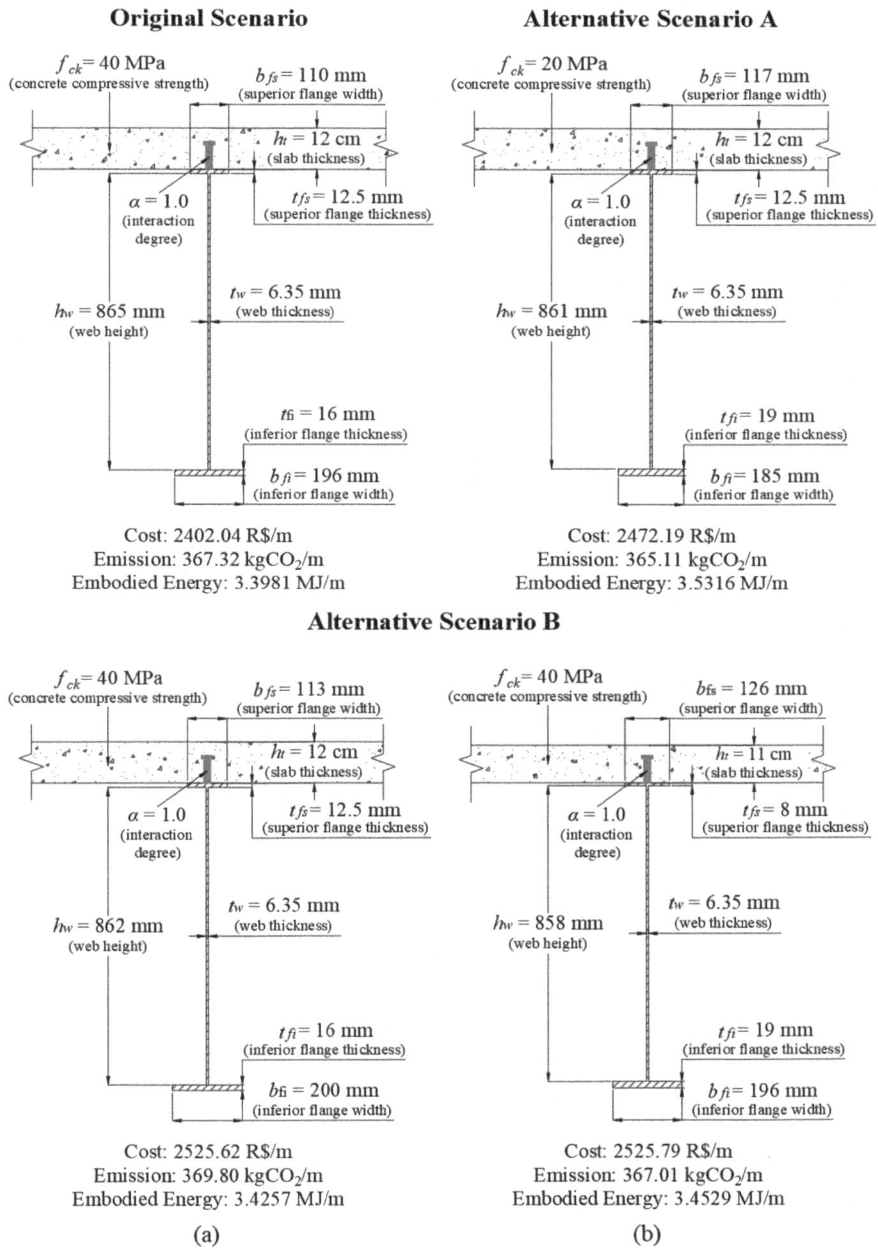

Fig. 14 Pareto optimal solution for each scenario

4 Conclusions

In this research, a multi-objective optimization of a footbridge with a steel–concrete composite structure was used to minimize the structure's cost and environmental impact. This study presents an application of LCA to quantify the environmental impacts of the footbridge, and from this, enable the sustainable design of the structure. Two LCA databases were applied: an international database for the steel of the girders and a database based on aspects of reinforced concrete production in the region where the footbridge is located. The environmental impact of the structure was evaluated based on CO_2 emissions and embodied energy. Different correlations between the optimization objectives were identified based on the results and the Pareto front obtained by the optimization. The most notable findings of the study are:

1. There is a direct linear relationship between the cost, CO_2 emission and embodied energy of the structure, so that solutions that minimize cost also have a good performance in terms of environmental impacts.
2. For every 6.56 saved per meter of the structure, 1 kg of CO_2 is prevented from being emitted; 1 MJ of embodied energy is decreased for each R$ 0.70 saved per meter of the footbridge.
3. Despite the fact that varied study scenarios produce minor variations in the Pareto front, the optimal cross sections are remarkably similar.
4. Employing higher-strength concrete for the slab is effective to improve the composite beam resistance, allowing reduction of steel consumption and resulting in a cheaper structure, besides the higher unit cost of the higher-strength concrete.
5. The LCA databases were successfully applied to quantify the environmental impacts produced by the materials employed in the structure, proving to be of great value in assessing the sustainability of the footbridge.
6. Structural optimization, when combined with LCA studies, has shown to be a successful approach for assisting in the design of more sustainable structures.

These assertions are especially pertinent in the context of developing countries such as Brazil, where infrastructure is seriously lacking. In this sense, it is feasible to contribute to the region's sustainable development through structural designs that have lower environmental impacts while yet being commercially desirable. In this context, LCA studies of the materials applied in structures are of utmost importance in allowing the assessment of the environmental impacts caused by the structure and, by extension, its sustainability. To continue with this research, the social effect of the structure may be added as an optimization objective, and the three pillars of sustainability can be considered simultaneously. It could also take into consideration the whole life cycle of the footbridge. Finally, the methods implemented may be easily extended to various types of structures, such as buildings and bridges.

Acknowledgements The authors are grateful for the financial support received by the Coordination for the Improvement of Higher Education Personnel—Brazil (Capes)—Finance Code 001; the Brazilian National Council for Scientific and Technological Development (Grant CNPq

305484/2023-0); and the Grant PID2023-150003OB-I00 funded by the Spanish Ministry of Science, Innovation and Universities MICIU/AEI/10.13039/501100011033 and by the European Regional Development Fund (ERDF/EU).

References

CAIXA (2023) Price and Costs References. https://www.caixa.gov.br/site/Paginas/downloads.aspx

Collette Y, Siarry P (2004) Multiobjective optimization: principles and case studies. Springer-Verlag Berlin Heidelberg, 1 ed. https://doi.org/10.1007/978-3-662-08883-8

Deb K, Pratap A, Agarwal S, Meyarivan T (2002) A fast and elitist multiobjective genetic algorithm: NSGA-II. IEEE T Evolut Comput 6(2):182–197. https://doi.org/10.1109/4235.996017

García-Segura T, Yepes V (2016) Multiobjective optimization of post-tensioned concrete box-girder road bridges considering cost, CO2 emissions, and safety. Eng Struct 125:325–336. https://doi.org/10.1016/j.engstruct.2016.07.012

Geem ZW, Kim JH, Loganathan GV (2001) A new heuristic optimization algorithm: Harmony Search. SIMULATION 76(2):60–68. https://doi.org/10.1177/003754970107600201

Gervásio H, Da Silva LS (2008) Comparative life-cycle analysis of steel-concrete composite bridges. Struct Infrastruct Eng 4(4):251–269. https://doi.org/10.1080/15732470600627325

Largaros ND (2018) He environmental and economic impact of structural optimization. Struct Multidisc Optim 58:1751–1768. https://doi.org/10.1007/s00158-018-1998-z

Lu P, Shi Q, Li D, Wu Y (2022) Optimization design on the section of curved composite box beam bridges. Struct Concr 24(1):1035–1052. https://doi.org/10.1002/suco.202100911

Mahdavi M, Fesanghary M, Damangir E (2007) An improved harmony search algorithm for solving optimization problems. Appl Math Comput 188(2):1567–1579. https://doi.org/10.1016/j.amc.2006.11.033

Martínez-Muñoz D, García J, Martí JV, Yepes V (2022a) Optimal design of steel–concrete composite bridge based on a transfer function discrete swarm intelligence algorithm. Struct Multidiscip Optim 65:312. https://doi.org/10.1007/s00158-022-03393-9

Martínez-Muñoz D, García J, Martí JV, Yepes V (2023) Hybrid swarm intelligence optimization methods for low-embodied energy steel-concrete composite bridges. Mathematics 11(1):140. https://doi.org/10.3390/math11010140

Martínez-Muñoz D, Martí JV, Yepes V (2021) Comparative life cycle analysis of concrete and composite bridges varying steel recycling ratio. Materials 14(15):4218. https://doi.org/10.3390/ma14154218

Martínez-Muñoz D, Martí JV, Yepes V (2022b) Social impact assessment comparison of composite and concrete bridge alternatives. Sustainability 14(9):5186. https://doi.org/10.3390/su14095186

Medeiros GF, Kripka M (2014) Optimization of reinforced concrete columns according to different environmental impact assessment parameters. Eng Struct 59:185–194. https://doi.org/10.1016/j.engstruct.2013.10.045

Mei L, Wang Q (2021) Structural optimization in civil engineering: a literature review. Materials 11(2):66. https://doi.org/10.3390/buildings11020066

Milani CJ, Yepes V, Kripka M (2020) Proposal of sustainability indicators for the design of small-span bridges. Int J Environ Res Public Health 17:4488. https://doi.org/10.3390/ijerph17124488

Molina-Pérez D, Portilla-Flores EA, Vega-Alvarado E, Calva-Yañez MB, Sepúlveda-Cervantes G (2021) A novel multi-objective harmony search algorithm with pitch adjustment by genotype. Appl Sci 11(19):8931. https://doi.org/10.3390/app11198931

NBR 5884 (2013) Electric Arc Welded Steel Structural I Profile—General Requirements

NBR 6118 (2014) Design of Concrete Structures—Procedure

NBR 6120 (2019) Actions for calculating building structures

NBR 6123 (1988) Wind forces on buildings

NBR 7187 (2021) Design of Concrete Bridges, Viaducts, and Footbridges
NBR 7188 (2013) Road and Pedestrian Live Loads on bridges, viaducts, footbridges and other structures
Niu Y, Fink G (2018) Life Cycle Assessment on modern timber bridges. Wood Mat Sci Eng 14(4):212–225. https://doi.org/10.1080/17480272.2018.1501421
NBR 8800 (2008) Design of Steel Structures and Composite Structures of Steel and Concrete for Buildings
Orcesi A, Cremona C, Ta B (2018) Optimization of design and life-cycle management for steel-concrete composite bridges. Struct Eng Int 28:185–195. https://doi.org/10.1080/10168664.2018.1453763
Penadés-Plà V, Martínez-Muñoz D, García-Segura T, Navarro IJ, Yepes V (2020) Environmental and social impact assessment of optimized post-tensioned concrete road bridges. Sustainability 12(10):4265. https://doi.org/10.3390/su12104265
Ricart J, Hüttermann G, Lima J, Barán B (2011) Multiobjective harmony search algorithm proposals. Electron Notes Theor Comput Sci 281:51–67. https://doi.org/10.1016/j.entcs.2011.11.025
Santoro JF, Kripka M (2020) Minimizing environmental impact from optimized sizing of reinforced concrete elements. Computers and Concrete 25(2):111–118. https://doi.org/10.12989/cac.2020.25.2.111
Sivasubramani S, Swarup S (2011) Multi-objective harmony search algorithm for optimal power flow problem. Int J Electr Power Energy Syst 33(3):745–752. https://doi.org/10.1016/j.ijepes.2010.12.031
Tormen AF, Pravia ZMC, Busato FR, Kripka M (2020) Optimization of steel-concrete composite beams considering cost and environmental impact. Steel and Composite Structures 34(3):409–421. https://doi.org/ https://doi.org/10.12989/scs.2020.34.3.409
Tres Junior FL, Yepes V, Medeiros GF, Kripka M (2023) Multi-objective optimization applied to the design of sustainable pedestrian bridges. Int J Environ Res Public Health 20(4):3190. https://doi.org/10.3390/ijerph20043190
United Nations (2015) Transforming our world: the 2030 Agenda for Sustainable Development. https://sdgs.un.org/2030agenda
UNEP (2022) 2022 Global Status Report for Buildings and Construction. https://www.unep.org/resources/publication/2022-global-status-reportbuildings-and-construction
Worldsteel (2023) Our performance: Sustainability Indicators. https://worldsteel.org/steel-topics/sustainability/sustainability-indicators-2023-report/
Yepes V, Dasí-Gil M, Martínez-Muñoz D, López-Desfilis VJ, Martí JV (2019) Heuristic techniques for the design of steel-concrete composite pedestrian bridges. Appl Sci 9(16):3253. https://doi.org/10.3390/app9163253

Pursuing Comparative Analysis of Life Cycle Assessment and Circular Economy Integration in Architectural Education: A Case Study of Universities in Belarus and China

Siarhei Zenchanka, Anna Gorbatcheva, and Zhang Pengfei

Abstract This paper presents a comparative analysis of the integration of Life Cycle Assessment (LCA) and Circular Economy (CE) principles in architectural education, focusing on universities in Belarus and China. The research addresses the growing importance of sustainability in the architectural field and the role of education in preparing future architects for environmentally responsible practice. Through a case study approach, the paper investigates the feedback of students mastering architecture in China and Belarus, teaching methods, and practical applications related to LCA and CE in architectural programs at selected universities in Belarus and China. The study aims to identify similarities, differences, challenges, and best practices in integrating LCA and CE into architectural education, considering both countries' cultural, economic, and regulatory contexts. The findings provide insights into the current state of sustainability education in architecture and offer recommendations for advancing the integration of LCA and CE principles in architectural curricula. This comparative analysis contributes to the ongoing dialogue on promoting sustainable design practices and fostering a culture of environmental stewardship within architectural education globally.

Keywords Life Cycle Assessment (LCA) · Circular economy · Architectural education · Comparative analysis · Belarus · China universities

S. Zenchanka (✉)
Minsk Branch of Plekhanov, Russian University of Economics, Moscow, Belarus
e-mail: zench@tut.by

A. Gorbatcheva
Faculty of Architecture, Institute of Urban Design and Landscaping, National Technical University, Minsk, Belarus

Z. Pengfei
Academic Affairs of International Institute of Management and Business, Minsk, Belarus
e-mail: pengfei19891115@gmail.com

L. Brandli et al. (eds.), *The Contribution of Life Cycle Analyses and Circular Economy to the Sustainable Development Goals*, World Sustainability Series,
https://doi.org/10.1007/978-3-031-85300-5_23

1 Introduction

The relationship between contemporary architecture, Life Cycle Assessment (LCA), and Circular Economy (CE) is multifaceted and symbiotic. Contemporary architecture increasingly integrates LCA and CE principles, emphasizing sustainability as a core aspect of design.

Architects are leveraging LCA to assess the environmental impacts of their designs and make informed decisions to minimize these effects (França 2021). Concurrently, CE principles guide architects to prioritize resource efficiency, waste reduction, and circularity, aligning with the broader sustainability goals of contemporary architecture. Material selection in modern architecture reflects the influence of LCA and CE, as architects consider the entire life cycle of materials—focusing on environmentally friendly, sustainable options such as recycled, renewable, and locally sourced materials (Vandervaeren et al. 2022). This approach resonates with the principles of CE, underpinning the emphasis on mindful and ecologically responsible material choices in contemporary architectural practice.

Integrating LCA and CE fosters design innovation in contemporary architecture, propelling architects to explore inventive approaches such as modular construction, adaptive reuse, and cradle-to-cradle design (Sanchez et al. 2019). This innovation mirrors contemporary architecture's dynamic and forward-thinking nature, as architects seek to address sustainability challenges through creative design solutions.

Compliance with increasingly stringent sustainability regulations drives contemporary architecture to embrace LCA and CE. Architects must demonstrate that their designs optimize resource use, reduce environmental impact, and contribute to a circular economy to meet these regulatory demands. Client demand and market trends also propel the adoption of LCA and CE in contemporary architecture, reflecting the increasing preference for environmentally responsible buildings that minimize operating costs and enhance occupant comfort. Finally, contemporary architecture education and research integrate LCA and CE concepts, equipping future architects with the knowledge and skills to design environmentally responsible buildings (Sanchez and Haas 2018). Ultimately, the interplay between contemporary architecture, LCA, and CE underscores the symbiotic relationship, with each element influencing the other to drive sustainable design, innovation, and regulatory compliance in the built environment. This integration facilitates the creation of architecturally stunning, functional, and environmentally sustainable buildings, reflecting a commitment to sustainability and responsible design.

Integrating LCA and CE principles into architectural education is important for several reasons. Firstly, it raises sustainability awareness by sensitizing students to the environmental impacts of architectural design and construction processes. Secondly, it enhances professional competence by equipping students with the knowledge and skills to design and construct environmentally conscious buildings. It also aligns with global efforts to promote sustainable development in the built environment.

Stimulating the innovation in architectural design and construction methods, integrating LCA and Circular Economy concepts fosters interdisciplinary collaboration and prepares graduates to comply with evolving sustainability regulations. This approach promotes responsibility, ethical design, and prioritization of sustainable practices (Filho et al. 2024a). Finally, integrating these principles not only equips students with essential knowledge and skills and cultivates a culture of sustainability, innovation, and responsible design in the next generation of architects, addressing pressing environmental challenges worldwide (Pieroni et al. 2019).

The discussion on the relationship between contemporary architecture, LCA and CE primarily addresses several United Nations Sustainable Development Goals (SDGs), including (UN n/d):

SDG 4: Quality Education: The incorporation of LCA and CE concepts into architectural education contributes to promoting inclusive and equitable quality education and fostering lifelong learning opportunities for all, aligning with the targets of SDG 4 related to quality education and sustainable development.

SDG 9: Industry, Innovation, and Infrastructure: This discussion aligns with the goal foster building resilient infrastructure promoting sustainable industrialization, and fostering innovation. Integrating LCA and CE principles in contemporary architecture represents a commitment to innovative and sustainable infrastructure development.

SDG 11: Sustainable Cities and Communities: The incorporation of LCA and CE in contemporary architecture contributes to the creation of sustainable, resilient, environmentally responsible cities and communities, in line with the targets of SDG 11.

SDG 12: Responsible Consumption and Production: The emphasis on mindful material selection, resource efficiency, and waste reduction in contemporary architecture reflects a commitment to promote sustainable consumption and production practices, addressing the objectives of SDG 12.

SDG 13: Climate Action: By integrating LCA and CE principles, contemporary architecture contributes to mitigating climate change and its impacts, aligning with the targets of SDG 13 related to climate action and environmental sustainability.

SDG 16: Peace, Justice, and Strong Institutions: The adoption of sustainable design practices, informed by LCA and CE principles, contributes to the promotion of peaceful and inclusive societies, reflecting the interconnectedness of sustainable development with peace, justice, and strong institutions.

Integrating LCA and CE principles into contemporary architecture education and practice contributes to advancing multiple SDGs related to sustainable infrastructure, climate action, responsible consumption, education, and the creation of sustainable, inclusive, and resilient communities (Merli et al. 2018).

This paper aims to comparatively analyze the integration of LCA and CE components in architectural education, focusing on universities in Belarus and China, to identify key similarities, differences, and best practices.

The Tasks of this paper are:

- To evaluate the curricula of architectural education programs in selected universities in Belarus and China to identify the extent of integration of LCA and CE principles. This involves comparing course offerings, modules, and learning objectives related to LCA and CE across the two countries.
- To conduct interviews and surveys of faculty members, students, and industry professionals in architectural education to gain insights into the implementation and reception of LCA and CE concepts. This task aims to gather qualitative data on the perception, challenges, and successes of integrating LCA and CE in architectural education in both countries.
- Based on the comparative analysis and stakeholder feedback, develop informed recommendations and best practices for enhancing the integration of LCA and CE in architectural education. Identify specific strategies, curricular enhancements, and potential areas of collaboration between universities in Belarus and China to promote sustainability education within the architectural curriculum further.

The subject of research is the integration of LCA and CE in architectural education programs at universities in Belarus and China. The object of this research is to analyze and compare the approaches, challenges, and outcomes of integrating LCA and CE in architectural education in the context of Belarusian and Chinese universities, with the aim of identifying best practices and areas for improvement in sustainability education within the architectural curriculum.

2 Literature Review

Interdisciplinary collaboration and action are key important for promoting sustainable development (Nerini et al. 2019). Complex sustainability challenges can be addressed more comprehensively by integrating knowledge from diverse industries and disciplines. This approach fosters innovation, creativity, and the development of holistic solutions that consider social, economic, and environmental aspects. It also promotes a deeper understanding of the interconnectedness of global challenges, leading to the co-creation of practical strategies and policies. Embracing interdisciplinary action is essential for achieving sustainable development that balances human well-being with environmental protection and economic prosperity (Giesenbauer and Mueller-Christ 2020).

Fostering collaboration among staff from various faculties is a powerful way to address complex global challenges and provide comprehensive solutions. By bringing diverse expertise and perspectives together, interdisciplinary teams can tackle multifaceted issues with innovative and blended approaches (Bassachs et al. 2020). This collaborative effort enables synthesizing knowledge from different disciplines, leading to more robust and sustainable solutions for existing and emerging problems. Through this collective approach, organizations and institutions can enhance their capacity to adapt, innovate, and effectively address the intricate challenges of our world.

In the context of architectural education, embracing interdisciplinary collaboration and action plays a crucial role in preparing future architects to address the complexities of sustainable design and development. Embracing interdisciplinary action in architectural education can enhance curriculum enrichment by incorporating diverse perspectives and knowledge into the learning experience, enabling students to develop a more comprehensive understanding of sustainability issues and innovative solutions.

Promoting teamwork among students from different faculties can simulate real-world professional collaborations, challenging students to communicate across disciplines and work towards common goals, mirroring the collaborative nature of architectural practice (Veckalne and Tambovceva 2022). Interdisciplinary research initiatives in architectural education can lead to groundbreaking insights and innovations in sustainable design and construction, enriching the learning experience and pushing the boundaries of sustainable architectural practices. Architectural education that emphasizes interdisciplinary collaboration also prepares students for the evolving demands of the profession, as architects increasingly need to work with experts from different fields to address complex sustainability challenges.

Embracing interdisciplinary action fosters a culture of innovation and adaptability, essential for addressing future challenges in the built environment. By integrating diverse perspectives and encouraging creative problem-solving, architectural programs can cultivate a mindset that embraces sustainability, resilience, and inclusive design, empowering students to become effective agents of change in creating a more sustainable and harmonious world through their architectural practice (Bertel et al. 2021).

Interdisciplinary action enhances curriculum enrichment by bringing together diverse perspectives from various fields, fostering a more holistic understanding of complex issues, and encouraging innovation in problem-solving. Through exposure to multiple disciplines, students develop critical thinking skills and analytical agility, enabling them to evaluate complex problems from different angles. This approach also bridges the gap between academic knowledge and real-world application, better preparing students for the complexities of modern challenges and the collaborative nature of professional environments (Butt and Dimitrijevic 2022). Furthermore, interdisciplinary collaboration promotes personal and professional growth by cultivating communication skills, teamwork abilities, and a broader mindset. By integrating interdisciplinary perspectives, educational institutions create a dynamic and impactful learning environment that equips students to thrive in an interconnected world.

Integrating LCA and CE concepts into interdisciplinary collaboration and curriculum enrichment aligns with holistic and sustainable problem-solving and design principles. These concepts encourage a comprehensive view of products and systems, considering their entire lifecycle and aiming to minimize environmental impacts (Filho et al. 2024b). Incorporating LCA and CE into interdisciplinary collaboration exposes students to diverse disciplinary perspectives, fostering a deeper understanding of sustainability and innovative solutions that consider environmental, social, and economic factors. This integration inspires creativity across disciplines, seeking integrated solutions incorporating sustainable principles and considering the full lifecycle of products and systems. By bridging academic knowledge with real-world application, this approach prepares students for careers in architecture and related fields by instilling a mindset that values sustainable, interdisciplinary, and collaborative approaches to design and construction, aligning with the demands of a rapidly evolving industry (Viera Trevisan et al. 2024).

Integrating LCA principles into architectural design processes provides a comprehensive understanding of the environmental impacts of building materials, construction practices, and operational considerations (Hollberg and Ruth 2016). Consequently, LCA empowers architects to make informed decisions regarding material selection, energy efficiency, and overall resource use, creating of environmentally sustainable and resource-efficient buildings. LCA encourages design innovation, supports regulatory compliance and certification, facilitates life cycle costing, and enhances client education and engagement. By integrating LCA into architectural design processes, architects can optimize the environmental performance of buildings, foster greater awareness of sustainability issues among stakeholders, and position their projects as environmentally responsible endeavors that contribute to a more sustainable built environment overall.

Integrating LCA into architectural education and design processes closely aligns with the CE approach (Joensuu et al. 2020). Both methodologies emphasize resource efficiency, waste reduction, and circular material flows, contributing to the creation of environmentally sustainable and resource-efficient buildings. LCA enables architects to quantify environmental impacts, design for disassembly, and prioritize durable, low-impact materials, aligning with CE's goal of reducing waste and promoting a circular approach to material use (Benachio et al. 2020). Furthermore, both LCA and CE encourage a focus on product longevity, systems thinking, collaboration, innovation, sustainable design strategies, and stakeholder engagement, fostering a holistic perspective and innovative approaches to sustainable architectural design. By integrating LCA into architectural practices, architects can effectively align their projects with the circular economy principles, creating buildings that minimize environmental impacts and contribute to a more sustainable built environment.

Regional perspectives on the integration of LCA and CE into architectural education and practice in China and Belarus provide a nuanced understanding of the challenges and opportunities faced in each country.

In China, there is a growing emphasis on sustainable development and green building practices driven by factors such as rapid urbanization, increasing environmental awareness, and government policies promoting green building initiatives (Fan and Fang 2020). The integration of LCA and CE into architectural education is gaining traction, with universities incorporating sustainability principles into their curricula and research activities. However, challenges such as the scale of construction projects, the impact of rapid urbanization on natural resources, and the need for standardized tools and methodologies for LCA and CE implementation remain significant hurdles (Wang et al. 2020). Promoting interdisciplinary collaboration, industry partnerships, and governmental support are crucial for further advancing the integration of LCA and CE in architectural practice in China.

In Belarus, there is a strong emphasis on sustainability and environmental protection, with government initiatives promoting energy efficiency and green technologies in the construction sector (Zenchanka et al. 2020). The integration of LCA and CE into architectural education is influenced by the country's commitment to sustainable development goals and green building standards. Belarusian universities are incorporating LCA and CE principles into their architectural programs, fostering a holistic approach to design that considers environmental impacts and resource efficiency (Gorbatchev and Zenchanka 2020). Challenges in Belarus include the need for increased awareness of LCA and CE concepts among stakeholders, the adoption of innovative sustainable design strategies, and the implementation of practical solutions in architectural practice. Enhancing industry collaboration, knowledge sharing, and capacity building are essential for further promoting the integration of LCA and CE in architectural education and practice in Belarus.

While both China and Belarus are making strides in integrating LCA and CE into architectural education and practice, each country faces unique challenges and opportunities that require tailored strategies and collaborative efforts to advance sustainable design practices for a more environmentally conscious built environment.

3 Methodology

The methodology for this study adopts a mixed-methods approach, combining qualitative and quantitative data collection methods. Purposive sampling will be used to select experienced architectural professionals for in-depth semi-structured interviews, focusing on their perspectives regarding the integration of LCA and CE in architectural education and practice.

The rationale for employing a mixed-methods approach in this study is to achieve a comprehensive understanding of the integration of LCA and CE in architectural education and practice. By combining qualitative interviews with architectural professionals and a quantitative questionnaire survey involving students, educators, and architects, the research can capture diverse perspectives and experiences on sustainable design practices in architecture. This approach allows for a more robust interpretation of the data through triangulation, ensuring the validity and reliability of the findings. The qualitative data from interviews offers in-depth insights and contextual understanding, complementing the quantitative survey data that provides broader insights and statistical analysis. By integrating qualitative and quantitative methods, the study aims to produce nuanced outcomes that address the complexities of sustainable architectural practices, enhancing the depth and validity of the research findings.

In addition, a questionnaire survey employing a stratified random sampling approach will gather quantitative data from students, educators, and practicing architects, exploring their understanding, perceptions, and experiences related to LCA and CE in architectural education and practice. Thematic analysis was applied to the interview transcripts, while descriptive and inferential statistical analysis will be used for the quantitative survey data. Ethical considerations include obtaining informed consent from participants, ensuring confidentiality and anonymity, and using the data solely for research purposes. Through this comprehensive methodology, the study aims to provide a nuanced understanding of the perspectives on integrating LCA and CE in architectural education and practice while upholding ethical standards.

The selection for questionnaire includes the 3rd and the 4th year architecture students from Belarus National Technical University (BNTU) and master students of "Project management" of School of Business Belarusian State University from various provinces of China that who previously studied architecture at China at bachelor level. This selection is justified by the participants' relevant expertise, continuity of architectural education, and diverse cultural backgrounds. The 3rd and 4th year students at BNTU and master students at Belarusian State University are expected to have a solid foundation in architectural education and exposure to sustainability concepts. Additionally, including individuals who previously studied architecture in China and are enrolled in a different master's program allows for an exploration of the transfer of sustainable knowledge from their undergraduate studies. The diverse educational levels and cultural backgrounds contribute to a holistic understanding of the integration of LCA and CE in architectural education and practice. This approach ensures a comprehensive representation of perspectives, enriching the survey data and facilitating a multifaceted analysis of sustainable architectural practices within the selected demographic.

4 Overview of LCA and CE Components and Outcomes in Architectural Education in Belarus and China

In Belarus, architectural education comprises two levels: colleges and universities—each catering to distinct aspects of architectural training and competencies. At the secondary specialized education level in colleges, students in the "Architecture" specialty undergo technical profile training, focusing on creating architectural and construction drawings, perspective projections, and solving basic compositional challenges. They also learn urban planning concepts, use normative literature, master architectural graphics, and develop interior elements. Currently, there are 4 colleges providing courses in architecture (in Minsk, Grodno, Mogilev, and Vitebsk).

On the other hand, the higher education first-stage specialty leading to an "Architect" qualification with an engineering profile emphasizes competencies in integrated territorial organization schemes, general and detailed city planning, and construction supervision. Professional activities encompass a wide range of architectural and urban planning projects, including buildings, landscapes, infrastructure, and small architectural forms, highlighting Belarus' commitment to nurturing skilled architectural professionals across various domains. There are 4 universities providing courses in architecture (in Minsk, Brest, Gomel, and Polotsk). For engineering profile courses, the duration of study is 5 years. There are two profiles for graduate programs: architecture and architectural design.

The curriculum, learning programs, and outcomes are adjusted with educational standards, which were adopted in 2018 (Standard, 2018). The universities are using the standard curriculum, which is approved by the national educational authority. The standard curriculum for "Architecture" includes a state component with twenty-one disciplines grouped into 7 educational modules, a university variable component with 24 disciplines grouped into 6 modules, and 6 elective disciplines. In addition, students complete fourteen course projects and 7 coursework. Also provided are educational and introductory, artistic, construction and technological, design, and diploma internships in the amount of 19 weeks (28 credits).

The educational standard stipulates the outcomes, which should be provided in curriculum courses. Outcomes include 17 universal (UC) and 12 basic professional competencies (BPC).

There are no direct connections between universal outcomes and LCA and CE concepts. However, some of them may relate to these concepts indirectly (Table 1).

Based on the analysis of the competencies outlined in the educational standards of higher education for Architecture in Belarus, it can be concluded that while direct references to Life Cycle Assessment (LCA) and Circular Economy (CE) concepts may be limited, there are indirect connections between the competences of architecture students and the principles of LCA and CE.

The competencies emphasizing critical thinking, interdisciplinary integration, innovation, and leadership in the design process indirectly align with the holistic and sustainable approach advocated by LCA and CE practices. Students with these competencies are potentially equipped to consider the environmental impact, resource optimization, and sustainability aspects essential to LCA and CE in their architectural education and professional activities.

With the interdisciplinary nature of LCA and CE, the competencies related to the use of construction technologies, materials, and life support systems, as well as the coordination of diverse design tools and interdisciplinarity, also indirectly support the incorporation of LCA and CE principles in architectural education. It can be inferred that the development of these competencies in architecture students may contribute to their ability to apply LCA and CE concepts in their future professional practice.

While the direct inclusion of LCA and CE concepts in the educational standards may be limited, the competencies outlined in the architectural educational standards indirectly support the integration of LCA and CE principles into the education and professional development of architecture students.

As per master programs in architecture, they are adjusted with educational standard adopted in 2023 (Standard 2023). The standard includes 20 universal competencies, 12 basic professional competencies and 4 universal professional competencies (UPC). The relations between outcomes according to educational standard and LCA and CE concepts for master equivalent programs for architectural education in Belarus are available in Table 2.

Table 1 Correlations between competencies and outcomes according to educational standard and LCA and CE concepts for bachelor equivalent programs for architects in Belarus

Competencies and outcomes according to educational standard	Relations with LCA and CE concepts
UC-4: Possess a culture of thinking, ability to perceive, summarize, and analyze information, including philosophical, ideological, and psychological-pedagogical problems in the field of interpersonal relationships and professional activities	This competence aligns with LCA and CE principles by emphasizing critical thinking, analysis, and consideration of the psychological and philosophical aspects related to sustainability, resource management, and environmental impact assessment
UC-8: Apply knowledge of related and adjacent disciplines when developing projects, act innovatively and technically competently when using construction technologies, materials, design-universal competencies, life support systems, and information and computer tools	This competence relates to LCA and CE by emphasizing the integration of knowledge from various disciplines, innovation in construction technologies, and the consideration of life support systems, reflecting the interdisciplinary nature and innovation required in LCA and CE practices
UC-10: Be able to mutually coordinate various design tools and factors, integrate diverse forms of knowledge and skills in developing design solutions, coordinate interdisciplinary goals, think creatively, initiate innovative solutions, and exercise leadership in the design process	This competence aligns with LCA and CE principles by emphasizing the coordination of interdisciplinary goals, creative thinking, and the initiation of innovative solutions, all of which are essential in the application of LCA and CE principles to architectural design
BPC 7: Be able to optimize the architectural environment, using the basic concepts and prerequisites for the formation of scientific ecological and urban planning theory and a multi-level approach to solving environmental problems of urban planning and landscape architecture	This competence reflects an understanding of ecological principles and a multi-level approach to addressing environmental issues, which aligns with the sustainability principles of LCA and CE in architectural design
BPC 10: Know the main methods of production, transportation, storage of energy resources, and ways to increase the efficiency of their consumption, taking into account the main directions of state policy in the field of energy saving	Understanding energy efficiency and sustainable resource management is essential in LCA and CE practices, as they involve assessing and optimizing resource use throughout a building's life cycle, aligning with principles of energy saving and sustainable practices

Source Developed by authors using competencies listed in (Standard 2018)

Upon analyzing the current versions of the Belarusian State Educational Standards for bachelor and master equivalent programs in architecture, several competencies align with LCA and CE concepts. These competencies emphasize the importance of sustainability, innovation, interdisciplinary collaboration, and the long-term development of the architectural environment.

The competencies highlighted in Tables 1 and 2 underscore the significance of initiating innovative design solutions, integrating diverse knowledge and skills, and using interdisciplinary methods in architectural projects. These competencies

Table 2 Correlations between competencies and outcomes according to educational standard and LCA and CE concepts for master equivalent programs for architects in Belarus

Competencies and outcomes according to educational standard	Relations with LCA and CE concepts
UC-8: Apply knowledge of related and related disciplines when developing architectural projects, act innovatively and technically competent when using construction technologies, materials, structures, life support systems and information and computer tools	This competence indirectly relates to LCA and CE by emphasizing the integration of knowledge from related disciplines, innovativeness in using construction technologies and materials, and the consideration of life support systems, reflecting the interdisciplinary nature and innovation required in LCA and CE practices
UC-9: Be ready to cooperate with colleagues and specialists in related fields, work in a creative team, knowing the principles and methods of organizing and managing small teams	Collaboration and teamwork are essential for integrating LCA and CE principles into architectural projects. The ability to cooperate with colleagues and specialists in related fields aligns with the interdisciplinary nature of LCA and CE practices, where collaboration and teamwork are crucial for considering life cycle impacts and developing circular design solutions
UC-10: Initiate innovative solutions and act as a leader in the design process, mutually coordinate various design tools and factors, integrate diverse forms of knowledge and skills when developing design solutions, think creatively, coordinate interdisciplinary goals	This competence directly aligns with the principles of Life Cycle Assessment (LCA) and Circular Economy (CE) by emphasizing the initiation of innovative solutions, leadership in the design process, coordination of interdisciplinary goals, and the integration of diverse knowledge and skills. These are essential aspects in applying LCA and CE principles to architectural design and construction practices
BPC-2: Create architectural projects according to structural and technical requirements and initiate innovative design solutions	This competence directly aligns with the principles of Life Cycle Assessment (LCA) and Circular Economy (CE) by emphasizing the initiation of innovative design solutions. This includes the consideration of materials, construction methods, and the impacts on the life cycle of architectural projects

(continued)

Table 2 (continued)

Competencies and outcomes according to educational standard	Relations with LCA and CE concepts
BPC-7: Optimize the architectural environment, using the basic concepts and prerequisites for the formation of scientific environmental and urban planning theory and a multi-level approach to solving environmental problems of urban planning, landscape architecture, and design	This competence directly relates to LCA and CE principles by emphasizing the optimization of the architectural environment and the integration of scientific environmental and urban planning theory. This involves considerations for sustainability, resource efficiency, and the environmental impact of architectural and urban design solutions
UPC-3: Apply the principles of integrated creation, transformation, preservation, and long-term development of the architectural environment and its components, including those of an innovative (conceptual), interdisciplinary, and specialized nature	This competence directly aligns with the principles of Life Cycle Assessment (LCA) and Circular Economy (CE) by emphasizing the holistic and long-term development of the architectural environment. It highlights the importance of considering sustainability, innovation, and interdisciplinary approaches when designing architectural solutions, which are key aspects of LCA and CE practices

Source Developed by authors using competencies listed in (Standard 2023)

directly support the principles of LCA and CE, which require innovative thinking, interdisciplinary collaboration, and the integration of diverse expertise to promote sustainable and circular design practices.

Some of the competencies are related to optimizing the architectural environment, considering environmental impacts, and applying principles for long-term development that align well with the goals of LCA and CE. These competencies stress the importance of sustainability, resource efficiency, and the environmental impact of architectural and urban design solutions, which are core principles of LCA and CE frameworks.

The competencies in the Master Standard advocate for a holistic approach to architectural design, encompassing integrated creation, preservation, and transformation of the architectural environment. This holistic perspective resonates with the cradle-to-cradle philosophy of LCA and the circular design principles of CE, emphasizing the comprehensive consideration of the life cycle competencies outlined in the Master Standard for Architecture programs in Belarus reflect a strong alignment with LCA and CE concepts. These competencies emphasize the need for sustainable, innovative, and interdisciplinary approaches in architectural design, highlighting the importance of integrating LCA and CE principles into architectural education and practice.

Architectural education in China typically involves an undergraduate and graduate program in architecture at colleges or universities. Students may focus on traditional Chinese architecture, as well as modern international architecture and sustainable architecture. In 2024 the international university rating system QS included fourteen architectural education programs in top 240 with Tsinghua University on the 8th

position in the World Rating (https://www.gotouniversity.com/universities-ranking/subject/architecture/china/qs).

Bachelor programs are available in several fields of architecture such as: arts, landscaping, urban planning, civil engineering etc. The program syllabus is more flexible and includes many electives, allowing for tailored student specialisation. The 2022 program of the Faculty of Architecture of Hong Kong University was analyzed as a typical syllabus. Core courses include Faculty Interdisciplinary (FIC) courses, which are a year-long initiative aimed at first-year undergraduate students. These courses are developed in collaboration with all specialized departments of the faculty and are aimed at collaborative thinking and group activities. These courses provide twelve credits and the last two semesters. Most courses provide 6 credits, which include 24–36 contact hours. The courses are grouped depending on area of study: History and Theory; Urbanization and Habitation; Technology and Sustainability; Digital Media and Fabrication; Practice and Management.

Links between LCA and CE concepts in elective courses are available in Table 3.

Table 3 Links between LCA and CE concepts in elective courses in the bachelor syllabus of the Faculty of Architecture of Hong Kong University

Courses of bachelor program	Links with LCA and CE
Architecture and the City	This course offers a platform to integrate LCA and CE principles by discussing sustainable urban development, eco-friendly infrastructure, and design considerations that embrace circularity such as incorporating recycled materials and designing for disassembly
Sustainable Building Systems	This course directly addresses sustainability within the built environment, offering an ideal platform to incorporate LCA and CE principles. Topics may include life cycle analysis of building materials, energy efficiency, and the adoption of circular economy practices in building system design
Design Research on Architecture and the Environment	This course can serve as a bridge to integrating LCA and CE by focusing on environmental impact assessment, sustainable architectural design, and the consideration of material life cycle aspects in design decisions
Building Structures and Systems	This course provides an opportunity to intertwine LCA and CE through discussions on sustainable material usage, energy-efficient building systems, and the integration of circular design principles in structural and systems engineering
Urban Studies and Urbanism Field Workshop	These courses offer possibilities to integrate LCA and CE by exploring sustainable urban planning, circular urban systems, and the environmental impacts of urban development, including the life cycle analysis of urban infrastructure and the implementation of circular economy practices in urban design

Source Developed by authors using syllabuses for the degree of bachelor of arts in architectural studies of university of Hong Kong (Syllabuses 2022)

The syllabus also includes a special module focused on sustainability (Technology and Sustainability). Table 4 suggests links with LCA and CE concepts in this module.

In recent years, the field of architecture has shifted towards sustainability, driven by the urgent need to address environmental challenges. As a response, the integration of LCA and CE outcomes into master programs for architects in China has emerged as a pivotal strategy to cultivate a new generation of environmentally conscious and innovative architectural professionals. This integration equips students with the knowledge and skills to employ LCA methodologies for evaluating the environmental impacts of architectural designs and to embrace circular economy principles for promoting resource efficiency, waste reduction, and sustainable material use. The following discussion articulates the significance of integrating LCA and CE outcomes into master programs for architects in China and its implications for shaping the future of sustainable architectural practice. The role of the Communist Party of China (CPC) in the integration of LCA and CE outcomes into educational programs for architects in China is multifaceted. The CPC plays a crucial

Table 4 Links between courses and LCA and CE outcomes from curriculum of bachelor of arts in architectural studies of university of Hong Kong

Discipline	Links with LCA and CE
Topics in architectural history and theory	This course may offer opportunities to explore historical and theoretical aspects of sustainable architecture's, potentially integrating LCA and CE through discussions on the historical context and evolution of sustainable design principles and circular economy concepts
Inter cities	This course, although focused on practice and management, can potentially incorporate LCA and CE when addressing inter-city planning and sustainable urban development, possibly emphasizing circular economy practices in urban management and the application of LCA in inter-city infrastructure decisions
Designing care in commons	This module presents opportunities to integrate LCA and CE through the exploration of sustainable and circular design principles within communal spaces, potentially including discussions on the life cycle impact of materials and the implementation of circular economy practices in communal design
Nordic latitudes	This course provide avenues for discussions on sustainable design practices and eco-friendly approaches in architectural design, potentially integrating LCA and CE through explorations of sustainable design principles derived from Nordic environmental conditions and the application of circular design concepts
Impossible architecture	This module might offer opportunities to explore innovative and sustainable architectural designs, potentially incorporating LCA and CE through discussions on the feasibility and life cycle implications of "impossible" architecture and the application of circular principles in architectural innovation

Source Developed by authors using Syllabuses for the Degree of Bachelor of Arts in Architectural Studies of University of Hong Kong (Syllabuses 2022)

role in formulating national policies and guidelines related to education and environmental sustainability. Through its policies, the CPC can set the agenda for integrating sustainability concepts, including LCA and CE outcomes, into educational curricula in alignment with national development goals, such as the promotion of ecological civilization and green development. Some of correlations between the courses of master curriculum of Tsinghua University is available in Table 5.

The integration of LCA and CE outcomes into master programs for architects in China holds substantial promise for advancing sustainable architectural practice and fostering a new cadre of professionals equipped to address complex environmental challenges. By systematically embedding LCA methodologies and CE principles into the curriculum, students gain a holistic understanding of the environmental implications of architectural decisions, learn to leverage sustainable material choices, and adopt circular strategies that minimize waste and resource depletion. Furthermore, this integration prompts students to approach architectural design as a conduit for environmental stewardship, emphasizing the interconnectedness of buildings with natural systems and the imperative to mitigate their ecological footprint.

As a result, graduates emerge with the expertise to envision and implement architecture that not only meets the present needs of society but also conserves resources

Table 5 Correlations between the courses of master curriculum of Tsinghua University

Courses of Master program	Correlation with LCA and CE
Building design and design methodology	During the design process, students can integrate LCA methodologies to assess the environmental impacts of building materials and design choices. Circular economy principles can also be incorporated into the design methodology to promote sustainable material use and end-of-life strategies for buildings
Site architectural design and site development	Students can apply LCA to evaluate the environmental performance of site development strategies and incorporate circular economy principles to minimize waste generation and promote efficient resource use on the site
City integrated design/Eco cities	In this studio, students can utilize LCA to assess the environmental impacts of city-scale design interventions and apply circular economy principles to promote resource efficiency and closed-loop systems within eco city designs
Systems design/Green building	LCA can be integrated into the design process to evaluate the life cycle environmental impacts of green building systems and materials. Circular economy principles can also be employed to enhance the sustainability of systems design and promote the reuse and recycling of building components
Green and sustainable building methods	This course directly aligns with LCA and CE outcomes by focusing on sustainable building methods. Students can learn to apply LCA to assess the environmental performance of green building strategies and integrate circular economy principles to promote sustainable material use and resource efficiency

Source Developed by authors using Tsinghua University. School of Architecture. Curriculum of English Master program in architecture (Curriculum n/d)

for future generations. Moreover, through the cultivation of a sustainability-oriented mindset, they are primed to become catalysts for positive change within the architectural field, influencing industry practices, policy frameworks, and societal perceptions of the built environment. The integration of LCA and CE outcomes into master programs for architects in China represents a pivotal step towards fostering a generation of architects who are committed to sustainability and are capable of redefining the trajectory of the architectural profession in alignment with the imperatives of a circular and ecologically responsible future.

5 Results of Survey and Interviews

For the purposes of this study, a group of 3rd and 4th year students of architecture from National Technical University (Belarus), as well as a group of Chinese master students studying "Project Management" at the School of Business of Belarusian State University, which previously got bachelor diploma in China were surveyed. The questionnaire with a concentration on LCA and CE outcomes included 26 questions grouped into 6 sections:

- General section
- Curriculum analysis
- Teaching and learning methods
- Student and Alumni Perspectives
- Industry Relevance and Employability
- Recommendations and Future Directions.

The selection included 60 students from Belarus representing National Technical University and 26 master students from China, who previously studied architecture at Chinese universities. All students mentioned that they had courses, that directly or indirectly included principles of LCA and CE.

Based on the responses to the question "How are sustainability principles integrated into core architectural courses, beyond dedicated LCA and CE courses", it is evident that sustainability principles are effectively woven into the curriculum through various means. The survey results show sustainability principles are being integrated into core architectural courses. It's clear that there is a multifaceted approach to this integration, as half of the students emphasized the integration of sustainability principles through topics such as passive design strategies, energy-efficient building systems, and sustainable material selection (40 votes), while the other half highlighted the significance of topics like biophilic design, green building certifications, and the consideration of life cycle thinking in architectural decision-making processes in addition to dedicated LCA and CE courses (39 votes).

Another indicative question was "Are there interdisciplinary or collaborative elements in the curriculum that expose architectural students to LCA and CE concepts in conjunction with other fields (e.g., engineering, environmental science, economics)?" Over 90% of respondents mentioned that curriculum in their universities incorporates interdisciplinary elements by offering joint modules with engineering and environmental science faculties, where students are exposed to LCA and CE concepts within the context of building physics, sustainable construction materials, and environmental impact assessments, fostering a collaborative approach to understanding and addressing sustainability challenges."

The results regarding learning methods employed to facilitate students' understanding and application of LCA and CE concepts within architectural design studio courses were mixed (see Table 6).

The high percentages in both Belarus (73%) and China (86%) indicate that design projects are a popular and effective method for integrating LCA and CE principles. This hands-on approach allows students to actively apply sustainability concepts in real-world design scenarios, enhancing their practical understanding. The disparity in responses between Belarus (16%) and China (73%) regarding interactive critiques highlights a potential area for improvement in the Belarusian educational context. Interactive critiques play a crucial role in fostering critical thinking and

Table 6 Results of surveying the students from Belarus and China on the question "What learning methods are employed to facilitate students' understanding and application of LCA and CE concepts within architectural design studio courses?" (multi selection allowed)

Questions	Belarus (%)	China (%)
Within architectural design courses, learning methods such as design projects, where students are encouraged to develop architectural solutions that integrate LCA and CE principles, provide opportunities for hands-on exploration and application of sustainability concepts in real-world design projects	73	86
Interactive critiques and feedback sessions are utilized in architectural design courses, allowing students to present and defend design proposals incorporating LCA and CE principles, fostering critical thinking and iterative refinement of sustainable design solutions	16	73
The integration of guest lectures by industry professionals, who share their experiences and best practices related to LCA and CE in architectural design, enhances students' understanding of how sustainability principles can be actualized within the design studio setting	53	100
Collaborative projects with engineering and environmental science disciplines, within the context of architectural design studio courses, provide students with exposure to interdisciplinary perspectives on LCA and CE concepts, promoting a holistic understanding of sustainability in building design and implementation	16	73
Design studio courses incorporate the use of advanced software and tools that enable students to conduct life cycle assessments, and analyse environmental impacts, and explore circular economy strategies within their design projects, offering practical applications of LCA and CE in architectural ed'ucation	73	86

Source Developed by the authors

refining sustainable design solutions through constructive feedback sessions. The overwhelming agreement in China (100%) on the importance of guest lectures signifies a strong endorsement of industry insights in enhancing students' understanding of LCA and CE. In Belarus (53%), there is still room for growth in leveraging industry expertise to enrich sustainability education within architectural design studios. The majority agreement in both Belarus (73%) and China (86%) underlines the importance of incorporating advanced software for conducting life cycle assessments and exploring circular economy strategies in design studio courses. Practical applications of LCA and CE through technology enhance students' skills and readiness for sustainable architectural practice.

Another important result of this survey is that lecturers and professors note increasing of student's interest in selecting the bachelor project of master thesis topics related with SDGs in general, and LCA and CE in particular. According to current practice in Belarus, universities suggest students a list of topics depending on sphere of expertise of supervisors. Topics covering LCA and CE outcomes for bachelor and master programs are 20–25% of total list of 40–50 suggested topics. The results of surveying are available on Fig. 1.

Both Belarus and China show a moderate increase in student interest in selecting topics related to SDGs, LCA, and CE over the three years covered in the survey. While the percentages fluctuate slightly, the overall trend suggests a growing inclination among students towards sustainability-focused research and projects. However, approximately another 25% of students select topics related with Building Information Modeling (BIM). BIM allows for the creation of a digital representation of a building throughout its entire lifecycle. By incorporating LCA into BIM models, architects can analyze the environmental impacts of materials and construction

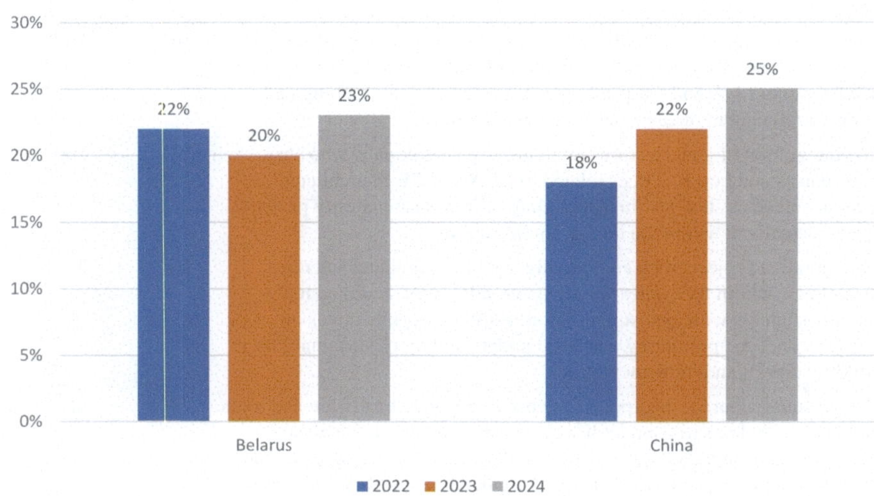

Fig. 1 Student's interest in selecting the project of master thesis topics related with LCA and CE in particular

processes from inception to demolition, enabling more informed decisions to reduce overall environmental footprints.

6 Discussion and Suggestions

Integrating LCA and CE concepts into interdisciplinary collaboration and curriculum enrichment is a proactive approach that aligns with the principles of holistic and sustainable problem-solving and design. These concepts promote a comprehensive view of products and systems, emphasizing the importance of considering their entire lifecycle and striving to minimize environmental impacts. By incorporating LCA and CE into educational settings and fostering interdisciplinary collaboration, institutions can effectively prepare students to address sustainability challenges and contribute to a more sustainable built environment. This approach also underscores the significance of equipping future architects and designers with the knowledge and skills necessary to create environmentally responsible and resilient solutions.

The analysis of syllabus in this paper discovered regional perspectives on the integration of LCA and CE into architectural education and practice in China and Belarus offer valuable insights into the unique challenges and opportunities present in each country. By examining the approaches and trends in sustainability education and implementation in these regions, a nuanced understanding of the advancements and obstacles can be gained. In China, rapid urbanization poses challenges in sustainable infrastructure development, necessitating a robust integration of LCA and CE in architectural education to address resource efficiency and environmental impacts. The country's scale and diversity of architectural projects require innovative approaches for effectively integrate LCA and CE across a wide spectrum of building types and contexts.

Belarus has shown a commendable level of comprehension and awareness of LCA and CE concepts among students of architecture. This understanding reflects a positive trend towards the integration of sustainability principles into architectural education, empowering future architects to approach design and construction with a holistic and environmentally conscious mindset. The evident grasp of LCA and CE concepts among Belarusian architecture students bodes well for the cultivation of a new generation of professionals equipped to address sustainability challenges and contribute to the creation of more environmentally responsible built environments.

The analysis of educational standards in Belarus has revealed a notable alignment between certain competencies and the principles of LCA and CE. Specifically, competencies related to optimizing the architectural environment, considering environmental impacts, and applying principles for long-term development resonate well with the goals of LCA and CE. These competencies underscore the significance of sustainability, resource efficiency, and the environmental impact of architectural and urban design solutions, emphasizing core principles that are intrinsic to LCA and CE frameworks. This synergy between educational competencies and sustainable principles further emphasizes the potential for Belarusian architectural education to cultivate a cohort of professionals well-versed in implementing sustainable, lifecycle-centric approaches in architectural and urban development practices.

On the other hand, the analysis of syllabus has discovered deep integration of LCA and CE outcomes into bachelor and master programs for architects in China holds significant promise for advancing sustainable architectural practice. By systematically embedding LCA methodologies and CE principles into the curriculum, students are poised to gain a comprehensive understanding of the environmental implications of architectural decisions. This was confirmed with the result of Chinese student's surveying, that discovered deep understanding of LCA and CE concepts.

Based on analysis above, the following recommendations can be introduced to Belarussian and Chinese educators on enhancing the integration of LCA and CE concepts into educational programs in architecture:

1. Curriculum Enhancement. Integrate specific courses or modules dedicated to LCA and CE within the general compulsory courses. This would ensure that students receive in-depth education on these concepts and their practical application in architectural design and construction from the early start of educational programs.
2. Practical Application. Incorporate real-world case studies and practical projects that require students to apply LCA and CE principles in architectural design in a form of CE oriented study projects
3. Faculty Training and Development. Universities should provide training and professional development opportunities for faculty members to enhance their understanding of LCA and CE, ensuring that they are equipped to effectively teach and mentor students in these areas. This hands-on approach can help students understand the practical implications and challenges related to sustainable architectural practices.
4. Collaboration with Industry Partners. As students noted in survey, guest lectures are an important piece in LCE and CE oriented courses. Universities should foster collaborations with architectural firms, construction companies, and sustainability-focused organizations to provide students with exposure to real-world projects and initiatives that integrate LCA and CE principles.
5. Interdisciplinary Collaboration. Encourage interdisciplinary collaboration with other relevant fields such as engineering, environmental science, and urban planning to provide a more holistic understanding of LCA and CE within the context of architectural education.

The integration of LCA and CE concepts into educational programs in architecture represents a crucial step towards nurturing a new generation of architects equipped to tackle complex environmental challenges with sustainable and innovative solutions. By enhancing curriculum, promoting practical application, empowering faculty, fostering industry collaboration, encouraging research and innovation, raising awareness, and facilitating interdisciplinary collaboration, educational institutions can effectively advance the integration of LCA and CE concepts. This concerted effort in Belarus can elevate the educational experience for architecture students but also to shape the future of architectural practice, positioning the industry to lead the way in sustainable and environmentally conscious design principles.

The next steps for the research include a deeper analysis of students' and educators' surveys, an analysis of course outcomes and their relations, an analysis of guest lecture statistics, and a further analysis of LCA and CE-oriented graduation project selection for a deeper understanding of the reasons for such choice from the side of both students and supervisors.

7 Conclusions

This paper was aimed at pursuing a comparative analysis of LCA and CE integration in architectural education based on a case study of universities in Belarus and China. Initially, it was linked with the following SDGs: SDG 4: Quality Education; SDG 9: Industry, Innovation, and Infrastructure; SDG 11: Sustainable Cities and Communities; SDG 12: Responsible Consumption and Production; SDG 13: Climate Action; SDG 16: Peace, Justice, and Strong Institutions. It was concluded that integrating LCA and CE principles into contemporary architecture education and practice contributes to advancing multiple SDGs related to sustainable infrastructure, climate action, responsible consumption, education, and the creation of sustainable, inclusive, and resilient communities.

The research included several steps: evaluation of the curricula of architectural education programs in selected universities in Belarus and China to identify the extent of integration of LCA and CE principles; conducting the interviews and survey of faculty members, students, and industry professionals in architectural education to gain insights into the implementation and reception of LCA and CE concepts; suggesting recommendations and best practices for enhancing the integration of LCA and CE in architectural education.

The comparative evaluation of curricula in architectural education programs in Belarus and China has unveiled varying degrees of integration of LCA and CE principles. This analysis comprehensively compares course offerings, modules, and learning objectives related to LCA and CE, highlighting distinct approaches to sustainability education in the two countries. While in China, there are special modules oriented on LCA and CE, in Belarus these outcomes are not directly included into curricula and are available only as outcomes of specific courses.

Interviews and surveys conducted among faculty members, students, and industry professionals have provided invaluable insights into implementing and receiving LCA and CE concepts within architectural education. Qualitative data gathered from these stakeholders offers a multifaceted understanding of the challenges, successes, and practical implications of integrating sustainability principles into architectural curricula.

A crucial outcome of this research is the development of recommendations and best practices based on comparative analysis and stakeholder feedback. These recommendations are derived from concrete findings and aim to enhance the integration of LCA and CE in architectural education, address specific challenges, and foster substantial improvements in sustainability education within the architectural curriculum.

The research has identified potential areas of collaboration between universities in Belarus and China, with a focus on fostering sustainable architectural education. Strategies, curricular enhancements, and collaborative initiatives between these institutions promise to further promote sustainability education and facilitate the exchange of best practices in integrating LCA and CE principles.

The research outcomes underscore the imperative of continually evolving and refining the integration of LCA and CE in architectural education. This imperative is vital for preparing future professionals to effectively address global environmental challenges and advance sustainable architectural practices. The insights gained from this comparative analysis provide a foundation for ongoing advancements in sustainability education within the architectural curriculum, with far-reaching implications for the industry and society.

References

Bassachs M, Cañabate D, Serra T, Colomer J (2020) Interdisciplinary cooperative educational approaches to foster knowledge and competences for sustainable development. Sustainability 12:8624. https://doi.org/10.3390/su12208624

Benachio G, Freitas MC, Tavares S (2020) Circular economy in the construction industry: a systematic literature review. J Cleaner Prod 260:121046. https://doi.org/10.1016/j.jclepro.2020.121046

Bertel L, Winther M, Routhe H, Kolmos A (2021) Framing and facilitating complex problem-solving competences in interdisciplinary megaprojects: an institutional strategy to educate for sustainable development. Int J Sustain Higher Educ. ahead-of-print. https://doi.org/10.1108/IJSHE-10-2020-0423

Butt, Dimitrijevic (2022) Multidisciplinary and transdisciplinary collaboration in nature-based design of sustainable architecture and urbanism. Sustainability 14:10339. https://doi.org/10.3390/su141610339

Curriculum (n/d) Tsinghua University. School of Architecture. Curriculum of English Master program in architecture. http://www.arch.tsinghua.edu.cn/column/Curriculum

Faculty of Architecture. University of Hong Kong. Review 2023 and prospectus 2024. https://foa-media.arch.hku.hk/media/upload/2024/03/HKU_Architecture_Prospectus24_MAR24.pdf

Fan, Fang (2020) Circular economy development in China-current situation, evaluation and policy implications. Environ Impact Assess Rev 84:106441. https://doi.org/10.1016/j.eiar.2020.106441

Filho W, Dibbern T, Ruiz de Maya S, Alarcón-del-Amo M-C, Rives L (2024a)The contribution of universities towards education for sustainable development. https://doi.org/10.1007/978-3-031-49853-4

Filho W, Dinis M, Salvia A, Sierra J, Vasconcelos H, Henderson-Wilson C, Diatta S, Kumar T, Meirelles M, Carvalho F (2024b) Assessing climate change and health provisions among staff in higher education institutions: a preliminary investigation. PLOS ONE 19. https://doi.org/10.1371/journal.pone.0304019

França W, Barros M, Salvador R, Francisco A, Moreira M, Piekarski C (2021) Integrating life cycle assessment and life cycle cost: a review of environmental-economic studies. Int J Life Cycle Assess. https://doi.org/10.1007/s11367-020-01857-y

Giesenbauer, Mueller-Christ (2020) University 4.0: promoting the transformation of higher education institutions toward sustainable development. Sustainability 12:3371. https://doi.org/10.3390/su12083371

Gorbatchev, Zenchanka (2020) Current approaches to waste management in Belarus. In: Leal Filho W, Borges de Brito P, Frankenberger F (eds) International business, trade and institutional sustainability. World Sustainability Series. Springer, Cham. https://doi.org/10.1007/978-3-030-26759-9_9

Hollberg, Ruth (2016) LCA in architectural design—a parametric approach. Int J Life Cycle Assess 21. https://doi.org/10.1007/s11367-016-1065-1

Joensuu T, Edelman H, Saari A (2020) Circular economy practices in the built environment. J Cleaner Prod 276:124215. https://doi.org/10.1016/j.jclepro.2020.124215

Merli R, Preziosi M, Acampora A (2018) How do scholars approach the circular economy? A systematic literature review. J Cleaner Prod 178:703–722. https://doi.org/10.1016/j.jclepro.2017.12.112

Nerini F, Hughes N, Cozzi L, Cosgrave E, Howells M, Tavoni M, Tomei J, Zerriffi H, Milligan B (2019) Connecting climate action with other sustainable development goals. Nat Sustain 2. https://doi.org/10.1038/s41893-019-0334-y

Pieroni M, McAloone T, Pigosso D (2019) Business model innovation for circular economy and sustainability: a review of approaches. J Cleaner Prod 215. https://doi.org/10.1016/j.jclepro.2019.01.036

Sanchez B, Rausch C, Haas C (2019) Deconstruction programming for adaptive reuse of buildings. Autom Constr 107:102921. https://doi.org/10.1016/j.autcon.2019.102921

Sanchez, Haas (2018) A novel selective disassembly sequence planning method for adaptive reuse of buildings. J Cleaner Prod 183. https://doi.org/10.1016/j.jclepro.2018.02.201

Standard (2018) Educational standard of higher education. First stage. Speciality 1-69 01 02 Architecture. Available at: https://www.bsut.by/images/MainMenuFiles/Obrazovanie/Studentam/umkd/umo/os/os_1_69_01_01_211218.pdf (in Russian)

Standard (2023) Educational standard of special higher education. Specialty 7-07-0731-01 Architecture. Qualification Architect Degree Master. Available at: https://www.bsut.by/images/MainMenuFiles/Obrazovanie/Studentam/umkd/umo/os/os_7-07-0732-01_281223.pdf (in Russian)

Syllabuses (2022) Syllabuses for The Degree of Bachelor of Arts in Architectural Studies of University of Hong Kong. Available at: https://foa-media.arch.hku.hk/media/upload/2015/01/Revisions-in-BAAS-Syl-from-2022-intake-and-thereafter-clean-copyamended-on-20231113-with-track-changes-2023.12.07-clean-copy.pdf

Tsinghua (n/d) Tsinghua University. School of Architecture. English Program for Masters in Architecture. https://yz.tsinghua.edu.cn/__local/4/F5/A3/C95B313626B6DEBBDD13634AA7B_75634F02_196F995.pdf

UN (n/d) Sustainable development goals. Available at: https://www.undp.org/sustainable-development-goals. Accessed on 28/05/2024

Vandervaeren C, Galle W, Stephan A, Temmerman N (2022) More than the sum of its parts: considering interdependencies in the life cycle material flow and environmental assessment of demountable buildings. Resources Conserv Recycl 177:106001. https://doi.org/10.1016/j.resconrec.2021.106001

Veckalne, Tambovceva (2022) The role of digital transformation in education in promoting sustainable development. Virtual Econ 5:65–86. https://doi.org/10.34021/ve.2022.05.04(4)

Viera Trevisan L, Filho W, Pedrozo E (2024) Transformative organisational learning for sustainability in higher education: a literature review and an international multi-case study. J Cleaner Prod 447:141634. https://doi.org/10.1016/j.jclepro.2024.141634

Wang H, Schandl H, Wang X, Ma F, Yue Q, Wang G, Wang Y, Wei Y, Zhang Z, Zheng R (2020) Measuring progress of China's circular economy. Resources Conserv Recycl 163:105070. https://doi.org/10.1016/j.resconrec.2020.105070

Zenchanka S, Antsipenka N, Busygin D, Korshuk E, Sidsaph H (2020) Circular economy: can belarus close the loop? Environ Res Eng Manage 76:7–19. https://doi.org/10.5755/j01.erem.76.4.24433

The Role of Venture Businesses and Startups in Life-Cycle Assessment and the Circular Economy: A Case Study of China

Hantongzhou Chen, Kirill Savrassov, Nikolai Gorbatchev, and Pengfei Zhang

Abstract The European Union (EU) has a long history of supporting the circular economy (CE) and life-cycle assessment (LCA), with progressive policies and regulations. The EU actively collaborates with developing countries, such as China, to exchange ideas and promote capacity building. The Chinese government is supportive in adopting CE concepts and incorporating LCA in policy frameworks and standards. It becomes a trend that many venture businesses and startups are driving LCA and CE practices as their prevailing business models and inspire companies to reevaluate processes and offerings by developing novel solutions and sparking discussions on the importance of sustainability and circularity across industries. In 2022 to 2023, the amount of the initial public offerings (IPOs) by venture businesses and startups aimed at supporting CE and LCA approach in China, considerably exceeded this indicator in the EU. Based on stock exchange data and surveying the companies' representatives, this paper compares the policies and investments currently made into the CE and LCA via stock markets. A comparison of investments in China and EU reveals market trends or a growing interest in companies that prioritize CE and LCA principles, showcasing increased recognition of their importance in achieving long-term sustainability. In discussion, authors consider that despite some caveats (not all companies go public; different companies may have dissimilar sizes and sustainability impacts; fluctuations in market conditions etc.), tracking investments

H. Chen (✉)
Belarusian State University, Minsk, Belarus
e-mail: chtz945@gmail.com

K. Savrassov
Phoenix CRetro Ltd, Chesham, UK

N. Gorbatchev
International Institute of Management and Business, Minsk Branch of Plekhanov Russian University of Economics, Minsk, Belarus

P. Zhang
International Institute of Management and Business, Minsk, Belarus

L. Brandli et al. (eds.), *The Contribution of Life Cycle Analyses and Circular Economy to the Sustainable Development Goals*, World Sustainability Series,
https://doi.org/10.1007/978-3-031-85300-5_24

provides insights into sustainability and growth in this area. The paper is aimed on analysis of current achievements and challenges of CE and LCA policies in China.

Keywords Circular economy (CE) investments · Life-cycle assessment (LCA) · Developing countries · Venture businesses · Startups

1 Introduction

Since the inception of the Sustainable Development Goals (SDGs), there has been a collective global endeavor to integrate sustainability into various sectors worldwide (UN 2015). The SDGs recognize the importance of sustainable consumption and production practices, as well as the need to minimize environmental impacts throughout the life cycle of products and services (Leal Filho et al. 2023). Several SDG's targets specifically aim at addressing aspects related to LCA and the CE:

- SDG 12 (Responsible Consumption and Production): Target 12.2 focuses on achieving sustainable management and efficient use of natural resources, and Target 12.5 aims to substantially reduce waste generation through prevention, reduction, recycling, and reuse.
- SDG 9 (Industry, Innovation, and Infrastructure): Target 9.4 promotes sustainable industrialization and encourages the adoption of environmentally friendly technologies and practices, and Target 9.8 aims to promote sustainable resource use and efficient production processes.
- SDG 13 (Climate Action): Target 13.3 emphasizes the implementation of policies and measures to reduce greenhouse gas emissions, and Target 13.2 highlights the integration of climate change measures into national policies, strategies, and planning.
- SDG 14 (Life below Water): Target 14.1 focuses on preventing and reducing marine pollution, including minimizing the impacts of land-based activities on oceans through sustainable consumption and production.
- SDG 15 (Life on Land): Target 15.2 aims to promote sustainable management and conservation of forests, including reducing deforestation and forest degradation.

LCA and CE are valuable tools and approaches that support the implementation of the SDGs (Ortiz-de-Montellano et al. 2023). By facilitating a better understanding of environmental impacts, promoting sustainable practices, and contributing to responsible consumption and production patterns, LCA and CE contribute to the broader sustainability agenda outlined by the SDGs (Backes and Traverso 2022).

The integration of LCA and the CE is crucial for assessing and improving the environmental performance of products and processes. LCA provides a scientific approach to quantify the impacts (Wagner et al. 2021), while the CE guides the transformation towards a more sustainable and resource-efficient system (Kirchherr et al. 2017). By applying LCA principles, the CE can identify areas of improvement and help prioritize actions to reduce the environmental footprint of economic activities. It is important to clarify that while many venture businesses and startups are increasingly aware of and embracing the principles of CE and LCA, it would be inaccurate to claim that all of them are active drivers of these concepts. Rather, a growing number of innovative ventures are recognizing the importance of integrating sustainability and LCA/CE principles into their business models, product development, and investment strategies. This trend is manifesting through various initiatives and efforts aimed at creating more sustainable and resource-efficient solutions, fostering collaborations with researchers and experts in the field, and actively seeking to address environmental challenges through their products and operations. Venture businesses and startups, through their innovation, sustainability focus, and collaborative networks, play a crucial role in advancing LCA and driving the transition towards the CE (Meskers et al. 2024).

The aim of this paper is to analyze the role and impact of venture businesses and startups in promoting LCA and CE principles with focus on China. Its main objectives include:

- Review the current research in field of CE, LCA and SDGs and their influence on investor behavior and initial public offerings (IPOs) of venture businesses promoting CE and LCA principles, with a focus on stock exchanges with the focus on China.
- Analysis of relationship between the growth of IPOs of businesses focused on LCA and CE projects and investor demand, market sentiment, and the increasing interest in sustainable initiatives, underscoring the market's emphasis on environmentally responsible practices and circular approaches to resource management on example of selected companies in China.
- Analysis of the findings of the survey conducted among Chinese business professionals at the China-Arab States Expo (CAS 2023) to understand their perspectives on the implementation of Life Cycle Assessment (LCA) and Circular Economy (CE) principles in venture businesses

2 Literature Review

The relationship between CE and LCA and their influence on Sustainable Development (SD) emphasizes how the principles of the CE can contribute to achieving the SDGs. Although many indicators in the CE domain claim to assess sustainability, they often overlook the three pillars of sustainability (De Oliveira and Oliveira 2023). Therefore, when considering the adoption of the CE as a strategy, it is crucial to thoughtfully evaluate its potential to contribute to sustainability performance. This evaluation should include the use of CE indicators that align with the chosen strategies, providing valuable support for professionals and decision-makers. To add a reference or two to this paragraph.

The concept of the CE encompasses a wide range of ideas, but it can be defined by three core principles. The first principle aims to protect and enhance natural capital by managing finite stocks and maintaining a balance in the use of renewable resources. The second principle focuses on maximizing the productive utilization of resources by fostering the circulation of products, components, and materials at their highest levels of utility, whether technical or biological. Lastly, the third principle entails enhancing the efficiency of the system by identifying and eliminating negative externalities from the outset (Ellen MacArthur Foundation 2022).

Moreover, CE and LCA principles can be applied at different levels, including individual, corporate, and global levels (Rigamonti and Mancini 2021; Samani 2023). At the individual level, the CE involves the adoption of circular practices by individual consumers. This include reducing waste through mindfull consumption and purchasing decisions, reusing or repurposing products, and recycling materials. Individuals can also engage in sharing or swapping platforms, repair services, and buying second-hand items to extend the life cycle of products (Ferreira et al. 2024). By embracing circular principles at an individual level, consumers can contribute to minimizing resource consumption, reducing waste generation, and fostering a more sustainable lifestyle.

At the corporate level, the CE refers to the integration of circular practices within organizations and businesses (Scarpellini et al. 2020). This includes implementing strategies such as eco-design, where products are designed for durability, reparability, and recyclability. Companies can also adopt business models like product-as-a-service, where customers pay for the use or performance of a product rather than owning it. Additionally, corporate CE initiatives involve implementing closed-loop supply chains, where waste is minimized through recycling, remanufacturing, and reverse logistics (Kazançoğlu et al. 2021). By implementing circular practices at the corporate level, businesses can reduce their environmental impact, enhance resource efficiency, and create new economic opportunities (Barros et al. 2021).

At the global level, the CE encompasses the coordination and collaboration among governments, international organizations, and stakeholders to promote circularity on a larger scale (Lacy et al. 2020). This includes developing policies and regulations that incentivize circular practices, fostering cross-sector collaborations, and promoting knowledge sharing and best practices at the international level (Ogunseitan et al. 2022). Global initiatives aim to create a supportive environment for circular business models, encourage sustainable consumption and production patterns, and address waste management challenges. By embracing circularity at the global level, societies can work towards a more sustainable and resilient economy, reducing resource depletion and environmental degradation (OECD 2021).

Venture businesses and startups aimed at supporting CE and LCA businesses are a growing trend in the industry (Han et al. 2023). They integrate the concepts of CE and LCA into its business model, manufacturing processes, supply chain management, product design, and overall decision-making (Nakano and Hirao 2011). These businesses prioritize sustainability, resource efficiency, waste reduction, and environmental impact mitigation in alignment with CE and LCA principles. Such businesses focus on developing products or services that follow circular economy principles, strive to minimize environmental impacts throughout their life cycle, and actively work towards achieving sustainability goals (Dieterle and Viere 2021). They also engage in conducting life cycle assessments for their own products, offering consulting services to other businesses looking to transition to a circular economy model, or providing expertise in LCA methodologies to support sustainable decision-making. These businesses play a crucial role in promoting sustainability and circular practices by providing innovative solutions to help businesses? transition towards more sustainable and environmentally friendly practices (Försterling et al. 2023). One area of focus is recycling technologies. These venture businesses develop and implement advanced recycling technologies that enable the efficient and effective recycling of various materials (BASF 2023). This also include developing new methods for recycling plastics, metals, and other waste materials, as well as creating advanced sorting and separation technologies to improve recycling rates.

Another area where these venture businesses are making a significant impact is in the development of biodegradable materials (Ali et al. 2023). This involves work on creating alternative materials that are renewable, compostable, and eco-friendly, reducing reliance on traditional materials that are harmful to the environment. This includes the development of bioplastics, biodegradable packaging, and other sustainable materials.

Furthermore, these venture businesses are also involved in optimizing supply chains and improving the overall efficiency of operations (He et al. 2020). They help companies identify ways to reduce waste, minimize energy consumption, and optimize resource utilization throughout the product lifecycle. This can involve implementing technologies such as blockchain for traceability, analytics for data-driven decision-making, and automation for streamlining processes.

Initial public offerings (IPOs) provide capital for the startups and venture businesses to invest in research and development, expand their operations, and hire talented professionals. Investors are attracted to these IPOs due to the potential for high returns, as many of these companies have disruptive technologies and innovative business models (Wang et al. 2022). The connection between IPOs and research related to CE, LCA and the SDGs may not be immediately apparent but is based on several interrelated factors. Research and innovation in CE, LCA, and sustainable development are increasingly shaping investor preferences. IPOs represent a critical stage for companies seeking investment, and those demonstrating alignment with CE, LCA, and SDGs may attract greater investor interest and support (Numa et al. 2023). Businesses integrating CE and LCA principles are often better prepared to address environmental, social, and governance (ESG) risks and regulatory requirements. This positive positioning can enhance market resilience, appealing to IPO investors seeking sustainable and long-term value. Analysis of IPO volumes of startups and venture businesses can provide valuable insights for CE, LCA, and SDG research (Malen and Marcus 2017). For instance, it can reveal the impact of venture investments on firm growth and maturity, the role of IPOs in exit strategies, and the contribution of startups to sustainable development. Additionally, the study of IPO volumes can help identify patterns and trends in the startup ecosystem, such as the concentration of IPOs in certain industries or geographic regions.

The major stock exchange locations for IPOs of startups and venture businesses can vary depending on several factors, which could include the country of origin and the target market. Several stock exchanges in the EU provide platforms for smaller and growing companies to conduct IPOs. Examples are Euronext Growth in Paris (Euronext, n/d), and AIM Italia in Milan that offer listing opportunities for startups and venture businesses with sustainable business models (AIM Italia, n/d). In China Shanghai Stock Exchange (SSE) is one of the world's largest stock exchanges by market capitalization and is frequently selected by Chinese companies for their IPOs (SSE, n/d). In recent years the Chinese government has implemented policies that encourage domestic companies to list on domestic stock exchanges (Tan 2023). These policies include streamlined approval processes, relaxed listing requirements, and various incentives aimed at supporting the growth of the domestic capital market.

Thus, comparing the venture businesses and investor's behavior as well as IPOs of SDG-friendly businesses at stock exchanges could provide valuable insights into the IPO landscape for SDG-friendly businesses in the EU and China, including the regulatory environment, investor sentiment, market performance, and growth potential. This can also help to assess the importance associated with sustainable and responsible business practices. In this paper the data collection was focused on investors and venture businesses, which promote LCA and CE principles.

3 Methodology

The methodology employed in this chapter incorporates a comprehensive mix of qualitative and quantitative research methods. To begin with, an extensive review of existing research findings was conducted, focusing on key aspects pertaining to the concepts of LCA, the CE, and the SDGs both on a global scale and within the context of China. This step was instrumental in informing the subsequent stages of the research. In order to meet the tasks of the research, the mission, vision and annual reports of selected companies were inspected that they promote the LCA and CE principles.

Upon it, open-source data on selected companies obtained from reliable sources, including official web-sites of stock exchanges was collected and analyzed to reflect current amounts of investments made at main IPO platforms in China. This part of the research included a comparative analysis of IPO approaches in USA, EU and China. This allows for the determination of the preferred industries for investors with higher IPO valuations, and to identify emerging trends and niche areas within LCA and CE venture businesses that have garnered substantial investor attention.

To conduct the quantitative part of the research, a survey was formulated leveraging the aforementioned research findings. The primary goal of the research survey was to evaluate the extent of awareness, comprehension, and implementation of LCA, CE in startups and venture businesses with focus on China. Its objective was to delve into the viewpoints, actions, and perceptions of the respondents concerning these concepts. The survey encompassed the selection of a representative sample of participants from diverse regions and industries from China. The questionnaire included 36 questions grouped into the following sections:

1. Introduction and personal information of respondents
2. Knowledge and awareness about concepts of SDG, LCA, CE
3. The role of government in CE and LCE
4. Relations with SDGs
5. Correlations between IPO and venture businesses related to CE and LCA
6. Future outlook and global perspectives of LCA and CE related startups and venture businesses

The survey was thoughtfully designed in alignment with the research objectives. Each question furnished options from which respondents could select their choices. Examples of questions from different categories are available in Table 1.

The respondents for the survey were identified at the China-Arab States Expo (CAS 2023), that took place in Yinchuan, Ningxia province, China in September 2023. This Expo focused on fields as financial investment, oil and gas, the digital economy, wind power and photovoltaics, and clean energy (Hu 2023). The survey gathered responses from 83 business professionals, regarding the implementation LCA and the CE in venture businesses, with a focus on China. The survey was translated into Chinese to make it accessible by local businesses, and all respondents represent China. The survey responses were entered into a structured format of electronic tables, allowing for efficient data management and analysis. After cleaning the data to remove any inconsistencies, errors, missing values, descriptive statistics, such as mean, median, mode, standard deviation, and frequency distributions have been calculated to summarize the responses and provide an overview of the survey findings. These statistics help in understanding the central tendencies and variability of the data.

The Chinese respondents were distributed between the provinces of China in the following way: Beijing (5%), Guangdong (2%), Guangxi (2%), Hebei (5%), Henan (5%), Hunan (2%), Jiangxi (3%), Liaoning (2%), Neimenggu (Inner Mongolia) (2%), Ningxia (49%), Shandong (5%), Shanghai (10%), Shanxi (5%), Sichuan (2%), and Tianjin (1%). These provinces represent more the half of the population of China and generate over 50% of China's GDP.

The selection is substantiated with the following facts:

- This forum attracts participants who are involved or interested in business activities between Arab countries having high investing potential and China with its technological potential.
- The forum brings together stakeholders who can provide valuable insights into the role of venture businesses and startups in promoting LCA and CE within these contexts.
- The participants represent various industries, backgrounds, and perspectives related to business and sustainability. By targeting this diverse group of Chinese business professionals, it is possible capture a range of experiences and perspectives, which can enhance the robustness and representativeness.

Table 1 Examples of questions by sections from the survey

Questions	Suggested variants
Knowledge and awareness about concepts of SDG, LCA, CE	
How familiar are you with the concept of LCA and CE?	Very familiar … No idea
Have you come across any venture businesses or startups that focus on sustainable solutions in the EU or China?	Yes/no
How important do you think venture businesses and startups are in the transition to a more sustainable economy in China compared to other actors, such as governments or established corporations?	Very important … No at all
Do you agree that venture businesses and startups can effectively drive sustainable consumption patterns?	Strongly agree … strongly disagree
The role of government in CE and LCE	
How effective do you perceive the government of China's collaboration with venture businesses and startups in promoting sustainable practices and innovations related to LCA and CE?	Very effective … No at all
How familiar are you with the policies and initiatives implemented by the Chinese government to support venture businesses and startups involved in LCA and CE?	To high extend … No at all
Relations with SDGs	
To what extent do you believe venture businesses and startups in the EU and China are actively contributing to the achievement of the SDGs through their involvement in LCA and CE practices?	To high extend … No at all
How do you perceive the alignment between the activities of venture businesses and startups in the EU and China, focusing on LCA and CE, and specific SDGs outlined by the United Nations?	Strong alignment … No opinion/Don't know
Correlations between IPO and venture businesses related to CE and LCA	

(continued)

Table 1 (continued)

Questions	Suggested variants
How does the transition of venture businesses involved in CE and LCA practices from private funding to IPO affect their ability to scale their operations and contribute to the achievement of sustainability goals?	Increased access to capital Enhanced visibility and credibility Regulatory compliance and reporting standards Potential market-driven pressures
Are there any notable differences or trends in the IPO activity and valuation of venture businesses engaged in CE and LCA, compared to businesses operating in traditional industries, and what implications does it have on the broader adoption of sustainable practices?	Increasing investor interest and valuation of CE and LCA Differentiated risk perception Sector-specific regulatory and policy support Impact on wider adoption of sustainable practices Investor demand for ESG investments Other

Source Developed by the authors

However, it is important to consider any limitations associated with this sampling choice. The participants' responses may not be representative of the broader population, and some bias may exist due to self-selection or specific event-related factors.

4 Current Support to LCA/CE-Oriented Startups and Venture Businesses

In the EU, there is significant support for startups and venture businesses focused on LCA and the CE (von Kolpinski et al. 2024). The European Commission has been actively promoting and supporting initiatives related to sustainability, resource efficiency, and the CE (Briguglio et al. 2021). One example of support is the Horizon Europe program (Horizon, n/d), which includes funding opportunities for research and innovation projects in various fields, including LCA and the CE. This program offers grants to companies, researchers, and other organizations working on projects that advance the SDGs, including CE practices. Moreover, the EU provides associated financial support through various funding mechanisms and initiatives. For instance, the European Investment Bank offers financing and advisory services to support CE projects (EIB, n/d). Additionally, regional development funds and business support programs in EU member states often prioritize projects that contribute to sustainable development and CE practices. Furthermore, the EU has set specific targets for waste reduction, recycling, and resource efficiency (Frilant et al. 2021). The Waste Framework Directive and the CE Action Plan provide a policy framework to drive the transition towards the CE. By implementing regulations and supporting initiatives, the EU aims to foster innovation and investment in LCA and the CE in various sectors, including manufacturing, construction, and agriculture.

On the other hand, China has shown a growing interest in learning from the experience and practices of the EU when it comes to matters as financing LCA and CE projects (Zhu et al. 2018). The EU has been at the forefront of promoting sustainable development and CE practices, and China recognizes the value of adopting similar approaches.

The Chinese government has been actively promoting sustainability and CE practices since 2000s (Ma et al. 2022). They provide financial support and incentives for LCA and CE projects through programs such as the National Circular Economy Pilot Projects and the Special Funds for Environmental Protection. China has been exploring different funding mechanisms used by the EU to finance LCA and CE projects. This includes research on the EU's use of grants, subsidies, low-interest loans, and other financial instruments to encourage businesses and organizations to adopt sustainable practices. Development banks like the China Development Bank and the Industrial and Commercial Bank of China, as well as other financial institutions, offer financing options for LCA and CE projects (Hao et al. 2020). These institutions support sustainable development efforts, including CE initiatives, and provide loans or investment capital to eligible projects. Furthermore, China has an active venture and private equity market. There are funds and investment businesses in China that focus on sustainable businesses and technologies, including those related to LCA and the CE. These entities provide funding and resources to support the growth of innovative projects in the CE space.

China has been actively participating in international collaborations and partnerships with the EU and other countries to advance sustainable finance and the CE agenda. Through these collaborations, China can learn from the EU's experience, share knowledge and best practices, and leverage international funding opportunities.

5 Venture Capital and Private Equity Markets for Supporting LCA and CE Initiatives

LCA and CE are closely related concepts that can work together to promote sustainability and reduce environmental impact. A LCA is a method used to evaluate the environmental impact of a product or service throughout its entire life cycle, from raw material extraction to disposal. A CE, on the other hand, is an economic system that aims to eliminate waste and pollution by keeping products and materials in use for as long as possible. LCA methodology can be used to support the transition to a circular economy by identifying opportunities for improvement and helping to inform decision-making. LCA helps to evaluate the environmental impact of products or services throughout their life cycle, providing insights into material and energy flows and informing eco-design strategies to minimize environmental impacts. By integrating LCA into the design process, products can be developed with circular principles in mind, such as durability, reparability, and recyclability.

The concept of LCA originated in the 1960s as a methodology to assess the environmental impacts of products and processes throughout their life cycle. It gained popularity in the 1970s and has since become a standard tool for sustainable product development and decision-making (Wagner et al. 2021). On the other hand, the CE concept has deeper historical roots, but the modern interpretation emerged in the late twentieth century. The term gained traction in the early 2000s and has since evolved into a key focus area for sustainability and resource management. Both LCA and CE have continued to develop and gain importance in the fields of environmental management, sustainable business practices, and policy development over the past few decades (Figge et al. 2023).

Nowadays, it is becoming increasingly common for venture businesses to be aware of and prioritize concepts of LCA and CE in their operations and investments. As sustainability and environmental consciousness become significant considerations for consumers, regulators, and investors, venture businesses are recognizing the importance of integrating LCA and CE principles into their business strategies. By embracing these concepts, venture businesses can position themselves as innovators in sustainable practices, differentiate their products or services in the market, and appeal to a growing segment of environmentally conscious consumers.

Venture capital and private equity play crucial roles in supporting LCA and CE initiatives in China, the EU, as well as the United States of America (USA) (Kanda et al. 2024). The USA has a mature and well-established venture capital and private equity market, particularly in technology and innovation-driven sectors. Many renowned venture capital and private equity businesses are based in regions like Silicon Valley, New York, and Boston, making sizable investments across various industries (Guerra and Leite 2021). The EU is home to a diverse and dynamic venture capital and private equity market. It has a robust ecosystem with a wide range of funds, investors, and investment networks. European countries, such as the Germany, France, Poland and Sweden, have particularly active venture capital and private equity sectors (Mhatre et al. 2020). Main differences in processes, requirements and rules for businesses for IPO in the USA, EU and China are listed in Table 2.

The generalizations provided in Table 1 offer a high-level overview of current IPO approaches in the USA, EU, and China. It's important to note that these generalizations may not encompass the unique circumstances of every individual IPO, and other specific considerations may apply. Furthermore, it's crucial to recognize that financial markets are dynamic, and the approaches outlined in the table are subject to ongoing reforms and developments. Therefore, the information presented should be interpreted with an understanding of the limitations inherent in the procedures followed to obtain it, as well as the evolving nature of financial regulations and practices.

An important step for the IPO market in China was the creation of the Shanghai Stock Exchange Science and Technology Innovation Board (STAR Market—http://star.sse.com.cn/), which is also called the Science and Technology Innovation Council. This project started in 2019, is part of the Shanghai Stock Exchange and specializes in listing and trading of innovative and fast-growing technology companies, many of which are directly related with LCA and CE businesses. Its main goal

Table 2 Main differences in processes, requirements and rules for businesses for IPO in the USA, EU and China

USA	EU	China
Under the direction of the Securities and Exchange Commission (SEC), the IPO process involves filing a registration statement (Form S-1), which is then verified	EU IPO rules are decentralized and subject to the rules of individual member countries and their securities regulators	The China Securities Regulatory Commission (CSRC) oversees the IPO. The process is lengthy and can take up to two years
Before the IPO, a roadshow is held to gauge investor interest	Disclosure requirements vary from country to country, but they generally follow the Prospectus Regulation and the Transparency Directive issued by the European Securities and Markets Authority (ESMA)	There is a significant level of government oversight that affect the IPO process and results
IPO pricing often involves lengthy negotiations between the underwriters and the company	In many EU countries, IPO underpricing is less common than in the US	Traditionally, China has used an "earnings forecast" model for IPOs. The company must have been profitable in the recent past and is expected to remain so
The US markets, especially NASDAQ and NYSE, are quite transparent	Cross-border IPOs in EU countries are quite common	In recent years, with the emergence of the STAR market, China has taken a more market-oriented approach, attempting to copy existing practices (especially from the US)

Source Developed by authors according to: (Elliot 2023), (IPOs, n/d), (Notice 2023)

is to support the development of China's technology sector by providing a platform for local and international investors.

The STAR Market differs from other Chinese stock markets due to its unique features. The first is the registration-based IPO system, which allows businesses to list more efficiently with fewer restrictions. This streamlined process aims to make it easier for qualified technology businesses to enter the market. Second, the STAR Market places more emphasis on evaluating technological innovation and growth potential than on profitability. This provides an opportunity for promising technology startups to access capital and grow their business. With the implementation of the registration system in February 2023 (Notice 2023), Weili Transmission, submitted a registration application to the Shanghai Stock Exchange Science and Technology Innovation Board on 15 May 2023 and received approval on May 24. Registration took only 9 days. In 2022, the average time required for an IPO from application to approval was 111 days (CSIA, n/d).

6 Examples of IPOs of LCA and CE Businesses in China

China has a vibrant startup ecosystem with numerous venture businesses that actively invest in startups across various industries (Xu 2024). Currently, China has a mix of public and private venture funds. Public funds are often established by government bodies or state-owned enterprises and are focused on supporting strategic sectors or industries aligned with national goals. Private funds include independent venture businesses and corporate venture funds, which invest in startups across various sectors. Venture funds in China typically focus on technology-driven startups, especially in areas such as artificial intelligence, biotechnology, fintech, e-commerce, and clean energy. The emphasis is often on high-growth potential sectors that align with China's economic priorities. Venture funds in China aim to generate returns on their investments through successful exits, such as IPOs or acquisitions. The Chinese stock exchanges and technology-focused stock markets like the Star Market in Shanghai provide opportunities for startups to go public.

Yinchuan Phoenix Tianyu Venture Capital Fund is one of the funds that invests in early-stage and growth-stage businesses? In various sectors. Its main goal is to identify and support startups with significant growth potential. This fund is actively involved in the development of its portfolio businesses?, offering strategic guidance, mentoring and consulting.

One of this fund's successful projects was the IPO of Yinchuan Weili Transmission Technology Co., Ltd. (hereinafter referred to as Weili Transmission), listing code 300904: CN. Founded in 2003, this business? Is engaged in the research, development and production of precision gearboxes. The main products include the research and development of wind turbine yaw reducers, wind turbine pitch reducers and new energy vehicle electric drive units, and the aim is to provide precision transmission solutions for the new energy industry (Yinchuan 2023).

This case was chosen as an example because, against the backdrop of deflation that occurred in China at the time of the initial public offering of this business, this IPO, along with two other IPOs (Bescient Technologies and Tianjin Guoan Mengguli New Materials Science and Technology), both taking place in July 2023, showed high growth from the start.

Yinchuan Weili Transmission Technology's (www.weili.com) commitment to innovation and the development of high-efficiency products aligns with principles of sustainability, energy efficiency, and environmental responsibility. The company has ISO 14001 certification, which demonstrates that the company has processes in place to monitor, manage, and continuously improve its environmental performance. Also, company has close relations with CE-responsible businesses as they became a supplier of Siemens Gamesa.

Tianjin Guoan Mengguli New Materials Science & Technology Co., Ltd. (www.htmgl.com.cn) focus on the research and development of high-performance materials for industries such as aerospace, automotive, and energy suggests a significant relationship with the principles of LCA and CE. The company's mission is "continually providing high-quality products and satisfactory services in the industry of new energy material, and achieving the harmony between human beings and the earth". The company utilizes LCA methodologies to assess and optimize the environmental performance of the new materials it develops by considering aspects such as energy consumption, emissions, and resource use across the entire life cycle of the products. The company has received a "National High-Tech Enterprise, Green Factory" certificate.

Bescient Technologies (en.bx-tec.com) is specializing in Environment, Health, and Safety (EHS) solutions in engineering and technology. Bescient Technologies provide LCA services to help clients assess the environmental impacts of their products or processes. Additionally, the company offer guidance on adopting circular economy practices, such as minimizing waste, promoting recycling and reuse of materials, and designing products for longevity and ease of disassembly. Given its focus on EHS and engineering solutions, Bescient Technologies intersects with LCA and CE principles, offering expertise and services to support clients in enhancing their environmental performance and sustainability efforts.

Based on the information presented in Table 3, we can conclude that the three businesses with the greatest growth in IPOs in China in July 2023 are also those related to environmental, LCA and CE projects. Such growth may be due to, among other things, investor demand, increased interest from portfolio investors, limited supply of shares available at the time of IPO, anticipation of future growth and potential profitability, especially if the company operates in a promising industry or offers innovative products and services.

This also indicates the importance of environmental, LCA, and CE projects in the current market and investor sentiment towards sustainable initiatives. It suggests that

Table 3 Top three companies with the greatest growth in IPOs in China for July 2023

Company name	IPO site, listing number	Share price at placement (RMB)	Share price at the end of the first day of trading (RMB)	% growth
Yinchuan Weili Transmission Technology (production of wind turbine elements)	SZSE, 300904:CN	35.41	128.01	261.5
Tianjin Guoan Mengguli New Materials Science and Technology (lithium battery cell production)	SZSE, 301487:CN	5.32	98.02	1742.5
Bescient Technologies (intelligent environmental monitoring and big data services for public safety)	SSEC, 688671:CN	36.12	70.07	94%

Source compiled by the authors based on data from investing.com (accessed 05/05/2024): https://www.investing.com/equities/yinchuan-weili-transmission
https://www.investing.com/equities/tianjin-guoan-mengguli-new
https://www.investing.com/equities/bescient-tech

investors are increasingly recognizing the value and potential profitability of businesses operating in these sectors. The connection between Yinchuan Weili Transmission Technology, Tianjin Guoan Mengguli New Materials Science and Technology, and Bescient Technologies, and the broader context of environmental, LCA and CE projects underscores the increasing importance of sustainable initiatives in the current market. Yinchuan Weili's focus on the production of wind turbine components indicates a dedication to renewable energy and reducing carbon emissions, involving LCA assessments to ensure minimal environmental impact and support for the circular economy through renewable energy utilization. Similarly, Tianjin Guoan Mengguli's specialization in lithium battery cell production for electric vehicles and renewable energy applications emphasizes the need for sustainable energy storage technologies, potentially involving LCA evaluations to address resource usage and waste management, thus contributing to circular economy principles. Furthermore, Bescient Technologies' provision of intelligent environmental monitoring and big data services reflects a commitment to leveraging technology for enhanced EHS practices and sustainability, aligning with LCA principles and supporting informed, data-driven strategies for environmental impact assessment and mitigation. Collectively, these companies exemplify the market's increasing focus on sustainable initiatives, LCA, and CE projects while reflecting positive investor sentiment towards environmentally responsible practices and the adoption of circular approaches to resource management and industrial processes.

7 Survey Among Business Professionals and Investors on LCA and the CE in Venture Businesses in China: Results and Discussion

The survey was conducted according to the methodology and with the use of the survey as set out in Sect. 3 of this chapter. The respondents for the survey were identified at the China-Arab States Expo (CAS 2023), that took place in Yinchuan, Ningxia province, China in September 2023. This Expo focused on fields as financial investment, oil and gas, the digital economy, wind power and photovoltaics, and clean energy (Hu 2023). The survey gathered responses from 83 business professionals from China only, regarding the implementation LCA and the CE in venture businesses, survey was provided in Chinese.

Section 1 contained general questions. The Chinese respondents were distributed between the provinces of China in the following way: Beijing (5%), Guangdong (2%), Guangxi (2%), Hebei (5%), Henan (5%), Hunan (2%), Jiangxi (3%), Liaoning (2%), Neimenggu (Inner Mongolia) (2%), Ningxia (49%), Shandong (5%), Shanghai (10%), Shanxi (5%), Sichuan (2%), and Tianjin (1%). These provinces represent more the half of the population of China and generate over 50% of China's GDP. The results obtained from the general section of survey showed that 68% of the respondents have more than three years of work experience, while 17% were business executives and 8% were founders of businesses.

Section 2 of the survey was concentrated on knowledge and awareness about concepts of SDG, LCA, CE. Analysis of the results obtained from the survey showed that there is a significant level of awareness and understanding of the principles of the CE among the respondents. The majority of respondents (over 80%) mentioned that they are familiar with the principles of the CE, indicating a good level of knowledge and understanding of this concept.

The survey furthermore revealed that 28.7% of the respondents from China have encountered venture businesses or start-ups working on sustainable solutions in China. The relatively low percentage of respondents who have come across such venture businesses suggests that there is still untapped potential, and room for growth and further development for sustainable solutions-related industries in both these regions. This could imply that sustainable business models and start-ups focusing on environmental and social issues are still emerging or have yet to gain widespread visibility and recognition in China. This finding also presents an opportunity for greater collaboration and knowledge exchange between the EU and China regarding sustainable solutions. By sharing best practices, experiences, and resources, both regions can foster innovation and accelerate the growth of sustainable industries.

Section 3 of the survey was related with the role of government in CE and LCE. Based on the survey results, it is evident that the majority of respondents (86.2%) agree that venture businesses and start-ups can effectively contribute to sustainable consumption patterns (Fig. 2). The high level of agreement among respondents suggests that venture businesses and start-ups are already perceived as important drivers of sustainable economic transition in the EU and China. This indicates that these entities are seen as vital players in fostering innovative and sustainable solutions to address environmental challenges and promote sustainable consumption patterns. Additionally, 82.8% of respondents believe that venture businesses and startups are very important in driving the sustainable economic transition in both the EU and China. These findings indicate that there is strong recognition and belief in the role of VC businesses and start-ups in promoting sustainability and contributing to the CE (Fig. 1).

A majority of respondents (70.1%) indicated that the government should provide more financial support and incentives to venture businesses and start-ups in the fields of LCA and the CE. This can include grants, subsidies, low-interest loans, or investment funds. These financial resources can help businesses in these fields to overcome initial barriers and stimulate further innovation and growth. Respondents (52.9%) also emphasized the importance of clear and consistent regulations and policies to support and encourage sustainable practices and innovations in LCA and CE. Such measures can provide a supportive environment for venture businesses and start-ups, including simplified licensing procedures and tax incentives for sustainable businesses. Clear guidelines and policies can enhance market certainty and facilitate sustainable business operations. Additionally, respondents (50.6%) believe that governments should promote greater collaboration and networking among venture business, start-ups, research institutions, and industry stakeholders.

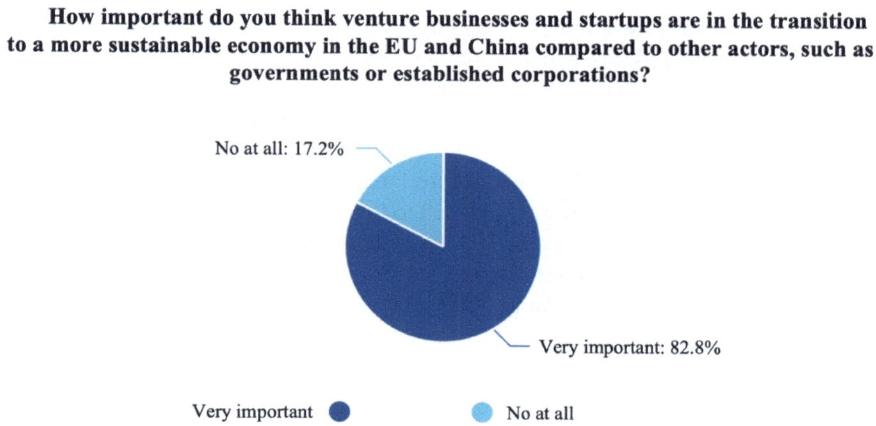

Fig. 1 Diagram of responses on the question "How important do you think venture businesses and startups are in the transition to a more sustainable economy in the EU and China compared to other actors, such as governments or established corporations?" *Source* Developed by the authors

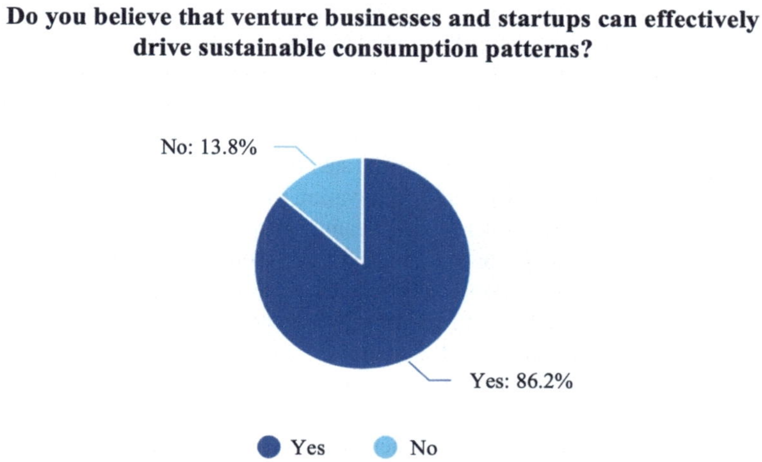

Fig. 2 Diagram of responses on the question "Do you believe that venture businesses and startups can effectively drive sustainable consumption patterns?" *Source* Developed by the authors

Section 4 was oriented on the relations with SDGs. The favorable outlook of survey participants regarding the contribution of venture businesses and start-ups to the shift towards a sustainable economy emphasizes the possibility of additional expansion and progress in this domain. Through fostering and backing the advancement of these enterprises, policymakers, investors, and industry stakeholders can leverage their inventive capabilities and expedite the transition to a more sustainable economy.

The survey results also emphasize the importance of promoting the CE, as venture businesses and start-ups are seen as important contributors in this endeavor. Encouraging circular business models, supporting start-ups focused on sustainability, and facilitating collaboration between venture businesses and established industries can all contribute to the promotion of the CE. Based on the survey findings (Fig. 2), it can be concluded that the efforts of venture businesses and start-ups in promoting sustainable practices and innovations related to LCA and the CE have been effective. However, it is also noteworthy that 41.4% of respondents believe that there is still room for policy optimization in this area.

The positive impact of relevant government policies and support measures on the development and success of venture businesses and start-ups in the field of LCA and the CE is acknowledged by the majority of respondents. Funding schemes, regulatory frameworks, and other initiatives have played a significant role in facilitating the growth of sustainable venture businesses and startups, and stimulating innovation in this domain. However, the acknowledgement that there is room for policy optimization indicates that there are still areas where governmental support can be further improved to enhance the effectiveness and efficiency of the policies and initiatives. This suggests that respondents may have specific recommendations or areas they believe could benefit from policy adjustments or enhancements. To optimize policies in this context, it would be beneficial for policymakers to engage with stakeholders, including venture businesses, start-ups, and industry experts, to identify areas for improvement and address any existing challenges or barriers. This collaborative approach can help refine existing policies, introduce new measures, and create an enabling environment that further accelerates the growth of sustainable venture businesses and startups and encourages innovation in LCA and the CE.

According to the survey results, a significant portion of the respondents (67.8%) believe that venture businesses and start-ups focusing on LCA and CE in China are aligned with the UN SDGs. These activities of these enterprises are perceived to contribute to goals such as SDG 12 (responsible consumption and production) and SDG 9 (industry, innovation and infrastructure). It is notable that these efforts are particularly recognized for their impact on sustainable production and consumption (SDG 12).

However, it is also mentioned that there is room for improvement in terms of aligning initiatives with the full set of SDGs, such as SDG 7 (affordable and clean energy) and SDG 13 (climate action). This suggests that there is potential for venture businesses and start-ups to further enhance their contributions to these specific SDGs, particularly in the areas of affordable and clean energy and climate action.

To optimize the alignment of these activities with the full set of SDGs, it is recommended for venture businesses and start-ups to consider integrating sustainable energy solutions and climate-related strategies into their operations. This involve exploring partnerships, investment in renewable energy sources, carbon reduction measures, and other environmentally sustainable practices. By doing so, venture businesses and start-ups can expand their impact and provide solutions that align more comprehensively with the broader sustainability agenda, contributing to a range of SDGs beyond those directly related to consumption and production.

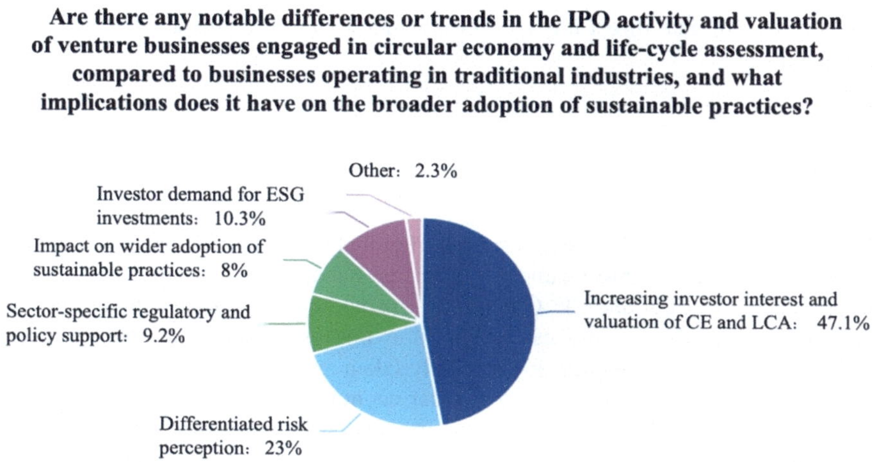

Fig. 3 Diagram of responses on the question "Are there any notable differences or trends in the IPO activity and valuation of venture businesses engaged in LCA and CE compared to businesses operating in traditional industries, and what implications does it have on the broader adoption of sustainable practices?" *Source* Developed by the authors

Section 5 was surveying for correlations between IPO and venture businesses related to CE and LCA. A significant percentage of respondents, 47.1%, indicated that there is increasing investor interest and valuation for venture businesses engaged in the CE and LCA. This suggests that investors are recognizing the value and potential profitability of businesses operating in sustainable practices, leading to higher investor demand and potentially higher valuations for these businesses during IPOs (Fig. 3).

Section 6 was surveying the opinion of respondents regarding future outlook and global perspectives of LCA and CE related businesses in China. A significant portion of respondents, 23%, mentioned that there is a differentiated risk perception when it comes to venture businesses in the CE and LCA compared to traditional industries. This suggests that investors perceive sustainable businesses as having lower risks or as having potential for long-term growth. Respondents also identified sector-specific regulatory and policy support as a factor influencing IPO activity and valuation in the CE and LCA. This highlights the importance of government support and favorable regulatory frameworks to encourage and incentivize sustainable practices in the business sector. A smaller percentage of respondents, 10.3%, mentioned investor demand for environmental, social, and governance (ESG) investments as a factor influencing IPO activity and valuation in the CE and LCA. This indicates that investors are seeking out businesses that align with their ESG values and are more likely to invest in businesses with strong sustainability practices. A relatively lower percentage of respondents, 8%, identified the impact on the broader adoption of sustainable practices as a factor influencing IPO activity and valuation in the CE and LCA. This suggests that while the growth of sustainable businesses during IPOs is recognized, it may have a limited direct impact on the wider adoption of sustainable practices across industries.

8 Discussion

In the recent years, China is becoming one of global leader in promoting and implementing the CE ideas into businesses. Initially started from copying the EU and US approaches to LCA and CE, China is now providing considerable governmental support to sustainable businesses. Initially drawing inspiration from approaches in the EU and the US, China has progressively taken significant strides in advancing sustainable practices and fostering a circular economy mindset. With a foundation rooted in resource efficiency, waste reduction, and sustainable production, China's commitment to sustainability is underscored by substantial governmental support for businesses embracing environmentally responsible strategies. This support spans financial incentives, policy frameworks, and collaboration initiatives to cultivate a conducive environment for sustainable enterprises. As China continues to align its economic development with ecological stewardship, its role as a global influencer in circular economy principles grows increasingly prominent.

The 14th Five-Year Plan of China for 2021–2025 (Plan 2021) demonstrates a comprehensive approach to environmental protection, encompassing a range of strategies and initiatives to address environmental challenges. The plan sets specific targets for reducing pollution, including air and water pollution, as well as greenhouse gas emissions. It emphasizes the transition to clean energy sources and promotes the development of renewable energy technologies. Furthermore, the plan places a strong focus on advancing the circular economy, encouraging resource efficiency, waste reduction, and sustainable production practices. It also prioritizes the protection and restoration of natural ecosystems, highlighting the importance of ecological conservation. There is an emphasis on fostering green technology and innovation to address environmental issues, alongside measures to strengthen environmental governance and regulation. The plan also underscores the importance of international cooperation on environmental matters, signaling China's commitment to global efforts in environmental protection.

Currently, the stock market of China is in a state of rapid growth. Between January and June 2023, Chinese companies raised $19.5 billion in IPOs (KPMG 2023). IPOs in China have received a boost from the lifting of stringent restrictions due to the COVID-19 pandemic and the new simplified listing regime on the Shanghai (SSEC) and Shenzhen Stock Exchanges (SZSE). The focus on regulatory changes and a simplified listing regime in China has heightened the appeal of local IPOs for companies seeking to go public. These regulatory adjustments are designed to streamline the listing process, alleviate regulatory constraints, and enhance Chinese companies' access to capital markets. While quantifying the precise number of IPOs tied to LCA and CE oriented businesses is challenging, this paper has identified three businesses as exemplars of successful IPOs that have sparked exceptional investor interest. These case studies aim to shed light on the growing traction of sustainable businesses within the Chinese IPO landscape and their ability to captivate investor enthusiasm.

The paper also included the survey of Chinese businessmen and investors on their perception of current trends in LCA and CE. The survey was structured into 6 sections, each of them was designed to analyze certain issues of approaches in China to CE:

Section 1 "Introduction and personal information of respondents" indicated that respondents represented 15 provinces of China (of 31 provinces, autonomous districts and municipalities). However, these provinces generate over 50% of Chinese GDP.

Section 2 "Knowledge and awareness about concepts of SDG, LCA, CE" demonstrated high awareness about sustainability. The fact that over 80% of the respondents expressed familiarity with the principles of CE indicates a robust level of knowledge and comprehension of this concept within the surveyed group. This high level of awareness bodes well for the integration and advancement of CE practices within the Chinese business landscape, suggesting a solid foundation for sustainable initiatives and investment in circular economy principles.

Section 3 "The role of government in CE and LCE" underlines the strong call from respondents for enhanced governmental support and a conducive policy environment to foster the growth and innovation of LCA and CE ventures in China. Regardless to governmental efforts proclaimed in the 14th 5-year plan (Plan 2021), this fact emphasizes the necessity for increased financial support and incentives, as indicated by the majority of respondents, suggests a recognition of the pivotal role that government-backed resources can play in propelling sustainable businesses past initial barriers and toward further advancements. Moreover, the highlighted significance of clear and consistent regulations and policies echoes the need for an enabling framework that can effectively underpin sustainable practices and innovation in LCA and CE.

Section 4 "Relations with SDGs" provides exploration of the relationship between LCA and CE ventures and the UN SDGs unveils the potential for these enterprises to extend their impact beyond specific domains into a more expansive and interconnected sustainability landscape. By adopting a holistic approach that encompasses not only responsible consumption and production but also clean energy and climate action, venture businesses and start-ups in China can significantly enhance their influence and contribute to a broader range of SDGs, aligning more fully with the global sustainability agenda.

Section 5 "Correlations between IPO and venture businesses related to CE and LCA" indicated that a significant percentage of respondents recognize a growing investor demand and potentially higher valuations for sustainable businesses during IPOs, which underscores the evolving perception of the value and potential profitability of enterprises operating within sustainable practices. This trend suggests a shifting investor landscape that increasingly values and prioritizes sustainability-driven ventures, reflecting a broader recognition of the financial viability and potential return on investment within the sustainable business sector. The alignment of investor interest with businesses focused on CE and LCA can signal an expanding market for sustainable initiatives, potentially incentivizing further innovation and growth within the sector.

The insights garnered from Sect. 6 "Future outlook and global perspectives of LCA and CE related startups and venture businesses" shed light on the differentiated risk perception of venture businesses in CE and LCA compared to traditional industries. The indication that a significant portion of respondents perceive sustainable businesses as having lower risks or as holding potential for long-term growth signifies an evolving investor mindset that acknowledges the long-term stability and growth potential of sustainable enterprises. This perception outlines the shifting investment landscape, reflecting an increasing recognition of the value and resilience inherent in sustainable business models.

The survey results highlight the significance of venture businesses and start-ups in driving sustainable practices and promoting the CE in China. While government policies have played a positive role, there is an opportunity for policymakers to continue optimizing and refining their support measures to further stimulate innovation and sustainable business growth.

It is important to consider that these observations are based on the specific survey data provided. Further research, incorporating a larger sample size and additional perspectives, could help to strengthen the understanding of the relationship between government policies, venture businesses, and start-ups in the context of LCA and CE.

Next steps for this research include a thorough policy analysis, engaging more relevant stakeholders such as government agencies, venture businesses, start-ups, industry associations, and environmental organizations, as well as benchmarking with international best practices. The next objective is to investigate the post-IPO long-term sustainability and financial performance of LCA and CE startups, assessing their ability to attract investments, generate positive financial returns, and determine potential correlations between sustainability metrics and financial performance.

9 Conclusion

This chapter presented the results of research on current trends in financing LCA and CE projects and their links with the SDGs. The main conclusions, which are based on a survey of LCA and CE business professionals and investors from mainly China and the analysis of current trends in IPOs at stock markets, include:

Recognition of the role of venture businesses and startups in China: The majority of respondents acknowledge the important role that venture businesses and startups play in promoting sustainable practices and innovations related to LCA and the CE. This recognition demonstrates the potential contribution of these entities in driving sustainable development.

Positive impact of government policies and support measures in China: Many respondents believe that relevant government policies and support measures have had a positive impact on the development and success of venture businesses and startups in the area of LCA and the CE. Funding schemes, regulatory frameworks, and other initiatives have been identified as facilitators of sustainable venture businesses and innovation. This highlights the importance of supportive policy environments in fostering sustainable entrepreneurship.

While government policies and support measures in China have made a positive impact, there is room for policy optimization. This suggests that there may be areas where policies can be further improved to better support the development of sustainable venture businesses and stimulate innovation. Identifying and addressing these opportunities can lead to more effective policy interventions in the future.

The intersection of CE, LCA and SDGs influences investor behavior and the performance of IPOs of venture businesses in China, which was confirmed with examples of 3 selected companies and results of survey. The increasing interest in sustainable initiatives and environmentally responsible practices may led to heightened investor demand and positive market sentiment, particularly for companies aligned with CE and LCA principles. This underscores the growing recognition of the value and potential profitability of businesses promoting sustainable practices, indicating a market emphasis on circular approaches to resource management.

The analysis of selected companies in China participating in IPOs and focused on LCA and CE projects illustrates a correlation between IPO growth and investor demand for businesses with a sustainability focus. The exceptional performance of these companies, as demonstrated by significant IPO valuation increases, reflects a market sentiment that favors environmentally responsible practices and circular economy initiatives. The success of these IPOs highlights the potential for sustainable initiatives to not only thrive in the market but also to attract substantial investor interest, impacting the broader adoption of sustainable practices across industries.

The findings from the survey conducted among Chinese business professionals at the China-Arab States Expo (CAS 2023) reveal a notable level of awareness and understanding of LCA and CE concepts among the respondents. Additionally, the survey demonstrates a strong belief in the role of venture businesses and startups in contributing to sustainable consumption patterns and driving the sustainable economic transition in both China and the EU. Furthermore, the survey identifies the importance of government support, financial incentives, and clear policy frameworks in fostering sustainable practices and innovations in LCA and CE. These insights underscore the potential for collaboration and knowledge exchange to further accelerate the growth of sustainable industries.

References

Agenda UN (2015) Transforming our world: the 2030 agenda for sustainable development. United Nations. A/RES/70/1. Resolution adopted by the General Assembly on 25 September 2015

AIM Italia (n/d) Listing on AIM Italia. Retrieved from: https://www.borsaitaliana.it/azioni/mercati/euronext-growth-milan/home/listingonaimitalia.en.htm. Accessed 9 Oct 2023

Ali GAM, Thalji MR, Makhlouf ASH (2023) Biodegradable materials: fundamentals, importance, and impacts. In: Ali GAM, Makhlouf ASH (eds) Handbook of biodegradable materials. Springer, Cham. https://doi.org/10.1007/978-3-031-09710-2_74

Backes J, Traverso M (2022) Life cycle sustainability assessment as a metrics towards SDGs agenda 2030. Curr Opin Green Sustain Chem 38:100683. https://doi.org/10.1016/j.cogsc.2022.100683

Barros MV, Salvador R, do Prado GF, de Francisco AC, Piekarski CM (2021) Circular economy as a driver to sustainable businesses. Cleaner Environ Syst 2:100006. ISSN 2666-7894. https://doi.org/10.1016/j.cesys.2020.100006

BASF (2023) BASF venture capital invests in startup DePoly SA. Retrieved from: https://www.basf.com/global/en/media/news-releases/2023/06/p-23-254.html. Accessed 13 Oct 2023

Briguglio M, Llorente-González LJ, Meilak C, Pereira Á, Spiteri J, Vence X (2021). Born or grown: enablers and barriers to circular business in Europe. Sustainability 13(24):13670. https://doi.org/10.3390/su132413670

CAS (2023) Official website of the China-Arab States Expo. Retrieved from: https://www.cas-expo.org.cn/en/index.html. Accessed 9 Oct 2023

CSIA (n/d) China Securities Industry Association, Comprehensive registration system process optimization, the time required for IPO company registration has been sharply reduced. Retrieved from: https://www.sac.net.cn/tzzzj/zxsd/xscyw/202306/t20230608_60377.html. Accessed 9 Oct 2023

De Oliveira CT, Oliveira GGA (2023) What circular economy indicators really measure? An overview of circular economy principles and sustainable development goals. Resources Conserv Recycl 190:106850. ISSN 0921-3449. https://doi.org/10.1016/j.resconrec.2022.106850

Dieterle M, Viere T (2021) Bridging product life cycle gaps in LCA & LCC towards a circular economy. Proc CIRP 98:354–357

EIB (n/d) European Investment Bank in the circular economy. Retrieved from: https://www.eib.org/en/about/initiatives/circular-economy/index.htm. Accessed 9 Oct 2023

Ellen MacArthur Foundation (2022) Ellen MacArthur Foundation, What is a circular economy? Retrieved from: https://ellenmacarthurfoundation.org/topics/circular-economy-introduction/overview. Accessed 9 Oct 2023

Elliot J (2023) Understanding the IPO process and how it works. https://www.uschamber.com/co/start/strategy/guide-to-the-ipo-process. Accessed 05 May 2024

Euronext (n/d) Euronext Paris. Retrieved from: https://www.euronext.com/en/markets/paris. Accessed 9 Oct 2023

Ferreira C, Gomes G, Castillo D, Bertassini A, Zaguetto C, Feltrin N, Rossi E, Carmo I, Vasconcellos J, Barboza L, Tonelli R, Dionisio G, Gerolamo M, De Mello A, Gomes L, Beal C, Ometto A (2024) Organizational practices, values, and mindsets as a basis for circular economy transition. https://doi.org/10.1007/978-3-031-55036-2_5

Figge F, Thorpe A, Gutberlet M (2023) Definitions of the circular economy: circularity matters. Ecol Econ 208:1–2. https://doi.org/10.1016/j.ecolecon.2023.107823

Försterling G, Orth R, Gellert B (2023) Transition to a circular economy in Europe through new business models: barriers, drivers, and policy making. Sustainability 15(10):8212. https://doi.org/10.3390/su15108212

Friant MC, Vermeulen WJV, Salomone R (2021) Analysing European Union circular economy policies: words versus actions. Sustain Prod Consump 27:337–353. https://doi.org/10.1016/j.spc.2020.11.001

Guerra BC, Leite F (2021). Circular economy in the construction industry: an overview of United States stakeholders' awareness, major challenges, and enablers. Resources Conserv Recycl 170:105617. ISSN 0921-3449. https://doi.org/10.1016/j.resconrec.2021.105617

Han D, Konietzko J, Dijk M, Bocken N (2023) How do circular start-ups achieve scale? Sustain Prod Consumption 40:363–375. ISSN 2352-5509. https://doi.org/10.1016/j.spc.2023.06.007

Hao Y, Wang Y, Wu Q, Sun S, Wang W, Cu M (2020). What affects residents' participation in the circular economy for sustainable development? Evidence from China. Sustain Dev 28. https://doi.org/10.1002/sd.2074

He L, Xue M, Gu B (2020). Internet-of-things enabled supply chain planning and coordination with big data services: certain theoretic implications. J Manage Sci Eng 5(1):1-22. ISSN 2096-2320. https://doi.org/10.1016/j.jmse.2020.03.002.

Horizon (n/d) Horizon Europe official web site. Retrieved from: https://research-and-innovation.ec.europa.eu/funding/funding-opportunities/funding-programmes-and-open-calls/horizon-europe_en. Accessed 9 Oct 2023

Hu D (2023) Ningxia attracts Arab countries' investments. Retrieved from: http://ningxia.chinadaily.com.cn/chinaarabstatesexpo/2023-02/09/c_857199.htm. Accessed 9 Oct 2023

IPOs (n/d) IPOs. Invest Europe. https://www.investeurope.eu/policy/key-policy-areas/ipos/. Accessed 05 May 2024

Kanda W, Klofsten M, Bienkowska D, Marvin H, Olof H (2024) Challenges of circular new ventures: an empirical analysis of 70 cases. J Clean Prod 442:2024. https://doi.org/10.1016/j.jclepro.2024.141103

Kazançoğlu İ, Sagnak M, Mangla S, Kazançoğlu Y (2021) Circular economy and the policy: a framework for improving the corporate environmental management in supply chains. Bus Strateg Environ 30:590–608. https://doi.org/10.1002/bse.2641

Kirchherr J, Reike D, Hekkert MP (2017) Conceptualizing the circular economy: an analysis of 114 definitions. Resour Conserv Recycl 127:221–232. https://doi.org/10.1016/j.resconrec.2017.09.005

KPMG (2023) KPMG report. Chinese Mainland and Hong Kong IPO markets 2023 mid-year review. Retrieved from: https://assets.kpmg.com/content/dam/kpmg/cn/pdf/en/2023/07/china-hk-ipo-2023-mid-year-review.pdf. Accessed 9 Oct 2023

Lacy P, Long J, Spindler W (2020) The circular economy handbook: realizing the circular advantage. https://doi.org/10.1057/978-1-349-95968-6

Leal Filho W, Kozlova V, Ávila LV (2023). Moving forward: visions on the future of sustainable development. In: Leal Filho W, Azul AM, Doni F, Salvia AL (eds) Handbook of sustainability science in the future. Springer, Cham. https://doi.org/10.1007/978-3-031-04560-8_111

Ma W, Hoppe T, de Jong M (2022) Policy accumulation in China: a longitudinal analysis of circular economy initiatives. Sustain Prod Consump 34(2022):490–504. https://doi.org/10.1016/j.spc.2022.10.010

Malen J, Marcus A (2017) Promoting clean energy technology entrepreneurship: the role of external context. Energy Policy 102:7–15. https://doi.org/10.1016/j.enpol.2016.11.045

Meskers C, Bartie NJ, Reuter MA (2024) Handbook of recycling. Chapter 47 life cycle assessment. Elsevier, pp 701–721

Mhatre P, Panchal R, Singh A, Bibyan S (2020) A systematic literature review on the circular economy initiatives in the European Union. Sustain Prod Consump 26. https://doi.org/10.1016/j.spc.2020.09.008

Nakano K, Hirao M (2011) Collaborative activity with business partners for improvement of product environmental performance using LCA. J Clean Prod 19:1189–1197. https://doi.org/10.1016/j.jclepro.2011.03.007

Notice (2023) Notice of transitional arrangements on related administrative licensing issues before and after full implementation of the registration based IPO system as of February 1, 2023 http://www.csrc.gov.cn/csrc/c100028/c7047622/content.shtml. Accessed 04 May 2024

Numa IAN, Wolf KE, Pastore GM (2023) FoodTech startups: technological solutions to achieve SDGs. Food Humanity 1(2023):358–369

OECD (2021) Towards a more resource-efficient and circular economy. The role of the G20. Retrieved from: https://www.oecd.org/environment/waste/OECD-G20-Towards-a-more-Resource-Efficient-and-Circular-Economy.pdf. Accessed 12 Oct 2023

Ogunseitan O, Nakamura S, Suh S, Kral U, Li J, Geng Y (2022) Reshaping global policies for circular economy. Circular Economy 1:100003. https://doi.org/10.1016/j.cec.2022.100003

Ortiz-de-Montellano C, Samani P, van der Meer Y (2023) How can the circular economy support the advancement of the Sustainable Development Goals (SDGs)? A comprehensive analysis. Sustain Prod Consump 40:352–362. https://doi.org/10.1016/j.spc.2023.07.003

Plan (2021) Main targets in the 14th 5 year plan. http://english.www.gov.cn/w/14thfiveyearplan/. Accessed 05 May 2024
Rigamonti L, Mancini E (2021) Life cycle assessment and circularity indicators. Int J Life Cycle Assess 26:1937–1942. https://doi.org/10.1007/s11367-021-01966-2
Samani P (2023) Synergies and gaps between circularity assessment and Life Cycle Assessment (LCA). Sci Total Environ 903:166611. ISSN 0048-9697. https://doi.org/10.1016/j.scitotenv.2023.166611
Scarpellini S, Marín-Vinuesa LM, Aranda-Usón A, Portillo P (2020) Dynamic capabilities and environmental accounting for the circular economy in businesses. Sustain Acc Manage Policy J. ahead-of-print. https://doi.org/10.1108/SAMPJ-04-2019-0150
SSE (n/d) SSE STAR Market. Retrieved from: http://star.sse.com.cn/en/. Accessed 9 Oct 2023
Tan C (2023) China has announced a slew of measures to bolster its economy. Retrieved from: https://www.cnbc.com/2023/07/24/china-policy-measures-to-bolster-economy.html. Accessed 9 Oct 2023
von Kolpinski C, Cagno E, Neri A (2024) Investigating the entrepreneurial orientation of circular-social hybrid start-ups by assessing their path toward sustainability: Evidence from 20 European cases. Sustain Prod Consump 47(2024):222–235. https://doi.org/10.1016/j.spc.2024.03.036
Wagner F, Murillo B, Rodrigo S, Antonio F, Moreira M, Piekarski, C (2021) Integrating life cycle assessment and life cycle cost: a review of environmental-economic studies. Int J Life Cycle Assess. https://doi.org/10.1007/s11367-020-01857-y
Wang Zh, Wang X, Xu Y, Cheng Q (2022) Are green IPOs priced differently? Evidence from China. In: Research in international business and finance, vol 61
Xu Y (2024) Harnessing venture capital in China. Socio-Econ Rev 22(1):349–371. https://doi.org/10.1093/ser/mwad037
Yinchuan (2023) Yinchuan Weili Transmission Technology Co, Ltd. Prospectus for initial public offering of shares and listing on the GEM. Retrieved from: http://www.szse.cn/disclosure/listed/bulletinDetail/index.html?3622cd19-a7bf-4dcd-8a3e-98b9bb708fbd. Accessed 9 Oct 2023
Zhu J, Fan Ch, Shi H, Shi L (2018) Efforts for a circular economy in China: a comprehensive review of policies. J Ind Ecol 23. https://doi.org/10.1111/jiec.12754

Valorization of Potato and Banana Waste in Bioethanol Production: A Life Cycle Assessment Perspective

Victória Dutra Fagundes, João Felipe Freitag, Viviane Simon, Ênio Leandro Machado, Rosana de Cássia de Souza Schneider, and Luciane Maria Colla

Abstract The aim was to conduct a Life Cycle Assessment (LCA) of bioethanol production from potato and banana waste, encompassing pretreatment, enzymatic hydrolysis, and fermentation steps. The LCA utilized Simapro 8.5.0.0 software with inventory data from Ecoinvent 3.4 and Impact 2002+ as the impact assessment methodology. Uncertainty analysis was performed using the Monte Carlo method. Two scenarios were analyzed: one considering energy consumption and the other without. Among the 15 impact categories evaluated by Impact 2002+, respiratory inorganics and global warming had the most significant impacts for both types of waste in both scenarios. In energy consumption scenarios, fermentation had the greatest impact on both banana and potato waste, while in scenarios without energy consumption, pretreatment had the most significant impact on both types of waste. Ethanol production from banana waste showed greater impacts than from potato waste. The study identified energy demand and the use of sodium phosphate in pretreatment as critical points in the LCA, associated with the largest impacts. Consequently, there is a pressing need to adopt sustainable energy sources and find more sustainable substitutes for sodium phosphate in the bioethanol production process.

Keywords LCA · Food waste · Impact 2002+ · Impact category · Global warming

V. D. Fagundes · J. F. Freitag · V. Simon · L. M. Colla (✉)
Program in Civil and Environmental Engineering, University of Passo Fundo, Campus I, Passo Fundo, Brazil
e-mail: lmcolla@upf.br

Ê. L. Machado · R. de C. de S. Schneider
Program in Environmental Technology, University of Santa Cruz Do Sul, Universitário, Santa Cruz Do Sul, Brazil

L. Brandli et al. (eds.), *The Contribution of Life Cycle Analyses and Circular Economy to the Sustainable Development Goals*, World Sustainability Series,
https://doi.org/10.1007/978-3-031-85300-5_25

1 Introduction

Currently, fossil fuels meet about 80% of global energy demand, with oil accounting for approximately 33% of the total energy supply (Sawarkar et al. 2022). According to the International Energy Agency (IEA) (2022), the transport sector is responsible for around 37% of total CO_2 emissions worldwide, originating mainly from the burning of fossil fuels.

Among the alternatives for replacing fossil fuels, bioethanol is considered one of the most promising liquid biofuels to replace these fuels, mainly gasoline, and mitigate their adverse environmental impacts and negative health effects (Kazemi Shariat Panahi et al. 2022). The production of bioethanol involves the fermentation of carbohydrates, a relatively fast biological process that offers several advantages, such as high energy use and the extraction of high-value byproducts (Panahi et al. 2019).

In 2019, global food waste production totaled 931 million tons (FAO 2021). Due to the high moisture content, this waste becomes a significant environmental concern, as it is difficult to transport, consumes more fuel when incinerated, and results in the generation of wastewater, unpleasant odors, and greenhouse gas emissions when disposed of in sanitary landfills (Kumar et al. 2022). Therefore, there is a need to value these wastes into products with high added value.

In general, the production of bioethanol depends heavily on the composition of the waste. Research often focuses on waste already known for its carbohydrate and free sugar content, such as fruit waste, vegetable waste, starchy bakery waste, and even kitchen waste (Arapoglou et al. 2010; Guerrero and Muñoz 2018; Byun et al. 2022; Narisetty et al. 2022).

The composition of food waste, mainly kitchen waste, contains 35.5–69% carbohydrates, 3.9–21.9% proteins, oils and fats, and organic acids, with the remainder being moisture (Uçkun Kiran and Liu 2015; Dhiman and Mukherjee 2021). With high percentages of carbohydrates, food waste can be hydrolyzed to obtain fermentable sugars and converted into bioethanol through fermentative processes (Singh et al. 2022).

The production of bioethanol from food waste involves a well-known process that includes: (a) collection and preparation of waste, focusing on selecting waste with higher carbohydrate content and treating it with processes such as crushing, grinding, or physical–chemical treatments to increase the surface area for enzymatic action; (b) enzymatic hydrolysis using specific enzymes to convert carbohydrate portions into simpler sugars; (c) fermentation of the sugars released during hydrolysis by yeast, such as *Saccharomyces cerevisiae*, which converts the sugars into ethanol and carbon dioxide; and (d) distillation, where the ethanol produced is purified to remove impurities and obtain high-concentration ethanol (Dhiman and Mukherjee 2021; Kumar et al. 2022; Singh et al. 2022; Bibra et al. 2022).

Although bioethanol production routes are already known, there are still challenges regarding the use of food waste to produce biofuel. Recent research has increasingly evaluated the use of food waste for the production of bioethanol, focusing mainly on optimizing technologies for converting this waste into biofuel (Byun et al. 2022; Narisetty et al. 2022). Significant challenges include the identification of ideal food waste, the optimization of enzyme use in the enzymatic hydrolysis stage due to the cost of these enzymes, and the scalability of production, as transitioning promising technologies from the laboratory to an industrial scale is challenging (Singh et al. 2022; Bibra et al. 2022.

However, some gaps still need to be filled in the effective use of food waste for bioethanol production. These include the improvement of waste pretreatment technologies to make them more efficient and economical, the optimization of the production process to maximize efficiency and sustainability, and the need for policies and incentives to promote the adoption of bioethanol production from food waste (Panahi et al. 2019; Singh et al. 2022; Bibra et al. 2022).

Furthermore, biofuels are currently evaluated for their environmental performance compared to their fossil counterparts (Konti et al. 2020). There is also a need to meet specific environmental criteria established by regulatory authorities that set environmental goals and criteria for the effective application of biofuels (Singh et al. 2022).

Thus, the most common tool for evaluating the environmental impact of a product or process is Life Cycle Analysis (LCA). LCA is an effective environmental management tool to deal with the complex interaction between the environment and a product or process, allowing the development and testing of improved environmental and socioeconomic scenarios for sustainable trade-offs (Liu et al. 2017; Król-Badziak et al. 2021). LCA, in addition to evaluating the environmental impacts associated with a process or product, allows for the identification of critical points and areas for improvement to make the process more efficient and sustainable (Konti et al. 2020; Król-Badziak et al. 2021).

The production of bioethanol from banana and potato waste is already known (Wang et al. 2013; Palacios et al. 2017; Sawarkar et al. 2022; Chauan et al. 2022). However, there is still a need for research aimed at identifying the environmental impacts associated with the production process, to evaluate both points for improvement and optimization and their applicability. Thus, the aim was to evaluate the life cycle of bioethanol production from banana and potato waste as a possibility of valuing this food waste, providing information on the critical points of its production process for process improvements and optimization.

2 Material and Methods

The LCA execution methodology involves defining the aim and scope, life cycle inventory, assessment, and interpretation of impacts as established by ISO 14.040 (2006a).

2.1 Goal and Scope Definition

The aim of the LCA is to examine the environmental impacts related to the production of bioethanol from food waste. The LCA was carried out based on the bioethanol production method described by Bender (2022), involving the stages of drying the waste for preparation, gelatinization as a preliminary treatment, enzymatic hydrolysis, and subsequent alcoholic fermentation. To achieve this objective, two distinct scenarios were defined:

(a) Scenario 01: Production of bioethanol using banana waste.
(b) Scenario 02: Production of bioethanol using potato waste.

LCA studies, mainly on a small scale, indicate that energy consumption represents a significant portion of environmental impacts, overshadowing the contribution of other important factors, such as the use of chemicals, enzymes, water, and other elements (Wang et al. 2013; Singh et al. 2022). Therefore, for a more comprehensive approach to the LCA of these wastes, each scenario was evaluated separately, excluding energy consumption generated by the use of equipment.

This work focuses on activities related to obtaining bioethanol from banana and potato waste, following a gate-to-gate approach, thereby limiting itself to the analysis that spans from obtaining the waste to the production of bioethanol. Figure 1 shows the system limits adopted for the LCA.

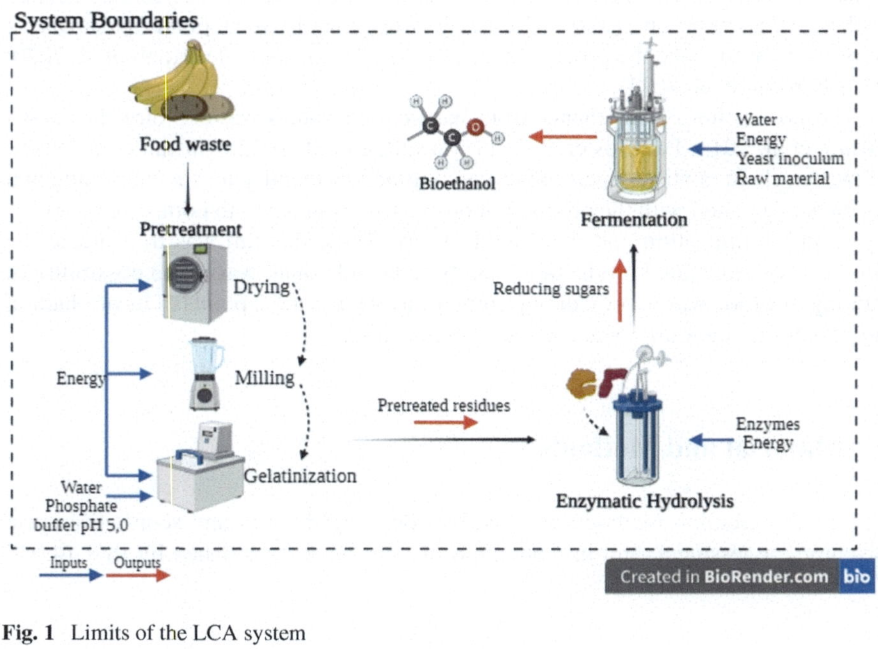

Fig. 1 Limits of the LCA system

The results of this study aim to raise awareness about the environmental implications associated with bioethanol production, identifying areas with potential for improvement and also demonstrating a way to value food waste by transforming it into high-value-added products. Within the established framework, the work aims to identify specific areas of intervention and improvement within the bioethanol production process. Furthermore, it will provide crucial information on the environmental impacts of the production method adopted, to assist decision-makers in choosing the most appropriate techniques for obtaining this biofuel.

The functional unit (FU) adopted in this study is 1 kg of waste, which means that all emissions, energy consumption, and materials used were calculated based on this quantity.

2.2 *Life Cycle Inventory*

At this stage, all inputs and outputs of the bioethanol production process were identified and quantified for each of the wastes evaluated. This includes the drying, grinding, and gelatinization operations of waste, as well as enzymatic hydrolysis with the application of different enzymes and the subsequent fermentation process.

Data collection for the inventory was conducted based on information from Letícia Bender's master's thesis (Bender 2022), carried out at the Postgraduate Program in Food Science and Technology at the University of Passo Fundo (PPGCTA-UPF). Additionally, data obtained from the literature describing similar production processes were used. Table 1 presents bioethanol production data obtained from the literature for banana and potato waste. The hydrolysis efficiency presented was obtained from Bender (2022), and these data were used for input and output calculations.

The bioethanol production process begins with drying banana and potato waste in a circulation oven for 48 h at 70 °C for banana waste and 50 °C for potato waste. After drying, the residues are ground in a blender and subsequently gelatinized for 10 min at 100 °C. In this step, a phosphate buffer solution (5% m/v) is added to adjust the pH to 5.5 during the pretreatment.

Table 1 Data used for the mass balance of LCA inputs and outputs

Food waste	Data	Value (%)	References
Banana	Waste moisture	86	Palacios et al. (2017)
	Hydrolysis efficiency	75	Bender (2022)
	Bioethanol yield	84	Ingale et al. (2014)
Potato	Waste moisture	85	Chauhan et al. (2022)
	Hydrolysis efficiency	93	Bender (2022)
	Bioethanol yield	88	Lareo et al. (2013)

In the enzymatic hydrolysis step, enzymes are added: α-amylase (LpHera) (Novozymes®) at 90 °C and subsequently amyloglucosidase (AMG) (Saczyme® Go 2x) at 50 °C. This step is carried out in a hydrolysis bioreactor for 6h. Finally, in the fermentation stage, the Saccharomyces cerevisiae yeast inoculum (10% v/v) is added, with fermentation lasting 72h for the banana waste and 31h for the potato waste.

The inputs and outputs were previously determined through a rigorous mass balance of the products. In scenarios that included energy consumption, the energy expenditure of the equipment was calculated based on the time of use. The Ecoinvent 3.4 database was used as the basis for the calculated inputs and outputs. Table 2 presents the calculated inputs and outputs for the production of bioethanol from banana and potato waste.

2.3 *Life Cycle Impact Assessment (LCIA)*

The impact classification and characterization phases, the results of the category indicators, as well as the weighting of the results, were carried out using the SimaPro 8.5.0.0 computer program. Impact2002+ was used as the LCIA methodology. This methodology proposes a viable implementation that combines impact assessment methods with approaches focused on both intermediate results (midpoint) and direct damages (endpoint) (Ecoinvent Support 2024). All LCI results are linked to the fifteen (15) midpoint categories, and these, in turn, are linked to the four (4) damage categories. Figure 2 presents the midpoint and endpoint categories evaluated by the Impact2002+ method.

When interpreting the data, the main impacts observed for the evaluated scenarios were assessed and discussed. As one of the focal points of the work is the identification of critical points and process improvement, only midpoints were considered for discussions. This decision was made due to the potential for identifying critical points in the estimated emissions for the midpoint impact categories.

Furthermore, to assess the reliability of the data, an uncertainty analysis was carried out using the Monte Carlo statistical method. This serves as a data reliability tool, allowing for a better understanding of the sources of uncertainty and contributing to more robust and efficient decision-making, thereby ensuring reliability regarding the applicability of the method. The SimaPro 8.5.0.0 software for each of the scenarios for the 15 impact categories was evaluated using the Impact 2002+ method. A normal distribution with 1000 iterations was used as the distribution function, and a 95% confidence interval was applied.

Table 2 Inputs and outputs calculated in the LCA

Process		Unit	Banana waste	Potato waste
			Quantities	
Phase	*Inputs*			
Pretreatment (Drying, milling and gelatinization)	Food waste	Kg	1	1
	Energy consumption	Kwh	48.81	48.81
	Distilled water	Kg	3.34	3.3
	Monobasic sodium phosphate	g	27.6	27.32
	Dibasic sodium phosphate	g	5.03	4.95
Enzymatic hydrolysis	Alpha-amylase	g	0.65	0.65
	Amyloglucosidase (AMG)	g	0.65	0.65
	Energy consumption	Kwh	6.5	8
Fermentation	*Saccharomyces cerevisiae*	Kg	0.33	0.33
	Pre inoculum preparation			
	Glucose	g	0.67	0.66
	Yeast extract	g	0.33	0.33
	Distilled water	g	33	33
	Energy consumption	Kwh	24	24
	Inoculum preparation			
	Glucose	g	0.06	0.06
	Yeast extract	g	0.008	0.0082
	KH_2PO_4	g	0.01	0.011
	Urea	g	0.08	0.082
	KCl	g	0.01	0.011
	$CaCl_2$	g	0.01	0.011
	Energy consumption	Kwh	139.5	74
	Distilled water	g	297	297
	Outputs			
Pretreatment	Post-pretreatment waste	g	167	165
Enzymatic hydrolysis	Yield of reducing sugars	g	148.4	136.4
Fermentation	Ethanol yield	$g\,L^{-1}$	19.2	19

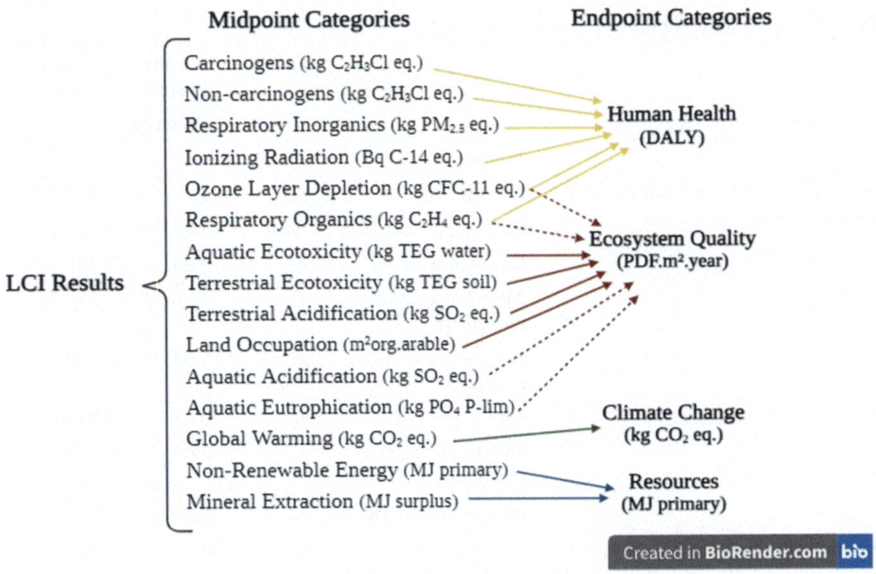

Fig. 2 Categories of environmental impacts and damages assessed using the Impact2022+ method. Adapted from Jolliet et al. (2003)

3 Results and Discussions

3.1 Impact Assessment

Figure 3 presents the normalization of the impact categories that were most relevant, evaluated by Impact 2002+ for banana waste considering or not considering energy consumption (Fig. 3a and b, respectively), and for potato waste considering and not considering energy consumption (Fig. 3c and d, respectively).

It is important to mention that in normalization, the impacts associated with a product and/or system are compared with reference values that describe the impacts relative to a reference system, which could be another product, a region, a country, or the entire globe (Crenna et al. 2019). According to ISO 14.044 (ISO 2006b), normalization is a step that allows the interpretation of characterized results in terms of the relative environmental relevance of impacts.

Thus, for banana waste, it is clear that when considering energy consumption, the fermentation stage presented the highest normalization scores for the categories presented, as seen in Fig. 3a. When energy consumption is not considered, the pretreatment stage has the highest scores, as shown in Fig. 3b.

As for potato waste, when considering energy consumption, the fermentation stage is associated with the greatest impact (Fig. 3c), while when not considering energy consumption, the pretreatment stage has the greatest impact (Fig. 3d).

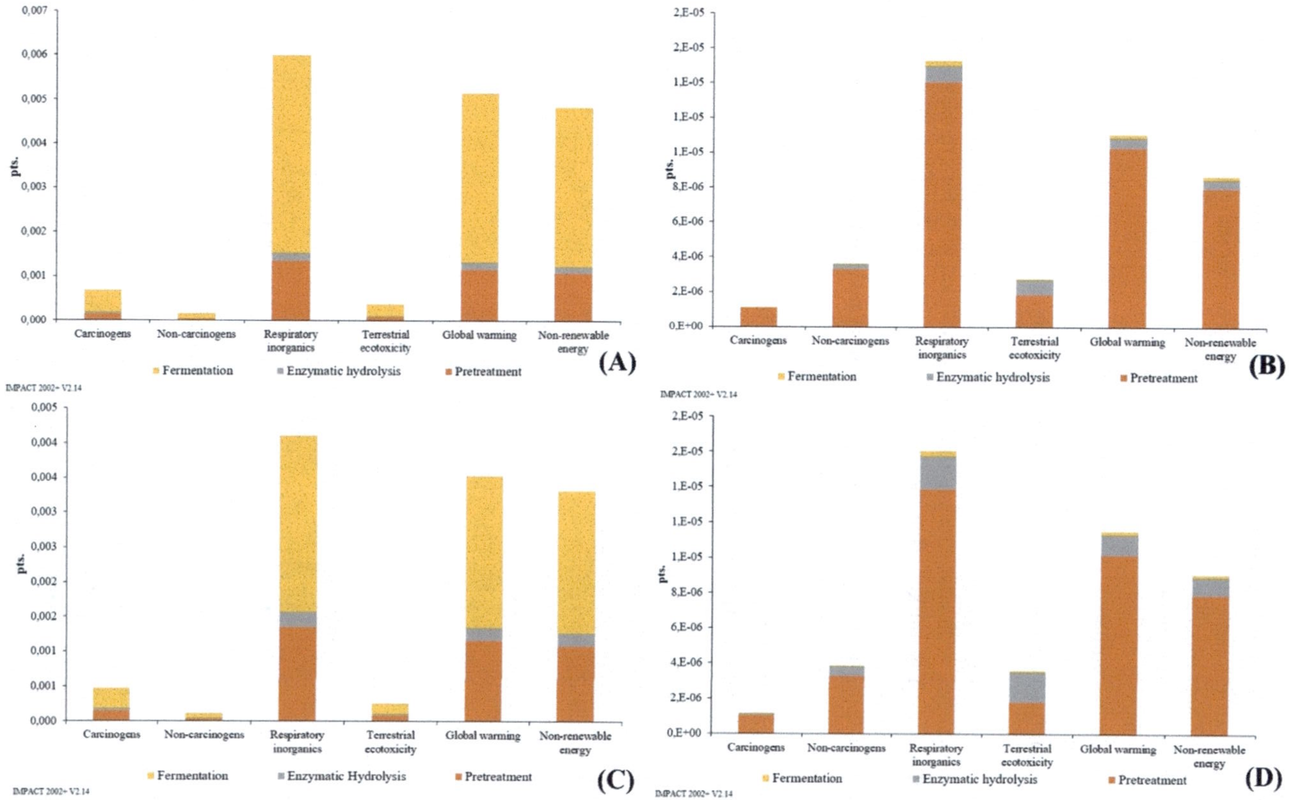

Fig. 3 Main impact categories obtained in the production of bioethanol from banana waste considering **a** and not considering **b** energy consumption and potato waste considering **c** and not considering **d** energy consumption, using Simapro 8.5.0.0 with Impact 2002+ method and Ecoinvent 3.4 database

Guerrero and Muñoz (2018) identified that the fermentation stage was responsible for around 24% of the impacts associated with the production of bioethanol from food waste, while the grinding steps and steam explosion pretreatment corresponded to 1% and 12% of impacts, respectively. Similar data was observed by Costa et al. (2018) in the LCA of bioethanol production from sweet potatoes in an experimental plant, where they found that the fermentation stage is responsible for around 47.3% of CO_2 emissions.

For all scenarios evaluated, including those not considering energy consumption, the most significant impact categories during normalization were respiratory inorganics (RI), global warming potential (GWP), and non-renewable energy (NRE), respectively.

The category of respiratory inorganics (RI) refers to respiratory effects caused by inorganic substances, with characterization factors provided only for emissions into the atmosphere (as it is unlikely that these pollutants will be emitted into the soil or water) (Humbert et al. 2012). The unit is represented by kg $PM_{2.5}$ eq, thus related to emissions of smog-forming gases such as carbon monoxide (CO), sulfur dioxide (SO_2), and nitrogen oxides (NO_X) (Li et al. 2019).

The global warming potential (GWP), also referred to as the life cycle carbon footprint, is an indicator that quantifies the influence of biofuels on climate change. This indicator is calculated using equivalent carbon dioxide (CO_2) emissions (Humbert et al. 2012).

The characterization factors for non-renewable energy (NRE) consumption, regarding the total primary energy extracted, are calculated based on higher calorific values, with the midpoint characterization factors expressed in MJ (megajoules) (Humbert et al. 2012).

Table 3 presents the estimated emissions for the most significant impact categories for banana and potato waste.

Table 3 Estimated emissions for the most relevant impact categories in the production of bioethanol from banana and potato waste

Impact category	Pretreatment		Hydrolysis		Fermentation	
	Banana	Potato	Banana	Potato	Banana	Potato
With energy						
GWP	11.45	11.45	1.52	1.87	38.02	21.61
NRE	164.93	164.92	21.88	26.98	548.44	311.70
RI	0.01	0.01	0.002	0.002	0.05	0.03
Without energy						
GWP	1.02E−01	1.01E−01	5.59E−03	1.12E−02	2.06E−03	2.03E−03
NRE	1.21E+0	1.20E+0	7.44E−02	1.49E−01	7.44E−02	2.73E−02
RI	1.42E−04	2.14E−05	9.52E−06	2.22E−06	3.15E−06	5.07E−07

GWP Global Warming Potential—kg CO_2 eq.; NRE Non-Renewable Energy—MJ primary; RI Respiratory inorganics—kg $PM_{2.5}$ eq

As mentioned, the fermentation stage was the stage associated with the greatest environmental impacts when considering energy consumption. The estimated emissions for the stage for the categories of respiratory inorganics (RI), global warming potential (GWP), and non-renewable inorganics were 0.05 kg $PM_{2.5}$ eq., 38.02 kg CO_2 eq. and 548.44 MJ primary for banana waste and 0.03 kg $PM_{2.5}$ eq., 21.61 kg CO_2 eq. and 311.70 MJ primary for banana waste, respectively.

It is noted that fermentation is the stage with the greatest energy demand, which contributes to its significant environmental impacts. This suggests that the fermentation process consumes significantly more energy and generates more greenhouse gas emissions, particularly in the GWP and NRE categories.

According to Costa et al. (2018), the fermentation stage is associated with the greatest impacts due to yeast production and the fermentation process itself, with its large production of CO_2 and high energy demand.

The use of energy and resources in the bioethanol production process significantly influences atmospheric emissions, such as carbon dioxide (CO_2), thus increasing the environmental impact associated with fermentation, mainly in the global warming potential (GWP) impact category (Roy et al. 2012; Singh et al. 2022).

The downstream processes of bioethanol production (fermentation, distillation, and dehydration) are energy-intensive. Obtaining high concentrations of bioethanol is crucial to achieving a positive net energy balance for the process (Singh et al. 2022). However, the environmental impacts related to energy use depend on the sources of the electrical grid mix and its related emissions (Ebner et al. 2014).

According to Costa et al. (2018), replacing grid electricity with locally generated renewable energy sources should be considered in an LCA to assess the reduction potential. Another recommendation is to improve both the equipment and the process itself to make bioethanol production more eco-efficient.

To reduce the impacts generated by electricity supply in small-scale laboratories, installing an electrical network from renewable energy sources could mitigate these adverse effects. However, this investigation must be specific to each country due to significant variations in energy supply and the location of experiments (Costa et al. 2018; Soleymani Angili et al. 2021).

The pretreatment stage was associated with the greatest impacts when energy consumption was not considered. The estimated emissions for the stage for the main impact categories were 1.42×10^{-4} kg $PM_{2.5}$ eq., 1.02^{-1} kg CO_2 eq. and 1.21 MJ primary for banana waste and 2.14×10^{-5} kg $PM_{2.5}$ eq., 1.01×10^{-1} kg CO_2 eq. and 1.20 MJ primary for potato waste for the respiratory inorganics (RI), global warming potential (GWP) and non-renewable inorganics categories, respectively.

It is important to mention that when energy is not considered, the environmental impacts of all stages decrease drastically. This highlights the importance of considering energy consumption over the full lifecycle to gain a more accurate view of environmental impacts.

The impacts generated when evaluating bioethanol production without considering energy consumption are mainly associated with the use of chemical products in bioethanol production, mainly in the pretreatment stage. Therefore, the use of sodium phosphate to adjust the pH in the pretreatment stage ends up being the input responsible for the greatest impacts. According to de Azevedo et al. (2017), the use of sodium phosphate contributed to the environmental impacts associated with the production of bioethanol from cattle manure.

The environmental impacts associated with phosphate are due to its complex production process, which mainly involves the phosphorus mining process commonly carried out through open-pit mining processes, in addition to not being a renewable resource with estimates of depletion of up to 100 years (Pradel and Aissani 2019; Issaoui et al. 2021). Thus, as it is a non-renewable resource, it ends up influencing the impacts associated with the non-renewable energy (NRE) category estimated for the scenario.

Issaoui et al. (2021) mention that high diesel consumption and the use of nitrate explosives during mining operations are highly related to higher CO_2 emissions, related to the global warming potential (GWP) impact category. According to Buratti et al. (2008), emissions of nitrogen oxides contribute 77% of the impacts associated with the category of respiratory inorganics (RI). Therefore, the use of nitrate explosives can be associated with impacts related to the use of sodium phosphate for the RI category.

De Azevedo et al. (2017) suggest that it is possible to improve bioethanol production, reducing its impacts, by modifying variables in various processes, such as pH adjustments. Furthermore, research into the development of materials from sustainable sources to replace conventional materials, as well as the practice of recycling and reusing chemicals/nutrients, can be considered in future studies as a strategy to mitigate possible environmental consequences (Costa et al. 2018).

Furthermore, analysis of the results reveals that the enzymatic hydrolysis stage has the lowest environmental impacts compared to the pretreatment and fermentation stages, both for banana and potato waste. This suggests that enzymatic hydrolysis may be a less resource- and energy-intensive step, which may have important implications for the sustainability of bioethanol production from these wastes.

Enzymatic hydrolysis is generally considered a key step in the conversion of lignocellulosic biomass into fermentable sugars due to its ability to break cellulose and hemicellulose bonds into simple sugars, which can then be fermented by microorganisms to produce ethanol (Amit et al. 2018; Singh et al. 2022).

However, it is important to note that results may vary depending on the specific conditions of each bioethanol production process and the characteristics of the waste used. Therefore, future research is needed to validate and further explore these findings, considering a broader range of variables such as enzyme type, reaction conditions, and biomass source.

When evaluating the inputs and outputs calculated for both wastes, presented in Table 2 in item 2.2, the inputs and outputs obtained demonstrate that banana waste is the residue with the highest input inputs, as well as the highest energy consumption, mainly in the fermentation phase with a consumption of 139.5 kWh, when compared to potato waste. However, it is observed that the yield of reducing sugars (148.4 g) and the yield of ethanol (19.2 g L^{-1}) are higher than those with potato waste.

The potato waste yields reducing sugars and ethanol at 136.4 g and 19 g L^{-1}, respectively, and energy consumption in the fermentation stage is 74 kWh. Bioethanol production from banana waste had greater impacts in all categories compared to potato waste. This may be due to differences in the composition of the waste and the efficiency of the processes involved.

The greatest production of bioethanol from banana waste occurs because this waste has high levels of sucrose, glucose, and other simple sugars (Utama et al. 2019). Furthermore, banana waste have a high source of energy and inverted sugar suitable for bioethanol fermentation, in addition to having higher total sugar contents than other residues (Dhanaseeli and Balasubramanian 2014; Vaitheki and Deepa 2016).

Bioethanol production from potato waste depends on physicochemical properties such as starch content (Chauhan et al. 2022) The high starch content of potato waste is documented for successful bioethanol production, and the low sugar-reducing needs more hydrolysis steps for enviable bioethanol production (Arapoglou et al. 2010; Khawla et al. 2014; Hossain et al. 2019).

Furthermore, the conversion of bioethanol from starch-based crops, such as potatoes, is more complex than sugar-based crops because they contain carbohydrates that must first be converted to simple sugars (glucose) and then fermented into bioethanol (Shapouri et al. 2002; Ensinas et al. 2009).

3.1.1 Uncertainty Analysis

Table 4 presents the study's uncertainty analysis, carried out using Monte Carlo statistical analysis, for banana and potato waste in the main impact categories. According to the IPCC, uncertainty is considered significant when the standard deviation divided by the mean (coefficient of variation) is greater than 0.3 (McMurray et al. 2017).

For both wastes, when considering energy consumption, the CV was less than 0.3. For the categories of global warming potential (GWP), non-renewable energy

Table 4 Uncertainty analysis by Monte Carlo simulation (1000 interactions and 95% confidence) of the impact results of the categories highlighted in the LCIA with the Impact 2002+ method

Impact category	Unit	Banana		Potato	
		Average	CV	Average	CV
With energy					
GWP	kg CO_2 eq	5.09E+01	0.03	3.61E+01	0.03
NRE	MJ primary	7.38E+02	0.09	5.21E+02	0.09
RI	kg $PM_{2.5}$ eq	6.09E−02	0.04	4.30E−02	0.04
Without energy					
GWP	kg CO_2 eq	1.10E−01	0.08	1.14E−01	0.08
NRE	MJ primary	1.32E+00	0.13	1.36E+00	0.11
RI	kg $PM_{2.5}$ eq	1.55E−04	0.12	1.61E−04	0.12

GWP Global Warming Potential; NRE Non-Renewable Energy; RI Respiratory inorganics

(NRE), and respiratory inorganics (RI), the CV was 0.03, 0.09, and 0.04, respectively, for banana and potato waste. Thus, there is greater reliability in the data obtained.

A CV < 0.3 was also observed when not considering energy consumption for banana and potato waste in the main impact categories. For the global warming potential (GWP) category, the CV was 0.08 for both wastes; for non-renewable energy (NRE), the CV was 0.13 and 0.11 for banana and potato waste, respectively; and for the respiratory inorganics (RI) category, the CV was 0.12 for both wastes. This also demonstrates greater reliability in the LCA data.

The study results indicate that the observed uncertainty is low, meaning there is greater reliability in the data. This increased reliability supports more secure decision-making and indicates consistency in the results. The low variation in the data suggests that the input data used are consistent and accurate.

According to Soleymani Angili et al. (2021), uncertainties derive from the lack of data or data inaccuracy in industrial processes, the selection of methodological elements (such as system boundary, allocation, impact categories, and characterization models), and the preparation of raw materials.

The production of bioethanol on a laboratory scale may present significant differences from the industrial scale due to experiments being conducted under strictly controlled conditions, which may not reflect the real conditions of large-scale production. Part of this uncertainty may be attributed to technology substitutions or materials used in real industrial facilities (Soleymani Angili et al. 2022).

Some authors mention that the reliability of data and results can be influenced by the level of technological progress. Therefore, there is a risk that the analysis of the effects of procedures occurs in the conceptual phases, in initial tests, and laboratories instead of being conducted on a commercial scale. This is due to the difficulty of evaluating the effectiveness of technologies in the early stages of their development (Spatari and MacLean 2010; Borrion et al. 2012).

4 Conclusions

Respiratory inorganics, global warming potential, and non-renewable energy are the most significant impact categories in the bioethanol production process from banana and potato waste, according to the Impact2002+ method. The fermentation phase has the largest impact on energy consumption, followed by the pretreatment phase when energy is not considered, for both scenarios.

These findings highlight the importance of addressing energy demand in bioethanol production, particularly focusing on the fermentation and pretreatment stages. Adopting renewable energy can mitigate these impacts, alongside optimizing the production process to reduce energy needs. Additionally, finding sustainable alternatives to sodium phosphate is crucial.

Bioethanol production from food waste offers a sustainable alternative to fossil fuels, aligning directly with UN Sustainable Development Goals (SDGs) like affordable and clean energy (SDG 7), consumption and sustainable production (SDG 12), and climate change mitigation (SDG 13).

Therefore. bioethanol from food waste plays a key role in promoting sustainability, reducing environmental impact, and advancing towards UN SDGs.

References

Amit K, Nakachew M, Yilkal B, Mukesh YJRJCE (2018) A review of factors affecting enzymatic hydrolysis of pretreated lignocellulosic biomass. Res J Chem Environ 22(7):62–67

Arapoglou D, Varzakas T, Vlyssides A, Israilides C (2010) Ethanol production from potato peel waste (PPW). Waste Manage 30(10):1898–1902. https://doi.org/10.1016/j.wasman.2010.04.017

Bender L (2022) Obtenção de bioetanol e ingrediente para ração animal a partir de resíduos alimentares. [Master's Thesis, Postgraduate Program in Food Science and Technology in University of Passo Fundo]

Bibra M, Samanta D, Sharma NK, Singh G, Johnson GR, Sani RK (2022) Food waste to bioethanol: opportunities and challenges. Fermentation 9(1):8. https://doi.org/10.3390/fermentation9010008

Borrion AL, McManus MC, Hammond GP (2012) Environmental life cycle assessment of lignocellulosic conversion to ethanol: a review. Renew Sustain Energy Rev 16(7):4638–4650. https://doi.org/10.1016/j.rser.2012.04.016

Buratti C, Barbanera M, Fantozzi F (2008) Environmental balance of bioethanol from corn grain: evaluation of different procedures of co-products allocation

Byun J, Kwon O, Kim J, Han J (2022) Carbon-negative food waste-derived bioethanol: a hybrid model of life cycle assessment and optimization. ACS Sustain Chem Eng 10(14):4512–4521. https://doi.org/10.1021/acssuschemeng.1c08300

Chauhan M, Dutt S, Manjul AS, Singh B, Garlapati VK (2022) A sustainable approach of turning potato waste towards bioethanol production using indigenous microbes of Himachal Pradesh, India. Chemosphere 299. https://doi.org/10.1016/j.chemosphere.2022.134429

Costa D, Jesus J, Virgínio e Silva J, Silveira M (2018) Life cycle assessment of bioethanol production from sweet potato (Ipomoea batatas L.) in an experimental plant. Bioenergy Res 11(3):715–725. https://doi.org/10.1007/s12155-018-9932-1

Crenna E, Secchi M, Benini L, Sala S (2019) Global environmental impacts: data sources and methodological choices for calculating normalization factors for LCA. Int J Life Cycle Assess 24(10):1851–1877. https://doi.org/10.1007/s11367-019-01604-y

de Azevedo A, Fornasier F, da Silva Szarblewski M, Schneider RCS, Hoeltz M, de Souza D (2017) Life cycle assessment of bioethanol production from cattle manure. J Clean Prod 162:1021–1030. https://doi.org/10.1016/j.jclepro.2017.06.141

Dhanaseeli PB, Balasubramanian V (2014) Studies of ethanol production from different fruit wastes using saccharomyces cerevisiae. Biosci Biotechnol Res Asia. 11(SepcialEdition 1):19–23. https://doi.org/10.13005/bbra/1386

Dhiman S, Mukherjee G (2021) Present scenario and future scope of food waste to biofuel production. J Food Process Eng 44(2). https://doi.org/10.1111/jfpe.13594

Ebner J, Babbitt C, Winer M, Hilton B, Williamson A (2014) Life cycle greenhouse gas (GHG) impacts of a novel process for converting food waste to ethanol and co-products. Appl Energy 130:86–93. https://doi.org/10.1016/j.apenergy.2014.04.099

Ecoinvent Support (2024) Impact assessment, ecoinvent support. https://support.ecoinvent.org/impact-assessment

Ensinas AV, Modesto M, Nebra SA, Serra L (2009) Reduction of irreversibility generation in sugar and ethanol production from sugarcane. Energy 34(5):680–688. https://doi.org/10.1016/j.energy.2008.06.001

Food and Agriculture Organization of the United Nations (FAO) (2021) 17% of total food available to consumers in 2019 was wasted. https://www.fao.org/platform-food-loss-waste/news/news-detail/17-of-total-food-available-to-consumers-in-2019-was-wasted/en

Guerrero AB, Muñoz E (2018) Life cycle assessment of second generation ethanol derived from banana agricultural waste: environmental impacts and energy balance. J Clean Prod 174:710–717. https://doi.org/10.1016/j.jclepro.2017.10.298

Hossain N, Razali AN, Mahlia TMI, Chowdhury T, Chowdhury H, Ong HC, Shamsuddin AH, Silitonga AS (2019) Experimental investigation, techno-economic analysis and environmental impact of bioethanol production from banana stem. Energies (Basel) 12(20). https://doi.org/10.3390/en12203947

Humbert S, De Schryver AM, Margni M, Jolliet O (2012) IMPACT 2002+: user guide. Draft for version Q2.2 (version adapted by Quantis). Quantis, Lausanne, Switzerland

Ingale S, Joshi SJ, Gupte A (2014) Production of bioethanol using agricultural waste: banana pseudo stem

IEA (2022) Energy system: transport, international energy agency. https://www.iea.org/energy-system/transport

International Organization for Standardization (2006a) Environmental management: life cycle assessment; principles and framework, vol 14040. ISO, Geneva, Switzerland

International Organization for Standardization (2006b) Environmental management: life cycle assessment; requirements and guidelines, vol 14044. ISO, Geneva, Switzerland

Issaoui R, Rösch C, Woidasky J, Schmidt M, Viere T (2021) Cradle-to-gate life cycle assessment of beneficiated phosphate rock production in Tunisia. Sustain Manag Foruml Nachhaltigkeits Manag Forum 29(2):107–118. https://doi.org/10.1007/s00550-021-00522-8

Jolliet O, Margni M, Charles R, Humbert S, Payet J, Rebitzer G, Rosenbaum R (2003) IMPACT 2002+: a new life cycle impact assessment methodology. Int J Life Cycle Assess 8(6):324–330. https://doi.org/10.1007/BF02978505

Kazemi Shariat Panahi H, Dehhaghi M, Guillemin GJ, Gupta VK, Lam SS, Aghbashlo M, Tabatabaei M (2022) Bioethanol production from food wastes rich in carbohydrates. Curr Opin Food Sci 43:71–81. https://doi.org/10.1016/j.cofs.2021.11.001

Khawla BJ, Sameh M, Imen G, Donyes F, Dhouha G, Raoudha EG, Oumèma NE (2014) Potato peel as feedstock for bioethanol production: a comparison of acidic and enzymatic hydrolysis. Ind Crops Prod 52:144–149. https://doi.org/10.1016/j.indcrop.2013.10.025

Konti A, Kekos D, Mamma D (2020) Life cycle analysis of the bioethanol production from food waste—a review. Energies (Basel) 13(19). https://doi.org/10.3390/en13195206

Król-Badziak A, Pishgar-Komleh SH, Rozakis S, Księżak J (2021) Environmental and socio-economic performance of different tillage systems in maize grain production: application of life cycle assessment and multi-criteria decision making. J Clean Prod 278. https://doi.org/10.1016/j.jclepro.2020.123792

Kumar V, Sharma N, Umesh M, Selvaraj M, Al-Shehri BM, Chakraborty P, Duhan L, Sharma S, Pasrija R, Awasthi MK et al (2022) Emerging challenges for the agro-industrial food waste utilization: a review on food waste biorefinery. Bioresour Technol 362. https://doi.org/10.1016/j.biortech.2022.127790

Lareo C, Ferrari MD, Guigou M, Fajardo L, Larnaudie V, Ramírez MB, Martínez-Garreiro J (2013) Evaluation of sweet potato for fuel bioethanol production: hydrolysis and fermentation. Springerplus 2(1):1–11. https://doi.org/10.1186/2193-1801-2-493

Li R, Chen W, Xiu A, Zhao H, Zhang X, Zhang S, Tong DQ (2019) A comprehensive inventory of agricultural atmospheric particulate matters (PM10 and PM2.5) and gaseous pollutants (VOCs, SO_2, NH_3, CO, NOx and HC) emissions in China. Ecol Indic 107. https://doi.org/10.1016/j.ecolind.2019.105609

Liu Y, Sun W, Liu J (2017) Greenhouse gas emissions from different municipal solid waste management scenarios in China: based on carbon and energy flow analysis. Waste Manage 68:653–661. https://doi.org/10.1016/j.wasman.2017.06.020

McMurray A, Pearson T, Casarim F (2017) Guidance on applying the monte Carlo approach to uncertainty analyses in forestry and greenhouse gas accounting. Guidance on applying the monte Carlo approach

Narisetty V, Nagarajan S, Gadkari S, Ranade VV, Zhang J, Patchigolla K et al (2022) Process optimization for recycling of bread waste into bioethanol and biomethane: a circular economy approach. Energy Convers Manage 266:115784. https://doi.org/10.1016/j.enconman.2022.115784

Palacios S, Ruiz HA, Ramos-Gonzalez R, Martínez J, Segura E, Aguilar M, Aguilera A, Michelena G, Aguilar C, Ilyina A (2017) Comparison of physicochemical pretreatments of banana peels for bioethanol production. Food Sci Biotechnol 26(4):993–1001. https://doi.org/10.1007/s10068-017-0128-9

Panahi HKS, Dehhaghi M, Kinder JE, Ezeji TC (2019) A review on green liquid fuels for the transportation sector: a prospect of microbial solutions to climate change. Biofuel Res J 6(3):995–1024. https://doi.org/10.18331/brj2019.6.3.2

Pradel M, Aissani L (2019) Environmental impacts of phosphorus recovery from a product life cycle assessment perspective: allocating burdens of wastewater treatment in the production of sludge-based phosphate fertilizers. Sci Total Environ 656:55–69. https://doi.org/10.1016/j.scitotenv.2018.11.356

Roy P, Orikasa T, Tokuyasu K et al (2012) Evaluation of the life cycle of bioethanol produced from rice straws. Bioresour Technol 110:239–244. https://doi.org/10.1016/j.biortech.2012.01.094

Sawarkar AN, Kirti N, Tagade A, Tekade SP (2022) Bioethanol from various types of banana waste: a review. Bioresour Technol Rep 18. https://doi.org/10.1016/j.biteb.2022.101092

Shapouri H, Duffield JA, Wang M (2002) The energy balance of corn ethanol: an update. [place unknown]

Singh A, Singhania RR, Soam S, Chen CW, Haldar D, Varjani S, Chang JS, Dong C Di, Patel AK (2022) Production of bioethanol from food waste: status and perspectives. Bioresour Technol 360. https://doi.org/10.1016/j.biortech.2022.127651

Soleymani Angili T, Grzesik K, Rödl A, Kaltschmitt M (2021) Life cycle assessment of bioethanol production: a review of feedstock, technology and methodology. Energies (Basel) 14(10). https://doi.org/10.3390/en14102939

Soleymani Angili T, Grzesik K, Salimi E, Loizidou M (2022) Life cycle analysis of food waste valorization in laboratory-scale. Energies (Basel) 15(19). https://doi.org/10.3390/en15197000

Spatari S, MacLean HL (2010) Characterizing model uncertainties in the life cycle of lignocellulose-based ethanol fuels. Environ Sci Technol 44(22):8773–8780. https://doi.org/10.1021/es102091a

Uçkun Kiran E, Liu Y (2015) Bioethanol production from mixed food waste by an effective enzymatic pretreatment. Fuel 159:463–469. https://doi.org/10.1016/j.fuel.2015.06.101

Utama GL, Sidabutar FEE, Felina H, Wira DW, Balia RL (2019) The utilization of fruit and vegetable wastes for bioethanol production with the inoculation of indigenous yeasts consortium

Vaitheki S, Deepa B (2016) A comparative study on the production of bioethanol from individual and mixed fruit wastes. Imperial J Interdisc Res (IJIR) 2(5):652–653

Wang M, Shi Y, Xia X, Li D, Chen Q (2013) Life-cycle energy efficiency and environmental impacts of bioethanol production from sweet potato. Bioresour Technol 133:285–292. https://doi.org/10.1016/j.biortech.2013.01.067

The Potential of Digital Product Passports as Data Providers for LCA—Data Mapping for Two Automotive Use Cases

Martin Popowicz, Nicolas J. Katzer, Antonia Pohlmann, Josef-Peter Schöggl, Katharina Berger, and Rupert J. Baumgartner

Abstract To enable a more sustainable future and achieve the United Nations Sustainable Development Goals (SDGs), it is important to quantify the impacts of processes, products, and services. Life cycle assessment (LCA) has become a widespread tool for assessing environmental impacts. The role of LCA is acknowledged by various decision-makers and is, thus, included as an important part of policies that aim to contribute to the achievement of the SDGs. Although LCA is a long-standing research topic and both the quality of obtained LCA results and the overall process of conducting LCAs have improved significantly over time, many challenges remain. LCA data collection is often time-consuming, the availability of robust data in the inventory analysis phase can be limited, and the assumptions made by producers and service providers for the subsequent life cycle phases (i.e., use and end-of-life) can be problematic. To overcome data-related LCA challenges, an emerging circular economy policy instrument, the digital product passport (DPP), offers a potential solution. Depending on its final design, a DPP can contain LCA-specific data relevant to all life cycle stages of a product. For example, a DPP may entail specific materials (comprising a share of secondary raw materials), may include information about manufacturers and suppliers of a product, and may anonymously track data related to the use phase or specific emissions and waste streams during the end-of-life phase. These data points are necessary for detailed LCA studies and constitutive policies; however, they are at present mostly available only as generic secondary data. To address this issue, this study focuses on the potential combination of DPPs and LCA by examining which data points may be incorporated into

M. Popowicz · A. Pohlmann · R. J. Baumgartner (✉)
Department of Environmental Systems Sciences, University of Graz, Graz, Austria
e-mail: rupert.baumgartner@uni-graz.at

N. J. Katzer · J.-P. Schöggl · K. Berger · R. J. Baumgartner
Christian Doppler Laboratory for Sustainable Product Management Enabling a Circular Economy, Department of Environmental Systems Sciences, University of Graz, Graz, Austria

J.-P. Schöggl
School of Sustainability, Governance, and Methods, Modul University Vienna, Vienna, Austria

L. Brandli et al. (eds.), *The Contribution of Life Cycle Analyses and Circular Economy to the Sustainable Development Goals*, World Sustainability Series,
https://doi.org/10.1007/978-3-031-85300-5_26

dynamic DPPs, and how this could directly provide the complete life cycle inventory (LCI; the calculation basis) to achieve more accurate LCAs. Differentiation between generic and potentially instance-specific data points (i.e., primary data of the specific product unit mostly stemming from automated tracking) for the LCA enables highlighting of some structural consequences for the DPP. Two use cases from the automotive industry are investigated, namely traction batteries of electric vehicles and turbochargers of combustion engines, to demonstrate possible configurations of DPP data points, their different LCA system models, and the implications of the results obtained. The automotive industry was considered an interesting use case, given the challenges it faces in becoming more sustainable and contributing to the SDGs. The results show the data mapping of the data points of the potential DPPs of both use cases to the corresponding life cycle inventories necessary for the calculation of an LCA. Consequently, this research serves as a starting point to further investigate the possibility of DPPs as data providers for LCIs and LCAs. Implementing this possible combination would facilitate the development of LCA as a reliable instrument to measure the environmental performance and progress of specific products towards the SDGs.

Keywords Digital product passport · Life cycle assessment · LCA · Instance-specific LCA

1 Introduction

Since their introduction in 2016, the Sustainable Development Goals (SDGs) have gained widespread attention as guiding principles for steering global sustainable development. Despite the widely acknowledged importance of sustainable development, achieving the SDGs remains a significant challenge. By 2023, half of the 140 quantifiable SDG targets exhibit moderate or severe deviations from the desired development pathway, with more than 30% showing no progress or regression below the 2015 baseline (United Nations Department of Economic and Social Affairs 2023). Overall, 15% of targets are considered on track, 48% are moderately or severely off track, and 37% demonstrate stagnation or regression (United Nations Department of Economic and Social Affairs 2023). This also holds for SDG12, which addresses sustainable production and consumption patterns. As outlined by the United Nations Department of Economic and Social Affairs (2023) in their 2023 SDGs Report, achieving SDG12 requires the effective implementation of integral policies that decouple economic growth from resource use to shift towards sustainable practices. This includes unequal responsibilities and consumption disparities, which must be uncovered by improved monitoring and reporting. It is crucial to understand the environmental impacts of varying consumption patterns and related products and services in order to ensure that the design of new products and services contribute meaningfully to achieving SDG 12 and other related objectives.

Life cycle assessment (LCA) has emerged as an important tool for evaluating the environmental impacts of different technologies, products, and services. LCA encompasses various environmental challenges and may serve as a decision-making tool for companies (Jolliet et al. 2015). Consequently, it can empower decision-makers within companies to choose technologies and products with lower environmental impacts, thereby contributing to achieving the SDGs. The support provided by LCA to decision-makers has fueled interest in the tool, leading to its widespread adoption on global scales (Curran 2014). Nevertheless, numerous challenges persist in applying LCAs, especially where data-related challenges play an important role. One major barrier to adopting LCA for environmental impact assessment lies in data collection; indeed, the gathering of data from suppliers, as well as internal data, are considered challenging and time-consuming (Testa et al. 2016). It is estimated that approximately 70–80% of the time and costs associated with conducting an LCA are attributed to data collection, the refitment of data inaccuracies, and addressing data gaps (Miah et al. 2018). These data gaps and inaccuracies can increase uncertainty in LCA models and challenge the representativeness of LCA results (Björklund 2002). Hence, data collection- and reliability-related challenges must be addressed to empower LCA as an effective means of measuring environmental impacts while contributing to achieving the SDGs of consumption patterns, products, and services. To address these challenges and provide accurate and timely data for conducting LCAs, the emerging circular economy policy instrument of digital product passports (DPPs) can offer a potential solution. A DPP can be described as a combination of a unique identifier of the product or service and corresponding data collected by actors along the value chain (Lemos 2020). These data may entail, for example, product characteristics and life cycle data (Lemos 2020) intended to facilitate information sharing among businesses operating within a specific value chain, while enabling companies to access relevant information to enhance their sustainability and circularity performance (European Commission 2022). Hence, dynamic DPPs can also incorporate data to facilitate the process of LCA, making it more accurate and reliable. A dynamic DPP can be understood to include data relevant to both the static and dynamic life cycles and, therefore, shares similarities with the concept of digital twins and with the newly emerging concept of data spaces (e.g., Catena-X (n.d.)). Given that the final design of DPPs and the different possible design options regarding the included data are not yet determined and that LCA probably plays an important role in achieving the SDGs, this contribution focuses on the potential role of the circular economy policy instrument of DPPs in facilitating and improving LCAs. To better showcase the potential of DPPs (and similar approaches such as digital twins and data spaces) for LCA, two cases from the automotive industry are considered: traction batteries from electric vehicles and turbochargers of internal combustion engines. The research question can hence be formulated as follows:

RQ: How can DPPs as data providers facilitate and improve the execution of LCAs?

To answer the research question, this study applies a technology-agnostic approach in identifying data points that could potentially be part of a dynamic DPP for traction batteries and turbochargers, then maps them to the life cycle inventories for both cases.

The remainder of this chapter is structured as follows: Sect. 2 addresses data-related LCA challenges, Sect. 3 introduces DPPs, and Sect. 4 presents the applied method. Section 5 shows the data mapping, and the paper closes with a discussion and conclusions section.

2 Data-Related Challenges in LCA

Life Cycle Assessment (LCA) is a globally recognized tool for assessing environmental impacts and performing environmental management. Its ISO standards 14040 and 14044 prescribe a "cradle to grave" perspective, incorporating the entirety of the life cycle—from the sourcing of materials, through their processing steps, until the manufacturing of parts, components, and final products, as well as the subsequent use of the product, and ultimately its end-of-life (ISO 2006a; ISO 2006b). Hence, LCA provides a comprehensive picture of the environmental impact of the assessed product or service and has been increasingly applied across various industries as a decision-supporting tool (Moutik et al. 2023).

However, due to its aim to provide a comprehensive picture of the environmental impact, the application of an LCA entails challenges related to measuring, collecting, and sufficiently modeling the physical and energetic input and output flows. The recommended approach is to use primary data, which is directly measured and compiled to represent one unit of the product or service (ISO 2006a). If measurement proves to be too resource-intensive or otherwise impossible, secondary data are used, in which case it is expected to rely on specialized life cycle inventory data providers. However, this approach relies solely on shared primary data, which are manually adapted and converged through extensive expert elicitation to represent process-specific average datasets (Ecoinvent 2024). It is thus understandable that practitioners often do not find appropriate datasets (Ciroth et al. 2016) that reflect the actual environmental impacts associated with all process-related parts of their assessment. This leads to the last resort: estimating missing data through best-practice modeling approaches.

Inheriting characteristics from its methodological evolution and applicability, conventional LCAs are also mostly attributional and static in order to reduce the data intensity required (Su et al. 2022), i.e., they model the accessible physical and energetical input and output flows by retrospectively aggregating them over the whole life cycle (Hauschild et al. 2018). A conventional LCA assumes constant resource consumption, emissions, and environmental effects throughout assessment and (mostly) neglects temporal- and spatial-dependent parameters. Thus, data providers and software tools have no or only rudimentary spatiotemporal capabilities (Ecoinvent 2024). This, in turn, limits or weakens the environmental relevance of the results and limits their applicability, as solely retrospective evaluations are possible (Su et al. 2022). Methodological developments of LCA, however, aim to enable additional fields of application, such as product development and strategic sustainable development (Bjørn et al. 2020; Van Der Giesen et al. 2020). However, an application in these fields requires evaluating the input and output streams over the entirety of the product's life cycle and related background systems even more granularly than in a conventional ex-post assessment, exacerbating the need for additional data. Theoretically, these data needs may be satisfied by leveraging the pillars of Industry 4.0 and 5.0.

Therefore, LCA faces the challenge of linking its concurrent methodological developments to emerging digital technologies and, specifically, to the DPP, in order to overcome its inherent data requirements.

3 Digital Product Passports

Digital product passports have recently become popular by virtue of their potential to contribute to a more sustainable and circular economy. DPPs are included in various regulatory frameworks and policies, for example, the circular economy action plan (European Commission 2020) and regulations on the establishment of a framework for the eco-design requirements of sustainable products (European Commission 2022), which mandates the development of DPPs for registering, processing, and sharing product-related information with supply chain actors, authorities, and consumers. Until now, the scientific community has focused on the following different DPP-related topical areas: Firstly, existing literature considers DPPs as general enablers of a circular economy (Adisorn et al. 2021; Walden et al. 2021). A second stream of literature focuses on the conceptualization of digital products passports, like for example Berger et al. (2022), who focused on the conceptualization of a battery passport, and Plociennik et al. (2022a), who introduced the concept of a DPP based on the Asset Administration Shell. A third identified literature stream deals with the definition of data requirements for DPPs. Examples include Plociennik et al. (2022b), focusing on the information requirements for a broadly applicable DPP in different sectors, Berger et al. (2023), who focus on the requirements of a battery passport, or also Jansen et al. (2023) and Jensen et al. (2023). Further DPP-related studies have focused on circular supply chain management systems based on a DPP (Koppelaar et al. 2023), the universal definition of a digital product passport ecosystem that applies for policy makers, technical, and industrial communities (King et al. 2023), or on investigations of DPPs from a data governance perspective (Ducuing and Reich 2023). Thus far, although Plociennik et al. (2022b) identified the potential of information included in DPPs for AI-based LCAs, no studies have focused on integrating DPPs and LCA.

4 Method

A four-step research approach was developed to investigate how information included in DPPs can be used to achieve improved and more precise LCA results. Firstly, based on previous research on the product passport of traction batteries (Berger et al. 2022), where the initial concept was continuously enhanced within the authors' research group, and on the conceptualization of a product passport for turbochargers (Pohlmann et al. 2024), lists of data points were derived and adapted to serve as the basis for the LCA-relevant data mapping. In the second step, the existing data points were analyzed to evaluate their suitability as LCIs for the LCA, irrelevant data points were deleted, existing points were enriched to work in the context of LCA, and new data points necessary for the execution of LCAs were added. In the third step, the resulting data points in both use cases were mapped onto a generic LCI based on previous automotive LCA studies. The generic life cycle inventory (LCI), including all sub-information categories, is shown in Fig. 1, which presents an own illustration based on Hauschild et al. (2018) and Berger et al. (2022). In the information category "generalized LCA and auxiliary information" data is summarized that can support the general LCA process in the case that a more generic LCA is conducted. Lastly, the prior steps were validated during an expert workshop (N = 7) to ensure coherence.

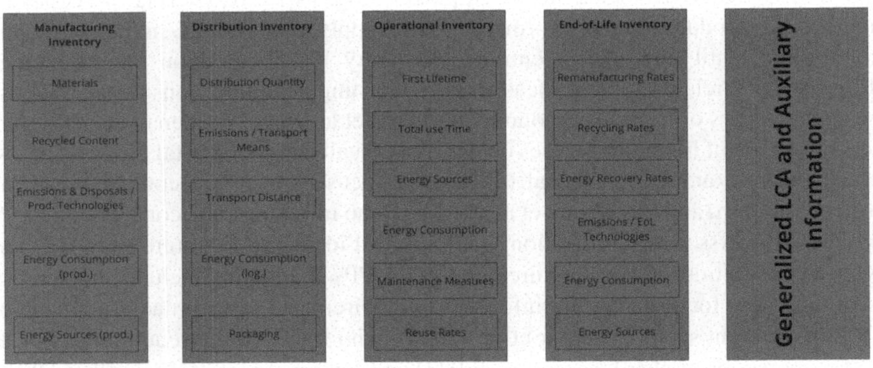

Fig. 1 Generic life cycle inventory. Illustration based on written context from Hauschild et al. (2018) and Berger et al. (2022)

5 DPP Information for LCA

In this section, the identified LCA-relevant data points for both use cases are introduced, then the mapping of the potential DPP data points on the different inventories for both use cases is presented.

5.1 Identified LCA-Relevant Data Points

Here, the identified data points relevant for both LCA use cases and their corresponding descriptions are introduced. The third column further classifies the role that the data point embodies for an LCA, distinguishing between *instance-specific* (i.e., primary data of the case-specific product unit mostly stemming from automated tracking), *generalized* (primary or secondary data of the general product line enabling a traditional LCA model), and *auxiliary* (key information for certain scenarios or coherence checks). The last column specifies the source of the data points, distinguishing between data points that are *regulatory requirements* (Regulation (EU) 1542/2023), data points based on the *DPP concept*s developed by Berger et al. (2022) (battery case) and Pohlmann et al. (2024) (turbocharger case), and data points considered *LCA-related* (i.e., newly added data points during this study). Table 1 shows the battery passport data points relevant for LCA.

Table 2 presents the data points of the turbocharger passport relevant for LCA accompanied by the corresponding categorization and description.

Table 1 Battery passport data points

Data point name	LCA relevant description	Role	Life cycle phase	Source
Weight	Refers to the weight (packs, modules, and cells)	Instance-specific or generalized	Manufacturing, logistics	Regulatory requirement
Full material disclosure	Detailed specification of the chemical composition of the materials (e.g., iron, unalloyed steel, stainless steel, etc.) used for the individual battery components, including its designation, material number, physical, and chemical properties	Instance-specific or generalized	Manufacturing	Regulatory requirement
Primary raw materials	Information on the manufacturing processes of primary raw materials (e.g., rimmed and hot rolled steel or killed and forged steel)	Instance-specific or generalized	Manufacturing	DPP concept

(continued)

Table 1 (continued)

Data point name	LCA relevant description	Role	Life cycle phase	Source
Secondary raw materials	Information on the manufacturing processes of secondary raw materials (e.g., mechanical separation of scrap combined with rimming and hot rolling of steel)	Instance-specific or generalized	Manufacturing	DPP concept
Emissions and disposals	Information about all outputs released (secondary products, emissions, disposals, etc.) next to the product	Instance-specific or generalized	Manufacturing, EoL	LCA-related
Production technologies	Information about the applied technologies during the production and manufacturing of the main components	Instance-specific or generalized	Manufacturing	LCA-related
Energy consumption along the lifecycle	Tracking the total energy (and potentially related carriers) used along the individual sub-phases of the battery's entire life cycle	Instance-specific or generalized	All life cycle phases	DPP concept
Transport specification	Next to the handling instructions during the transportation of the battery or parts of the battery, information about the packaging (including its materials and their quantity)	Instance-specific or generalized	Logistics	DPP concept
Logistics information	Information collection about the transport means, load, distance, and energy source along the value chain	Instance-specific or generalized	Logistics	LCA-related

(continued)

Table 1 (continued)

Data point name	LCA relevant description	Role	Life cycle phase	Source
Battery identifier	Refers to an identifier, such as serial number or batch number, used to track the item (potentially, the main components may also include identifiers)	Instance-specific	Operation	DPP concept
Maintenance history	Record collection about applied maintenance	Instance-specific or generalized	Operation	LCA-related
Circularity assessment-related information	Information on various R-strategy application rates (actual tracked maintenance, longevity, replacements, and recycling), including the calculated overall circularity indexes/ results (and potentially the components' IDs)	Instance-specific or generalized	EoL	DPP concept
Technologies—EoL	Information about applied EoL-technologies (potentially the last information added to the DPP)	Instance-specific or generalized	EoL	LCA-related
Expected range	Information regarding the expected distance that can be achieved after one charge with a new battery in the respective battery electric vehicle	Generalized		DPP concept
Expected lifetime	Information on the expected battery lifetime, expressed in full charge and discharge cycles	Generalized		Regulatory requirement

(continued)

Table 1 (continued)

Data point name	LCA relevant description	Role	Life cycle phase	Source
Number of total modules	Number of total modules	Generalized		DPP concept
Number of cells per module	Refers to the number of total cells per battery module	Generalized		DPP concept
Number of total cells	Refers to the number of total cells per battery pack	Generalized		DPP concept
Supplier information	Information about the suppliers of the battery and battery components such as the supplier of the battery cell and the cathode producer. This comprises the name of the suppliers and contact information as well as geographical details about the companies supplying materials, such as their location	Generalized		DPP concept
Manufacturer information	Information about the battery manufacturer, such as the name and contact information, as well as geographical details of the battery manufacturer, such as location, production site, and unique facility identifier	Generalized		Regulatory requirement
Information on fulfilled regulations and laws	List of compliance with regulations, standards, and laws that are relevant and mandatory for the product, such as the Declaration of Conformity (DoC)	Auxiliary		Regulatory requirement

(continued)

Table 1 (continued)

Data point name	LCA relevant description	Role	Life cycle phase	Source
Repair instructions	Information for repairing the battery or parts, e.g., instructions for carrying out an error analysis and for fixing an error or deficiency	Auxiliary		DPP concept
Dismantling instructions	Instructions or detailed description/ drawings of how the battery can be dismantled. This includes both disassembling the battery from the car and dismantling the battery itself and its individual parts. This entails, for example, the battery discharging manual	Auxiliary		Regulatory requirement
Refurbishment/ Remanufacturing instructions	Information for remanufacturing a used battery and making it available for redistribution either "like new" (remanufacturing) or "used" (refurbishment)	Auxiliary		DPP concept
Recycling instructions	Information on recycling-specific information for different components, e.g., melting points, compatible solvent agents, etc	Auxiliary		LCA-related

(continued)

Table 1 (continued)

Data point name	LCA relevant description	Role	Life cycle phase	Source
State of health	Information about the current state of health of the battery (as a percentage), denoting its ability to deliver the desired performance relative to its initial condition. Typically, a battery's state of health at the time of manufacture is 100% and decreases with time and usage	Auxiliary		Regulatory requirement
State of Certified Energy (SOCE)	Indicates the measured or on-board UBE performance at a specific point in its lifetime, expressed as a percentage of the certified usable battery energy	Auxiliary		DPP concept
State of Certified Range (SOCR)	Indicates the measured or on-board electric range at a specific point during its lifetime, expressed as a percentage of the certified range	Auxiliary		DPP concept
Long-term trend of the state of health	Information about the long-term trend of the state of health since the first use of the battery	Auxiliary		DPP concept
Maintenance instructions	Instructions for carrying out maintenance measures, such as inspection, maintenance, restoration, and improvement to maintain the target condition of the battery	Auxiliary		DPP concept

(continued)

Table 1 (continued)

Data point name	LCA relevant description	Role	Life cycle phase	Source
Maintenance work trigger	Refers to triggers of suggested or required maintenance work	Generalized		DPP concept

Table 2 Turbocharger passport data points

Data point name	LCA relevant description	Role	Life Cycle Phase	Source
Primary raw materials	Information about the manufacturing processes of primary raw materials (e.g., rimmed and hot rolled steel or killed and forged steel)	Instance-specific or generalized	Manufacturing	DPP concept
Secondary raw materials	Information on the manufacturing processes of secondary raw materials (e.g., mechanical separation of scrap combined with rimming and hot rolling of steel)	Instance-specific or generalized	Manufacturing	DPP concept
Material specification	Detailed specification for the chemical composition of the material (e.g., iron, unalloyed steel, stainless steel, etc.) used for the individual turbocharger components, including its physical and chemical properties	Instance-specific or generalized	Manufacturing	DPP concept
Material number	Clear designation of the materials of the individual components	Instance-specific	Manufacturing	DPP concept

(continued)

Table 2 (continued)

Data point name	LCA relevant description	Role	Life Cycle Phase	Source
Information about the production processes	Information about the technologies applied during the production and manufacturing of the main components	Instance-specific or generalized	Manufacturing	DPP concept
Specification of upstream processes	Information regarding upstream production processes (e.g., auxiliary materials, secondary products, emissions, disposals, etc.)	Instance-specific or generalized	Manufacturing	DPP concept
Design information	Next to the drawings of the product, this point also includes its main components and information about weights	Instance-specific or generalized	Manufacturing, logistics	DPP concept
Designation of the components	Exact listings/ designation (BOM) of the various components that make up the turbocharger	Instance-specific or generalized	Manufacturing	DPP concept
Energy consumption along the lifecycle	Information regarding the total energy consumption of the individual sub-phases of the entire life cycle. (potentially including relevant energy carriers during the different phases of the life cycle)	Instance-specific or generalized	All life cycle phases	DPP concept

(continued)

Table 2 (continued)

Data point name	LCA relevant description	Role	Life Cycle Phase	Source
Packaging information	Document is required to contain information about the packaging category, its materials, and their quantities	Instance-specific or generalized	Logistics	LCA-related
Logistics information	Specification of the delivery routes and distances covered by the product	Instance-specific or generalized	Logistics	DPP concept
Maintenance history	Records of any maintenance conducted	Instance-specific or generalized	Operation	LCA-related
Turbocharger identifier	Refers to an identifier, such as serial number or batch number, which is used to track the item (the main components may also feature identifiers)	Instance-specific	Operation	LCA-related
Specification of downstream processes	Information regarding the downstream EoL processes (e.g., auxiliary materials, secondary products, emissions, disposals, etc.)	Instance-specific or generalized	EoL	LCA-related
Technologies—EoL	Information about the applied EoL-technologies (potentially the last information added to the DPP)	Instance-specific or generalized	EoL	LCA-related

(continued)

Table 2 (continued)

Data point name	LCA relevant description	Role	Life Cycle Phase	Source
Circularity assessment related information	Information about various R-strategy application rates (actual tracked maintenance, longevity, replacements, and recycling), including the calculated overall circularity indexes/ results (and potentially the components' IDs)	Instance-specific or generalized	EoL	LCA-related
Supplier information	Information about the suppliers of components. This comprises the name of the suppliers and their contact information, as well as geographical details about the supplying companies, such as their location	Generalized		LCA-related
Manufacturer information	Information about the manufacturer, such as their name, contact information, and geographical details	Generalized		LCA-related
Expected lifetime	Information on the expected battery lifetime, expressed in full charge and discharge cycles	Generalized		LCA-related
Maintenance work trigger	Refers to triggers of suggested/required maintenance work	Generalized		LCA-related
Sensor data	Collection of sensors data (e.g., regarding operation diagnosis or maintenance scheduling)	Generalized		DPP concept

(continued)

Table 2 (continued)

Data point name	LCA relevant description	Role	Life Cycle Phase	Source
Information on fulfilled regulations and laws	List of compliance with regulations, standards, and laws that are relevant and mandatory for the product, such as the Declaration of Conformity (DoC)	Auxiliary		DPP concept
Remanufacturing instructions	Information required for remanufacturing the turbocharger (e.g., replacing the bearing)	Auxiliary		DPP concept
Overall condition of the system	Information regarding the current condition of the turbocharger, e.g., electronic parameters	Auxiliary		DPP concept
Disassembly instructions	Instructions or detailed description explaining how the turbocharger can be disassembled or dismantled	Auxiliary		DPP concept
Recycling instructions	Description of all necessary information for the potential recycling processes (e.g., hardness and melting points)	Auxiliary		DPP concept
Repair instructions	Information required for repairing the turbocharger or parts thereof, e.g., instructions for carrying out error analysis, maintenance, and servicing	Auxiliary		DPP concept

5.2 *Data Mapping of Life Cycle Inventories for Use Cases*

Here, data mapping for different inventories, starting with the manufacturing inventory, is presented for both cases.

5.2.1 Manufacturing Inventory

Figure 2 shows the mapping of the data points from the turbocharger and the battery use case onto the manufacturing inventory.

Regarding the turbocharger, the data points *primary raw materials, secondary raw materials, material specification, material number, specification of upstream processes, design information*, and *designation of components* are mapped onto the manufacturing inventory information category *materials. Material number, material specification*, and *secondary raw materials* are mapped onto *recycled content*. The data points *information about the production process* and *specification of upstream processes* are mapped onto the inventory information category *emissions & disposals/ production technologies*. The data point *energy consumption along the lifecycle* is mapped onto the information category *energy consumption (production)* and onto the information category *energy sources.*

In the case of the traction battery, the data points *weight, full material disclosure, primary raw materials,* and *secondary raw materials* are mapped onto the inventory information category *materials*. The data points *full material disclosure* and *secondary raw materials* are mapped onto *recycled content*. The data points *emissions & disposals* and *production technologies* are mapped to the inventory information category *emissions & disposals/production technologies*. The data point *energy consumption along the lifecycle* is mapped onto the information categories *energy consumption (production)* and *energy sources (production).*

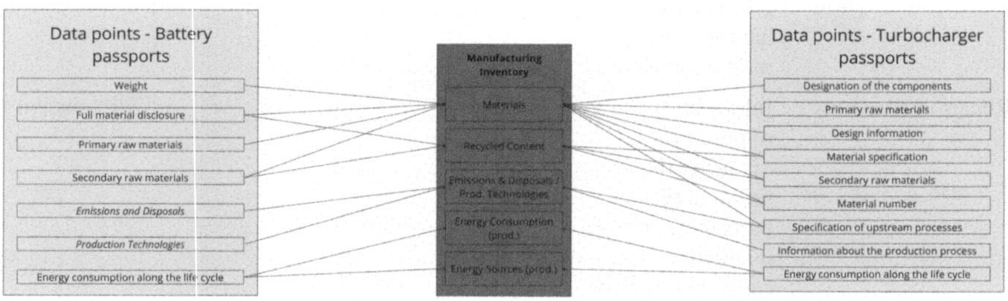

Fig. 2 Data mapping manufacturing inventory. Italic font denotes the newly added LCA-relevant datapoints

5.2.2 Distribution Inventory

The mapping of the data points from the turbocharger and the battery onto the distribution inventory is shown in Fig. 3.

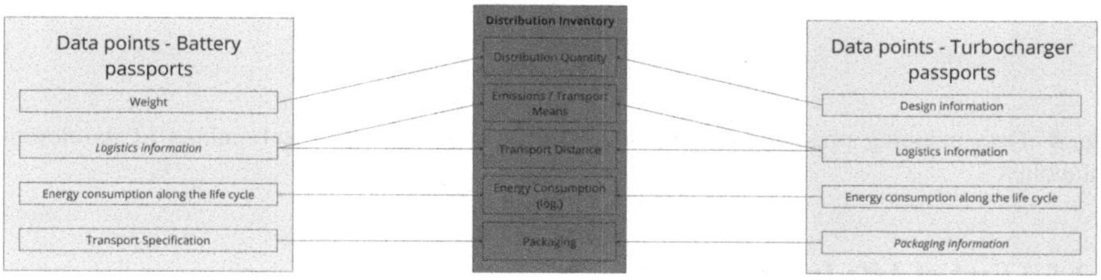

Fig. 3 Data mapping distribution inventory. Italic font denotes the newly added LCA-relevant datapoints

In the case of the turbocharger use case, the data point *design information* is mapped onto *distribution quantity*. The data point *logistics information* is mapped onto the distribution inventory information categories *emissions/transport means* and *transport distance*. The data point *energy consumption along the life cycle* is mapped onto the information category *energy consumption (logistics)*, and the data point *packaging information* is mapped onto the information category *packaging*.

In the case of the battery use case, the data point *weight* is mapped onto the *distribution quantity* and the *transport specification* data point is mapped onto *packaging*. The data point *logistics information* is mapped onto the distribution inventory information categories *emissions/transport means* and *transport distance*. Finally, the data point *energy consumption along the lifecycle* is mapped onto the information category *energy consumption (logistics)*.

5.2.3 Operational Inventory

The mapping of the data points from the turbocharger and the battery use case onto the operational inventory are shown in Fig. 4.

With regard to the operational inventory, the data points of relevance to the battery and the turbocharger are identical. Both comprise, firstly, the *maintenance history*, which is mapped onto *maintenance measures* and *reuse rates*. Secondly, both use cases' data points include *energy consumption along the life cycle*, mapped onto *energy sources*, and *energy consumption*. Furthermore, both the battery passport as well as the turbocharger passport contain the data point *battery identifier*, respective *turbocharger identifier* mapped onto *first lifetime*, and *total use time*.

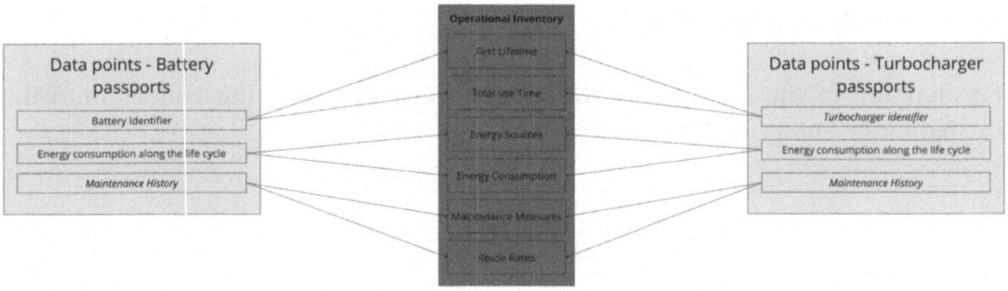

Fig. 4 Data mapping operational inventory. Italic font denotes the newly added LCA-relevant datapoints

5.2.4 End-of-Life Inventory

The mapping of the data points from the turbocharger and the battery use case onto the end-of-life inventory are shown in Fig. 5.

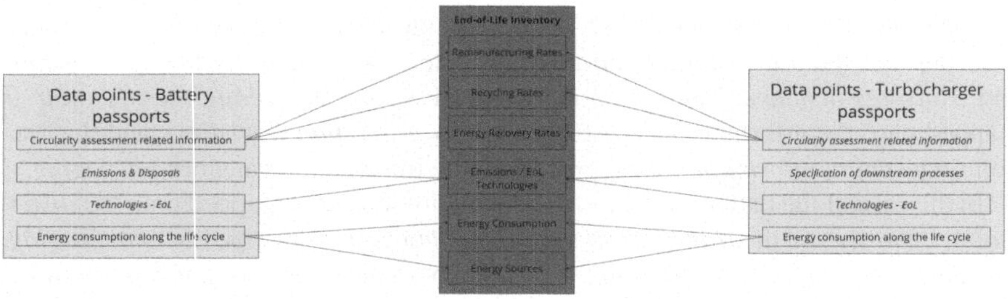

Fig. 5 Data mapping end-of-life inventory. Italic font denotes the newly added LCA-relevant datapoints

Concerning the end-of-life inventory, the relevant data points and their mappings are again similar for both use cases. Both contain the data point *technologies—EoL*, which is mapped onto *emissions/EoL technologies*. Similarly to the operational inventory, *energy consumption along the life cycle* is connected to the categories *energy consumption* and *energy sources* in both cases. The data point *circularity assessment related information* is mapped onto the categories *remanufacturing rates*, *recycling rates* and *energy recovery rates* in both use cases. The data point *emissions and disposals* of the battery passport is mapped onto the category *emissions/EoL technologies*. Finally, the data point *specification of downstream processes* of the turbocharger passport is mapped onto *emissions/EoL technologies*.

5.2.5 Data Points for Assumptions and Auxiliary Information

In the case that some data points described above are not designed to be instance-specific, a more traditional and generalized LCA approach requiring more data to model average impacts must be applied. They may be distinguished between *generalized* data points, which directly serve as rendering blocks within the LCA model, or *auxiliary* data points, which provide important coherence checks or enablers for certain scenario designs. Therefore, this section shows a first draft of the additional data points potentially necessary for traditional LCAs and corresponding to generalized assumptions. The lists, however, are not exhaustive but rather show the data points currently discussed for the DPPs. Due to the high number of relevant data points, in this case, there are two separate illustrations for both use cases. Figure 6 shows the relevant data points for the battery use case, for which the corresponding descriptions can be found in Table 1.

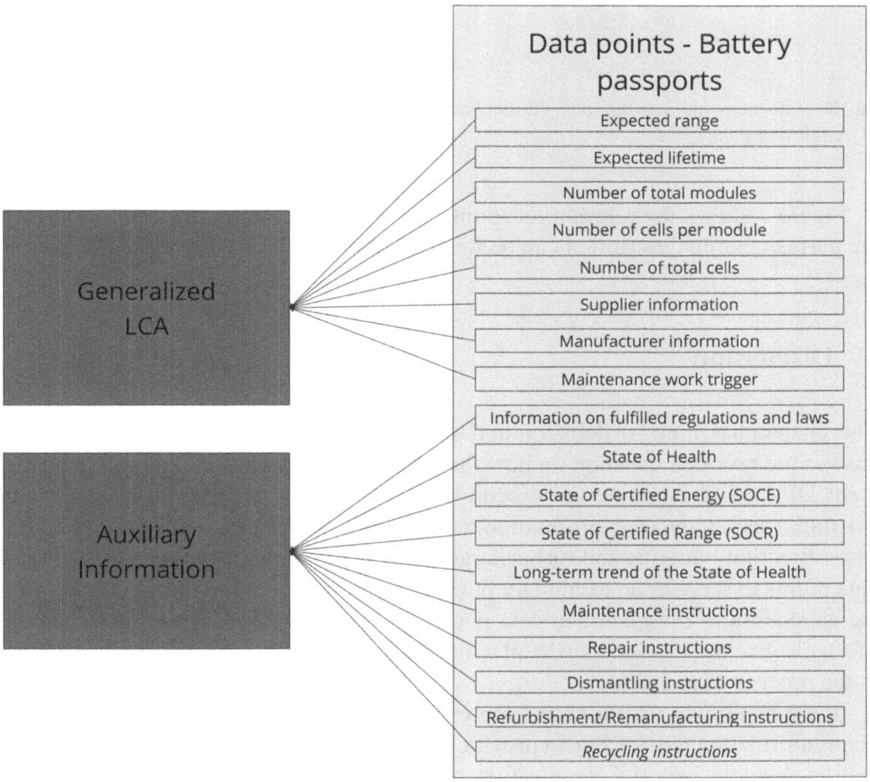

Fig. 6 Data mapping assumptions—battery use case. Italic font denotes the newly added LCA-relevant datapoints

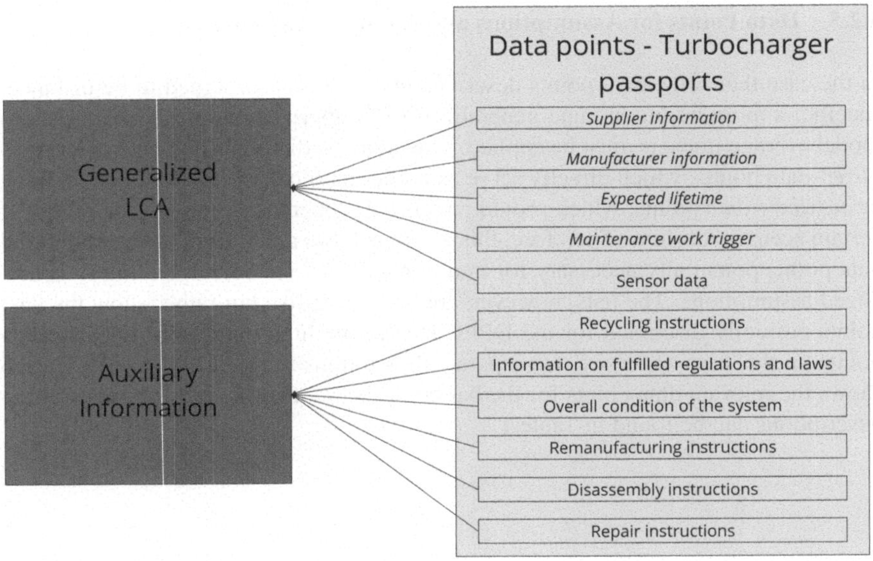

Fig. 7 Data mapping assumptions—turbocharger use case. Italic font denotes the newly added LCA-relevant datapoints

Figure 7 shows the relevant datapoints for the turbocharger use case, for which the corresponding description can be found in Table 2.

6 Discussion

This chapter has aimed to deliver initial insights into the use of DPP data for LCA. To answer the research question on how the emerging circular economy policy instrument DPP could facilitate the execution of LCAs, this research built on already existing research into the information requirements for DPPs in two automotive use cases (traction batteries and turbochargers), expanding and adapting the identified data points to serve as an exemplary but conclusive input set for an LCA. Following the adaptation of the identified data points, DPP data were mapped onto a generic life cycle inventory to show in detail which DPP data can be used for which purposes in the process of an LCA.

In the beginning-of-life phase, represented by the manufacturing inventory, the main information that DPPs can provide for LCA can be described as detailed information about the material composition, production process, including emissions and disposals, and information about the energy consumption and respective carriers. Therefore, based on the investigation of both use cases, DPPs offer great potential to improve the reliability and accuracy of LCAs by delivering accurate, instance-specific, and up-to-date primary data for the life cycle inventory. By adding instance-specific data, the DPP also offers the possibility to considerably increase the reliability of the LCA by substituting generic data from inventory databases. This holds true not only for the materials used but especially for the production processes, where DPP data allow companies that invest in more sustainable production technologies to account for them, reflecting an improvement relative to the use of generic data from databases, which might not allow for the possibility to consider environmental advantage appropriately.

In the middle-of-life phase, represented by the operational inventory, DPPs offer three main functions regarding data provision for LCAs. Firstly, for both use cases considered herein, DPPs can deliver information about the maintenance schedule and history of the product in question. By delivering specific information about the maintenance history of a battery or turbocharger, the LCA can include the materials used during maintenance and repair, as well as auxiliary materials in detail, without referring to generic maintenance rates or other assumptions based on experience or expert knowledge. Additionally, the detailed DPP data can help in accounting for the reuse of batteries or turbochargers in other first- or second-life product applications and, hence, increased product lifetime. Furthermore, DPPs can also deliver information relating to energy consumption and sources during instance-specific use and maintenance. While this is of lower priority in the turbocharger case, given that the energy input might only include energy used during the maintenance process, this is an essential improvement for LCAs of traction batteries. LCA results can be made significantly more accurate and reliable by accounting for the exact amount of energy consumed and the specific energy mix used for charging the battery instead of the generic country- or region-based energy mix.

During the end-of-life phase, DPPs can deliver the specific energy used for end-of-life processes and the corresponding energy sources for both use cases. Furthermore, the specific end-of-life technologies applied can be tracked, and their emissions and disposals can be accounted for in detail when the relevant information is included in DPPs. Both passports could also deliver information on recycling, remanufacturing, and energy recovery rates. The potential DPPs could provide the possibility to include specific and detailed end-of-life pathways for both use cases examined herein and determine the exact application of the end-of-life processes.

The distribution of the products can be seen as spanning all life cycle phases, given that all lifecycle phases include at least some degree of transportation. In this study, the distribution and transportation phases are represented by the distribution inventory. Although the data points in this case are named differently for both use cases, the principal information that they can deliver for LCA is similar. Firstly, the exact distribution quantity and the means by which the respective product is transported, the corresponding distances, and emissions can be tracked and are hence included in detail in the LCA. Secondly, a DPP could deliver the exact amount of energy consumed during the transport processes in the different life cycle stages. Finally, both DPPs could offer the possibility to deliver information regarding the packaging material used during the transportation process. The potential of DPPs for

LCA in this stage is the detailed tracking of the transportation processes, where high granularity primary data is then available and enables instance-specific or at least product-line-specific LCAs.

The generalized and auxiliary data points denote the first draft of additional data points potentially necessary whenever non-instance-specific modeling elements are applied to support the corresponding assumptions and scenarios. In principle, such traditional LCA modeling approaches may be applied to every LCA data point described, leading to an extensive range of possible implementations into DPPs. The generalized and auxiliary data points shown above have been formulated based on data points currently discussed for the DPPs and, hence, are not exhaustive. Additional generalized or auxiliary data points may be required depending on the final LCA system model application. Nonetheless, the considerations presented already provide a thorough introduction to the potential implementation pathways. Ultimately, it will be necessary to underline two aspects: (1) non-instance-specific LCA data points introduce complexity and uncertainty to the system model, but (2) they enable the retention of the data points necessary to conduct a full LCA, for which instance-specific tracking is not yet feasible.

This study also has methodological limitations. Firstly, the data mapping was solely literature-based and based on two use cases in which the DPPs were not explicitly developed for LCA. However, this can also be seen as close to reality, since DPPs are designed to support a circular economy and are not explicitly intended for LCA. Furthermore, data mapping was carried out from those DPP data points onto a generic life cycle inventory required for a full LCA, which may be structured differently. Nonetheless, mapping onto the generic inventory can offer valuable first insights into the combination of DPPs and LCAs. The study makes the first attempt to distinguish between the potentially instant-specific and traditional approaches of generalized DPP data points for an LCA. Although no analysis of the applicability of instance-specific data points, their potential implementation costs or their specific direct and indirect benefits compared to general LCA calculations has been conducted, it highlights the need for further research in this area. Given that the development of DPPs is not very advanced and its data points have not yet been selected, starting such studies would aid the development of DPPs and is ultimately essential to completely exploit their value. However, this is also strongly connected to the initial question that must be answered by the development of DPPs: which additional fields of application (such as product development, strategic sustainable development, or policy evaluation) should/must be incorporated into a circular economy policy instrument?

7 Conclusions

This research aimed to deliver the first insights into using DPPs to facilitate LCA. To describe the potential of combining a dynamic DPP with LCA, two existing product passport concepts developed by the authors' research group were adapted to serve as the basis for mapping DPP data points onto a generic life cycle inventory. The results show the usefulness of aligning different DPP data points of the product with those of a generic LCA model and identifying the different potentials of the data provision via DPPs to LCA. This research has wider implications. For researchers, it serves as a starting point for conducting focused research on the application of DPPs for LCA in different sectors and for different products, refining the proposed generic LCA data points. Furthermore, the mapping performed in this research serves as the basis for researching potential instant-specific LCA data points based on DPP data, further shaping the structure of DPPs. For policymakers, this research offers the opportunity to gain insights into how DPP can serve as a data provider for LCA and how policies could be shaped accordingly to enable this additional function. Finally, this paper offers practitioners the possibility to identify which information they would want to be included and accessible in the DPPs for the execution of an LCA.

References

Adisorn T, Tholen L, Götz T (2021) Towards a digital product passport fit for contributing to a circular economy. Energies 14(8):2289. https://doi.org/10.3390/en14082289

Berger K, Baumgartner RJ, Weinzerl M, Bachler J, Preston K, Schöggl J-P (2023) Data requirements and availabilities for a digital battery passport—a value chain actor perspective. Cleaner Prod Lett 4:100032. https://doi.org/10.1016/j.clpl.2023.100032

Berger K, Schöggl J-P, Baumgartner RJ (2022) Digital battery passports to enable circular and sustainable value chains: Conceptualization and use cases. J Clean Prod 353:131492. https://doi.org/10.1016/j.jclepro.2022.131492

Bjørn A, Chandrakumar C, Boulay A-M, Doka G, Fang K, Gondran N, Hauschild MZ, Kerkhof A, King H, Margni M, McLaren S, Mueller C, Owsianiak M, Peters G, Roos S, Sala S, Sandin G, Sim S, Vargas-Gonzalez M, Ryberg M (2020) Review of life-cycle based methods for absolute environmental sustainability assessment and their applications. Environ Res Lett 15(8):083001. https://doi.org/10.1088/1748-9326/ab89d7

Björklund AE (2002) Survey of approaches to improve reliability in LCA. Int J Life Cycle Assess 7(2):64–72. https://doi.org/10.1007/BF02978849

Catena-X (n.d.) Catena-X. Retrieved 23.05.2024 from https://catena-x.net/en

Ciroth A, Muller S, Weidema B, Lesage P (2016) Empirically based uncertainty factors for the pedigree matrix in ecoinvent. Int J Life Cycle Assess 21(9):1338–1348. https://doi.org/10.1007/s11367-013-0670-5

Curran MA (2014) Strengths and limitations of life cycle assessment. In: Klöpffer W (ed) Background and future prospects in life cycle assessment. LCA compendium—the complete world of life cycle assessment. Springer Dordrecht, pp 189–206. https://doi.org/10.1007/978-94-017-8697-3

Ducuing C, Reich RH (2023) Data governance: Digital product passports as a case study. Competition Regulation Netw Ind 24(1):3–23. https://doi.org/10.1177/17835917231152799

Ecoinvent (2024) Ecoinvent database (3.10) [Computer software]. https://ecoinvent.org

European Commission (2020) Communication from the Commission to the European Parliament, the Council, the European Economic and Social Committee and the Committee of the Regions: a new circular economy action plan

European Commission (2022) Communication from the Commission to the European Parliament, the Council, the European Economic and Social Committee and the Committee of the Regions: on making sustainable products the norm

Hauschild MZ, Rosenbaum RK, Olsen SI (eds) (2018) Life cycle assessment: theory and practice. Springer International Publishing. https://doi.org/10.1007/978-3-319-56475-3

ISO (2006a) ISO 14040:2006: Environmental management life cycle assessment principles and framework [Standard]. https://www.iso.org/standard/37456.html

ISO (2006b) ISO 14044:2006: environmental management life cycle assessment Requirements and guidelines [Standard]. https://www.iso.org/standard/38498.html

Jansen M, Meisen T, Plociennik C, Berg H, Pomp A, Windholz W (2023) Stop guessing in the dark: identified requirements for digital product passport systems. Systems 11(3):123. https://doi.org/10.3390/systems11030123

Jensen SF, Kristensen JH, Adamsen S, Christensen A, Waehrens BV (2023) Digital product passports for a circular economy: data needs for product life cycle decision-making. Sustain Prod Consump 37:242–255. https://doi.org/10.1016/j.spc.2023.02.021

Jolliet O, Saadé-Sbeih M, Shaked S, Jolliet A, Crettaz P (2015) Environmental Life Cycle Assessment (1st edn). CRC Press. https://doi.org/10.1201/b19138

King MRN, Timms PD, Mountney S (2023) A proposed universal definition of a Digital Product Passport Ecosystem (DPPE): worldviews, discrete capabilities, stakeholder requirements and concerns. J Clean Prod 384:135538. https://doi.org/10.1016/j.jclepro.2022.135538

Koppelaar RHEM, Pamidi S, Hajósi E, Herreras L, Leroy P, Jung H-Y, Concheso A, Daniel R, Francisco FB, Parrado C, Dell'Ambrogio S, Guggiari F, Leone D, Fontana A (2023) A digital product passport for critical raw materials reuse and recycling. Sustainability 15(2):1405. https://doi.org/10.3390/su15021405

Lemos P (2020) Towards a circular economy: presentation on product passport. Hambg Sustain Sess #2

Miah JH, Griffiths A, McNeill R, Halvorson S, Schenker U, Espinoza-Orias N, Morse S, Yang A, Sadhukhan J (2018) A framework for increasing the availability of life cycle inventory data based on the role of multinational companies. Int J Life Cycle Assess 23(9):1744–1760. https://doi.org/10.1007/s11367-017-1391-y

Moutik B, Summerscales J, Graham-Jones J, Pemberton R (2023) Life cycle assessment research trends and implications: a bibliometric analysis. Sustainability 15(18):13408. https://doi.org/10.3390/su151813408

Plociennik C, Pourjafarian M, Nazeri A, Windholz W, Knetsch S, Rickert J, Ciroth A, do Carmo Precci Lopes A, Hagedorn T, Vogelgesang M, Benner W, Gassmann A, Bergweiler S, Ruskowski M, Schebek L, Weidenkaff A (2022a) Towards a digital lifecycle passport for the circular economy. Proc CIRP 105:122–127. https://doi.org/10.1016/j.procir.2022.02.021

Plociennik C, Pourjafarian M, Saleh S, Hagedorn T, do Carmo Precci Lopes A, Vogelgesang M, Baehr J, Kellerer B, Jansen M, Berg H, Ruskowski M, Schebek L, Ciroth A (2022b) Requirements for a digital product passport to boost the circular economy. INFORMATIK 2022:1485–1494. https://doi.org/10.18420/INF2022_127

Pohlmann A, Popowicz M, Schöggl J-P, Bachler J, Keler J, Baumgartner RJ (2024) Conceptualization of a digital product passport to enable circular and sustainable automotive value chains—the combustion engine use case. Proc CIRP 122:169–174. https://doi.org/10.1016/j.procir.2024.01.025

Regulation (EU) 1542/2023 Regulation (EU) 2023/1542 of the European Parliament and of the Council of 12 July 2023 concerning batteries and waste batteries, amending Directive 2008/98/EC and Regulation (EU) 2019/1020 and repealing Directive 2006/66/EC

Su S, Ju J, Ding Y, Yuan J, Cui P (2022) A comprehensive dynamic life cycle assessment model: considering temporally and spatially dependent variations. Int J Environ Res Public Health 19(21):14000. https://doi.org/10.3390/ijerph192114000

Testa F, Nucci B, Tessitore S, Iraldo F, Daddi T (2016) Perceptions on LCA implementation: evidence from a survey on adopters and nonadopters in Italy. Int J Life Cycle Assess 21(10):1501–1513. https://doi.org/10.1007/s11367-016-1106-9

United Nations Department of Economic and Social Affairs (2023) The sustainable development goals report 2023: special edition. United Nations. https://doi.org/10.18356/9789210024914

Van Der Giesen C, Cucurachi S, Guinée J, Kramer GJ, Tukker A (2020) A critical view on the current application of LCA for new technologies and recommendations for improved practice. J Clean Prod 259:120904. https://doi.org/10.1016/j.jclepro.2020.120904

Walden J, Steinbrecher A, Marinkovic M (2021) Digital product passports as enabler of the circular economy. Chem Ing Tec 93(11):1717–1727. https://doi.org/10.1002/cite.202100121

Correction to: An Assessment of the Nexus Between the Ricardian Thoughts on Comparative Advantage and Circular Economy in Developing Economies

Oluwabunmi O. Adejumo, Joshua F. Obisanya, and Akintoye V. Adejumo

Correction to: Chapter 12 in: L. Brandli et al. (eds.), *The Contribution of Life Cycle Analyses and Circular Economy to the Sustainable Development Goals*, World Sustainability Series, https://doi.org/10.1007/978-3-031-85300-5_12

In the original version of the book, the following belated corrections have been incorporated: The author name "Oluwabunmi J. Adejumo" has been changed to "Oluwabunmi O. Adejumo" in the Frontmatter, Backmatter and in Chapter 12. The book and the chapters have been updated with the changes.

The updated version of this chapter can be found at
https://doi.org/10.1007/978-3-031-85300-5_12

L. Brandli et al. (eds.), *The Contribution of Life Cycle Analyses and Circular Economy to the Sustainable Development Goals*, World Sustainability Series,
https://doi.org/10.1007/978-3-031-85300-5_27

The manufacturer's authorised representative in the EU is Springer Nature Customer Service Centre GmbH, Europaplatz 3, 69115 Heidelberg, Germany. If you have any concerns regarding our products, please contact ProductSafety@springernature.com

Printed and bound by CPI Group (UK) Ltd, Croydon, CR0 4YY
30/09/2026
02225572-0019